GENETICS, GENOMICS AND BREEDING OF COOL SEASON GRAIN LEGUMES

Genetics, Genomics and Breeding of Crop Plants

Series Editor

Chittaranjan Kole
Department of Genetics and Biochemistry
Clemson University
Clemson, SC
USA

Books in this Series:

Published or in Press:

- Jinguo Hu, Gerald Seiler & Chittaranjan Kole: *Sunflower*
- Kristin D. Bilyeu, Milind B. Ratnaparkhe & Chittaranjan Kole: *Soybean*
- Robert Henry & Chittaranjan Kole: *Sugarcane*
- Kevin Folta & Chittaranjan Kole: *Berries*
- Jan Sadowsky & Chittaranjan Kole: *Vegetable Brassicas*
- James M. Bradeen & Chittaranjan Kole: *Potato*
- C.P. Joshi, Stephen DiFazio & Chittaranjan Kole: *Poplar*
- Anne-Françoise Adam-Blondon, José M. Martínez-Zapater & Chittaranjan Kole: *Grapes*
- Christophe Plomion, Jean Bousquet & Chittaranjan Kole: *Conifers*
- Dave Edwards, Jacqueline Batley, Isobel Parkin & Chittaranjan Kole: *Oilseed Brassicas*
- Marcelino Pérez de la Vega, Ana María Torres, José Ignacio Cubero & Chittaranjan Kole: *Cool Season Grain Legumes*
- Yi-Hong Wang, Tusar Kanti Behera & Chittaranjan Kole: *Cucurbit*
- Michael Pillay, George Ude & Chittaranjan Kole: *Banana*

GENETICS, GENOMICS AND BREEDING OF COOL SEASON GRAIN LEGUMES

Editors

Marcelino Pérez de la Vega
Area de Genética
Universidad de León
León, Spain

Ana María Torres
Area de Mejora y Biotecnología
IFAPA, Centro Alameda del Obispo
Córdoba, Spain

José Ignacio Cubero
Departamento de Genética
Universidad de Córdoba
Córdoba, Spain.

Chittaranjan Kole
Department of Genetics and Biochemistry
Clemson University
Clemson, SC, USA

CRC Press
Taylor & Francis Group
an **informa** business
www.crcpress.com

6000 Broken Sound Parkway, NW
Suite 300, Boca Raton, FL 33487
270 Madison Avenue
New York, NY 10016
2 Park Square, Milton Park
Abingdon, Oxon OX14 4RN, UK

Science Publishers
Jersey, British Isles
Enfield, New Hampshire

Published by Science Publishers, an imprint of Edenbridge Ltd.

- St. Helier, Jersey, British Channel Islands
- P.O. Box 699, Enfield, NH 03748, USA

E-mail: info@scipub.net Website: www.scipub.net

Marketed and distributed by:

CRC Press
Taylor & Francis Group
an informa business
www.crcpress.com

6000 Broken Sound Parkway, NW
Suite 300, Boca Raton, FL 33487
270 Madison Avenue
New York, NY 10016
2 Park Square, Milton Park
Abingdon, Oxon OX14 4RN, UK

ISBN 978-1-57808-765-5

Cover Illustration: Flowers and seeds of cool season grain legumes. From left to right: pea, faba bean, lentil, chickpea, and grass pea.

Library of Congress Cataloging-in-Publication Data

Genetics, genomics and breeding of cool season grain legumes / editors: Marcelino Pérez de la Vega ... [et al.]. -- 1st ed.
p. cm. -- (Genetics, genomics and breeding of crop plants)
Includes bibliographical references and index.
ISBN 978-1-57808-765-5 (hardcover)
1. Legumes--Genetics. 2. Legumes--Genome mapping. 3. Legumes--Breeding. I. Pérez de la Vega, M. (Marcelino) II. Series: Genetics, genomics and breeding of crop plants.
SB317.L43G46 2011
633.3--dc23

2011019797

Printed in the United States of America

Preface to the Series

Genetics, genomics and breeding has emerged as three overlapping and complimentary disciplines for comprehensive and fine-scale analysis of plant genomes and their precise and rapid improvement. While genetics and plant breeding have contributed enormously towards several new concepts and strategies for elucidation of plant genes and genomes as well as development of a huge number of crop varieties with desirable traits, genomics has depicted the chemical nature of genes, gene products and genomes and also provided additional resources for crop improvement.

In today's world, teaching, research, funding, regulation and utilization of plant genetics, genomics and breeding essentially require thorough understanding of their components including classical, biochemical, cytological and molecular genetics; and traditional, molecular, transgenic and genomics-assisted breeding. There are several book volumes and reviews available that cover individually or in combination of a few of these components for the major plants or plant groups; and also on the concepts and strategies for these individual components with examples drawn mainly from the major plants. Therefore, we planned to fill an existing gap with individual book volumes dedicated to the leading crop and model plants with comprehensive deliberations on all the classical, advanced and modern concepts of depiction and improvement of genomes. The success stories and limitations in the different plant species, crop or model, must vary; however, we have tried to include a more or less general outline of the contents of the chapters of the volumes to maintain uniformity as far as possible.

Often genetics, genomics and plant breeding and particularly their complimentary and supplementary disciplines are studied and practiced by people who do not have, and reasonably so, the basic understanding of biology of the plants for which they are contributing. A general description of the plants and their botany would surely instill more interest among them on the plant species they are working for and therefore we presented lucid details on the economic and/or academic importance of the plant(s); historical information on geographical origin and distribution; botanical origin and evolution; available germplasms and gene pools, and genetic and cytogenetic stocks as genetic, genomic and breeding resources; and basic information on taxonomy, habit, habitat, morphology, karyotype, ploidy level and genome size, etc.

Classical genetics and traditional breeding have contributed enormously even by employing the phenotype-to-genotype approach. We included detailed descriptions on these classical efforts such as genetic mapping using morphological, cytological and isozyme markers; and achievements of conventional breeding for desirable and against undesirable traits. Employment of the in vitro culture techniques such as micro- and megaspore culture, and somatic mutation and hybridization, has also been enumerated. In addition, an assessment of the achievements and limitations of the basic genetics and conventional breeding efforts has been presented.

It is a hard truth that in many instances we depend too much on a few advanced technologies, we are trained in, for creating and using novel or alien genes but forget the infinite wealth of desirable genes in the indigenous cultivars and wild allied species besides the available germplasms in national and international institutes or centers. Exploring as broad as possible natural genetic diversity not only provides information on availability of target donor genes but also on genetically divergent genotypes, botanical varieties, subspecies, species and even genera to be used as potential parents in crosses to realize optimum genetic polymorphism required for mapping and breeding. Genetic divergence has been evaluated using the available tools at a particular point of time. We included discussions on phenotype-based strategies employing morphological markers, genotype-based strategies employing molecular markers; the statistical procedures utilized; their utilities for evaluation of genetic divergence among genotypes, local landraces, species and genera; and also on the effects of breeding pedigrees and geographical locations on the degree of genetic diversity.

Association mapping using molecular markers is a recent strategy to utilize the natural genetic variability to detect marker-trait association and to validate the genomic locations of genes, particularly those controlling the quantitative traits. Association mapping has been employed effectively in genetic studies in human and other animal models and those have inspired the plant scientists to take advantage of this tool. We included examples of its use and implication in some of the volumes that devote to the plants for which this technique has been successfully employed for assessment of the degree of linkage disequilibrium related to a particular gene or genome, and for germplasm enhancement.

Genetic linkage mapping using molecular markers have been discussed in many books, reviews and book series. However, in this series, genetic mapping has been discussed at length with more elaborations and examples on diverse markers including the anonymous type 2 markers such as RFLPs, RAPDs, AFLPs, etc. and the gene-specific type 1 markers such as EST-SSRs, SNPs, etc.; various mapping populations including F_2, backcross, recombinant inbred, doubled haploid, near-isogenic and pseudotestcross; computer software including MapMaker, JoinMap, etc. used; and different types of genetic maps including preliminary, high-resolution, high-density, saturated, reference, consensus and integrated developed so far.

Mapping of simply inherited traits and quantitative traits controlled by oligogenes and polygenes, respectively has been deliberated in the earlier literature crop-wise or crop group-wise. However, more detailed information on mapping or tagging oligogenes by linkage mapping or bulked segregant analysis, mapping polygenes by QTL analysis, and different computer software employed such as MapMaker, JoinMap, QTL Cartographer, Map Manager, etc. for these purposes have been discussed at more depth in the present volumes.

The strategies and achievements of marker-assisted or molecular breeding have been discussed in a few books and reviews earlier. However, those mostly deliberated on the general aspects with examples drawn mainly from major plants. In this series, we included comprehensive descriptions on the use of molecular markers for germplasm characterization, detection and maintenance of distinctiveness, uniformity and stability of genotypes, introgression and pyramiding of genes. We have also included elucidations on the strategies and achievements of transgenic breeding for developing genotypes particularly with resistance to herbicide, biotic and abiotic stresses; for biofuel production, biopharming, phytoremediation; and also for producing resources for functional genomics.

A number of desirable genes and QTLs have been cloned in plants since 1992 and 2000, respectively using different strategies, mainly positional cloning and transposon tagging. We included enumeration of these and other strategies for isolation of genes and QTLs, testing of their expression and their effective utilization in the relevant volumes.

Physical maps and integrated physical-genetic maps are now available in most of the leading crop and model plants owing mainly to the BAC, YAC, EST and cDNA libraries. Similar libraries and other required genomic resources have also been developed for the remaining crops. We have devoted a section on the library development and sequencing of these resources; detection, validation and utilization of gene-based molecular markers; and impact of new generation sequencing technologies on structural genomics.

As mentioned earlier, whole genome sequencing has been completed in one model plant (Arabidopsis) and seven economic plants (rice, poplar, peach, papaya, grapes, soybean and sorghum) and is progressing in an array of model and economic plants. Advent of massively parallel DNA sequencing using 454-pyrosequencing, Solexa Genome Analyzer, SOLiD system, Heliscope and SMRT have facilitated whole genome sequencing in many other plants more rapidly, cheaply and precisely. We have included extensive coverage on the level (national or international) of collaboration and the strategies and status of whole genome sequencing in plants for which sequencing efforts have been completed or are progressing currently. We have also included critical assessment of the impact of these genome initiatives in the respective volumes.

Comparative genome mapping based on molecular markers and map positions of genes and QTLs practiced during the last two decades of the last century provided answers to many basic questions related to evolution, origin and phylogenetic relationship of close plant taxa. Enrichment of genomic resources has reinforced the study of genome homology and synteny of genes among plants not only in the same family but also of taxonomically distant families. Comparative genomics is not only delivering answers to the questions of academic interest but also providing many candidate genes for plant genetic improvement.

The 'central dogma' enunciated in 1958 provided a simple picture of gene function—gene to mRNA to transcripts to proteins (enzymes) to metabolites. The enormous amount of information generated on characterization of transcripts, proteins and metabolites now have led to the emergence of individual disciplines including functional genomics, transcriptomics, proteomics and metabolomics. Although all of them ultimately strengthen the analysis and improvement of a genome, they deserve individual deliberations for each plant species. For example, microarrays, SAGE, MPSS for transcriptome analysis; and 2D gel electrophoresis, MALDI, NMR, MS for proteomics and metabolomics studies require elaboration. Besides transcriptome, proteome or metabolome QTL mapping and application of transcriptomics, proteomics and metabolomics in genomics-assisted breeding are frontier fields now. We included discussions on them in the relevant volumes.

The databases for storage, search and utilization on the genomes, genes, gene products and their sequences are growing enormously in each second and they require robust bioinformatics tools plant-wise and purpose-wise. We included a section on databases on the gene and genomes, gene expression, comparative genomes, molecular marker and genetic maps, protein and metabolomes, and their integration.

Notwithstanding the progress made so far, each crop or model plant species requires more pragmatic retrospect. For the model plants we need to answer how much they have been utilized to answer the basic questions of genetics and genomics as compared to other wild and domesticated species. For the economic plants we need to answer as to whether they have been genetically tailored perfectly for expanded geographical regions and current requirements for green fuel, plant-based bioproducts and for improvements of ecology and environment. These futuristic explanations have been addressed finally in the volumes.

We are aware of exclusions of some plants for which we have comprehensive compilations on genetics, genomics and breeding in hard copy or digital format and also some other plants which will have enough achievements to claim for individual book volume only in distant future. However, we feel satisfied that we could present comprehensive deliberations on genetics, genomics and breeding of 30 model and economic plants, and their groups in a few cases, in this series. I personally feel also happy that I could work with many internationally celebrated scientists who edited the book volumes on the leading plants and plant groups and included chapters authored by many scientists reputed globally for their contributions on the concerned plant or plant group.

We paid serious attention to reviewing, revising and updating of the manuscripts of all the chapters of this book series, but some technical and formatting mistakes will remain for sure. As the series editor, I take complete responsibility for all these mistakes and will look forward to the readers for corrections of these mistakes and also for their suggestions for further improvement of the volumes and the series so that future editions can serve better the purposes of the students, scientists, industries, and the society of this and future generations.

Science publishers, Inc. has been serving the requirements of science and society for a long time with publications of books devoted to advanced concepts, strategies, tools, methodologies and achievements of various science disciplines. Myself as the editor and also on behalf of the volume editors, chapter authors and the ultimate beneficiaries of the volumes take this opportunity to acknowledge the publisher for presenting these books that could be useful for teaching, research and extension of genetics, genomics and breeding.

Chittaranjan Kole

Preface to the Volume

Cool season grain legumes (pea, faba bean, lentil, chickpea and grass pea) constitute a set of important leguminous species mainly for food as dry seeds, traditionally valued for their content of protein and carbohydrates. Nowadays, they are also valued for their low content in fats and high content in dietary fiber, thus contributing to healthy nutrition. Crop legumes have been traditionally classified into cool season and warm season legumes, reflecting their origin and agroecological requirements. Cool season legumes are included in two tribes, Vicieae (pea, faba bean, lentil and grass pea) and Cicereae (chickpea). A sister tribe to these two tribes is Trifolieae which includes the model legume species *Medicago truncatula* (barrel medic), the closest reference for comparative genetics and genomics for these crop species. As other legumes, cool season grain legumes enrich the soil by fixing atmospheric nitrogen in symbiosis with root nodule bacteria, and therefore are a traditional component of agriculture in temperate climates, generally in rotation with cereals and other crops. These five species were domesticated in the Fertile Crescent among other first domesticated plant species such as wheat and barley. The oldest archaeological remains are dated around 10,000 BP in aceramic Neolithic layers. The crops spread first to Europe, the rest of Asia and North Africa with the expansion of agriculture and were introduced into America and Australia after 1500 AD. They are mostly consumed as dry seeds, but immature pods and seeds of faba bean and pea are also consumed as vegetables. Although none of these species is individually included in the list of main crops on which human diet depends, their joint contribution to human nutrition is significant, and even more so considering that they are mainly consumed directly by humans. In fact they constitute, together with other pulses, one of the cheapest and healthiest sources of food proteins. According to FAOSTAT, the 2009 world production of these crops reached approximately 19 million tons for pea (dry and green), 9.8 for chickpea, 4.1 for faba bean and 3.6 for lentil. No individual record is available for grass pea. In spite of their traditional cultivation and consumption during millennia, some of these crops can be considered as neglected and underutilized species. However, many neglected and underutilized species are extremely important for food production in low-income food countries. The general tendency in the last decades for these cool season legume species is a slow increase in production and yield.

The genetics and genomics knowledge as well as the breeding advances in these species are unequal. Plant breeders have used traditional or conventional techniques to improve these species. In comparison with old classical early-domesticated crops, systematic breeding research started later. The exception is pea, which was the pioneering species in genetic research and was bred since the XIX century. The interest in these crops has increased in the past decades and research and the application of modern breeding techniques have increased recently. This volume, as others in the Series, deals with recent advances in cool season grain legume research and breeding, including genome mapping, molecular breeding and genomics and intends to bridge the gap between traditional breeding research and modern molecular developments. As can be seen in the different chapters, the "state of the art" for each individual crop differs; however, their phylogenetic proximity to *M. truncatula*, a legume model species, justifies their joint study, since most of the knowledge obtained in one species is useful to speed up the research and breeding in the other ones.

There are 12 chapters in total. The first five chapters are each devoted to a species (pea, faba bean, lentil, grass pea and chickpea, respectively), while the remaining chapters are devoted to horizontal common aspects which affect all these crops. When information from one chapter is needed or relevant for completing another, cross references are provided. Likewise, cross references within chapters are provided. The first five chapters introduce the basic knowledge of each species, then, describe classical genetics and traditional breeding, genetic mapping efforts and linkage map construction, identification and tagging of genes (both quantitative and qualitative), major biotic and abiotic threats, and the genomics advances, and their breeding applications as they exist. Chapter 6 reviews the functional genomics of cool season food legumes, though many insights into the biology of a specific legume derive from so-called model plants; describes the complex worlds of transcriptomics and some of the techniques frequently used in the study of these species. Because of the limited availability of large-scale genome sequences in cool season grain legumes, Chapter 7 focuses on comparative mapping between the most important cool season grain legumes at the macrosynteny level. Results achieved so far suggest that the knowledge gained from comparative mapping may have considerable utility to solve basic and applied agronomic questions of importance in the crop species. Comparative studies are particularly relevant in this set of related species. Chapter 8 focuses on proteomic and metabolomic techniques that have been successfully adopted for systems biology approaches especially in *Medicago*. Furthermore, applications of these novel strategies in the field of cool season legumes research is described. The studies on regeneration and genetic transformation in cool season legumes are described in Chapter 9. It summarizes the hurdles being faced in the efficient recovery of transgenic plants and presents a comparative account of explants used, mode of regeneration, gene delivery techniques and recovery of transgenic plants in these crops.

Legume nitrogen fixation plays a key role in the maintenance of crop production worldwide. However, this property is often not considered in new cultivar selection programs. Chapter 10 reviews the effectiveness and efficiency of the process and the structural and functional genomic studies on plant-bacteria interactions. The identification of symbiotic genes in legumes has been speed up by the use of model legume species and functional genomic technologies such as map-based gene cloning, transcriptomics, proteomics and metabolomics. The chapter also stresses the importance of breeding for enhancing nitrogen fixation. In legume breeding programs, more attention should be directed at understanding and evaluating plant mechanisms that control root nodule formation and activity, otherwise biological nitrogen fixation may be compromised. Chapter 11 is devoted to broomrape. Broomrapes are annual or perennial obligate parasitic plants which represent a major threat to cool season legumes and other crops in many temperate areas, and are one of the main limiting factors to crop legumes around the Mediterranean basin, one of the major traditional areas of production and consumption of these crops. Traditional plant breeding has failed to produce stable resistance; although genetic resistance exists and has been used in cool season legume breeding. Molecular mapping of broomrape resistance QTLs together with the development of MAS techniques are promising approaches to rapidly improve crop resistance. With the emergence of large-scale genomic tools, the combination of genetic mapping with gene expression studies, can offer an integrated approach to study resistance to parasitic plants. Finally, Chapter 12 summarizes the current status of genetic and genomic research and points to future developments and applications.

Each chapter has been written by several experts who have worked thoroughly in compiling the information available in their area of expertise. We are indebted to all of them for the dedication and effort devoted towards producing the book. We hope that this book is useful for researchers and breeders interested in cool season crop legumes, in particular those interested in a broader spectrum of legumes and crop plants.

The Editors
Marcelino Pérez de la Vega
Ana M. Torres
José I. Cubero
Chittaranjan Kole

Contents

List of Contributors

Michael J. Ambrose
John Innes Center, Colney, Norwich NR4 7UH, UK.
Phone: +44 1603 450630
Fax: +44 1603 450045
Email: *mike.ambrose@bbsrc.ac.uk*

Cesar Arrese-Igor
Departamento de Ciencias del Medio Natural. Universidad Pública de Navarra. Campus de Arrosadía, s/n. 31006 Pamplona, Spain.
Phone: +34 948 169119
Fax: +34 948 169169
Email: *cesarai@unavarra.es*

Carmen María Avila
IFAPA, Centro Alameda del Obispo, Area de Mejora y Biotecnología, Apdo. 3092, E-14080 Córdoba, Spain.
Phone: +34 957 016005
Fax: +34 957 016043
Email: *carmenm.avila@juntadeandalucia.es*

Judith Burstin
INRA, UMRLEG, BP 86510, F-21065 Dijon, France.
Phone: +33 3 80 69 31 59
Fax: +33 3 80 69 32 63
Email: *judith.burstin@dijon.inra.fr*

Hong-Kyu Choi
Dong-A University, Department of Genetic Engineering, Busan, South Korea.
Phone: + 82-51-200-7508
Fax: + 82-51-200-7505
Email: *hkchoi@dau.ac.kr*

María José Cobos
IFAPA, Centro Alameda del Obispo, Area de Mejora y Biotecnología, Apdo. 3092, 14080 Córdoba, Spain.
Phone: +34 957 016132
Fax: +34 957 016043
Email: *covamj@gmail.com*

Jose Ignacio Cubero
Departamento de Genética, Universidad de Córdoba, Campus de Rabanales, Edificio C5, 2ª planta, 14071 Córdoba, Spain.
Phone: + 34 957 499278
Fax: +34 957 499252
Email: *ge1cusaj@lucano.uco.es*

José Vicente Die
IFAPA, Centro Alameda del Obispo, Area de Mejora y Biotecnología, Apdo. 3092, 14080 Córdoba, Spain.
Phone: +34 957 016061
Fax: +34 957 016043
Email: *jose.die.ext@juntadeandalucia.es*

Claire Domoney
John Innes Centre, Norwich Research Park, Colney, Norwich NR4 7UH, UK.
Phone: +44 1603 450381
Fax: +44 1603 450045
Email: *claire.domoney@bbsrc.acuk*

Noel Ellis
John Innes Centre, Department of Crop Genetics, Norwich NR4 7UH, UK.
Phone: + 44 (0) 1603 450243
Fax: + 44 (0) 1603 450027
Email: *noel.ellis@bbsrc.ac.uk*

Richard M. Fratini
Area de Genética, Departamento de Biología Molecular, Universidad de León, 24071 León, Spain.
Phone: +34 987 291553
Fax: +34 987 291409
Email: *richard.fratini@unileon.es*

Juan Gil
Departamento de Genética, Univ. de Córdoba, Campus Rabanales Edif. C-5, 14071 Córdoba, Spain.
Phone: +34 957 218510
Fax: +34 957 218503
Email: *ge1gilij@uco.es*

Clara Isabel González-Verdejo
IFAPA, Centro Alameda del Obispo, Area de Mejora y Biotecnología, Apdo. 3092, 14080 Córdoba, Spain.
Phone: +34 957 016120
Fax: +34 957 016043
Email: *clarai.gonzalez@juntadeandalucia.es*

Colin D. Hanbury
Department of Agriculture and Food, Western Australia, 3 Baron Hay Court, South Perth 6151, Australia.
Phone: +61 893683251
Fax: +61 893683945
Email: *colin.hanbury@agric.wa.gov.au; colinhanbury@yahoo.com.au*

Günter Kahl
Molecular BioSciences, Biocenter, University of Frankfurt am Main, D-60438 Frankfurt am Main, Germany.
Phone: +49 69 95739 602/603
Fax: +49 69 95739 706
Email: *kahl@em.uni-frankfurt.de*

Péter Kaló
Agricultural Biotechnology Center, Szent-Györgyi Albert u. 4, Gödöllő, Hungary.
Phone: + 36 28-526104
Fax: +36 28-526101
Email: *kalo@abc.hu*

György B. Kiss
Agricultural Biotechnology Center, Szent-Györgyi Albert u. 4, Gödöllő, Hungary.
Phone: + 36 28-526111
Fax: +36 28-526101
Email: *gbkiss@abc.hu*

Fernand Lambein
Institute of Plant Biotechnology for Developing Countries (IPBO), Ghent University, K.L. Ledeganckstraat 35, 9000 Ghent, Belgium.
Phone: +32 92645196
Fax: +32 92648795
Email: *fernand.lambein@ugent.be*

Teresa Millán
Departamento de Genética, Universidad de Córdoba, Campus Rabanales, Edif. C-5, 14071 Córdoba, Spain.
Phone: + 34 957 218508
Fax: +34 957 218503
Email: *ge1mivat@uco.es*

Carlos Molina Medina
Unité de Recherche en Légumineuses, INRA-URLEG, 17 Rue Sully, 21000 Dijon, France.
Phone: +33-380693-430
Fax: +33-380693-263
Email: *carlos.molina@dijon.inra.fr*

Fred J. Muehlbauer
U.S. Department of Agriculture, Agriculture Research Service, 303 Johnson Hall, Washington State University, Pullman, WA 99164-6434. U.S.A.
Phone: +1 509 335 9521
Fax: +1 509 335 7692
Email: *muehlbau@wsu.edu*

Salvador Nadal
IFAPA, Centro Alameda del Obispo, Area de Producción Agraria, Apdo. 3092, 14080 Córdoba, Spain.
Phone: +34 957 016120
Fax: +34 957 016043
Email: *salvador.nadal@juntadeandalucia.es*

Sergio Ochatt
INRA, Centre de Recherches de Dijon, UMRLEG, BP 86510, 21065 Dijon Cedex, France.
Phone: + 33 380693161
Fax: +33 380693263
Email: *ochatt@dijon.inra.fr*

José Olivares
Departamento de Microbiología del Suelo y Sistemas Simbióticos, Estación Experimental del Zaidín, CSIC, Prof. Albareda 1, 18008 Granada, Spain.
Phone: +34 958 181600 Ext. 258
Fax: +34 958 129600
Email: *olivares@eez.csic.es*

Marcelino Pérez de la Vega
Area de Genética, Departamento de Biología Molecular, Universidad de León, 24071 León, Spain.
Phone: +34 987 291550
Fax: +34 987 291409
Email: *m.perez.delavega@unileon.es*

Belén Román
IFAPA, Centro Alameda del Obispo, Area de Mejora y Biotecnología, Apdo. 3092, 14080 Córdoba, Spain.
Phone: +34 957 016125
Fax: +34 957 016043
Email: *mariab.roman@juntadeandalucia.es*

Diego Rubiales
Instituto de Agricultura Sostenible, CSIC, Apartado 4084, 14080 Córdoba, Spain.
Phone: +34 957499215
Fax: + 34 957499252
Email: *diego.rubiales@ias.csic.es*

Josefa Rubio
IFAPA, Centro Alameda del Obispo, Area de Mejora y Biotecnología, Apdo. 3092, 14080 Córdoba, Spain.
Phone: +34 957 016132
Fax: +34 957 016043
Email: *josefam.rubio@juntadeandalucia.es*

María L. Ruiz Sánchez
Area de Genética, Departamento de Biología Molecular, Universidad de León, 24071 León, Spain.
Phone: + 34 987 291548
Fax: + 34 987 291409
Email: *mlruiz@unileon.es*

Juan Sanjuán
Departamento de Microbiología del Suelo y Sistemas Simbióticos, Estación Experimental del Zaidín, CSIC, Prof. Albareda 1, 18008 Granada, Spain.
Phone: +34 958 181600 Ext. 259
Fax: +34 958 129600
Email: *juan.sanjuan@eez.csic.es*

María J. Soto
Departamento de Microbiología del Suelo y Sistemas Simbióticos, Estación Experimental del Zaidín, CSIC, Prof. Albareda 1, 18008 Granada, Spain.
Phone: +34 958 181600 Ext. 150
Fax: +34 958 129600
Email: *mariajose.soto@eez.csic.es*

Fred L. Stoddard
Department of Agricultural Sciences, University of Helsinki, PO Box 27 (Latokartanonkaari 5), FIN-00014 Helsingin Yliopisto, Finland.
Phone: +358 (0)9 191 58 342
Fax: +358 (0)9 191 58 582
Email: *frederick.stoddard@helsinki.fi*

Ana María Torres
IFAPA, Centro Alameda del Obispo, Area de Mejora y Biotecnología, Apdo. 3092, E-14080 Córdoba, Spain.
Phone: +34 957 016178
Fax: +34 957 016043
Email: *anam.torres.romero@juntadeandalucia.es*

Marijke Van Moorhem
Department of Basic Medical Sciences, Faculty of Medicine and Health Sciences, Ghent University, De Pintelaan 185 (Blok B), 9000 Ghent, Belgium.
Phone: +32 93323366
Fax: +32 93323059
Email: *marijkevanmoorhem@hotmail.com*

Francisca Vaquero
Area de Genética, Departamento de Biología Molecular, Universidad de León, 24071 León, Spain.
Phone: + 34 987 291547
Fax: + 34 987 291409
Email: *fvaqr@unileon.es*

Maria Carlota Vaz Patto
Instituto de Tecnologia Química e Biológica (ITQB), Universidade Nova de Lisboa (UNL), Apartado 127, 2781-901 Oeiras, Portugal.
Phone: +351 21 4469461
Fax: +351 21 4411277
Email: *cpatto@itqb.unl.pt*

Stefanie Wienkoop
Department of Molecular Systems Biology, University of Vienna, Althanstrasse 14, 1090 Vienna, Austria.
Phone: +43 1 4277 577 03
Fax: + 43 1 4277 9577
Email: *stefanie.wienkoop@univie.ac.at*

Peter Winter
GenXPro GmbH, Frankfurt Innovation Center Biotechnology (FIZ), Altenhöferallee 3, D-60438 Frankfurt am Main, Germany.
Phone: +49(0)69 95739705
Fax: +49(0)69-95739706
Email: *pwinter@genxpro.de*

Abbreviations

2-DE	Two-dimensional gel electrophoresis
ABA	Abscisic acid
ACE	Angiotensin converting enzyme
ACGM	Amplified consensus gene markers
AFLP	Amplified fragment length polymorphism
AM	Arbuscular mycorrhiza
AMV	Alfalfa mosaic virus
BAC	Bacterial artificial chromosome
BBI	Bowman-Birk inhibitor
BBSV	Broad bean stain virus
BLAST	Basic local alignment search tool
BNF	Biological nitrogen fixation
BSA	Bulked segregant analysis
BYMV	Bean yellow mosaic virus
CA	Carbonic anhydrase
CAGE	Cap analysis of gene expression
CAPS	Cleaved amplified polymorphic sequence
CCaMK	Calcium/calmodulin-dependent protein kinase
CCD	Charge-Coupled Device
cDNA	Complementary DNA
ciRNA	Circular interfering RNA
cM	CentiMorgan
CMV	Cucumber mosaic virus
CP	Viral coat protein
CPR	Chlorophenol Red
CSIC	Consejo Superior de Investigaciones Científicas (Spain)
DAF	DNA Amplification Fingerprinting
DAO	Diamine oxidase
DAPI	4′,6-diamidino-2-phenylindole
dCAPS	Derived cleaved amplified polymorphic sequence
DH	Double haploid
DS	Disease severity
EC	Enzyme Commission
emPCR	Emulsion PCR

EMS	Ethyl methane sulphonate
EPRG	Expressed potential resistance genes
eQTL	Expression QTL
ERA-PG	European Research Area on Plant Genomics
EST	Expressed sequence tag
ESTP	EST polymorphism
FAO	Food and Agriculture Organization of the United Nations
FAOSTAT	FAO Statistical tables and series
FBNYV	Faba bean necrotic yellows virus
FISH	Fluorescence in situ hybridization
GABA	Gamma aminobutyric acid
GC-MS	Gas chromatography coupled mass spectrometry
GCxGC-MS	Two-dimensional gas-chromatography mass spectrometry
GI	Glycemic index
GLIP	Grain Legume Integrated Project
GO	Gene ontology
GS	Glutamine synthetase
GUS	β-glucuronidase
IBPGR	International Board for Plant Genetic Resources
ICA	Independent component analysis
ICARDA	International Center for Agricultural Research in the Dry Areas
ICRISAT	International Crop Research Institute for the Semi-Arid Tropics
IFAPA	Instituto de Investigación y Formación Agraria y Pesquera de Andalucía (Spain)
IGS	Ribosomal intergenic spacer
IMGAG	International *Medicago* gene annotation group
INRA	Institut National de la Recherche Agronomique (France)
IPK	Leibniz Institute of Plant Genetics and Crop Plant Research
IRLC	Inverted repeat-lacking clade
ISSR	Inter simple sequence repeat
IT	Intron targeted
ITAP	Intron-targeted amplified polymorphic
ITS	Internal transcribed spacer
ITPGRFA	International Treaty on Plant Genetic Resources for Food and Agriculture
ITy	Infection type
Lb	Leghemoglobin
LC-MS	Liquid chromatography-mass spectrometry

LC-MS/MS	Liquid chromatography-tandem quadrupole mass spectrometry
LCO	Lipochitooligosaccharides
LEGRESIST	Exploiting Genetic Variability of Resistance Genes in major European Food Legumes to Improve Varieties for Sustainable Agriculture (an ERA-PG project)
LG	Linkage group
LIS	Legume Information Service
LPS	Lipopolysaccharides
LRR	Leucine-rich repeat
LysM	N-acetyl-glucosamine-binding lysine motif
MALDI	Matrix-assisted laser desorption ionization
MALDI-TOF	Time of Flight mass spectrometry
MAPA	Mass accuracy precursor alignment
MAS	Marker-assisted selection
MeJA	Methyl jasmonate
methQTL	Methylation quantitative trait locus
MIQE	Minimum information for publication of quantitative real-time PCR Experiments
miRNA	MicroRNA
MPSS	Massively parallel signature sequencing
MRM	Multiple reaction monitoring
MS	Mass spectrometry
Mt	Million tonnes
MYA	Million years ago
NCBI	National Center for Biotechnology Information
NBS	Nucleotide binding site
NF	Nod factor
NGS	Next generation sequencing
NIL	Near isogenic line
NOR	Nucleolar organizing region
NTS	Non-transcribed spacer
ODAP	β-N-oxalyl-L-α,β-diaminopropionic acid (β-ODAP)
ODB	Oxigen diffusion barrier
PA	Pea seed albumins
PAGE	Polyacrylamide gel electrophoresis
PCR	Polymerase chain reaction
PEBV	Pea early browning virus
PEMV	Pea enation mosaic virus
PEPC	Phosphoenolpyruvate carboxylase
PGE	Polyethylene glycol
PGIP	Polygalacturonase-inhibiting protein
PI	Propidium iodide

PIC	Polymorphic information content
pQTL	Protein quantitative trait locus
PSbMV	Pea seed-borne mosaic virus
pSym	Symbiotic plasmid
qPCR	Quantitative PCR
qRT-PCR	Quantitative real-time PCR
QTL	Quantitative trait locus/loci
RACE	Rapid amplification of cDNA ends
RAPD	Random amplified polymorphic DNA
rDNA	Ribosomal-DNA
RFDD-PCR	Restriction fragment differential display PCR
RFLP	Restriction fragment length polymorphism
RGA	Resistance gene analogue
RNAi	Interference RNA
RIL	Recombinant inbred line
RLK	Receptor-like kinases
ROS	Reactive oxygen species
RT–PCR	Real-time PCR
SA	Salicylic acid
SAAT	Sonication-assisted *Agrobacterium*-mediated transformation
SABE	Serial analysis of binding elements
SACO	Serial analysis of chromatin occupancy
SAGE	Serial analysis of gene expression
SCAR	Sequence characterized amplified region
SGC	Symbiotic genome compartments
SNP	Single nucleotide polymorphism
SOD	Superoxide dismutase
SOLiD™	Sequencing by oligonucleotide ligation and detection
sRNA	Small RNA
SS	Sucrose synthase
SSAP	Sequence-specific amplification polymorphism
SSH	Suppressive subtractive hybridization
SSR	Simple sequence repeat
sstDNA	Single-stranded template DNA
SuperSAGE	Super serial analysis of gene expression
STMS	Sequence-tagged microsatellite site
STS	Sequence-tagged site
TAF	Transcription activation factor
TAR	Transcriptionally active region
TBP	TATA-box binding protein
TC	Tentative consensus sequence
TF	Transcription factor

TFBS	Transcription factor binding site
TI	Tripsin inhibitor
TIGR	The Institute for Genome Research
TILLING	Targeted induced local lesions in genomes
TIU	Trypsin inhibitor units
TOF	Time of flight
TPR	Transcript-poor region
tQTL	Transcriptional QTL,
TRR	Transcript-rich Region
TSS	Transcription start site
UBF	Upstream binding factor
UDP-glucose:	Uridine diphosphate glucose
USDA-ARS	United States Department of Agriculture—Agricultural Research Service
UTR	Untranslated region
WANA	West Asia and North Africa

1

Pea

Diego Rubiales,[1,*] *Michael J. Ambrose,*[2] *Claire Domoney*[2] and *Judith Burstin*[3]

ABSTRACT

Pea is a cool season legume crop produced worldwide, mainly in temperate regions. In this chapter we will review and critically discuss current and future strategies on pea breeding. Conventional breeding methods have been successful in improving pea germplasm towards development of superior cultivars through introduction of novel traits from wild germplasm and landraces as well as pyramiding multiple positive alleles in adapted genetic backgrounds. Detailed genetic maps have been generated, placing many genes of interest using a variety of molecular techniques, and marker-assisted selection (MAS) is possible for some desired traits that have been associated with DNA markers. Nevertheless, the improvement of several important agronomic characters such as disease resistance and stability of seed yield and composition is difficult. Breeding success will depend on availability of consistent resistance genes or quantitative trait loci (QTLs) within or outside the species, and adequate molecular markers flanking these genes or QTLs that can be used to trace resistance in breeding populations. However, the long genetic distance existing in most cases between the identified genetic markers and the QTL, the common lack of codominant markers and the general lack of knowledge on biochemistry of the trait greatly limit the use of genetic markers in marker-assisted breeding. Effectiveness of MAS might soon increase with the adoption of new improvements in marker technology together with the integration of comparative mapping and functional genomics.

Keywords: biotechnology, breeding, comparative mapping, genetic resources, pea, quality, resistance

[1]Institute for Sustainable Agriculture, CSIC, Apdo. 4084, 14080 Córdoba, Spain;
e-mail: *diego.rubiales@ias.csic.es*
[2]John Innes Center, Colney, Norwich NR4 7UH, UK.
[3]INRA, UMRLEG, 17 rue de Sully, BP 86510, F-21065 Dijon, France.
*Corresponding author

1.1 Introduction

In common with other grain legumes, pea (*Pisum sativum* L.) plays a critical role in crop rotation (Nemecek et al. 2008). Because the pea plant is able to fix atmospheric nitrogen through symbiosis with nitrogen-fixing soil bacteria from the *Rhizobium* genus, its cultivation does not require N-fertilizer, reduces the need of N-fertilizer of the following crop, and thus significantly reduces energy costs. It also permits a higher diversification of crop rotations, improving soil physical conditions and decreasing the influence of pathogen and weed populations. Finally, it provides protein-rich raw material complementary to cereals for animal and human nutrition. Energetically and economically, it is an advantage to utilize nitrogen inputs from this legume source for securing carbon sequestration into food, feed and biofuel crops (Nemecek et al. 2008).

Pea is a cool season legume crop produced worldwide, mainly in the temperate regions. Dry pea production amounted to 9.8 Mt produced over 5.9 Mha worldwide in 2008, with 22% produced in Europe (1.3 Mha), 29% produced in Asia (1.7 Mha), and 32% produced in North America (1.9 Mha). This production is less than soybean (230 Mt) and dry bean (20.4 Mt) (FAOSTAT: http://faostat.fao.org). The main dry pea producers are Canada, China, India, Russian Federation, the USA, and France. The contribution of these different countries has dramatically changed in the recent decades. The pea crop area has significantly increased in Canada, from 259 kha in 1992 to 1.6 Mha in 2008, making this country the world's leading producer and exporter of dry pea. This rapid increase in production was prompted by the rotational advantages of pea (Lupwayi and Kennedy 2007) and the availability of cultivars adapted to dry land conditions of the northern Great Plains. Conversely, the pea crop area decreased significantly in western Europe during the same period. In France, the pea cultivated area dropped from 700 kha in 1992 to 100 kha in 2008. European agriculture relies extensively on nitrogen fertilizer and chemical pesticides and lacks diversity. It is thus highly important to re-introduce leguminous crops into farming systems and raise their profile through the revision of agricultural policies towards achieving this aim. Fresh pea production amounted to 8.3 Mt in 2008 worldwide. The main fresh pea producers are China, India, the USA, France and Egypt (FAOSTAT: http://faostat.fao.org).

Pea seeds are used in a wide range of forms. Approximately half of the dry pea world production is fed to livestock while the remaining half is used for human food, primarily in developing countries. Dry mature seeds are used in diverse food specialties but can also be industrially processed to produce starch and protein concentrates. Immature seeds can be eaten fresh, canned, or frozen. Pea seed nutritional value relies on a balanced composition: dry pea seeds contain 18–33% proteins, 35–50%

starch and 4–7% fibres. Pea seed proteins are deficient in sulphur amino acids, but contain relatively high levels of lysine, making pea a good dietary complement to cereals. No major antinutritional factor is present in pea. Tannins, trypsin inhibitors, and lectins are relatively minor or can be easily counter-selected, and thus do not adversely affect crop use. Colour, appearance, flavour and dietary proprieties are significant criteria for peas used for human consumption.

Pea is generally considered to have originated in the Near East region and its domestication may have occurred as early as 7000–6000 BC (Smartt 1990). Peas were disseminated towards Europe, together with wheat and barley ca 5500 BC. They also spread towards East Africa through Egypt and Ethiopia ca 5000 BC, and towards India ca 2000 BC. The *Pisum sativum* species includes an extraordinary variety of forms and uses. Its taxonomy has long been a subject of debate. Based on several lines of evidence, a consensus has emerged (Ben Ze'ev and Zohary 1973; Smartt 1990; Maxted and Ambrose 2001). The *Pisum* genus includes two species, *P. sativum* and *P. fulvum*, which can be crossed to produce viable hybrids only when *P. fulvum* is used as the male parent (Ben Ze'ev and Zohary 1973). Wild pea forms from which cultivated types originated (i.e., former *P. elatius* and *P. humile*) are now considered as one subspecies *P. sativum* ssp. *elatius*. Cultivated forms that are also very diverse morphologically and that became differentiated following the action of both natural and human selection (formerly *P. arvense*, *P. hortense*, *P. abyssinicum*, and *P. transcaucasicum*) are now commonly grouped into one *P. sativum* subspecies *sativum*. This diversity of form prompted early experimenters in plant hybridization to analyze pea segregating populations and allowed Gregor Mendel's initial discovery of inheritance (Mendel 1866).

Pea is an annual plant, mainly autogamous and cleistogamous even though cross-pollination can occur in nature through the action of insect pollinators. Pea genetic improvement began with domestication and has continued throughout history. Since that time, breeders discovered and introduced novel traits to adapt the indeterminate, tall, slender, bushy or climbing plants with small and coloured seeds to a short, determinate field combining crop with large seeds without tannins. Significant improvement was made progressively using genes with major effects: for example, indehiscent pods through the use of *dpo*, palatability through tannin removal (*a*), dwarfism (*le*) and, more recently, sugary taste (*r* and *rb*) or enhanced gripping through tendril replacement of leaflets (*af*). Conventional breeding methods have been successful in improving pea germplasm towards development of superior cultivars through introduction of novel traits from wild germplasm and landraces as well as pyramiding multiple positive alleles in adapted genetic backgrounds. Detailed genetic maps have been generated for pea, placing many genes of interest using a variety of molecular techniques, and marker-assisted selection (MAS) is possible for some desired traits that have been associated with DNA markers. A series of problems have held back progress in the development of genomic resources in pea as a result of its relatively large genome size comprising an estimated 90% of of repetitive sequences (Ellis and Poyser 2002). This is however being addressed through the increasing use of genomic resources generated in other legumes, notably *Medicago truncatula* with which the pea genetic map shares a high degree of colinearity. Nevertheless, the improvement of several important agronomic characters in pea such as disease resistance and stability of seed yield and composition is difficult and molecular markers are lacking. The development of genomic resources in pea is needed for further breeding progress.

1.2 Taxonomy, Genetic Resources and Diversity

The primary centre of diversity for pea is in the eastern Mediterranean and the Near and Middle East where wild forms of *P. fulvum* and *P. sativum* ssp. *elatius* can still be found growing today. The distribution of *P. fulvum* is restricted to the eastern Mediterranean (Syria, Lebanon, Israel, Palestine and Jordan), whereas *P. sativum* ssp. *elatius* is distributed more widely across from Spain through to the Middle East (Maxted and Ambrose 2001). Carbonized remains found at neolithic farming villages in northern Iraq, southern and southeastern Turkey and Syria indicate the cultivation and use of pea as food as early as 7000–6000 BC. Soon after this, pea was found in remains at sites in southern Europe (Zohary and Hopf 1973). Outside the primary centre, two important secondary centres of diversity for pea are recognized. First is the highland Asiatic region of the Hindu Kush where the distinct long vined "Afghan" types through to the shorter statured Tibetan ecotype are grown at high altitudes. A further secondary centre of diversity includes the central highland region of Ethiopia and uplands of southern Yemen, which covers the currently known distributional range of *P. sativum* ssp. *abyssinicum* where it grows sympatrically with *P. sativum*. The taxon is well described and distinct from all other *P. sativum* forms for a range of morphological characters including foliage, flower colour and seed shape. Molecular diversity studies have confirmed the narrow genetic base within known germplasm of this form but also its distinctness from all other cultivated forms, which strongly suggest this taxon to be an independent domestication event to that of *P. sativum* (Lu et al. 1996; Vershinin et al. 2003). The continual drive to identify new variation to assist with crop improvement means there is constant screening and evaluation of germplasm. While the bulk of this occurs in more cultivated material (primary genepool), the absence of relevant variation for specific traits focuses attention on screening more exotic and even wild germplasm.

This is especially true in the case of disease and pest resistance (Maxted and Kell 2009).

The worldwide distribution of peas coupled with their widespread popularity under cultivation has contributed to the extensive genetic diversity within this genus. This alongwith their autogamous nature and diploid status makes them relatively easy to maintain as fixed inbred lines and has been instrumental in making pea a model system in the study of plant genetics. A large number of *ex situ* germplasm collections for pea have been developed to provide long term conservation and ready access to the broad range of diversity from wild crop relatives, landraces, elite cultivars, mutants, mapping populations and host differentials for use by research and to underpin the requirements of breeders (Ambrose 2008a). These *ex situ* collections have a long history of active collaboration (Ambrose and Green 1991). A working group for grain legumes exists as part of the European Cooperative Program for Crop Genetic Resources, which brings together the formal and informal sectors to collaborate on areas of common interest (ECP/GR; http://www.ecpgr.cgiar.org/Networks/Grain_legumes/grain_legumes.htm). A recent output from this group is a web searchable compilation of passport data for the holdings of 43 collections from 26 European countries representing some 32,500 accessions (Ambrose 2008b). This initiative is being further expanded to include other key international pea collections to form a world inventory as an activity organized by the International Consortium for Pea Genetic Resources (PeaGRIC), that links together key collections within Europe, the USA, ICARDA and Australia (Furman et al. 2006).

The community of pea geneticists has a long history of collaboration and exchange of novel variants and mutant stocks. Since the work of Vilmorin in Paris in the 1890s through to key geneticists such as Lamprecht, Lamm, Gottschalk, Marx and Blixt, the collection of classical mutants in pea has grown into a rich resource that now extends to mapping populations and near-isogenic lines (PGene: http://www.jic.ac.uk/GERMPLAS/pisum/Zgc4g.htm). Other notable collections include the G.A. Marx Pea genetic stocks collection (http://www.ars.usda.gov/Main/docs.htm?docid=15144) and the more recently formed TILLING population (Dalmais et al. 2008).

Investigations into the structure of diversity within cultivated pea and the relationship to wilder forms have been the subject of a growing number of studies for which there are several drivers; to understand and refine phylogenetic relationships within the genus, to clearly delineate differences between different cultivated forms (Amurrio et al. 1995) and to support the structuring, management and sampling of germplasm collections. These studies explore the relationships between different cultivated forms, the taxonomic structure and the organization of germplasm collections and assess the relationship between the different cultivated forms and wild germplasm. Several molecular diversity studies in recent years have had a significant impact on our understanding of the underlying nature of the diversity within pea germplasm, highlighting the importance of ecogeographical factors (Burstin et al. 2001; Baranger et al. 2004; Tar'an et al. 2005; Kosterin and Bogdanova 2008; Zong et al. 2009; Kosterin et al. 2010) and the significance that introgression has played in the development of population stucture (Lu et al. 1996; Ellis et al. 1998; Pearce et al. 2000; Vershinin 2003; Baranger et al. 2004; Jing et al. 2007). Further improvement in the reliability of marker systems and the ability to develop them as high throughput systems (Flavell et al. 2003), together with enhanced database software capabilities for storing and analyzing large datasets, make this a very dynamic area of germplasm exploration. It is now entirely feasible to screen entire germplasm collections with markers which are opening up new opportunities to study the underlying population structure of the genus. The first such example in pea is the application of retrotransposon element markers to the entire John Innes *Pisum* collection (TEGERM; Jing et al. 2010). The level of activity in exploring the structure of diversity within the genus is such that there is a concerted effort to ensure that common markers and reference germplasm accessions are used to enable both a better integration of data sets and a combined effort in establishing a world *Pisum* core collection (Smykal et al. 2008).

1.3 Genetics and Genomics

1.3.1 From the Origin of Pisum Genetic Studies to Recent Molecular Maps

The *P. sativum* genome is large (ca 5,000 Mbp) and diploid. The chromosome number of the *Pisum* genus is $2n = 14$ (Ben Ze'ev and Zohary 1973) and the pea karyotype includes two sub-metacentric chromosomes (1 and 2) and five acrocentric chromosomes (3, 4, 5, 6, and 7) showing a secondary constriction at the 45S rRNA gene cluster (Ellis and Poyser 2002). Several linkage re-arrangements, such as translocations between non-homologous chromosomes, have been reported in different pea accessions (Hammarlund 1923, 1929, cited by Blixt 1972). Moreover, the pea genome is rich in retrotransposons. A recent paper by Macas et al. (2007) analyzed the repetitive DNA in the pea genome using 454 sequencing. It revealed the predominance of Ty3-*gypsy* retroelements, representing an estimated proportion of ca 24% of the genome, with *Ogre*-like retrotransposons making up ca 20% of the genome, followed by Ty1-*copia* retroelements (ca 7% of the genome), rDNA (ca 1% of the genome) and DNA transposons (ca 0.4% of the genome).

Pea is a rather polymorphic species, presenting extensive phenotypic variability. Hybridization experiments between different pea morphotypes have been performed since the XIXth century and genetic analyses began in 1865 with Mendel's study of the inheritance of seven major effect genes: *i*, *a*, *le*, *v* or *p*, *fa*, *r* and *gp* present on five or six linkage groups (Fig. 1-1A) with the corresponding phenotypic variation shown in Figure 1.1B. In 1912, Vilmorin and Bateson described linkage relationships among different genes in pea (cited by Blixt et al. 1972), a first mutation map was published by Wellensiek in 1925 and the first pea genetic map comprising seven linkage groups (LGs) and 37 morphological markers was published by Lamprecht in 1948 (cited by Rosov et al. 1999). A multitude of conventional morphological markers has been described in pea, at first from spontaneous mutations and then from induced mutations. In 1972, Blixt published a mutation map including

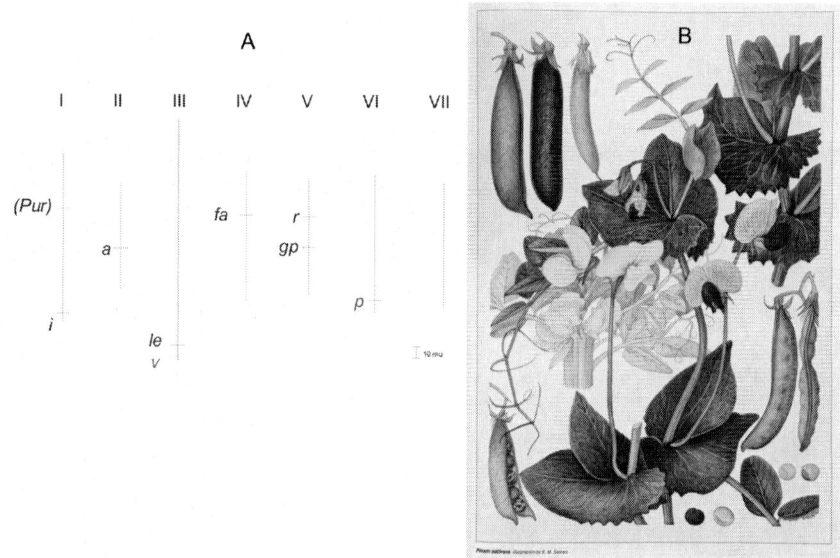

Figure 1-1 (A) The positions of the genes recorded by Mendel on five of the seven linkage groups in pea, where six of the seven traits can be assigned with certainty (blue). In grey are the positions of two genes controlling parchment (content of sclerenchyma) in the pod wall, a trait also studied by Mendel. Since *p* and *v* have similar phenotypes, Mendel could have studied either one. *Pur* (brackets), controlling purple pods, is a trait that Mendel observed but he did not study its inheritance. Genetic information supplied by Noel Ellis, John Innes Centre, Norwich NR4 7UH, UK. (B) The phenotypes of the traits studied by Mendel, reproduced from "Legumes of the World" (G. Lewis, B. Schrire, B. Mackinder, M. Lock eds.), Kew Publishing, 2005, by kind permission of the artist, Margaret Stones. Phenotypic variation at *i* (yellow or green cotyledon colour), *a* (purple or white flower colour), *le* (long or short internodes), *fa* (normal or fasciated stem), *r* (round or wrinkled seeds), *gp* (green or yellow pod colour) and *p* or *v* (parchmented or reduced parchment pods) loci is shown. *Pur* is a distinct locus (see Fig. 1-1A).

Color image of this figure appears in the color plate section at the end of the book.

169 morphological markers. Since then, pea genetic mapping has exploited new morphological (for example, Rozov et al. 1999), isozyme (Weeden and Marx 1984, 1987), restriction fragment length polymorphism (RFLP, Ellis et al. 1992; Dirlewanger et al. 1994), random amplified polymorphic DNA (RAPD, Laucou et al. 1998), retro-transposon based (Ellis et al. 1998), amplified fragment length polymorphism (AFLP, Timmerman-Vaughan et al. 1996; Irzykowska et al. 2001), microsatellite (Loridon et al. 2005) and expressed sequence tag (EST) or gene-derived markers (Gilpin et al. 1997; Aubert et al. 2006; Deulvot et al. 2010), to obtain more precise and reliable genetic maps. In all cases, common markers have allowed data to be related and integrated consensus maps to be developed. Similarly, the cytological analyses of the pea chromosomes improved, together with the use of more precise techniques. From the use of translocation lines (Blixt 1959; Lamm and Miravalle 1959; Hall et al. 1997) to fluorescent *in situ* hybridization (Fuchs et al. 1998), the correspondence between LGs and chromosome became more refined.

A linkage map including more than 200 morphological and isozyme markers was published by Weeden et al. in 1993 and was updated in 1998 to integrate morphological, isozyme, RFLP and EST-derived markers (Weeden et al. 1998). Based on complementary DNA marker data obtained from three different recombinant inbred line (RIL) populations, Ellis and Poyser (2002) further refined the pea classical marker linkage map. Indeed, inherent difficulties are linked to the use of morphological markers, such as the effect of pleiotropy, epistasis or the environment on the expression of phenotypic traits, and the limited number of variants in any given segregating population. Additionally, the report of frequent translocations among pea accessions had blurred some linkage analysis results. Ellis and Poyser (2002) specified the linkage relationship and chromosome assignments of 120 classical markers described by Blixt (1972). It was concluded that LG I corresponds to chromosome (Chr.) 2, LG II to Chr. 6, LG III to Chr. 5, LG IV to Chr. 4, LG V to Chr. 3, LG VI to Chr. 1 and LG VII to Chr. 7.

The advent of new generation easy-to-use PCR markers facilitated a significant increase of linkage mapping capacity in pea. At first, anonymous non-specific marker systems such as RAPD and/or AFLP were used to rapidly saturate linkage maps dedicated to gene or quantitative trait locus (QTL) mapping (Timmerman-Vaughan et al. 1996; Laucou et al. 1998; Pilet-Nayel et al. 2002; Tar'an et al. 2003; Prioul et al. 2004; Fondevilla et al. 2008b). Because the transferability of these markers from one mapping population to another is poor, though possible, other marker systems largely used in other species were developed in pea. Highly polymorphic microsatellite or simple sequence repeat (SSR) markers were developed with the view of connecting mapping data obtained for different mapping populations (Burstin et al. 2001; Loridon et al. 2005). A reference map was built by Loridon et al. (2005) based on segregation patterns observed in three RIL populations and included 229 microsatellite markers. This set of microsatellite markers has been used successfully in many other mapping populations (Prioul et al. 2004; Ek et al. 2005; Wang et al. 2008; Mishra et al. 2009). Gene-based markers were also developed in order to provide candidates for gene identification. Functional maps were developed (Aubert et al. 2006; Bordat et al. 2011), using a range of different methods to reveal size or single nucleotide polymorphism (SNP) in the mapped genes. More specific candidate genes, for example disease-resistance genes (Timmerman-Vaughan et al. 2000; Prioul-Gervais et al. 2007) were also mapped. With the advent of high-throughput sequencing and SNP genotyping technologies, new sets of SNP markers will soon be available in pea for high-throughput genotyping and consensus mapping (Deulvot et al. 2010).

1.3.2 Comparative Mapping

Linkage maps, especially those that include gene-based markers, allow comparative mapping among related species. Linkage maps of lentil (Weeden et al. 1992) and chickpea (Simon and Muehlbauer 1997) showed some colinearity with the pea map, suggesting that genome synteny may be conserved among cool-season legumes. Linkage maps of *Medicago truncatula* and *M. sativa* were also compared with that of pea. The pea genome is 10 times larger than the *M. truncatula* genome. The major interest in identifying the molecular determinants of legume specific traits, such as the *Rhizobium* symbiosis, has boosted the development of genomics resources and the sequencing of the gene-space of legume model species with small diploid genomes (Sato et al. 2007). The sequence of euchromatic regions of *M. truncatula* is almost completed (http://medicago.org.genome/). To assess the conservation of synteny between pea and *M. truncatula*, sets of bridge markers were mapped in both *M. truncatula* and pea genomes. Gutierrez et al. (2005) tested the cross amplification of EST-derived microsatellite markers from *M. truncatula* to three major pulse crops, including *P. sativum*: while ca 37% of primers designed in *M. truncatula* succeeded in amplifying a fragment of the same size in pea, the level of conservation and polymorphism of the embedded microsatellite motif was very low, preventing the use of these sequences for comparative mapping. Choi et al. (2004) designed gene-specific primers in exon regions flanking introns in order to sequence the introns of these genes in the parents of mapping populations from several legume species to be compared to *M. truncatula*. Fifty-seven gene polymorphisms were mapped in both pea and *M. truncatula* and revealed a good conservation of synteny among the two species. Aubert et al. (2006) compared the map positions of 45 gene markers, of which 43 were different from the genes analyzed by Choi et al. (2004). They confirmed the conservation of synteny between pea and *M. truncatula,* with a possible inversion in the middle of pea LG II relative to *M. truncatula* LG 1. Kaló et al. (2004) compared the map positions of 74 gene sequences between the diploid *M. sativa* and the pea genomes: extensive colinearity was observed. From these three studies, we can deduce that pea LG I corresponded to *Medicago* LG 5, LG II to LG 1, LG III to LG 3 and bottom of LG 2, LG IV to LG 8, LG V to LG 7, LG VI to LG 6 and top of LG 2, and LG VII to LG 4 (see Chap 7). Refined syntenic information is now available thanks to new pea gene maps and to the published genome sequences of *M. truncatula, L. japonicus,* and soybean (Bordat et al. 2011).

The conservation of genome structure between pea, *Medicago,* lentil, and chickpea, irrespective of the genome size of these species, suggests that genome size variation is unlikely to be due to large segmental duplications but rather to retrotransposon evolution. Interestingly, Macas et al. (2007) compared the repetitive DNA genome composition of pea, soybean and *M. truncatula.* Little repetitive sequence similarity was observed between pea and soybean, except for rDNA sequences. Conversely, repetitive sequences were more similar between pea and *M. truncatula* but show different genomic proportions.

1.3.3 Mapping, Tagging, and Identifying Simply Inherited Traits

Mapping major genes affecting important target traits has been a constant activity since the beginning of pea genetics. Numerous major genes controlling disease resistance (for example: *er1*, *er2* and *Er3*, Timmerman-Vaughan et al. 1994; Janila and Sharma 2004; Fondevilla et al. 2007c; *En*, Yu et al. 1995; BYMV, Yu et al. 1996; *Fwf*, Coyne et al. 2000), seed composition (*Tri*, Page et al. 2002), flowering time and plant architecture (Rameau et al. 1998; von Stackelberg et al. 2003) have been mapped and tagged. Many results of gene tagging have been published and regularly posted in the *Pisum* gene list (http://data.jic.bbsrc.ac.uk/cgi-bin/pgene/).

Identifying the genes behind mutations is a challenging task in a species with such a large genome as pea. To overcome the difficulty, the use of candidate genes whether based on their function or because they are orthologs of genes identified in another species has proved quite successful. Several genes used by Mendel have been identified: *Le*, the gene controlling internode length, was identified as an enzyme from the gibberellin biosynthetic pathway (Lester et al. 1997). Mendel's green cotyledon gene was identified by Sato et al. (2007) as being an ortholog of a rice stay-green gene. The *r* mutation and four other independent mutations determining the wrinkled seed trait in pea were all identified in genes encoding enzymes involved in starch biosynthesis: *r* encodes the starch branching enzyme I (Bhattacharyya et al. 1990), *rb* encodes ADP-glucose pyrophosphorylase

(Martin and Smith 1995), *rug3* encodes phosphoglucomutase (Harrison et al. 1998), *rug4* encodes a sucrose synthase (Craig et al. 1999), and *rug5* encodes starch synthase II (Craig et al. 1998). Major genes involved in the control of pea-*Rhizobium* symbiosis were identified by way of orthologous genes in *M. truncatula* (*sym9*, Levy et al. 2004) or *Lotus japonicus* (*sym29*, Krusell et al. 2002). Foucher et al. (2003) identified *det* and *lf* as homologs of the TFL1 family controlling flowering time in several plant species. *Stamina pistilloida* was identified as the ortholog of *Fim* and *UFO* (Taylor et al. 2001). However, in some cases, positional cloning has been inevitable but was aided by the availability of genomic sequence of closely related model species. Comparative mapping of *L. japonicus* and pea *Cycloidea* genes helped to identify *K* and *LST1*, two genes controlling floral zygomorphy in pea (Wang et al. 2008). Hofer et al. (1997) cloned *Unifoliata*, a gene regulating leaf and flower morphogenesis. *Tendril-less* was isolated by cloning a fast neutron deletion mutant of a Class I HDZIP transcription factor (Hofer et al. 2009).

All together, this report shows that gene discovery is possible in pea, is greatly enhanced by the comparative approach with model species, and will take advantage of the wealth of mutations identified in pea.

1.3.4 Mapping ComplexTraits

QTLs have been mapped for a wide range of agronomic traits in pea. Results are summarized in Table 1-1. Most QTL mapping studies were carried out on RIL populations, allowing replicated measurements in a range of environments. Provided that common molecular markers are placed on the maps serving QTL detection, it is possible to identify genomic regions consistently involved in the variation of target traits in different populations for different environments. *Aphanomyces* partial resistance QTLs have been mapped in one pea population, in the field and the greenhouse after inoculation with two pathogen isolates (Table 1-1). Three QTLs (*Aph1*, *Aph2*, and *Aph3*) were consistently mapped in these three conditions (Pilet-Nayel et al. 2005). Ascochyta blight partial resistance QTLs have been mapped in six different pea populations, both in field and growth chamber experiments (Table 1-1). Common Ascochyta blight partial resistance QTLs were identified on LG II, LG III, LG V, and LG VII in two, three or four populations (Timmerman-Vaughan et al. 2004, 2002; Prioul et al. 2004; Fondevilla et al. 2008b). However, more common markers should be added to the respective maps to refine these colocations. Seed trait QTLs were mapped in upto six different populations and in several field environments (Timmerman-Vaughan et al. 1996, 2005; Irzykowska and Wolko 2004; Tar'an et al. 2005; Burstin et al. 2007). Here also, the lack of common markers makes it difficult to relate the data. However, both

Table 1-1 Summary of QTLs detected in pea for biotic and abiotic stress resistance, seed traits, plant development and growth.

Reference	Trait	Environment	Cross	Population size	N° of QTLs	R^{2a}
Biotic stress resistance						
Pilet-Nayel et al. 2002	*Aphanomyces euteiches* partial resistance	Field (4 envir.)	Puget x 90-2079	127 RIL	7	0.06–0.47
Pilet-Nayel et al. 2005	"	Greenhouse Strain SP7	"	"	6	0.08–0.16
"	"	Greenhouse Strain Ae108	"	"	6	0.07–0.14
Dirlewanger et al. 1994	Ascochyta blight partial resistance	Field (1 envir.)	Erygel x 661	102 $F_{2:3-4}$	3	0.71
Tar'an et al. 2003	"	Field (11 envir.)	Carneval x MP1401	88 RIL	3	0.08–0.14
Timmerman-Vaughan et al. 2002	"	Field (3 envir.)	3148-A88 x Rovar	133 $F_{2:3-4}$	13	0.09–0.35
Timmerman-Vaughan et al. 2004	"	Field (4 envir.)	3147-A26 x Rovar	148 $F_{2:3-4}$	11	0.07–0.21
"	"	Field (4 envir.)	3148-A88 x Rovar	133 $F_{2:3-4}$	14	0.06–0.36
Prioul et al. 2004	"	Field (adult stage)	JI296 x DP	135 RIL	10	0.06–0.29
"	"	Growth chamber (seedling)	"	"	6	0.05–0.18
Fondevilla et al. 2008b	"	Field (adult stage)	*P. s.* ssp *syriacum* x Messire	111 RIL	4	0.09–0.52
"	"	Growth chamber (seedling)	"	"	4	0.09–0.29
Weeden and Porter 2007	Fusarium root rot partial resistance	Greenhouse	CMG x PI 220174	275 RIL	3	-
Valderrama et al. 2004	*Orobanche crenata* resistance	Field (1 envir.)	*P. s.* ssp *syriacum* x Messire	115 F_3	2	0.10–0.11

Fondevilla et al. 2010b	″	Field (2 envir.)	″	111 RIL	4	0.11–0.17
″	″	Minirizotrons	″	111 RIL	7	0.13–0.37
Abiotic stress resistance						
Lejeune-Hénaut et al. 2008	Winter Frost tolerance	Field (11 envir.)	Champagne x Térèse	164 RIL	6	0.03–0.49
Dumont et al. 2009	″	Field (2 envir.)	″	78 RIL	4	0.06–0.46
″	Frost Damage	Growth chamber (GC)	″	″	4	0.11–0.30
″	Raffinose content	Field and GC (3 envir.)	″	″	2	0.09–0.25
″	Glucose content	Field and GC (3 envir.)	″	″	1	0.12
″	RubisCO content	Field and GC (3 envir.)	″	″	1	0.16
″	Electrolyte leakage	Field and GC (3 envir.)	″	″	1	0.03–0.12
Seed traits						
Timmerman-Vaughan et al. 1996	Seed weight	Field (1 envir.)	Primo x OSU442-15	102 F_2	3	0.13–0.26
″	″	Greenhouse (1 envir.)	JI1794 x Slow	51 RIL	7	-
Irzykowska and Wolko 2004	Seed Yield	Field (1 envir.)	Wt10245 x Wt11238	114 F_4	6	0.09–0.47
″	Seed Number	″	″	″	5	0.10–0.55
	Seed weight (X1000)	″	″	″	5	0.10–0.24
″	Seed Protein content	″	″	″	5	0.13–0.25
Tar'an et al. 2004	Seed yield	Field (13 envir.)	Carneval x MP1401	88 RIL	4	0.39
″	Seed Protein content	″	″	″	3	0.45

Table 1-1 contd....

Table 1-1 contd....

Reference	Trait	Environment	Cross	Population size	N° of QTLs	R^{2a}
Timmerman-Vaughan et al. 2005	Seed yield	Field (3 envir.)	Primo x OSU442-15	227 $F_{2:3-4}$	5	0.05–0.15
"	Seed Number	"	"	"	9	0.03–27
"	Seed weight (X1000)	"	"	"	10	0.03–0.19
Burstin et al. 2007	Seed yield	Field (5 envir.)	Térèse x K586	139 RIL	6	0.09–0.53
"	Seed Protein content	"	"	"	8	0.10–0.46
"	Seed Number	"	"	"	5	0.10–0.44
"	Seed weight (X1)	"	"	"	9	0.09–0.32
"	Number of cotyledon cells	"	"	"	2	0.19–0.27
"	Volume of cotyledon cells	"	"	"	4	0.20–0.32
Plant phenology and morphology						
Tar'an et al. 2003	Lodging resistance	Field (11 envir.)	Carneval x MP1401	88 RIL	2	0.18–0. 51
"	Plant Height	"	"	"	3	0.06–0.47
Tar'an et al. 2004	Days to maturity	Field (13 envir.)	"	"	4	0.35
Timmerman-Vaughan et al. 2005	Number of nodes	Field (3 envir.)	Primo x OSU442-15	227 $F_{2:3-4}$	6	0.06–0.33
"	First flowering node	"	"	"	4	0.06–0.20

″	Number of flowering nodes	″	″	″	5	0.07–0.12
″	Harvest index	″	″	″	4	0.06–0.20
Burstin et al. 2007	Beginning of flowering	Field (5 envir.)	Térèse x K586	139 RIL	8	0.12–0.67
″	End of flowering	″	″	″	8	0.11–0.47
″	Plant Height	″	″	″	4	0.09–0.90
″	Number of basal branches	″	″	″	7	0.09–0.63
″	Harvest index	″	″	″	7	0.11–0.66
Plant growth and physiology						
Burstin et al. 2007	Nitrogen harvest index	Field (5 envir.)	Térèse x K586	139 RIL	8	0.09–0.34
″	Nitrogen Nutrition Index	″	″	″	3	0.08–0.64
″	Plant Biomass (DM at BSF)	″	″	″	6	0.10–0.45
″	Plant %N at BSF	″	″	″	6	0.10–0.76
″	Plant Vegetative Biomass at Harvest	″	″	″	7	0.09–0.57
″	Plant %N at harvest	″	″	″	5	0.09–0.66

[a] Part of phenotypic variance explained by the QTLs.

Timmerman-Vaughan et al. (2005) and Burstin et al. (2007) suggested a significant role of developmental genes in seed yield and protein content variation. Burstin et al. (2007) further used the QTL approach in order to identify the physiological determinants of seed yield and protein content, by using indicators of plant source capacity (for example, nitrogen nutrition efficiency) and sink potential (for example, seed cotyledon cells). This study suggested a predominant role of plant source capacity in determining seed protein content and yield.

The definition of a common set of easy-to-use markers is necessary to unify QTL results from different laboratories and identify consistent QTL regions. SSR markers are highly polymorphic and provide a useful tool towards this aim. Forthcoming SNP marker sets could also be adapted to this aim. As for major genes, the identification of the underlying molecular determinants can greatly benefit from model species genome sequences and tools. However, for most of the traits considered here, the effect of environment on their expression is high and the precise evaluation of phenotypes is a real challenge.

1.3.5 Structural and Functional Genomics Resources

Genomic resources available in pea are scarce. Around 30,000 pea sequences are available in the GenBank database. These include EST-sequences obtained from suppressive subtractive hydridization (SSH) libraries developed from seedlings or plants submitted to different stresses (J Burstin, I Lejeune-Hénaut, A. Baranger, C. Rameau, X Sarda, unpubl. results), and from pea shoot apical meristem (Wong et al. 2008). Other sequences obtained from seed EST libraries are available at IPK Gatersleben (http://pgrc.ipk-gatersleben.de/cr-est/index.php). Some post-genomic approaches have been undertaken in pea. Transcriptomic analyses were carried out using a 6k oligo-array (Ps6kOLI1) established mainly for seed development analyses (Weigelt et al. 2008, 2009). Proteomic reference maps of roots, leaves and seeds under different stresses were produced (Castillejo et al. 2004, 2010a; Schiltz et al. 2004; Curto et al. 2006; Dumont et al. 2009; Bourgeois et al. 2009; Islam et al. 2009) but the paucity of sequence data renders proteomic analyses difficult.

For gene identification, pea bacterial artificial chromosome (BAC) libraries were developed at USDA Pullman (3.2×, genotype PI269818, Coyne et al. 2007) and at INRA URGV (9×, cv Cameor, A Bendhamane pers. comm.). For forward and reverse genetics, large mutant populations are available, such as the pea TILLING mutant population of ca 5,000 lines developed recently from the cv Cameor (Dalmais et al. 2008; http://www.legumbase.fr). Some transgenic peas have been described (for example, Charlton et al. 2004; Islam et al. 2009) but no efficient and reliable transformation system is available to routinely express protein fusions or for RNA-interference (RNAi).

1.4 Breeding

1.4.1 Breeding for Quality: Exploiting Natural and Induced Variation in Seed Quality Traits for Improved End Use

The promotion of legume crops as significant contributors to sustainable agriculture is linked to many factors, including aspects of seed quality. Any definition of seed end-use quality traits relates to the various requirements of feed and food industries that, at least for pea, can pose different or opposing challenges. These diverse requirements are united through the common scientific goal of identifying and understanding aspects of seed development related to product accumulation, and combining these with genomic and genetic data to enable MAS in breeding programs. In addition, such information can also benefit the exploitation of seeds for non-food use, which is considered below. The development of molecular markers facilitates breeding programs for seed traits in particular, where early screening can reduce population sizes, thus saving on space and costly seed assays.

Higher seed protein content is desired for the use of pea in animal feed in some, but not all, EU countries, while the relevance of various antinutrients also differs among countries. For food use, there are important traits that may be either specific, or common, to the use of seeds as immature fresh, canned or mature dried products. It is clear that some of the so-called antinutrients, defined in the context of optimizing animal feed, may act as health-promoting agents for humans. All of these traits require a definition of their genetics, ideally coupled with an understanding of the underlying biochemistry, to provide and optimize the use of tools and resources that may be exploited for a broad range of end-uses by the feed and food industries.

1.4.1.1 Manipulating Seed-Expressed Genes for Optimal Feed Use

Several proteins have been defined as either being poorly digested themselves, or interfering directly with digestive processes. The pea seed albumin 2 (PA2) and the trypsin/chymotrypsin inhibitor (TI) proteins fall into these two respective classes. PA2 is a seed protein that has been shown to resist digestion in piglets and chickens and it can persist through the digestive tract (Crévieu et al. 1997; Salgado et al. 2003; Le Gall et al. 2007). A naturally occurring variant line, JI 1345, that lacks PA2 was identified in the John Innes *Pisum* germplasm collection that appeared to lack most of the PA2 genes present in other pea lines. The null mutation was introgressed into a cultivar (cv Birte) through development of an RI population and associated backcrossing (Fig. 1-2A; Vigeolas et al. 2008). Studies of the RI and backcross (BC) lines showed that a loss of PA2 did not impact negatively on seed protein content, but rather the PA2-deficient lines showed higher seed protein, despite their reduced albumin content (Vigeolas et al. 2008). This study demonstrated that variation in the amount of a second albumin, PA1, is not associated directly with a compensatory increase in seed protein in lines lacking PA2. Expression analyses using the Ps6kOLI1 microarray have identified genes that are more highly expressed in the absence of PA2 gene expression in mutant lines. These data pinpoint genes whose expression may be associated more generally with higher seed protein content, and this is under investigation. There are several genetic loci associated with

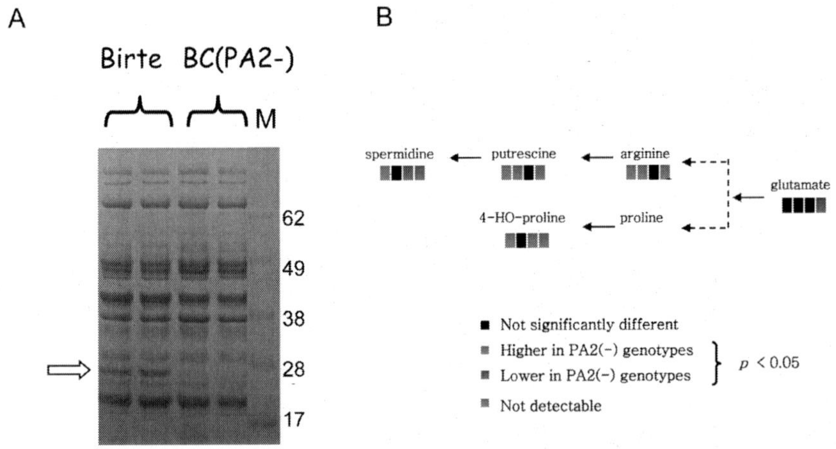

Figure 1-2 (A) Seed protein profiles of mature seeds from the cv Birte and lines lacking pea albumin 2 (PA2, open arrow), derived from two backcrosses (BC) of a selected mutant RIL with cv Birte. (B) The metabolite changes in seeds of the BC lines are indicated for the polyamine pathway, where a higher content of a metabolite in the mutant is indicated in red. The four coloured boxes indicated for every metabolite are based on measurements made at four stages (10, 20 and 30 days after flowering, and in mature seeds, left to right) of seed development (Vigeolas et al. 2008).

Color image of this figure appears in the color plate section at the end of the book.

quantitative variation in seed nitrogen content in pea, ranging from three QTLs and two mapped genomic regions (Tar'an et al. 2004) to five QTLs on three LGs (Irzykowska and Wolko 2004) and eight genomic regions corresponding to 14 QTLs (Burstin et al. 2007). Some of these regions are associated with the locations of genes whose functions are relevant to many aspects of seed development (Burstin et al. 2007). One of the QTLs mapped to LG V (Irzykowska and Wolko 2004) may correspond to the *Vc-2* locus, where variation in the number of functional genes has been demonstrated (Chinoy et al. 2011).

Earlier work showed that metabolic changes in seed starch and sugar content were associated with alterations in protein content, which in part reflect a reduction in expression of a major class of legumin genes (Turner et al. 1990). In such pea mutants, the higher seed sugar content is a direct consequence of a lesion in a starch biosynthetic gene (Bhattacharyya et al. 1990). Similar data have been obtained more recently for induced wrinkled-seeded pea mutants transformed with RNAi constructs. Here alterations in starch and protein biosynthesis have been investigated thoroughly at the level of gene transcription and metabolite content (Weigelt et al. 2009). The elevated seed protein content in PA2 mutant lines was investigated further at the metabolite level by profiling seeds of cv Birte and the BC PA2-deficient lines during development (Vigeolas et al. 2008). Significant changes were noted in a number of metabolic pathways, most notably that leading to polyamine compounds (Fig. 1-2B). Specifically, at mid-late stages of development, there were differences in the amounts of several individual amino acids: levels of Ala, Phe, Cys, Arg, and Gln were increased, while others, notably Ile, Trp and Glu, were lower in the mutant. A lack of PA2 led to a decrease in the level of spermidine, while the level of the precursor of spermidine synthesis, Arg, increased (Fig. 12B). Further analyses of enzymatic activities revealed a lower arginine decarboxylase activity in mutant lines, providing an explanation for some of the observed effects (Vigeolas et al. 2008).

The decrease in spermidine and spermine in lines lacking PA2 was accompanied by changes in the relative amounts of sugars, and organic and amino acids, indicative of an increase in the organic N content and a decrease in the C to N ratio in seeds. The metabolic changes documented for PA2-deficient lines are compatible with a regulatory role for PA2, consistent with its structural relatedness to mammalian regulatory proteins that contain hemopexin domains (Vigeolas et al. 2008). The crystal structure of the homologous protein (LS-24) from seeds of *Lathyrus sativus* (grass pea) shows clearly that this is a hemopexin fold protein (Gaur et al. 2010). The structure provides insights into the function of LS-24 and suggests that mutually exclusive binding of heme and spermine may play a role in sensing oxidative stress through a ligand-regulated monomer-dimer transition switch (Gaur et al. 2010). Apart from direct roles in development and stress responses, polyamines are also likely to be involved in regulating metabolism in developing seeds. The introgression of the PA2 mutation into pea breeding material is underway, and field tests will reveal if there is any further phenotype associated with this mutation.

The relevance of TI to digestibility of feed has been shown in studies comparing the effects of near-isogenic lines that differ at least 5-fold in their content of TI (Wiseman et al. 2003, 2006). Here the coefficient of apparent ileal digestibility was significantly higher when the near-isolines having the lower TI activity were included in broiler diets (Fig. 1-3). These data suggest that lowering TI activity further could provide more benefit. Novel variants for TI have been isolated by TILLING (Domoney et al. in collaboration with Dalmais et al. 2008) (http://urgv.evry.inra.fr/UTILLdb). Besides their potential for improving digestibility in animal feed, lines with reduced activity provide tools for examining the physiological role of seed TI, and for assessing the contribution of such proteins to human health (see next section).

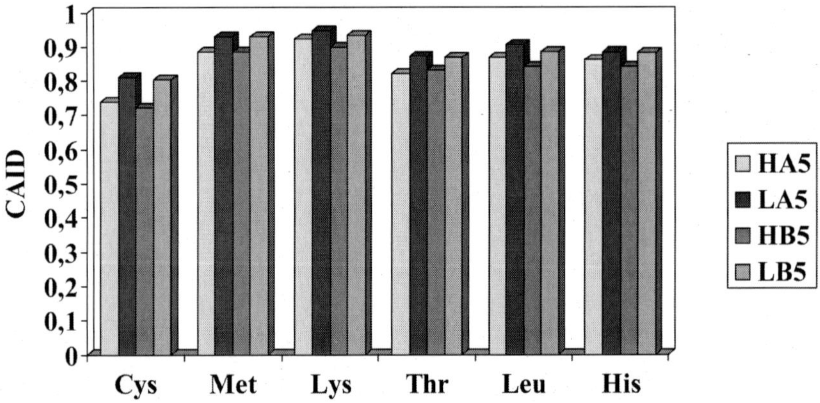

Figure 1-3 The coefficient of apparent ileal digestibility (CAID) determined for seeds of two sets of near-isogenic pea lines, HA5/LA5 and HB5/LB5, where A and B have different genetic backgrounds. Within each pair of near-isolines, H denotes a high seed trypsin inhibitor activity, whereas L denotes a low seed trypsin inhibitor activity. The CAID determined for six amino acids is shown (Wiseman et al. 2003).

Color image of this figure appears in the color plate section at the end of the book.

1.4.1.2 Manipulating Seed-Expressed Genes for Optimal Food Use

It has been shown that PA2 and the equivalent protein in chickpea (*Cicer arietinum*) can behave as potential allergens in humans (Vioque et al. 1998) and so a reduced content of this protein may provide benefit in diets for certain groups of people, in addition to benefits for improved feed as outlined above.

The positive contribution that TI may make to human health as anticarcinogenic proteins is under investigation. A clear reduction in the growth of HT29 colon cancer cells has been shown when cells are grown *in vitro* in the presence of pea TI (Clemente et al. 2005). These observations have acquired a greater significance, since it has been demonstrated that the homologous TI from soybean (STI) can retain activity following exposure to the digestive tract (Clemente et al. 2008; Marín-Manzano et al. 2009).

Although earlier work had suggested that the chymotrypsin inhibitory domain may be more effective at inhibiting the growth of cancer cells, recent data have shown that both chymotrypsin and trypsin inhibitory domains are significant in terms of their inhibitory effects on cell growth *in vitro* (Clemente et al. 2010). Furthermore, normal fibroblast cells were shown to be unaffected by exposure to STI, while chemically inactivated STI proteins showed little or no effect on HT29 cells (Fig. 1-4). Serine proteases have been identified that are active in epithelial cells, are associated with cell membranes, and are good candidates for significant involvement in processes associated with cell migration and metastasis. These proteases include the type II transmembrane serine protease, matriptase, where interaction with legume TI warrants further investigation (Bhatt et al. 2006). If validated by *in vivo* studies, genotypes with high TI activity may be selected positively for the development of health-promoting foodstuffs. It is already the case that many of the lines used for marrowfat pea products have a high TI activity, which may have been selected positively but unintentionally, alongwith selection for the "marrowfat trait" and associated visual and sensory quality determinants.

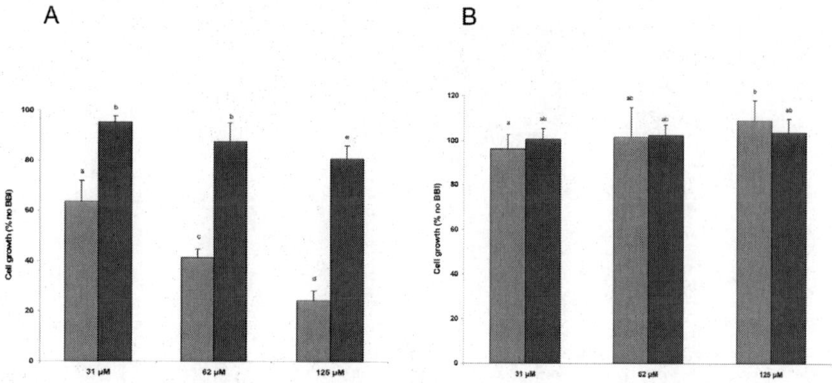

Figure1-4 Effects of soybean trypsin inhibitor (STI) on cell growth in vitro of the cancer cell line, HT29 (A) or the control fibroblast cell line, CCD-18Co (B), where native STI showed a reduced growth of the HT29 cells (blue bars), in contrast to chemically inactivated STI (red bars). The growth of fibroblast cells (B) was not affected by either form of the protein (Clemente et al. 2010). Means not sharing superscript letters differ significantly (p <0.05; Bonferroni's test).

Currently there are widespread concerns over food security and, within this climate, health issues may recede somewhat in priority listings. However, the definition of compounds involved in the generation of food that is nutritious, desirable and beneficial to health has already benefited from metabolomic studies that permit simultaneous detection and quantification of many small compounds. When linked to genetic analyses, these studies will unravel biochemical and genetic determinants of metabolite concentration from environmental effects. Metabolite analyses have been used to identify the downstream changes to pathways and compounds that have occurred as a consequence of the manipulation of genetic background in pea, in studies exploiting near-isogenic lines (Vigeolas et al. 2008; see previous section). Other work has documented changes that can be attributed to the environment (Charlton et al. 2008). In the latter study, the environment was shown to have a much greater effect on the leaf metabolome, when compared with the effects of transgenes. The further identification of seed metabolites linked to food quality will provide a guide to pathways and markers for improved selection processes.

Traits affecting the general appearance of vegetables include economically significant quality parameters, which in turn influence consumer choice, price, return to the farmer, and ultimately crop choices for rotations. These are all particularly relevant to the adoption and extended use of sustainable crops, and efficient nitrogen use. Included here is control of the shape and colour of seeds. The general appearance of seeds is important and often relates to pest control. As an example, severe damage to pea and bean seeds can be delivered by bruchid beetle larvae, and here progress is being made with the genetic basis of resistance (Byrne et al. 2008; Clement et al. 2009; see below). Aspects of seed colour are being investigated and variant lines used to develop RILs with associated maps to define genetic loci involved in seed colour determination, its stability and degradation (C Chinoy, C Domoney et al. in prep.).

1.4.1.3 Manipulating Seed-Expressed Genes for General Industrial Use

Throughout the studies ou tlined above, the selection of novel germplasm is being used alongside the generation of new mutants, for example by TILLING, to identify relevant genetic loci and candidate genes. Transformation of pea remains a long process, with a low success rate, that has not been accepted widely as a routine research tool. Nonetheless, the protection of pea seeds from bruchid beetle attack in storage and in the field, following transformation with an α-amylase inhibitor gene from another legume species, was a notable success story (Schroeder et al. 1995). However differential post-translational modification of the transgene product in pea (Islam et al. 2009) possibly elicited a pre-immune response in mice, meaning that the seeds were rejected by the regulatory authorities and not developed commercially (see Editorial 2006). These developments have not encouraged the exploitation of pea seeds for the over-expression of valuable products for industrial use.

It is clear that some pea seed proteins have innate insecticidal properties, which can be exploited for the protection of other crops. Among these are the TI, where related proteins from other legume species have been shown to inhibit the development of larvae of lepidopteran but not coleopteran insects (Domoney 1999). A distinct class of pea seed albumin, however, is responsible for the resistance of pea to insects of the *Sitophilus* species that can destroy cereal seeds during storage. The pea albumin 1 (PA1) peptide, PA1b, has been associated specifically with strong insecticidal properties and acute mortality of these insects, and genes encoding PA1b and related peptides have been identified in some but not all subfamilies of the Leguminoseae (Louis et al. 2007). In addition to the protective role played by PA1b, secondary metabolites in pea seeds have been implicated additionally in antifeeding activity against *Sitophilus* (Louis et al. 2007) and it has been speculated that legume storage proteins in general may play important defence roles (Sales et al. 2000).

Transgenic pea lines have been generated as laboratory tools to underpin the study of metabolic pathways and their manipulation, either as marker lines to monitor relative plant responses (Charlton et al. 2008) or to examine directly the consequences of up- and down-regulating specific genes (Radchuk et al. 2006; Weigelt et al. 2008, 2009). These last studies have provided insights into specific interactions between N and C metabolism in seed development, focussing on amino acid transport (Weigelt et al. 2008), starch biosynthetic lesions (Weigelt et al. 2009) and the role of a kinase in the transition phase from cell proliferation to differentiation, a stage that is critical for storage product accumulation (Radchuk et al. 2006). These insights facilitate the rational manipulation of seed products for a wide range of industrial use, including "molecular pharming" to provide antibodies and vaccines for medical and veterinary use, where pleiotropic effects of the transgenes on plant growth and development might be minimized. Transgenic pea seeds expressing an antibody that protects chickens from infection by *Eimeria*, and resulting coccidiosis, have been developed (Zimmermann et al. 2009), that illustrate the potential of this approach for the future. It is notable that, in the study of Mikschofsky et al. (2009), pea seeds proved to be a superior expression system for the production of a vaccine that provided protection in rabbits against haemorrhagic disease virus.

1.4.2 Breeding for Resistance to Biotic Stresses

Abiotic and biotic stress agents have impacted on pea production since domestication; however, it was not until the last century that many stress resistance genes were identified and characterized. Many of these genes have since been incorporated into modern cultivars, but several others, including those which are controlled quantitatively have been more difficult to manipulate. Development of improved cultivars with resistance to a single stress is often straightforward if a good source of resistance is available and an efficient, easily controlled and practical screening procedure exists to provide good selection pressure. Unfortunately, this is seldom the case.

Breeding for stress resistance involves diagnosis, development of efficient screening methods, availability of resistance sources, and the development of effective breeding methodology. Screening methods have been developed and refined for most stresses, and a number of sources of resistance identified (Infantino et al. 2006 for root pathogens; Rubiales et al. 2006 for parasitic plants; Sillero et al. 2006 for biotrophic pathogens; Edwards and Singh 2006 for insect pests; Stoddard et al. 2006 for abiotic stresses; Tivoli et al. 2006 for necrotrophic pathogens).

Breeding success will depend on availability of consistent resistance genes or QTLs within or outside the species, and adequate molecular markers flanking these genes or QTLs that can trace resistance in breeding populations. However, the long genetic distance existing in most cases between the identified genetic markers and the resistance QTL, the common lack of codominant markers and the general lack of knowledge on resistance mechanisms limit greatly the use of genetic markers to confer resistance to pea.

Unfortunately, for some pathogens and for most pests, genetic resistance in the genus *Pisum* is scarce or simply unknown. Genes for resistance to *Bruchus pisorum* and to *Erysiphe pisi* have been transferred from *P. fulvum* to *P. sativum* through natural hybridization (Byrne and Hardie 2004; Fondevilla et al. 2007c). Resistance to *Mycosphaerella pinodes* and *Orobanche crenata* has been detected in some *P. fulvum* accessions, and crosses were made to transfer the genes to pea (Rubiales et al. 2005; Fondevilla et al. 2008b). Certain resistance can also be found in species from the tertiary pea genepool. For example, *Lathyrus sativus* is resistant to *M. pinodes* (Gurung et al. 2002), where resistance is quantitative and controlled by two QTLs (Skiba et al. 2004). Valuable resistance to pea rust, powdery mildew and broomrape is available is *Lathyrus* spp. (Vaz Patto et al. 2006, 2007, 2009; Sillero et al. 2005; Vaz Patto and Rubiales 2009; Fernández-Aparicio et al. 2009). However it is still difficult to use these species as gene sources for pea breeding. The potential use of *in vitro* culture, embryo rescue and protoplast fusion to circumvent this drawback seems rather low, since pea has been shown to be largely recalcitrant to regeneration techniques. Ultimately, these genes could be transferred to pea if procedures such as protoplast fusion were developed as stated by Durieu and Ochatt (2000).

Major progress is likely to be made from the use of *M. truncatula* as a model for pea disease studies. *M. truncatula* is affected by many of the pathogens and pests limiting pea yield and thus offers great opportunities to improve knowledge of resistance mechanisms against pea pathogens and to identify effective resistance genes against them. In parallel, the transcriptomic and proteomic approaches developed for this model legume are being used to understand the molecular components and identify candidate genes involved in the defence of *M. truncatula* against these pathogens. For instance, suppression subtractive hybridization libraries allowed the identification of candidate genes for *M. truncatula* resistance to *Aphanomyces euteiches* (Nyamsuren et al. 2003) or *O. crenata* (Die et al. 2007). Comparison of the proteomic profile of *M. truncatula* lines with varying levels of resistance also identified proteins potentially involved in *A. euteiches* (Colditz et al. 2005), powdery mildew (*E. pisi*) (Curto et al. 2008), broomrape (*O. crenata*) (Castillejo et al. 2009) or rust resistance (*Uromyces striatus*) (Castillejo et al. 2010b; Madrid et al. 2010; Rubiales et al. 2011). Microarray analysis of different *M. truncatula* genotypes with different defence mechanisms against *E. pisi* (Foster-Hartnett et al. 2007) or *O. crenata* (Dita et al. 2009) allowed a set of genes involved in these defence mechanisms to be determined. Post-genomic approaches are also being applied to tackle other fungal diseases such as *M. pinodes* (Fondevilla et al. 2011).

1.4.2.1 Resistance to Fungal Diseases

Fusarium root rot is a soilborne disease incited by *Fusarium solani* f.sp. *pisi* (also known as *Nectria haematocca*). It causes brownish-black lesions, which coalesce and ultimately destroy the roots. Control is only possible by long crop rotations (Kraft 2001), and by the use of tolerant pea cultivars recently released (Infantino et al. 2006). Resistance is quantitatively inherited (Knavel 1967) with three QTLs identified so far together with sequence-tagged microsatellite sites (STMS) markers for use in marker-assisted breeding (Coyne et al. 2004). More recently, a candidate resistance gene, *DRR230* was mapped to the same pea linkage group regions as one of the QTLs on LG III (Coyne and Pilet-Nayel 2008).

Fusarium wilt is a widespread soilborne disease caused by the ascomycete *F. oxysporum* f.sp. *pisi*. The pathogen penetrates the fibrous roots, colonizes the vascular tissues, spreads through the xylem into the stem causing wilting and yellowing of the foliage, followed by plant death. Control is primarily through the use of genetically resistant cultivars, but crop rotation is also used to reduce disease incidence. The disease can be spred through contaminated seed and pre-plant seed treatment is recommended (Bradford 1996). Four races of Fusarium wilt (races 1, 2, 5, 6) were first described using host-pathogen response, and these differ in vegetative compatibility (Correll 1991) and molecular fingerprint profiles (Grajal-Martin et al. 1993). New variants of the pathogen continue to emerge (Bødker et al. 1993; Kraft 2001), and this indicates that *F. oxysporum* f.sp. *pisi* is an evolving pathogen. Resistance to race 1 is controlled by a single dominant gene, *Fw*, that is bred into most cultivars grown currently. The gene was localized on LG III in pea (Grajal-Martin and Muehlbauer 2002).

RFLP, AFLP, sequence characterized amplified region (SCAR) and STMS markers have been reported (Dirlewanger et al. 1994; McPhee et al. 1999; McClendon et al. 2002; Watt et al. 2004). Resistance to race 2 is also conferred by a single dominant gene, *Fwn*, (Kraft 2001) although it has recently been suggested that more genes might be involved (Infantino et al. 2006). Resistance to race 5 is controlled by a single dominant gene *Fwf* located on LG II (Coyne et al. 2000). A RAPD marker has been identified for *Fwf* (Okubara et al. 2002).

Ascochyta blight of pea is caused by a complex of three fungal pathogens: *Mycosphaerella pinodes*, *Phoma medicaginis* var. *pinodella* and *Ascochyta pisi*. The disease affects all aerial parts of the plant, including stems, leaves and pods (Bretag and Ramsey 2001), and also causes foot rot symptoms, particularly when infected seed is planted. Ascochyta blight can have a significant effect on yield (Beasse et al. 1999). *A. pisi* can be divided into five pathotype groups based on the reaction of a differential set of *P. sativum* host lines (Darby et al. 1986). Genetic studies have shown that major genes conferring resistance to specific pathotypes are present in pea germplasm and minor genes may also be involved in determining host plant resistance. Resistance against *A. pisi* is conditioned by a single dominant genetic locus, *Rap2*, possibly modified by a minor gene or polygenic system (Darby et al. 1985). A major QTL for race C resistance was mapped to LG IV (Dirlewanger et al. 1994).

There is no agreement on the existence of pathotypes for *M. pinodes* (Tivoli et al. 2006). A number of studies suggest the existence of pathotypes (Ali et al. 1978; Clulow et al. 1991a; Nasir and Hoppe 1991), whereas others denying them (Wroth 1998; Xue et al. 1998; Onfroy et al. 1999). Some levels of incomplete resistance against *M. pinodes* and *P. medicaginis* have been reported (Ali et al. 1978; Bretag 1989; Kraft 1998; Xue and Warkentin 2001a; Prioul et al. 2003; Fondevilla et al. 2005, 2008b). The majority of the genetic studies concerning resistance to *M. pinodes* in pea has considered resistance as a quantitative trait and have concluded that resistance is under polygenic control (Zlamal 1984; Wroth 1999; Fondevilla et al. 2007b; Rubiales and Fondevilla 2010). As an exception, Clulow et al. (1991b) used a qualitative approach and reported four major genes: *Rmp1* and *Rmp2* for stem resistance and *Rmp3* and *Rmp4* for leaf resistance. The genes identified by Clulow have not been verified by any further study. In contrast, QTL mapping studies have resulted in the identification of numerous genomic regions involved in the control of the resistance, confirming the polygenic nature of resistance. Timmerman-Vaughan et al. (2002, 2004) identified 19 QTLs associated with resistance, six of which were common for two crosses. Tar'an et al. (2003) identified three QTLs specific for resistance to *M. pinodes* under field conditions. Prioul et al. (2004) reported six QTLs associated with seedling resistance to *M. pinodes* under controlled conditions and 10 for adult plant field resistance, with four being common to both stages. More recently, Fondevilla et al. (2008b) identified six QTLs (*mp1–mp6*) associated with resistance to *M. pinodes* in a cross of the cultivar Messire with *P. sativum* subsp. *syriacum*. Prioul-Gervais et al. (2007) reported three candidate genes co-locating with QTLs previously described for resistance to *M. pinodes*.

Powdery mildew is a worldwide distributed pea disease caused by the biotroph pathogen *Erysiphe pisi*. Three genes for resistance to powdery mildew, named *er1*, *er2* and *Er3* (Heringa et al. 1969; Fondevilla and Rubiales 2011; Fondevilla et al. 2007a, 2010a) have been described so far, although additional resistance is available in *Pisum* germplasm (Fondevilla et al. 2007a). The gene *er1* is in wide use in pea breeding programs (Heringa et al. 1969; Tiwari et al. 1997a, b; Fondevilla et al. 2006). Resistance conferred by this gene has proven to be stable based on a barrier to pathogen penetration (Fondevilla et al. 2006) and has recently been reported to be *mlo* type, due to loss-of-function mutation of *PsMLO1* (Humphry et al. 2011). The gene *er2* (Heringa et al. 1969) is not commonly in commercial use. The gene confers a high level of resistance in some locations but is ineffective in others (Tiwari et al. 1997a, b; Fondevilla et al. 2006). Expression of *er2* is strongly influenced by temperature and leaf age. The *er2* governed resistance is based mainly on post-penetration cell death, complemented by a reduction of percentage penetration success in mature leaves (Fondevilla et al. 2006). The gene *Er3* was identified recently in *P. fulvum* and has been introduced successfully into adapted *P. sativum* material by sexual crossing (Fondevilla et al. 2007a, 2010a). Resistance conferred by *Er3* is due to a high frequency of cell death that occurs both as a rapid response to attempted infection and a delayed response that follows colony establishment (Fondevilla et al. 2007a, c).

The first attempts to identify molecular markers linked to the *er1* locus were carried out by Dirlewanger et al. (1994), who found an RFLP marker located at 9.8 cM from this resistance locus and by Timmerman-Vaughan et al. (1994) who mapped a RAPD/SCAR marker 2.1 cM from the same locus. Three additional RAPD markers were later identified by Tiwari et al. (1998) who converted two of them into SCARs. Janila and Sharma (2004) identified two new RAPDs, and Ek et al. (2005) mapped five SSRs, all of them more than 10 cM apart from the *er1* locus. More recently, Pereira et al. (2010) reported markers located closer to the *er1* locus, three of which were converted into SCARs. AFLPs linked to the powdery mildew resistance gene *er2* have been developed using JI 2480 as the source of resistance (Tiwari et al. 1999). More recenlty, RAPD markers closely linked to *er2* have been converted to a SCAR (Katoch et al. 2010). Similarly, RAPD markers tightly linked to *Er3* have been identified and converted into SCARs using a *P. fulvum* accession as a source of resistance (Fondevilla et al. 2008a).

Pea rust has become an important pathogen of dry pea from the mid-1980s particularly in regions with warm, humid weather. Pea rust has been reported to be caused either by the fungus *Uromyces viciae-fabae* (syn. *U. fabae*) or *U. pisi*. *U. viciae-fabae* is the principal causal agent of pea rust in tropical and subtropical regions such as India and China, where warm humid weather is suitable for the appearance of both the uredial and the aecidial stage (Kushwaha et al. 2006). However, in temperate regions, it has been observed that, although pea seedlings can be infected by *U. viciae-fabae*, the fungus hardly gets established or progresses under field conditions (Barilli et al. 2009c). Many efforts have been made to identify sources of resistance in pea against *U. viciae-fabae* (Pal et al. 1980; Singh and Ram 2001; Xue and Warketin 2001b; Vijayalakshmi et al. 2005; Kushwaha et al. 2006). The studies revealed that most pea genotypes studied were infected, although genotypic differences in the rust intensity were identified and some genotypes described as slow rusting types (Chand et al. 2006). To date, the dominant nature of the partial resistance against faba bean rust *U. viciae-fabae* in pea, recorded as reduced infection frequency, has been justified as the expression of a single major gene, for which the symbol *Ruf* was proposed (Katiyar and Ram 1987; Vijayalakshmi et al. 2005). Two RAPD markers have been reported flanking *Ruf*, but neither marker is close enough to the gene to allow a dependable MAS for rust resistance (Vijayalakshmi et al. 2005). Additional minor genes have been proposed (Singh and Ram 2001).

Resistance to *U. pisi* has recently been reported (Barilli et al. 2009a). Only incomplete resistance has been identified that has not been associated so far with host cell death (Barilli et al. 2009a, b). A QTL explaining a high percentage (63%) of the phenotypic variance has been located in LG III (Barilli et al. 2010).

1.4.2.2 Resistance to Diseases Incited by Oomycete

Aphanomyces root rot is a soilborne disease caused by *Aphanomyces euteiches*. It causes soft honey-brown rotting of the roots and epicotyls resulting in yellowing and stunting of aerial plant parts, and then death of the plant. There is no efficient and reliable cultural or chemical method to manage the pathogen. Genetic resistance to *A. euteiches* in pea is known to be quantitative and largely influenced by interactions with environmental conditions (Shehata et al. 1983). Pilet-Nayel et al. (2002) identified seven QTLs associated with partial field resistance. Three QTLs were consistently detected in at least two environmental conditions and in controlled conditions (Pilet-Nayel et al. 2005). Weeden et al. (2000) identified a major gene controlling tolerance to Aphanomyces root rot that might correspond to *Aph1*.

Downy mildew of pea is casued by *Peronospora viciae* f.sp. *pisi*. It is widely distributed all over the world, causing systemic infection of seedlings, local infections on leaves and pod infections. Oospores in the soil are the most important primary inoculum. Conidia can be distributed over long distances. Treatment of seeds with acylalanine fungicides is very effective against systemic seedling infection (Stegmark 1994). There is variation in resistance to downy mildew among pea varieties (Matthews and Dow 1976). Some varieties have race-specific resistance but there is evidence of this resistance breaking down (Thomas et al. 1999). Partial resistance has also been reported. The cultivar "Dark Skin Perfection" has been reported to carry partial resistance that has proven durable for over 30 years (Stegmark 1994). One breeding line of pea with a high level of partial resistance was described by Stegmark (1988). This line exhibited low susceptibility to all isolates of the fungus, but never complete resistance to any isolate (Stegmark 1990). The partial resistance is determined by more than a single gene. A combination of dominant, recessive and intermediate resistance genes are involved (Stegmark 1992).

1.4.2.3 Resistance to Bacterial Diseases

Pea bacterial blight caused by *Pseudomonas syringae* pv. *pisi* is a widely distributed seed borne disease. It may affect all above ground plant parts. Under cool and wet conditions, favourable for the disease, lesions are initially discrete, shiny and water-soaked. Defoliation, blasting of blossoms and pods and unsightly pods can lower yield, quality and the value of the crop. Key measures for disease control are disease avoidance through seed testing and deployment of resistant cultivars. Seven races are currently recognized (Taylor et al. 1989) but a new race 8 has recently been postulated (Martín-Sanz et al. 2011). Races 2, 4 and 6 are predominant in Europe but race 6 has become increasingly important (Reeves et al. 1996), possibly due to the employment in Europe of cultivars resistant to races 2 and 4.

Stem resistance to races 2, 3 and 4 is controlled by single dominant genes, *Ppi1* (for race R2), *Ppi3* (R3) and *Ppi4* (R4), although there is an incomplete expression of R4 in some cases (Bevan et al. 1995, Hunter et al. 2001). No resistance to race 6 is available in pea cultivars, but resistance has been identified in *P. abyssinicum*. This resistance is controlled by a major recessive gene together with a number of modifiers (Elvira-Recuenco and Taylor 2001). Three markers cosegregating with resistance and three cosegregating with susceptibility were identified using a Ty1-*copia* group retrotransposon of pea (PDR1) and sequence-specific amplification polymorphisms (SSAP) markers (Elvira-Recuenco et al. 2003). Under field conditions, susceptibility in the leaf to race 6 is relatively low and susceptibility to race 4 is moderate in the leaf with some stem snapping. Race 4 is known to be a very aggressive race, and both race specific and race non-specific resistance seems not to be fully effective. An additional benefit from the use of *P. abyssinicum* is its frost tolerance, since frost damage is a factor that favours infection.

1.4.2.4 Resistance to Viral Diseases

Pea seed-borne mosaic virus (PSbMV) is a pathogen of peas, broad beans, lentils and chickpeas and causes major crop losses worldwide. Since PSbMV is seed transmitted, losses result both from reduced productivity/yield and reduced seed quality. The disease can only be managed by identifying and discarding batches of infected seed. Resistance to PSbMV has been reported (Hagedorn and Gritton 1973; Khetarpal et al. 1990) and several recessive resistance genes have been identified, namely *sbm1*, *sbm2*, *sbm3* and *sbm4* (Provvidenti and Alconero 1988a, b). Through linkage of these loci with morphological markers the introgression of the resistance genes into commercial lines without virus testing was possible (Provvidenti et al. 1991). Efficiency of selection will be enhanced now through use of available molecular markers (Gao et al. 2004). Hence many commercial varieties of pea remain susceptible to the more common strains of PSbMV.

Pea enation mosaic virus (PEMV) can be a very devastating viral disease. It causes stunted growth and possibly plant death, vein clearing or "windowing" on the leaves and stipules, distorted pod development, and reduced seed size. Resistance to PEMV is conferred by a single dominant gene (*En*) (Yu et al. 1995) and has been incorporated into recently released varieties (McPhee and Muehlbauer 2002).

Resistance genes against white lupin mosaic virus (WLMV) (Provvidenti and Hampton 1993) and bean yellow mosaic virus (BYMV) (Schroeder and Provvidenti 1971; Yu et al. 1996) have also been reported.

1.4.2.5 Resistance to Parasitic Plants

Pea cultivation is strongly hampered in Mediterranean and Middle East farming systems by the occurrence of crenate broomrape (*Orobanche crenata*), causing significant yield losses (Rubiales et al. 2003). Strategies for broomrape control have been developed including cultural practices and chemical control. However, all have met with limited success and are either not feasible, uneconomical, hard to achieve or resulting in incomplete protection (Rubiales et al. 2009b; Rubiales and Fernández-Aparicio 2011; Rubiales and Heide-Jørgensen 2011). Breeding for broomrape resistance is a difficult task considering the scarce and complex nature of resistance in legumes in general, and in pea in particular (Rubiales et al. 2003, 2006, 2009a). Similarly, only incomplete resistance was found in faba bean to *O. crenata* that has been successfully accumulated by breeding, allowing the release of resistant faba bean cultivars (Pérez-de-Luque et al. 2010). A similar effort in breeding pea for broomrape resistance was not made in the past, however.

Levels of resistance have been reported in accessions of *P. sativum* ssp. *sativum, abyssinicum, arvense* and *elatius* and in *P. fulvum* (Rubiales et al. 2005). All these resistant accessions have been successfully crossed with commercial pea varieties and a breeding program for broomrape resistance is underway (Rubiales et al. 2009a). Resistance is the result of several mechanisms acting at different stages of the infection process, including low stimulation of broomrape seed germination, unsuccessful penetration of host roots, delay in post-attachment tubercle development and necrosis of the attached tubercles (Pérez-de-Luque et al. 2005; Rubiales et al. 2005). In the identified resistant accessions the broomrape parasite is stopped in the host cortex before reaching the central cylinder, associated with protein cross-linking and accumulation of H_2O_2, peroxidases, β-glucanases and callose (Castillejo et al. 2004; Pérez-de-Luque et al. 2006).

QTLs conferring resistance to *O. crenata* have been detected in pea using a map developed from F_2 plants derived from the cross *P. sativum* ssp. *syriacum* x *P. sativum* (Valderrama et al. 2004). Field screenings allowed identification of two QTLs for broomrape resistance that explained an insufficient proportion of the observed variation to be used efficiently in MAS. More recent studies with RILs from the same cross, and an improved map, allowed the identification of additional QTLs (Fondevilla et al. 2010b). Four genomic regions associated with field resistance, assessed as the number of emerged broomrape shoots per pea plant under field conditions were identified but each of them explained a low to moderate proportion of the phenotypic variation. A more accurate screening of this RIL population, assaying different phases of the parasite cycle using a petri dish technique, enabled identification of QTLs governing specific mechanisms of resistance. Thus, QTLs for low induction of *O. crenata* seed germination, lower numbers of established tubercles per host root length unit, and slower development of tubercles were identified. Each of these QTLs explained from 8 to 37% of the variation observed, and altogether from 38 to 59%, depending on the trait. The effect of a low plant vigour, which is an undesirable trait, on the number of broomrapes emerged per plant should be taken into account and avoided in field screenings. In relation to that, simultaneous mapping of resistance and vigour traits performed in that study could be helpful as it would allow selection of those QTLs governing resistance mechanisms, but not linked to low biomass, for MAS.

1.4.2.6 Resistance to Insects

Pea aphid (*Acyrthosiphon pisum*) is a serious pest (Wilkinson and Douglas 1998). Insecticides are used to control aphid populations and also to minimize the spread of virus particles in many production areas (McPhee 2003). Sources of host plant resistance in pea have been reported (Soroka and Mackay 1990, 1991). Also the modification of plant morphology in pea to enhance predator effectiveness on *A. pisum* has recently been investigated and reported to have a positive influence (Rutledge et al. 2003). There is evidence for parallel QTL effects on host-plant acceptance and fecundity on the accepted host, which might produce assortative mating (Via and Hawthorne 2002). A mutant with reduced wax in the stems, leaves, stipules, and pods has been shown with potential for biological control of *A. pisum* as it favours acceptance by predators (Eigenbrode et al. 1998).

Pea bruchid (*Bruchus pisorum*) is a major constraint to the production of dry pea in many areas of the world (Clement et al. 2000). Contact insecticides can be used to control adults of *B. pisorum* before they can lay eggs on the green pods (Horne and Bailey 1991). There is little resistance in pea germplasm, apart from lines carrying the "Neoplastic pod" wild allele that causes formation of callus tissue in response to the presence of *B. pisorum* eggs on pods resulting in a reduced larval survival (Berdnikov et al. 1992). The highest resistance to pea bruchid has been found in *P. fulvum* (Hardie et al. 1995; Clement et al. 2002). Resistance from *P. fulvum* is conferred by three genes (Byrne and Hardie 2004). A number of polymorphic AFLP bands associated with resistance have been identified (Byrne and Hardie 2004; Byrne et al. 2008). *B. pisorum* might also be controlled by the use of a transgenic pea line containing a gene from common bean that inhibits α-amylase in the digestive system and effectively causes starvation of the weevil larvae (see also Section 1.4.1.3; Morton et al. 2000, Sousa-Majer et al. 2004).

Pea leafminers (*Chromatomyia horticola* and *Melanagromyza phaseoli*) can be controlled by insecticide applications (Khajuria and Sharma 1995). Varying levels of resistance against *C. horticola* and *M. phaseoli* have been reported in pea (Brar et al. 1993; Gupta and Pandey 1998; Mehta et al. 1998).

Pea weevil (*Sitona lineatus*) adults affect young pea shoots while larvae can affect root nodules, resulting in a severe reduction in the nitrogen content of the seeds and in the final yield. Genetic variation in pea germplasm for resistance to *S. lineatus* is available. Resistance to adult feeding on seedlings is related to either early vigour or large leaflets (Nouri-Ghanbalani et al. 1978). There was also evidence of antibiosis in lines which produced the fewest nodules in the field and greenhouse and supported the lowest population of larvae (Nouri-Ghanbalani et al. 1978). Management is also possible by a transgenic approach expresssing insecticidal activity in *Rhizobium* containing the delta-endotoxin gene cloned from *Bacillus thuringiensis* (Skot et al. 1990; Quinn and Bezdicek 1996).

1.4.2.7 Resistance to Nematodes

Pea is affected by a number of nematodes like root-lesion nematode (*Pratylenchus* spp.), root-knot nematode (*Meloidogyne* spp.) and pea cyst nematode (*Heterodera goettingiana*) to which little resistance is available. No pea cultivars with resistance to *H. goettingiana* are available (Inglis 1998) but some accessions show a reduced number of cysts per root and oxidative enzyme pattern, suggesting that they might control *H. goettingiana* (Di Vito and Perrino 1978; Bleve-Zacheo et al. 1990). Pea is resistant to *P. neglectus* and *P. thornei*, but not to *P. penetrans* (Vanstone et al. 2004).

1.4.3 Breeding for Resistance to Abiotic Stresses

Potential yields of major crops are usually reduced due to environmental constraints in addition to biotic constraints. Moreover, global climate change predictions suggest new scenarios with larger arid areas and extreme climatic events. Thus, it is essential to understand how plants respond to different abiotic stresses in order to improve crop performance.

1.4.3.1 Drought Stress

The severity of drought stress is unpredictable as it depends on many factors such as the occurrence and distribution of rainfall, evaporative demands of the atmosphere and moisture storing capacity of the soils (Wery et al. 1994). Drought escape occurs when phenological development is successfully matched with periods of soil moisture availability, and where the growing season is short and terminal drought stress predominates (Turner 1986). The success of increasing pea production in drier regions prone to terminal drought largely depends on the development of short season varieties that enable the crop to escape severe soil water deficits. Drought and high temperatures occur together in many environments important for legume production, and both stresses can interact to reduce yields. Pea is less sensitive to soil moisture deficit than faba bean that wilts at higher water potentials (McDonald and Paulsen 1997).

A decrease in relative water content is a main factor resulting in reduced growth in response to osmotic stress in pea (Alexieva et al. 2001). Under drought stress, sensitive pea genotypes were more affected by the decline in relative water content than tolerant ones (Upreti et al. 2000). Physiological parameters related to decrease in canopy temperature, such as epicuticular wax content and leaf colour, are beneficial to seed yield in pea under drought conditions (Manzanares et al. 1998). Osmotic adjustment is important for drought tolerance (Sánchez et al. 1998). In 49 pea cultivars, the stimulation of sugar levels by drought was proportional to osmotic adaptation and related to turgor maintenance (Sánchez et al. 1998). Cultivars that maintained turgor also exhibited higher harvest index. Proline content increased under drought stress in peas (Sánchez et al. 1998; Alexieva et al. 2001), and cultivars with high proline accumulation showed lower water contents upon turgor loss.

The few studies on carbon isotope discrimination show its value as a marker for water-use efficiency. In field peas, better water use efficiency of conventional tall cultivars compared with semi-dwarf lines was shown by means of increased ^{13}C (Armstrong et al. 1994). Under drought stress, the content of abscisic acid (ABA), which is associated with stomatal closure, increased in peas, more so in the tolerant genotype than in the susceptible genotype (Upreti and Murti 1999).

Selection for deep and extensive root system has been advocated to increase productivity of food legumes under moisture deficit conditions as it can optimize the capacity to acquire water. Turner et al. (2001) identified rooting depth and density as a main drought avoidance trait in grain legumes for use in terminal-drought environments. Grzesiak et al. (1997) showed that drought resistant pea cultivars had extensive and prolific root systems.

Alterations in the structure and function of cell membranes are early effects of stress injury and have been attributed to oxidative damage from active oxygen species. Membrane stability of plant tissues, mostly leaves, is often determined by electrolyte leakage measured as electrical conductivity. Membrane damage was not only found after exposure to drought stress in pea (Alexieva et al. 2001), but was also found to be a suitable screening technique for drought tolerance rating in legume species and cultivars (Grzesiak et al. 1996). Under conditions of drought, the electron transport chain was down-regulated as a response to stomatal closure and substrate limitation in pea (Cornic and Fresneau 2002). Recent studies have identified a number of metabolites that are altered in relative concentration in leaves from drought-stressed pea plants (Charlton et al. 2008). These include trigonelline that may have a protective function.

1.4.3.2 Winter Tolerance

Most cultivars are very susceptible to cold. Winter peas are not at present very successful, due to inadequate winter hardiness (Weightman 2005). Winter hardiness is reported to be a quantitatively inherited trait (Markarian and Andersen 1966; Auld et al. 1983; Cousin 1997; Lejeune-Henaut et al. 2008). Selection can be made under field conditions and under controlled conditions (Cousin 1997). QTLs for frost resistance have been identified, the most important of which co-segregates with the gene for delayed flowering, thus emphasizing the importance of appropriate flowering time for successful overwintering (Lejeune-Henaut et al. 2004, 2008).

Under chilling stress, pea leaves accumulated high quantities of sucrose, glucose-6-phosphate, fructose-6-phosphate and mannose-6-phosphate (Streb et al. 2003) and a close relationship was found between soluble sugar concentration of leaves just before the frost and the degree of freezing tolerance obtained by the different genotypes (Bourion et al. 2003). During freeze-thaw cycles, exogenous glycine betaine prevented changes in relative proportion and antenna sizes of PSII in pea (Busheva and Apostolova 1997). Exogenous ABA treatment improved freezing tolerance, but did not replace cold hardening in peas (Welbaum et al. 1997). More recently, QTLs for frost tolerance were associated with QTLs controlling the expression of proteins potentially involved in frost response (Dumont et al. 2009).

References

Alexieva V, Sergiev I, Mapelli S, Karanov E (2001) The effect of drought and ultraviolet radiation on growth and stress markers in pea and wheat. Plant Cell Environ 24: 1337–1344.

Ali SM, Nitschke LF, Dube AJ, Krause MR, Cameron MR (1978) Reaction of pea lines to pathotypes of *Ascochyta pinodes*, *A. pisi* and *Phoma medicagines* var *pinodella*. Aust J Agri Res 29: 841–849.

Ambrose M (2008a) Garden Pea. In: J Prohens, F Nuez (eds) Vegetables II: Fabaceae, Liliaceae, Umbelliferae, and Solanaceae. Springer, New York, USA.

Ambrose MJ (2008b) European *Pisum* central crop database (EPDB). *Pisum* Genet 40: 21

Ambrose MJ, Green FN (1991) A review of *Pisum* genetic resources and germplasm utilisation. Asp Appl Biol 27: 243–251.

Amurrio, JM, de Ron AM, Zeven AC (1995) Numerical taxonomy of Iberian pea landraces based on quantitative and qualitative characters. Euphytica 82: 195–205.

Armstrong EL, Pate JS, Tennant D (1994) The field pea crop in southwestern Australia—patterns of water use and root growth in genotypes of contrasting morphology and growth habit. Aust J Plant Physiol 21: 517–532.

Aubert G, Morin J, Jacquin F, Loridon K, Quillet MC, Petit A, Rameau C, Lejeune-Hénaut I, Huguet T, Burstin J (2006) Functional mapping in pea, as an aid to the candidate gene selection and for investigating synteny with the model legume *Medicago truncatula*. Theor Appl Genet 112: 1024–1041.

Auld DL, Dittedine RL, Murray GA, Swensen JB (1983) Screening peas for winter hardiness under field and laboratory conditions. Crop Sci 23: 85–88.

Baranger A, Aubert G, Arnau G, Lainé AL, Deniot G, Potier J, Weinachter C, Lejeune-Hénaut I, Lallemand J, Burstin J (2004) Genetic diversity within *Pisum sativum* using protein and PCR-based markers. Theor Appl Genet 108: 1309–1321.

Barilli E, Sillero JC, Fernández-Aparicio M, Rubiales D (2009a) Identification of resistance to *Uromyces pisi* (Pers.) Wint. in *Pisum* spp. germplasm. Field Crops Res 114: 198–203.

Barilli E, Sillero JC, Rubiales D (2009b) Characterization of resistance mechanisms to *Uromyces pisi* in pea. Plant Breed 128: 665–670.

Barilli E, Sillero JC, Serrano A, Rubiales D (2009c) Differential response of pea (*Pisum sativum*) to rusts incited by *Uromyces viciae-fabae* and *U. pisi*. Crop Prot 28: 980–986.

Barilli E, Zatovic S, Rubiales D, Torres AM (2010) Mapping of quantitative trait loci controlling partial resistance against rust incited by *Uromyces pisi* (Pers.) Wint. in a *Pisum fulvum* L. intraspecific cross. Euphytica 175: 151–159.

Beasse C, Ney B, Tivoli B (1999) Effects of pod infection by *Mycosphaerella pinodes* on yield components of pea (*Pisum sativum*). Ann Appl Biol 135: 359–367.

Ben Ze'ev N, Zohary D (1973) Species relationships in genus *Pisum*. Isr J Bot 2: 73–91.

Berdnikov VA, Trusov YA, Bogdanova VS, Kosterin OE, Rozov SM, Nedel'kina SV, Nikulina YN (1992) The neoplastic pod gene (*Np*) may be a factor for resistance to the pest *Bruchus pisorum* L. *Pisum* Genet 24: 37–39.

Bevan JR, Taylor JD, Crute IR, Hunter PJ, Vivian A (1995) Genetics of specific resistance in pea (*Pisum sativum*) cultivars to seven races of *Pseudomonas syringae* pv. *pisi*. Plant Pathol 44: 98–108.

Bhatt A-S, Takeuchi T, Ylstra B, Ginzinger D, Albertson D, Shuman M-A, Craik C-S (2006) Quantitation of membrane type serine protease 1 (MT-SP1) in transformed and normal cells. Biol Chem 384: 257–266.

Bhattacharyya MK, Smith AM, Ellis THN, Hedley C, Martin C (1990) The wrinkled-seed character of pea described by Mendel is caused by a transposon-like insertion in a gene encoding starch-branching enzyme. Cell 60: 115–122.

Bleve-Zacheo T, Melillo MT, Zacheo G (1990) Syncytia development in germplasm pea accessions infected with *Heterodera goettingiana*. Nematol Medit 18: 93–102.

Blixt S (1959) Cytology of *Pisum* III Investigation of five interchange lines and coordination of linkage groups with chromosomes. Agri Hort Genet 17: 47–75.

Blixt S (1972) Mutation genetics in *Pisum*. Agri Hort Genet 30: 1–293.

Bordat A, Savois V, Nicolas M, Salse J, Chauveau A, Bourgeois M, Potier J, Houtin H, Rond C, Murat F, Marget P, Aubert G, Burstin J (2011) Translational genomics in legumes allowed placing in silico 5460 Unigenes on the pea functional map and identified candidate genes in Pisum sativum L. Genes, Genomes, Genetics, in press.

Bourgeois M, Jacquin F, Savois V, Sommerer N, Labas V, Henry C, Burstin J (2009) Dissecting the proteome of pea mature seeds reveals the phenotypic plasticity of seed protein composition. Proteomics 9: 254–271.

Bourion V, Lejeune-Henaut I, Munier-Jolain N, Salon C (2003) Cold acclimation of winter and spring peas: carbon partitioning as affected by light intensity. Eur J Agron 19: 535–548.

Bødker L, Lewis BG, Coddington A (1993) The occurrence of a new genetic variant of *Fusarium oxysporum* f.sp. *pisi*. Plant Pathol 42: 33–838.

Bradford D (1996) *Fusarium oxysporum* f. sp. *pisi*. Mycopathologia 133: 57–59.

Brar KS, Dhillon GS, Singh M (1993) Field resistance in early sown pea to stem fly, *Ophiomyia phaseoli* Tryon. J Insect Sci 6: 119–120.

Bretag TW (1989) Resistance of pea cultivars to Ascochyta blight caused by *Mycosphaerella pinodes, Phoma medicaginis* and *Ascochyta pisi*. Ann Appl Biol 114: 156–157.

Bretag TW, Ramsey M (2001) Foliar diseases caused by fungi: *Ascochyta* spp. In: JM Kraft, PM Pfleger (eds) Pea Disease Compendium. American Phytopathological Society (APS) Press, St. Paul, MN, USA, pp 24–28.

Burstin J, Deniot G, Potier J, Weinachter C, Aubert G, Baranger A (2001) Microsatellite polymorphism in *Pisum sativum*. Plant Breed 120: 311–317.

Burstin J, Marget P, Huart M, Moessner A, Mangin B, Duchene C, Desprez B, Munier-Jolain N, Duc G (2007) Developmental genes have pleiotropic effects on plant morphology and source capacity, eventually impacting on seed protein content and productivity in pea. Plant Physiol 144: 768–781.

Busheva M, Apostolova E (1997) Influence of saccharides and glycine betaine on freezing of photosystem 2 enriched particles: a chlorophyll fluorescence study. Photosynthetica 34: 591–594.

Byrne O, Hardie D (2004) Incorporation of pea weevil resistance into a cultivar field pea. Grains Research and Development Corporation Project Report No UWA314SR, CLIMA, University of Western Australia.

Byrne OM, Hardie DC, Khan T, Yan G (2008) Genetic analysis of pod and seed resistance to pea weevil in a *Pisum sativum* x *P. fulvum* interspecific cross. Aust J Agri Res 59: 854–862.

Castillejo MA, Amiour N, Dumas-Gaudot E, Rubiales D, Jorrín J (2004) A proteomic approach to studying plant response to crenate broomrape (*Orobanche crenata*) in pea (*Pisum sativum*). Phytochemistry 65: 1817–1828.

Castillejo MA, Maldonado AM, Dumas-Gaudot E, Fernández-Aparicio M, Susin R, Rubiales D, Jorrín J (2009) Differential expression proteomics to investigate responses and resistance to *Orobanche crenata* in *Medicago truncatula*. BMC Genom 10: 294.

Castillejo MA, Curto M, Fondevilla S, Rubiales D, Jorrín JV (2010a) Two-dimensional electrophoresis based proteomic analysis of the pea (*Pisum sativum*) in response to *Mycosphaerella pinodes*. J Agri Food Chem 58: 12822–12832.

Castillejo MA, Susín R, Madrid E, Fernández-Aparicio M, Jorrín JV, Rubiales D (2010b) Two-dimensional electrophoresis based proteomic analysis of the *Medicago truncatula* -rust (*Uromyces striatus*) interaction. Ann Appl Biol 157: 243–257.

Chand R, Srivastava CP, Singh BD, Sarode SB (2006) Identification and characterization of slow rusting components in pea (*Pisum sativum* L.). Genet Resour Crop Evol 53: 219–224.

Charlton A, Allnut T, Holmes S, Chisholm J, Bean S, Ellis N, Mullineaux P, Oehlschlager S (2004) NMR profiling of transgenic peas. Plant Biotechnol J 2: 27–35.

Charlton AJ, Donarski JA, Harrison M, Jones SA, Godward J, Oehlschlager S, Arques JL, Ambrose M, Chinoy C, Mullineaux PM, Domoney C (2008) Responses of the pea (*Pisum sativum* L.) leaf metabolome to drought stress assessed by nuclear magnetic resonance spectroscopy. Metabolomics 4: 312–327.

Chinoy C, Welham T, Turner L, Moreau C, Domoney C (2011) The genetic control of seed quality traits: effects of allelic variation at the *Tri* and *Vc-2* genetic loci in *Pisum sativum* L. Euphytica, DOI 10.1007/s10681-011-0363-8.

Choi HK, Mun JH, Kim DJ, Zhu H, Baek JM, Mudge J, Roe B, Ellis N, Doyle J, Kiss GB, Young ND, Cook DR (2004) Estimating genome conservation between crop and model legume species. Proc Natl Acad Sci USA 101: 15289–15294.

Clement SL, Wightman JA, Hardie DC, Bailey P, Baker G, McDonald G (2000) Opportunities for integrated management of insect pests of grain legume. In: R Knight (ed) Linking Research and Marketing Opportunities for Pulses in the 21st Century. Kluwer, Dordrecht, The Netherlands, pp 467–480.

Clement SL, Hardie DC, Elberson LR (2002) Variation among accessions of *Pisum fulvum* for resistance to pea weevil. Crop Sci 42: 2167–2173.

Clement SL, McPhee KE, Elberson LR, Evans MA (2009) Pea weevil, *Bruchus pisorum* L. (Coleoptera: Bruchidae), resistance in *Pisum sativum* x *Pisum fulvum* interspecific crosses. Plant Breed 126: 478–485.

Clemente A, Gee JM, Johnson IT, MacKenzie DA, Domoney C (2005) Pea (*Pisum sativum* L.) protease inhibitors from the Bowman-Birk class influence the growth of human colorectal adenocarcinoma HT29 cells *in vitro*. J Agri Food Chem 53: 8979–8986.

Clemente A, Jiménez E, Marín-Manzano MC, Rubio LA (2008) Functional Bowman-Birk inhibitors survive gastrointestinal digestion at the terminal ileum of cannulated pigs fed chickpea-based diets. J Sci Food Agri 88: 523–531.

Clemente A, Moreno FJ, Marín-Manzano MC, Jiménez E, Domoney C (2010) The cytotoxic effect of Bowman-Birk isoinhibitors, IBB1 and IBBD2, from soybean (*Glycine max*) on HT29 human colorectal cancer cells is related to their intrinsic ability to inhibit serine proteases. Mol Nutr Food Res 54: 396–405.

Clulow SA, Matthews P, Lewis BG (1991a) A pathotype classification of *Mycosphaerella pinodes*. J Phytopathol 131: 322–332.

Clulow SA, Matthews P, Lewis BG (1991b) Genetical analysis of resistance to *Mycosphaerella pinodes* in pea seedlings. Euphytica 58: 183–189.

Colditz F, Braun HP, Jacquet C, Niehaus K, Krajinski F (2005) Proteomic profiling unravels insights into the molecular background underlying increased *Aphanomyces euteiches*-tolerance of *Medicago truncatula*. Plant Mol Biol 59: 387–406.

Cornic G, Fresneau C (2002) Photosynthetic carbon reduction and carbon oxydation cycles are the main electron sinks for photosystem II activity during a mild drought. Ann Bot 89: 887–894.

Correll JC (1991) The relationship between formae speciales, races and vegetative compatibility groups in *Fusarium oxysporum*. Phytopathology 81: 1061–1064.

Cousin R (1997) Peas (*Pisum sativum* L.). Field Crops Res 53: 111–130.

Coyne CJ, Pilet-Nayel ML (2008) Candidate genes associated with QTL controlling resistance to fusarium root rot in pea. In: Plant Anim Genome XVI Conf, San Diego, CA, USA: http://www.intl-pag.org/16/abstracts/PAG16_W18_133.html

Coyne CJ, Inglis DA, Whitehead SJ, Muehlbauer FJ (2000) Chromosomal location of *Fwf* in pea. *Pisum* Genet 32: 20–22.

Coyne CJ, Grünwald NJ, Inglis DA, McPhee KE, Pilet-Nayel ML (2004) Inheritance of fusarium root rot resistance in pea using RILs. Proc 5th Eur Conf on Grain Legumes, Dijon, AEP, Paris, France, p 340.

Coyne CJ, McClendon MT, Walling J, Timmerman-Vaughan GM, Murray S, Meksem K, Lightfoot DA, Shultz JL, Keller KE, Martin RR, Inglis DA, Rajesh PN, McPhee KE, Weeden NF, Grusak MA, Li CM, Storlie EW(2007) Construction and characterization of two bacterial artificial chromosome libraries of pea (*Pisum sativum* L.) for the isolation of economically important genes. Genome 50: 871–875.

Craig J, Lloyd JR, Tomlinson K, Barber L, Edwards A, Wang TL, Martin C, Hedley CL, Smith AM (1998) Mutations in the gene encoding starch synthase II profoundly alter amylopectin structure in pea embryos. Plant Cell 10: 413–426.

Craig J, Barratt P, Tatge H, Déjardin A, Handley L, Gardner CD, Barber L, Wang TL, Hedley C, Martin C, Smith AM (1999) Mutations at the *rug4* locus alter carbon and nitrogen metabolism of pea plants through an effect on sucrose synthase. Plant J 17: 353–362.

Crévieu I, Carré B, Chagneau AM, Quillien L, Guéguen J, Bérot S (1997) Identification of resistant pea (*Pisum sativum* L.) proteins in the digestive tract of chickens. J Agri Food Chem 45: 1295–1300.

Curto M, Camafeita E, López JA, Maldonado AM, Rubiales D, Jorrín JV (2006) A proteomic approach to study pea (*Pisum sativum*) responses to powdery mildew (*Eysiphe pisi*). Proteomics 6: 163–174.

Curto M, Gil C, Gutiérrez M, Rubiales D, Maldonado AM, Jorrín JV (2008) *Medicago truncatula* resistance to powdery mildew (*Erysiphe pisi*): a proteomic study. Proteómica 1: 135.

Dalmais M, Schmidt J, Le Signor C, Moussy F, Burstin J, Savois V, Aubert G, Brunaud V, de Oliveira Y, Guichard C, Thompson R, Bendahmane A (2008) UTILLdb, a *Pisum sativum* in silico forward and reverse genetics tool. Genome Biol 9: R43.

Darby P, Lewis BG, Matthews P (1985). Inheritance and expression of resistance to *Ascochyta pisi*. In: PD Hebblethwaite, MC Heath, TC Dawkins (eds) The Pea Crop: A Basis for Improvement. Butterworths, London, UK, pp 231–236.

Darby P, Lewis BG, Matthews P (1986) Diversity of virulence within *Ascochyta pisi* and resistance in the genus *Pisum*. Plant Pathol 35: 214–223.

Deulvot C, Charrel H, Marty A, Jacquin F, Donnadieu C, Lejeune-Hénaut I, Burstin J, Aubert G (2010) Highly-multiplexed SNP genotyping for genetic mapping and germplasm diversity studies in pea. BMC Genom 11: 468.

Die JV, Dita MA, Krajinski F, González-Verdejo CI, Rubiales D, Moreno MT, Román B (2007) Identification by suppression subtractive hybridization and expression analysis of *Medicago truncatula* putative defence genes in response to *Orobanche crenata* parasitization. Physiol Mol Plant Pathol 70: 49–59.

Dirlewanger E, Isaac P, Ranade S, Belajouza M, Cousin R, Devienne D (1994) Restriction fragment length polymorphism analysis of loci associated with disease resistance genes and developmental traits in *Pisum sativum* L. Theor Appl Genet 88: 17–27.

Di Vito M, Perrino P (1978) Reazione di *Pisum* spp. agli attacchi di *Heterodera goettingiana*. Nematol Medit 6: 113–118.

Dita MA, Die JV, Román B, Krajinski F, Küster H, Moreno MT, Cubero JI, Rubiales D (2009) Gene expression profiling of *Medicago truncatula* roots in response to the parasitic plant *Orobanche crenata*. Weed Res 49: 66–80.
Domoney C (1999) Inhibitors of legume seeds. In: PR Shewry, R Casey (eds) Seed Proteins. Kluwer Academic Publishers, Dordrecht, The Netherlands, pp 635–655.
Dumont E, Fontaine V, Vuylsteker C, Sellier S, Bodèle S, Voedts N, Devaux R, Frise M, Avia K, Hilbert JL, Bahrman N., Hanocq E, Lejeune-Henaut I, Delbreil B (2009) Associattion of sugar content QTL and PQL with physiological traits relevant to frost damage resistance in pea under field and controlled conditions. Theor Appl Genet 118: 1561–1571.
Durieu P, Ochatt SJ (2000) Effecient intergeneric fusion of pea (*Pisum sativum* L) and grass pea (*Lathyrus sativus* L) protoplasts. J Exp Bot 51: 1237–1242.
Editorial (2006) Genetically modified mush. Nat Biotechnol 24: 2.
Edwards O, Singh KB (2006) Resistance to insect pests: What do legumes have to offer? Euphytica 147: 273–285.
Eigenbrode S, White C, Rohde M, Simon J (1998) Epicuticular wax phenotype of the wel mutation and its effects on pea aphid populations in the greenhouse and in the field. *Pisum* Genet 29: 13–17.
Ek M, Eklund M, von Post R, Dayteg C, Henriksson T, Weibull P, Ceptilis A, Issac P, Tuvesson S (2005) Microsatellite markers for powdery mildew resistance in pea (*Pisum sativum* L.). Hereditas 142: 86–91.
Ellis THN, Poyser SJ (2002) An integrated and comparative view of pea genetic and cytogenetic maps. New Phytol 153: 7–25.
Ellis THN, Turner L, Hellens RP, Lee D, Harker CL, Enard C, Domoney C, Davies DR (1992) Linkage maps in pea. Genetics 130: 649–663.
Ellis THN, Poyser SJ, Knox MR, Vershinin AV, Ambrose MJ (1998) Polymorphism of insertion sites of Ty1-*copia* class retrotransposons and its use for linkage and diversity analysis in pea. Mol Gen Genet 200: 9–19.
Elvira-Recuenco M, Taylor JD (2001) Resistance to bacterial blight (*Pseudomonas syringae* pv. *pisi*) in Spanish pea (*Pisum sativum*) landraces. Euphytica 118: 305–311.
Elvira-Recuenco M, Bevan JR, Taylor JD (2003) Differential responses to pea bacterial blight under glasshouse and field conditions. Eur J Plant Pathol 109: 555–564.
Fernández-Aparicio M, Flores F, Rubiales D (2009) Field response of *Lathyrus cicera* germplasm to crenate broomrape (*Orobanche crenata*). Field Crops Res 113: 321–327.
Flavell AJ, Bolshakov VN, Booth A, Jing R, Russell J, Ellis THN, Isaac P (2003) A microarray-based high throughput molecular marker genotyping method- the tagged microrray (TAM) marker approach. Nucl Acids Res 31: e115.
Fondevilla S, Avila CM, Cubero JI, Rubiales D (2005) Response to *Mycosphaerella pinodes* in a germplasm collection of *Pisum* spp. Plant Breed 124: 313–315.
Fondevilla S, Carver TLW, Moreno MT, Rubiales D (2006) Macroscopic and histological characterisation of genes *er1* and *er2* for powdery mildew resistance in pea. Eur J Plant Pathol 115: 309–321.
Fondevilla S, Carver TLW, Moreno MT, Rubiales D (2007a) Identification and characterisation of sources of resistance to *Erysiphe pisi* Syd. in *Pisum* spp. Plant Breed 126: 113–119.
Fondevilla S, Cubero JI, Rubiales D (2007b) Inheritance of resistance to *Mycosphaerella pinodes* in two wild accessions of *Pisum*. Eur J Plant Pathol 119: 53–58.
Fondevilla S, Torres AM, Moreno MT, Rubiales D (2007c) Identification of a new gene for resistance to *Erysiphe pisi* Syd. in pea. Breed Sci 57: 181–184.
Fondevilla S, Rubiales D, Moreno MT, Torres AM (2008a) Identification and validation of RAPD and SCAR markers linked to the gene *Er3* conferring resistance to *Erysiphe pisi* DC in pea. Mol Breed 22: 193–200.
Fondevilla S, Rubiales D, Zatovic S, Torres AM (2008b) Mapping of quantitative trait loci for resistance to *Mycosphaerella pinodes* in *Pisum sativum* subsp. *syriacum*. Mol Breed 21: 439–454.

Fondevilla S, Cubero JI, Rubiales D (2010a) Confirmation that the *Er3* gene, conferring resistance to *Erysiphe pisi* in pea, is a different gene from *er1* and *er2* genes. Plant Breed 130: 281–282.
Fondevilla S, Fernández-Aparicio M, Satovic Z, Emeran AA, Torres AM, Moreno MT, Rubiales D (2010b). Identification of quantitative trait loci for specific mechanisms of resistance to *Orobanche crenata* Forsk. in pea (*Pisum sativum* L.). Mol Breed 25: 259–272.
Fondevilla S, Küster H, Krajinski F, Cubero JI, Rubiales D (2011). Identification of genes differentially expressed in a resistant reaction to *Mycospherella pinodes* in pea using microarray technology. BMC Genomics 12: 28.
Fondevilla S, Rubiales D (2011). Powdery mildew control in pea. A review. Agron Sust Dev, in press. DOI 10.1007/s13593-011-0033-1
Foster-Hartnett D, Danesh D, Penuela S, Sharopova N, Endre G, Vandenbosch KA, Young ND, Samac DA (2007) Molecular and cytological responses of *Medicago truncatula* to *Erysiphe pisi*. Mol Plant Pathol 8: 307–319.
Foucher F, Morin J, Courtiade J, Cadioux S, Ellis N, Banfield MJ, Rameau C (2003) *DETERMINATE* and *LATE FLOWERING* are two *TERMINAL FLOWER1/CENTRORADIALIS* homologs that control two distinct phases of flowering initiation and development in pea. Plant Cell 15: 2742–2754.
Fuchs J, Kuhne M, Schubert I (1998) Assignement of linkage groups to pea chromosomes after karyotyping and gene mapping by fluorescent *in situ* hybridization. Chromosoma 107: 272–276.
Furman BJ, Ambrose MJ, Coyne CJ, Redden B (2006) Formation of PeaGRIC: An international consortium to co-ordinate and utilize the genetic diversity and agro ecological distribution of major collections of *Pisum*. *Pisum* Genet 38: 32–34.
Gao Z, Eyers S, Thomas C, Ellis N, Maule A (2004) Identification of markers tightly linked to *sbm* recessive genes for resistance to Pea seed borne mosaic virus. Theor Appl Genet 109: 488–494.
Gaur V, Qureshi IA,. Singh A, Chanana V, Salunke DM (2010) Crystal structure and functional insights of hemopexin fold protein from grass pea. Plant Physiol 152: 1842–1850.
Gilpin BJ, McCallum JA, Frew TJ, Timmerman-Vaughan GM (1997) A linkage map of the pea *Pisum sativum* L. genome containing cloned sequences of known function and expressed tags ESTs. Theor Appl Genet 95: 1289–1299.
Grajal-Martin MJ, Muehlbauer FJ (2002) Genomic location of the *Fw* gene for resistance to fusarium wilt race 1 in peas. J Hered 93: 291–293.
Grajal-Martin MJ, Simon CJ, Muehlbauer FJ (1993) Use of random amplified polymorphic DNA (RAPD) to characterize race 2 of *Fusarium oxysporum* f.sp. *pisi*. Phytopathology 83: 612–614.
Grzesiak S, Filek W, Skrudlik G, Niziol B (1996) Screening for drought tolerance: Evaluation of seed germination and seedling growth for drought resistance in grain legumes. J Agron Crop Sci 177: 245–252.
Grzesiak S, Iijima M, Kono Y, Yamauchi A (1997) Differences in drought tolerance between cultivars of field bean and field pea. A comparison of drought-resistant and drought-sensitive cultivars. Acta Physiol Plant 19: 349–357.
Gupta RN, Pandey R (1998) Field reaction of new selections/varieties in early maturity group of table pea for resistance to stemfly. Indian J Pulses Res 11: 149–151.
Gurung AM, Pang ECK, Taylor PWJ (2002) Examination of *Pisum* and *Lathyrus* species as sources of ascochyta blight resistance for field pea (*Pisum sativum*). Australas Plant Pathol 31: 41–45.
Gutierrez MV, Vaz Patto MC, Huguet T, Cubero JI, Moreno MT, Torres AM (2005) Cross species amplification of *Medicago truncatula* microsatellites across three major pulse crops. Theor Appl Genet 110: 1210–1217.
Hagedorn DJ, Gritton ET (1973) Inheritance of resistance to the pea seed-borne mosaic virus. Phytopathology 63: 1130–1133.

Hall KJ, Parker JS, Ellis THN (1997) The relationship between genetic and cytogenetic maps of pea. I. Standard and translocation karyotypes. Genome 40: 744–754.

Hardie DC, Baker GJ, Marshall DR (1995) Field screening of *Pisum* accessions to evaluate their susceptibility to pea weevil (Coleoptera: Bruchidae). Euphytica 84: 155–161.

Harrison CJ, Hedley CL, Wang TL (1998) Evidence that the rug3 locus of pea (*Pisum sativum* L.) encodes a plastidial phosphor glucomutase confirms that the imported substrate for starch synthesis in pea amyloplasts is glucose-6-phosphate. Plant J 13: 753–762.

Heringa RJ, Van Norel A, Tazelaar MF (1969) Resistance to powdery mildew (*Erysiphe polygoni* D.C.) in peas (*Pisum sativum* L.). Euphytica 18: 163–169.

Hofer J, Turner L, Hellens R, Ambrose M, Matthews P, Michael A, Ellis N (1997) *UNIFOLIATA* regulates leaf and flower morphogenesis. Curr Biol 7: 581–567.

Hofer J, Turner L, Moreau C, Ambrose M, Isaac P, Butcher S, Weller J, Dupin A, Dalmais M, Le Signor C, Bendahmane A, Ellis N (2009) *Tendril-less* regulates tendril formation in pea leaves. Plant Cell 21: 420–428.

Horne J, Bailey P (1991) *Bruchus pisorum* L. (Coleoptera, Bruchidae) control by a knockdown pyrethroid in field peas. Crop Protec 10: 53–56.

Hunter PJ, Ellis N, Taylor JD (2001) Association of dominant loci for resistance to *Pseudomonas syringae* pv. *pisi* with linkage groups II, VI and VII of *Pisum sativum*. Theor Appl Genet 103: 129–135.

Humphry M, Reninstädler A, Ivanov, S, Bisselingm T, Panstruga R (2011) Durable broad-spectrum powdery mildew resistance in pea *er1* plants is conferred by natural loss-of-function mutations in *PsMLO1*. Mol Plant Pathol DOI:10.1111/J.1364-3703.2011.00718.

Infantino A, Kharrat M, Riccioni L, Coyne CJ, McPhee KE, Grünwald NJ (2006) Screening techniques and sources of resistance to root diseases in legumes. Euphytica 147: 201–221

Inglis DA (1998) Pea cyst nematode: biology and prevention. UWS Extension Bulletin EB1872.

Irzykowska L, Wolko B (2004) Interval mapping of QTLs controlling yield-related traits and seed protein conte in *Pisum sativum*. J Appl Genet 45: 297–306.

Irzykowska L, Wolko B, Swiecicki WK (2001) The genetic linkage map of pea (*Pisum sativum* L) based on molecular, biochemical, and morphological markers. *Pisum* Genet 33: 13–18.

Islam N, Campbell PM, Higgins TJV, Hirano H, Akhurt RJ (2009) Transgenic peas expressing an α-amylase inhibitor gene from beans show altered expression and modification of endogeneous proteins. Electrophoresis 30: 1863–1868.

Janila P, Sharma B (2004) RAPD and SCAR markers for powdery mildew resistance gene *er* in pea. Plant Breed 123: 271–274.

Jing R, Johnson R, Seres A, Kiss G, Ambrose MJ, Knox MR, Ellis THN, Flavell A (2007) Gene-based sequence diversity analysis of field pea (*Pisum*). Genetics 177: 2263–2275.

Jing R, Vershinin A, Grzebyta J, Shaw P, Smýkal P, Marshall D, Ambrose MJ , Ellis THN, Flavell AJ (2010) The genetic diversity and evolution of field pea (*Pisum*) studied by high throughput retrotransposon based insertion polymorphism (RBIP) marker analysis. BMC Evol Biol 10:44: doi: 10.1186/1471-2148-10-44.

Kaló P, Seres A, Taylor SA, Jakab J, Kevei Z, Kereszt A, Endre G, Ellis THN, Kiss GB (2004) Comparative mapping between *Medicago sativa* and *Pisum sativum*. Mol Genet Genom 272: 235–246.

Katiyar RP, Ram RS (1987) Genetics of rust resistance in pea. Indian J Genet 47: 46–48

Katoch V, Sharma S, Pathania S, Banayal DK, Sharma SK, Rathour R (2010) Molecular mapping of pea powdery mildew resistance gene *er2* to pea linkage group III. Mol Breed 25: 229–237.

Khajuria DR, Sharma JP (1995) Efficacy of insecticides in controlling pea leafminer (*Chromatomyia horticola*) on seed crop of pea (*Pisum sativum*). Indian J Agri Sci 65: 381–384.

Khetarpal RK, Maury Y, Cousin R, Burghofer A, Varma A (1990) Studies on resistance of pea to pea seed borne mosaic virus pathotypes. Ann Appl Biol 116: 297–304.

Knavel DE (1967) Studies on resistance to fusarium root rot *Fusarium solani* f. sp. *pisi* (F.R. Jones) in *Pisum sativum*. Am Soc Hort Sci 75: 510–516.

Kosterin OE, Bogdanova VS (2008) Relationship of wild and cultivated forms of *Pisum* L. as inferred from an analysis of three markers, of the plastid, mitochondrial and nuclear genomes. Genet Resour Crop Evol 55: 735–755.

Kosterin OE, Zaytseva OO, Vera S, Bogdanova VS, Ambrose MJ (2010) New data on three molecular markers from different cellular genomes in Mediterranean accessions reveal new insights into phylogeography of *Pisum sativum* L. subsp. *elatius* (Bieb.) Schmalh. Genet Resour Crop Evol 57: 733–739.

Kraft JM (1998) A search for resistance in peas to *Mycosphaerella pinodes*. Plant Dis 82: 251–253.

Kraft JM (2001) Fusarium Wilt. In: JM Kraft, FL Pfleger (eds) Compendium of Pea Diseases and Pests. American Phytopahological Society (APS) Press, St. Paul, MN, USA, pp 14–17.

Krusell L, Madsen LH, Sato S, Aubert G, Genua A, Szczyglowski K, Duc G, Kaneko T, Tabata S, de Bruijn F, Pajuelo E, Sandal N, Stougaard J (2002) Shoot regulation of nodule and root development is mediated by a serine/threonine receptor kinase. Nature 420: 422–426.

Kushwaha C, Chand R, Srivastava C (2006) Role of aeciospores in outbreaks of pea (*Pisum sativum*) rust (*Uromyces fabae*). Eur J Plant Pathol 115: 323–330.

Lamm R, Miravalle RJ (1959) A translocation tester set in *Pisum*. Hereditas 45: 417–440.

Laucou V, Haurogné K, Ellis N, Rameau C (1998) Genetic mapping in pea. 1. RAPD-based genetic linkage map of *Pisum sativum*. Theor Appl Genet 97: 905–915.

Le Gall M, Quillien L, Sève B, Guéguen J, Lallès JP (2007) Weaned piglets display low gastrointestinal digestion of pea (*Pisum sativum* L.) lectin and albumin pea albumin 2. J Anim Sci 85: 2972–2981.

Lejeune-Henaut I, Morin J, Fontaine V, Etévé G, Devaux R, Thomas M, Boilleau M, Stempniak JJ, Petit A, Rameau C, Baranger A (2004) Towards genes to breed for freezing resistance in pea. In: Proc 5th Eur Conf on Grain Legumes, Dijon, AEP, Paris, France, p 127.

Lejeune-Henaut I, E. Hanocq, L. Bethencourt, Fontaine V, Delbreil B, Morin J, Petit A, Devaux R, Boilleau M, Stempniak JJ, Thomas M, Lainé AL, Foucher F, Baranger A, Burstin J, Rameau C, Giauffret C (2008) The flowering locus *Hr* colocalizes with a major QTL affecting winer frost tolerance in *Pisum sativum* L. Theor Appl Genet 116: 1105–1116.

Lester D, Ross JJ, Davies PJ, Reid JB (1997) Mendel's stem length gene (*Le*) encodes a gibberellin 3-Hydroxylase. Plant Cell 9: 1435–1443.

Levy J, Bres C, Geurts R, Chalhoub B, Kulikova O, Duc G, Journet EP, Ane JM, Lauber E, Bisseling T, Denarie J, Rosenberg C, Debelle F (2004) A putative Ca2+ and calmodulin-dependent protein kinase required for bacterial and fungal symbioses. Science 303: 1361–1364.

Loridon K, McPhee K, Morin J, Dubreuil P, Pilet-Nayel ML, Aubert G, Rameau C, Baranger A, Coyne C, Lejeune-Hènaut I, Burstin J (2005) Microsatellite marker polymorphism and mapping in pea (*Pisum sativum* L.). Theor Appl Genet 111: 1022–1031.

Louis S, Delobel B, Gressent F, Duport G, Diol O, Rahioui I, Charles H, Rahbé Y (2007) Broad screening of the legume family for variability in seed insecticidal activities and for the occurrence of the A1b-like knottin peptide entomotoxins. Phytochemistry 68: 521–535.

Lu J, Knox MR, Ambrose MJ, Brown JKM, Ellis THN (1996) Comparative analysis of genetic diversity in pea assessed by RFLP- and PCR-based methods. Theor Appl Genet 93: 1103–1111.

Lupwayi NZ, Kennedy AC (2007) Grain legumes in Northern Great Plains: impacts on selected biological soil processes. Agron J 99: 1700–1709.

Macas J, Neumann P, Navratilova A (2007) Repetitive DNA in the pea (*Pisum sativum* L.) genome: comprehensive characterization using 454 sequencing and comparison to soybean and *Medicago truncatula*. BMC Genom 8: 427.

Madrid E, Gil J, Rubiales D, Krajinski F, Schlereth A, Millán T (2010) Transcription factor profiling leading to the identification of putative transcription factors involved in the *Medicago truncatula–Uromyces striatus* interaction. Theor Appl Genet 121: 1311–1321.

Manzanares M, Sánchez FJ, de Andres EF, Tenorio JL, Eyerbe L (1998) Contribution of some physiological traits to drought tolerance in peas (*Pisum sativum* L.). In: Proc 3rd Eur Conf on Grain Legumes. Valladolid, Spain, AEP, Paris, France, pp 38–39.

Markarian D, Andersen RL (1966) The inheritance of winter hardiness in *Pisum*. Euphytica 15: 102–110.

Marín-Manzano MC, Ruiz R, Jiménez E, Rubio LA, Clemente A (2009) Anti-carcinogenic Bowman-Birk inhibitors from soybean survive fermentation in active form using faecal inoculum and do not affect the microbiota composition. Brit J Nutr 101: 967–971.

Martin C, Smith AM (1995) Starch biosynthesis. Plant Cell 7: 971–985.

Martín-Sanz A, Palomo JL, Pérez de la Vega M, Caminero C (2011) Identification of pathovars and races of *Pseudomonas syringae*, the main causal agent of bacterial disease in pea in North-Central Spain, and the search for disease resistance. Eur J Plant Pathol 129: 57–69.

Matthews P, Dow K (1976) Sources and inheritance of resistance to downy mildew of the pea, *Peronospora pisi*. Ann Appl Biol 84: 281.

Maxted N, Ambrose M (2001) Peas (*Pisum* L.). In: N Maxted, SJ Bennett (eds) Plant Genetic Resources of Legumes in the Mediterranean. Kluwer, Dordrecht, The Netherlands, pp 181–190.

Maxted N, Kell SP (2009) Establishment of a global network for the *in situ* conservation of crop wild relatives: status and needs. FAO Commission on Genetic Resources for Food and Agriculture, Rome, Italy.

McClendon MT, Inglis DA, McPhee KE, Coyne CJ (2002) DNA markers for fusarium wilt race 1 resistance gene in pea. J Am Soc Hort Sci 127: 602–607.

McDonald GK, Paulsen GM (1997) High temperature effects on photosynthesis and water relations of grain legumes. Plant Soil 196: 47–58.

McPhee K (2003) Dry pea production and breeding—A mini-review. J Food Agri Environ 1: 64–69.

McPhee KE, Muehlbauer FJ (2002) Registration of 'Franklin' green dry pea. Crop Sci 42: 1378.

McPhee KE, Tullu A, Kraft JM, Muehlbauer FJ (1999) Resistance to fusarium wilt race 2 in the *Pisum* core collection. J Am Soc Hort Sci 124: 28–31.

Mehta PK, Sharma TN, Chandel RS (1998) Resistance in pea (*Pisum sativum* var *arvense*) genotypes to leaf miner (*Chromatomyia horticola*). Indian J Agri Sci 68: 271–273.

Mendel G (1866) Veruche uber pflanzen-hybriden. Verhandlungen des Naturforschenden veereins in Brunn 4: 3–47.

Mikschofsky H, Schirrmeier H, Keil GM, Lange B, Polowick PL, Keller W, Broer I (2009) Pea-derived vaccines demonstrate high immunogenicity and protection in rabbits against rabbit haemorrhagic disease virus. Plant Biotechnol J 7: 537–549.

Mishra RK, Kumar A, Chaudhary S, Kumar S (2009) Mapping of the multifoliate pinna (*mfp*) leaf-blade morphology mutation in grain pea *Pisum sativum*. J Genet 88: 227–232.

Morton RL, Schroeder HE, Bateman KS, Chrispeels MJ, Armstrong E, Higgins TJV (2000) Bean alpha amylase inhibitor 1 in transgenic peas (*Pisum sativum*) provides complete protection from pea weevil (*Bruchus pisorum*) under field conditions. Proc Natl Acad Sci USA 97: 3820–3825.

Nasir M, Hoppe HH (1991) Studies on pathotype differentiation within *Mycosphaerella pinodes* (Berk & Bloxam) Vestergren, a component of the Ascochyta-disease–complex of peas (*Pisum sativum* L.). J Plant Dis Protec 98: 619–626.

Nemecek T, von Richthofen JS, Dubois G, Casta P, Charles R, Pahl H (2008) Environmental impacts of introducing grain legumes into European crop rotations. Eur J Agron 28: 380–393.

Nouri-Ghanbalani G, Auld DL, O'Keeffe LE, Campbell AR (1978) Inheritance of resistance to the adult pea leaf weevil in Austrian winter peas. Crop Sci 18: 858–860.

Nyamsuren O, Colditz F, Rosendahl S, Tamasloukht M, Bekel T, Meyer F,Küster H, Franken P, Krajinski F (2003) Transcriptional profiling of *Medicago truncatula* roots after infection with *Aphanomyces euteiches* (oomycota) identifies novel genes upregulated during this pathogenic interaction. Physiol Mol Plant Pathol 63: 17–26.

Okubara PA, Inglis DA, Muehlbauer FJ, Coyne CJ (2002) A novel RAPD marker linked to the fusarium wilt race 5 resistance gene (*Fwf*) in *Pisum sativum*. *Pisum* Genet 34: 6–8.

Onfroy C, Tivoli B, Coubiere R, Bouznad Z (1999) Cultural, molecular and pathogenic variability of *Mycosphaerella pinodes* and *Phoma medicaginis* var. *pinodella* isolates from dried pea (*Pisum sativum*) in France. Plant Pathol 48: 218–229.

Page D, Aubert G, Duc G, Welham T, Domoney C (2002) Combinatorial variation in coding and promoter sequences of genes at the *Tri* locus in *Pisum sativum* accounts for variation in trypsin inhibitor activity in seeds. Mol Genet Genom 267: 359–69.

Pal AB, Brahmappa HS, Rawal RD, Ullasa BA (1980) Field resistance of pea germplasm to powdery mildew (*Erysiphe polygoni*) and rust (*Uromyces fabae*). Plant Dis 64: 1085–1086.

Pearce SR, Knox MR, Ellis THN, Flavell AJ, Kumar A (2000) Pea Ty1-*copia* group retrotransposons: Transpositional activity and use as markers to study genetic diversity in *Pisum*. Mol Gen Genet 263: 898–907.

Pereira G, Marques C, Ribeiro R, Formiga S, Dâmaso M, Tavares-Sousa M, Farinhó M, Leitão JM (2010) Identification of DNA markers linked to an induced mutated gene conferring resistance to powdery mildew in pea (*Pisum sativum* L.). Euphytica 171: 327–335.

Pérez-de-Luque A, Jorrín JV, Cubero JI, Rubiales D (2005) *Orobanche crenata* resistance and avoidance in pea (*Pisum* spp.) operate at different developmental stages of the parasite. Weed Res 45: 379–387.

Pérez-de-Luque A, González-Verdejo CI, Lozano MD, Dita MA, Cubero JI, González-Melendi P, Risueño MC, Rubiales D (2006) Protein cross-linking, peroxidase and β-1,3-endoglucanase involved in resistance of pea against *Orobanche crenata*. J Exp Bot 57: 1461–1469.

Pérez-de-Luque A, Eizenberg H, Grenz JH, Sillero JC, Ávila CM, Sauerborn J, Rubiales D (2010) Broomrape management in faba bean. Field Crops Res 115: 319–328.

Pilet-Nayel ML, Muehlbauer FJ, McGee RJ, Kraft JM, Baranger A, Coyne CJ (2002) Quantitative trait loci for partial resistance to Aphanomyces root rot in pea. Theor Appl Genet 106: 28–39.

Pilet-Nayel ML, Muehlbauer FJ, McGee RJ, Kraft JM, Baranger A, Coyne CJ (2005) Consistent QTL in pea for partial resistance to *Aphanomyces euteiches* isolates from United States and France. Phytopathology 95: 1287–1293.

Prioul S, Onfroy C, Tivoli B, Baranger A (2003) Controlled environment assessment of partial resistance to *Mycosphaerella pinodes* in pea (*Pisum sativum* L.) seedlings. Euphytica 131: 121–130.

Prioul S, Frankewitz A, Deniot G, Morin G, Baranger A (2004) Mapping of quantitative trait loci for partial resistance to *Mycosphaerella pinodes* in pea (*Pisum sativum* L.), at the seedling and adult plant stages. Theor Appl Genet 108: 1322–1334.

Prioul-Gervais S, Deniot G, Receveur EM, Frankewitz A, Fourmann M, Rameau C, Pilet-Nayel ML, Baranger A (2007) Candidate genes for quantitative resistance to *Mycosphaerella pinodes* in pea (*Pisum sativum* L.). Theor Appl Genet 114: 971–984.

Provvidenti R, Alconero R (1988a) Inheritance of resistance to a lentil strain of pea seed-borne mosaic virus in *Pisum sativum*. J Hered 79: 45–47.

Provvidenti R, Alconero R (1988b) Inheritance of resistance to a third pathotype of pea seed-borne mosaic virus in *Pisum sativum*. J Hered 79: 76–77.

Provvidenti R, Hampton RO (1993) Inheritance of resistance to white lupin mosaic virus in common pea. HortScience 28: 836–837.

Provvidenti R, Hampton RO, Muehlbauer FJ (1991) G-1000, Marx's pea breeding line for multiviral resistance. *Pisum* Genet 23: 50–52.

Quinn M, Bezdicek D (1996) Effect of CryIIIA protein production in nodules on pea-pea leaf weevil (Coleoptera: Curculionidae) interactions. J Econ Entomol 89: 550–557.

Radchuk R, Radchuk V, Weschke W, Borisjuk L, Weber H (2006) Repressing the expression of the SUCROSE NONFERMENTING-1-RELATED PROTEIN KINASE gene in pea embryo causes pleiotropic defects of maturation similar to an abscisic acid-insensitive phenotype. Plant Physiol 140: 263–278.

Rameau C, Dénoue D, Fraval F, Haurogné K, Josserand J, Laucou V, Batge S, Murfet IC (1998) Genetic mapping in pea. 2. Identification of RAPD and SCAR markers linked to genes affecting plant architecture. Theor Appl Genet 97: 916–928.

Reeves JC, Hutchins JD, Simpkins SA (1996) The incidence of races of *Pseudomonas syringae* pv. *pisi* in UK pea (*Pisum sativum*) seed stocks, 1987–1994. Plant Var Seeds 9: 1–8.

Rosov SM, Kosterin O, Borisov AY, Tsyganov V (1999) The history of the pea gene map: last revolutions and the new symbiotic genes. *Pisum* Genet 31: 55–570.

Rubiales D, Fondevilla S (2010) Resistance of cool season food legumes to ascochyta blight. Field Veg Crop Res 47: 439–442.

Rubiales D, Pérez-de-Luque A, Cubero JI, Sillero JC (2003) Crenate broomrape (*Orobanche crenata*) infection in field pea cultivars. Crop Prot 22: 865–872.

Rubiales D, Moreno MT, Sillero JC (2005) Search for resistance to crenate broomrape (*Orobanche crenata*) in pea germplasm. Genet Resour Crop Evol 52: 853–861.

Rubiales D, Pérez-de-Luque A, Fernández-Aparicio M, Sillero JC, Román B, Kharrat M, Khalil S, Joel DM, Riches Ch (2006) Screening techniques and sources of resistance against parasitic weeds in grain legumes. Euphytica 147: 187–199.

Rubiales D, Fernández-Aparicio M, Pérez-de-Luque A, Prats E, Castillejo MA, Sillero J, Rispail N, Fondevilla S (2009a) Breeding approaches for crenate broomrape (*Orobanche crenata* Forsk.) management in pea (*Pisum sativum* L.). Pest Manag Sci 65: 553–559.

Rubiales D, Fernández-Aparicio M, Wegmann K, Joel DM (2009b) Revisiting strategies for broomrape (*Orobanche* and *Phelipanche* spp.) seed bank demise. Weed Res 49: 23–33.

Rubiales D, Castillejo MA, Madrid E, Barilli E, Rispail N (2011) Legume breeding for rust resistance—lessons to learn from the model *Medicago truncatula*. Euphytica, in press. DOI: 10.1007/s10681-011-0367-4.

Rubiales D, Fernández-Aparicio M (2011) Innovations in crop protection: parasitic weeds infecting legume crops as a case study. Agron Sust Dev, in press.

Rubiales D, Heide-Jørgensen HS (2011). Parasitic Plants. In: Enciclopedia of Life Sciences (ELS), John Wiley & Sons, Ltd: Chichester. DOI: 10.1002/9780470015902.a0021271.

Rutledge CE, Robinson AP, Eigenbrode SD (2003) Effects of a simple plant morphological mutation on the arthropod community and the impacts of predators on a principal insect herbivore. Oecologia 135: 39–50.

Sales MP, Gerhardt IR, Grossi-de-Sá MF, Xavier-Filho J (2000) Do legume storage proteins play a role in defending seeds against bruchids? Plant Physiol 124: 515–522.

Salgado P, Freire JPB, Ferreira RB, Teixera A, Bento O, Abreu MC, Toullec R, Lalles JP (2003) Immunodetection of legume proteins resistant to small intestinal digestion in weaned piglets. J Sci Food Agri 83: 1571–1580.

Sánchez FJ, Manzanares M, de Andres EF, Tenorio JL, Eyerbe L (1998) Turgor maintenance, osmotic adjustment and soluble sugar and praline accumulation in 49 pea cultivars in response to water stress. Field Crops Res 59: 225–235.

Sato Y, Morita R, Nishimura M, Yamaguchi H, Kusaba M (2007) Mendel's green cotyledon gene encodes a positive regulator of the chlorophyll degrading pathway. Proc Natl Acad Sci USA 104: 14169–14174.

Schiltz S, Gallardo K, Huart M, Negroni L, Sommerer N, Burstin J (2004) Proteome reference maps of vegetative tissues in *Pisum sativum*: an investigation of nitrogen mobilization from leaves during seed filling. Plant Physiol 135: 2241–2260.

Schroeder HE, Gollasch S, Moore A, Tabe LM, Craig S, Hardie DC, Chrispeels MJ, Spencer D, Higgins TJV (1995) Bean α-amylase inhibitor confers resistance to the pea weevil (*Bruchus pisorum*) in transgenic peas (*Pisum sativum* L.). Plant Physiol 107: 1233–1239.

Schroeder WT, Provvidenti R (1971). A common gene for resistance to bean yellow mosaic virus and watermelon mosaic virus 2 in *Pisum sativum*. Phytopathology 61: 846–848.

Shehata MA, Davis DW, Pfleger FL (1983) Breeding for resistance to *Aphanomyces euteiches* root rot and *Rhizoctonia solani* stem rot in peas. J Am Soc Hort Sci 108: 1080–1085.

Sillero JC, Cubero JI, Fernández-Aparicio M, Rubiales D (2005) Search for resistance to crenate broomrape (*Orobanche crenata*) in *Lathyrus*. Lathyrus Lathyrism Newsl 4: 7–9.

Sillero JC, Fondevilla S, Davidson J, Vaz Patto MC, Warkentin TD, Thomas J, Rubiales D (2006) Screening techniques and sources of resistance to rusts and mildews in grain legumes. Euphytica 147: 255–272.

Simon CJ, Muehlbauer FJ (1997) Construction of a chickpea linkage map and its comparison with maps of pea and lentil. J Hered 88: 115–119.

Singh R, Ram H (2001) Inheritance of days to flowering and rust resistance in peas. Res Crops 2: 414–418.

Singh PP, Yadav RP, Singh B (2001) Response of pea genotypes to pea stemfly (*Ophiomyia phaseoli* Tryon). Shashpa 8: 59–61.

Skiba B, Ford R, Pang ECK (2004) Construction of a linkage map based on a *Lathyrus sativus* backcross population and preliminary investigation of QTLs associated with resistance to ascochyta blight. Theor Appl Genet 109: 1726–1735.

Skot L, Harrison S, Nath A, Mytton R, Clifford B (1990) Expression of insecticidal activity in *Rhizobium* containing the delta-endotoxin gene cloned from *Bacillus thuringiensis* subsp. *tenebrionis*. Plant Soil 127: 285–295.

Smartt J (1990) Grain legumes: evolution and genetic resources. Cambridge University Press, Cambridge, UK 392 p.

Smykal P, Coyne CJ, Ford R, Redden R, Hybl M, Flavell AJ, Warkentin T, Burstin J, Duc G, Ambrose M, Ellis T (2008) Effort towards a world pea (*Pisum sativum* L.) germplasm core collection: The case for common markers and data compatibility. *Pisum* Genet 40: 11–14.

Soroka JJ, Mackay PA (1990) Seasonal occurrence of the pea aphid, *Acyrthosiphon pisum* (Harris) (Homoptera: Aphididae), on cultivars of field peas in Manitoba and its effects on pea growth and yield. Can Entomol 122: 503–513.

Soroka JJ, Mackay PA (1991) Antibiosis and antixenosis to pea aphid (Homoptera: Aphididae) in cultivars of field peas. J Econ Entomol 84: 1951–1956.

Sousa-Majer MJ, Turner NC, Hardie DC, Morton RL, Lamont B, Higgins TJV (2004) Response to water deficit and high temperature of transgenic peas (*Pisum sativum* L.) containing a seed- specific α-amylase inhibitor and the subsequent effects on pea weevil (*Bruchus pisorum* L.) survival. J Exp Bot 55: 497–505.

Stegmark R (1988) Downy mildew resistance of various pea genotypes. Acta Agri Scand 38: 373–379.

Stegmark R (1990) Variation for virulence among Scaninavian isolates of *Peronospora viciae* f. sp. *pisi* (pea downy mildew) and response of pea genotypes. Plant Pathol 39: 118–124.

Stegmark R (1992) Diallel analysis of the inheritance of partial resistance to downy mildew in peas. Plant Breed 108: 111–117.

Stegmark R (1994) Downy mildew on peas (*Peronospora viciae* f. sp. *pisi*). Agronomie 14: 641–647.

Stoddard FL, Balko C, Erskine W, Khan HR, Link W, Sarker A (2006) Screening techniques and sources of resistance to abiotic stresses in cool-season food legumes. Euphytica 147: 167–186.

Streb P, Aubert S, Gout E, Bligny R (2003) Cold- and light-induced changes of metabolite and antioxidant levels in two high mountain plant species *Soldanella alpina* and *Ranunculus glacialis* and a lowland species *Pisum sativum*. Physiol Planta 118: 96–104.

Tar'an B, Warkentin T, Somers DJ, Miranda D, Vandenberg A, Blade S, Woods S, Bing D, Xue A, DeKoeyer D, Penner G (2003) Quantitative trait loci for lodging resistance, plant height and partial resistance to mycosphaerella blight in field pea (*Pisum sativum* L.). Theor Appl Genet 107: 1482–1491.

Tar'an B, Warkentin T, Somers DJ, Miranda D, Vandenberg A, Blade S, Bing D (2004) Identification of quantitative trait loci for grain yield, seed protein concentration and maturity in field pea (*Pisum sativum* L.). Euphytica 136: 297–306.

Tar'an B, Zhang C. Warkentin T, Tullu A, Vandenberg A (2005) Genetic diversity among varieties and wild species accessions of pea (*Pisum sativum* L.) based on molecular markers, and morphological and physiological characters. Genome 48: 257–272.

Taylor JD, Bevan JR, Crute IR, Reader SL (1989) Genetic relationship between races of *Pseudomonas syringae* pv. *pisi* and cultivars of *Pisum sativum*. Plant Pathol 38: 364–375.

Taylor S, Hofer J, Murfet I (2001) Stamina pistilloida, the pea ortholog of Fim and UFO is required for normal developement of flowers, inflorescences, and leaves. Plant Cell 13: 31–46.

Thomas JE, Kenyon DM, Kightley SPJ (1999) Progress in the exploitation of disease resistance in oilseed rape, field peas and field beans. Asp Appl Biol 56: 67–74.

TEGERM (2000) Analysis and exploitation of germplasm resources using transposable element molecular markers: http://www.biocenter.helsinki.fi/bi/tegerm/ (Cited 29 Oct 2009).

Timmerman-Vaughan GM, Frew TJ, Weeden NF (1994) Linkage analysis of *er1*, a recessive *Pisum sativum* gene for resistance to powdery mildew fungus (*Erysiphe pisi* D.C). Theor Appl Genet 88: 1050–55.

Timmerman-Vaughan GM, McCallum JA, Frew TJ, Weeden NF, Russell AC (1996) Linkage mapping of quantitative trait loci controlling seed weight in pea (*Pisum sativum* L.). Theor Appl Genet 93: 431–439.

Timmerman-Vaughan GM, Frew TJ, Weeden NF (2000) Characterization and linkage mapping of R-gene analogous DNA sequences in pea (*Pisum sativum* L.). Theor Appl Genet 101: 241–247.

Timmerman-Vaughan GM, Frew TJ, Russell AC, Khan T, Butler R, Gilpin M, Murray S, Falloon K (2002) QTL mapping of partial resistance to field epidemics of Ascochyta blight of peas. Crop Sci 42: 2100–2111.

Timmerman-Vaughan GM, Frew TJ, Butler R, Murray S, Gilpin M, Falloon K, Johnston P, Lakeman MB, Russell A, Khan T (2004) Validation of quantitative trait loci for Ascochyta blight resistance in pea (*Pisum sativum* L.), using populations from two crosses. Theor Appl Genet 109: 1620–1631.

Timmerman-Vaughan GM, Mills A, Whitfield C, Frew T, Butler R, Murray S, Lakeman M, McCallum J, Russell A, Wilson D (2005) Linkage mapping of QTL for seed yield, yield components, and development traits in pea. Crop Sci 45: 1336–1344.

Tivoli B, Baranger A, Avila CM, Banniza S, Barbetti M, Chen W, Davidson J, Lindeck K, Kharrat M, Rubiales D, Sadiki M, Sillero JC, Sweetingham M, Muehlbauer FJ (2006) Screening techniques and sources of resistance to foliar diseases caused by major necrotrophic fungi in grain legumes. Euphytica 147: 223–253.

Tiwari KR, Penner GA, Warkentin TD, Rashid KY (1997a) Pathogenic variation in *Erysiphe pisi*, the causal organism of powdery mildew of pea. Can J Plant Pathol 19: 267–271.

Tiwari KR, Penner GA, Warkentin TD (1997b) Inheritance of powdery mildew resistance in pea. Can J Plant Sci 77: 307–310.

Tiwari KR, Penner GA, Warkentin TD (1998) Identification of coupling and repulsion phase RAPD markers for powdery mildew resistance gene *er-1* in pea. Genome 41: 440–444.

Tiwari KR, Penner GA, Warkentin TD (1999) Identification of AFLP markers for powdery mildew resistance gene *er2* in pea. *Pisum* Genet 31: 27–29.

Turner NC (1986) Crop water deficits: A decade of progress. Adv Agron 39: 1–51

Turner NC, Wright GC, Siddique KHM (2001) Adaptation of grain legumes (pulses) to water limited environments. Adv Agron 71: 193–231.

Turner SR, Barratt DHP, Casey R (1990) The effect of different alleles at the *r* locus on the synthesis of seed storage proteins in *Pisum sativum*. Plant Mol Biol 14: 793–803.

Upreti KK, Murti GGR (1999) Effect of polyamines on the changes in endogenous hormones in pea under water stress conditions. Indian J Plant Physiol 4: 1–5.

Upreti KK, Murti GGR, Bhatt RM (2000) Response of pea cultivars to water stress: changes in morpho-physiological characters, endogenous hormones and yield. Veg Sci 27: 57–61.

Valderrama MR, Román B, Satovic Z, Rubiales D, Cubero JI, Torres AM (2004) Locating genes associated with *Orobanche crenata* resistance in pea. Weed Res 44: 323–328.

Vanstone V, Kelly S, Hunter H (2004) Are field pea and faba bean really resistant to root lesion nematodes? GRDG-Research Update for Growers-Western Region, Perth, Australia, pp 81–82.

Vaz Patto MC, Rubiales D (2009) Identification and characterization of partial resistance to rust in a germplasm collection of *Lathyrus sativus*. Plant Breed 128: 495–500.

Vaz Patto MC, Fernández-Aparicio M, Moral A, Rubiales D (2006) Characterization of resistance to powdery mildew (*Erysiphe pisi*) in a germplasm collection of *Lathyrus sativus*. Plant Breed 125: 308–310.

Vaz Patto MC, Fernández-Aparicio M, Moral A, Rubiales D (2007) Resistance reaction to powdery mildew (*Erysiphe pisi*) in a germplasm collection of *Lathyrus cicera* from Iberian origin. Genet Resour Crop Evol 54: 1517–1521.

Vaz Patto MC, Fernández-Aparicio M, Moral A, Rubiales D (2009) Pre and posthaustorial resistance to rusts in *Lathyrus cicera*. Euphytica 165: 27–34.

Vershinin AV, Allnutt TR, Knox MR, Ambrose MJ, Ellis THN (2003) Transposable elements reveal the impact of introgression, rather than transposition in *Pisum* diversity, evolution and domestication. Mol Biol Evol 20: 2067–2075.

Via S, Hawthorne D (2002) The genetic architecture of ecological specialization: correlated gene effects on host use and habitat choice in pea aphids. Am Nat 159: 576.

Vigeolas H, Chinoy C, Zuther E, Blessington B, Geigenberger P, Domoney C (2008) Combined metabolomic and genetic approaches reveal a link between the polyamine pathway and albumin 2 in developing pea seeds. Plant Physiol 146: 74–82.

Vijayalakshmi S, Yadav K, Kushwaha C, Sarode SB, Srivastava CP, Chand R, Singh BD (2005) Identification of RAPD markers linked to the rust (*Uromyces fabae*) resistance gene in pea (*Pisum sativum*). Euphytica 144: 265–274.

Vioque J, Clemente A, Sánchez-Vioque R, Pedroche J, Bautista J, Millán F (1998) Comparative study of chickpea and pea PA2 albumins. J Agri Food Chem 46: 3609–3613.

Von Stackelberg M, Lindemann S, Menke M, Riesslemann S, Jacobsen HJ (2003) Identification of AFLP and STS markers closely linked to *def* locus in pea. Theor Appl Genet 106: 1293–1299.

Wang Z, Luo Y, Li X, Wang L, Xu S, Yang J, Weng L, Sato S, Tabata S, Ambrose M, Rameau C, Feng X, Hu X, Luo D (2008) Genetic control of floral zygomorphy in pea (*Pisum sativum* L.). Proc Natl Acad Sci USA 105: 10414–10419.

Watt C, Razai L, Coyne CJ (2004) Molecular diversity of USDA pea core collection. In: ASA-CSSA-SSSA Annual Meetings Abstracts [CD-ROM], p 1092.

Weeden NF, Marx GA (1984) Chromosome location of twelve isozyme loci in *Pisum sativum*. J Hered 75: 65–370.

Weeden NF, Marx GA (1987) Further genetic analysis and linkage relationships of isozyme loci in pea. J Hered 78: 153–159.

Weeden NF, Porter L (2007) The genetic basis of fusarium root rot tolerance in the 'Afghanistan' pea. *Pisum* Genet 39: 35–36.

Weeden NF, Muehlbauer FJ, Ladizinski G (1992) Extensive conservation of linkage relationship between pea and lentil genetic maps. J Hered 83: 123–129.

Weeden NF, Swiecicki WK, Ambrose M, Timmerman GM (1993) Linkage groups of pea. *Pisum* Genet 25: 4.

Weeden NF, Ellis THN, Timmerman-Vaughan GM, Swiecicki WK, Rozov SM, Berdnikov VA (1998) A consensus linkage map for *Pisum sativum*. *Pisum* Genet 30: 1–4.

Weeden NF, McGee RJ, Grau CR, Muehlbauer FJ (2000) A gene influencing tolerance to common root rot is located on linkage group IV. *Pisum* Genet 32: 53–55.

Weigelt K, Küster H, Radchuk R, Müller M, Weichert H, Fait A, Fernie AR, Saalbach I, Weber H (2008) Increasing amino acid supply in pea embryos reveals specific interactions of N and C metabolism, and highlights the importance of mitochondrial metabolism. Plant J 55: 909–926.

Weigelt K, Küster H, Rutten T, Fait A, Fernie AR, Miersch O, Wasternack C, Emery RJN, Desel C, Hosein F, Müller M, Saalbach I, Weber H (2009) ADP-glucose pyrophosphorylase-deficient pea embryos reveal specific transcriptional and metabolic changes of carbon-nitrogen metabolism and stress responses. Plant Physiol 149: 395–411.

Weightman RM (2005) The rise and fall of grain legumes and their predicted yields in UK agriculture. In: J Wiseman, R Sylvester-Bradley (eds) Yields of Farmed Species: Constraints and Opportunities in the 21st Century. Nottingham University Press, Nottingham, UK, pp 261–288.

Welbaum GE, Bian D, Hill DR, Grayson RL, Gunatilaka MK (1997) Freezing tolerance, protein composition and abscisic acid localization and content of pea epicotyl, shoot, and root tissue in response to temperature and water stress. J Exp Bot 48: 643–654.

Wery J, Silim SN, Knights EJ, Malhotra RS, Cousin R (1994) Screening techniques and sources and tolerance to extremes of moisture and air temperature in cool season food legumes. Euphytica 73: 73–83.

Wilkinson T, Douglas A (1998) Plant penetration by aphids (*Acyrthosiphon pisum*) of different plant range. Entomol Exp Appl 87: 43–50.

Wiseman J, Al-Mazooqi W, Welham T, Domoney C (2003) The apparent ileal digestibility, determined with young broilers, of amino acids in near-isogenic lines of peas (*Pisum sativum* L) differing in trypsin inhibitor activity. J Sci Food Agri 83: 644–651.

Wiseman J, Al-Marzooqi W, Hedley C, Wang TL, Welham T, Domoney C (2006) The effects of genetic variation at *r*, *rb* and *Tri* loci in *Pisum sativum* L. on apparent ileal digestibility of amino acids in young broilers. J Sci Food Agri 86: 436–444.

Wong CE, Bhalla P, Ottenhof H, Singh M (2008) Transcriptional profiling of the pea shoot apical meristem reveals processes underlying its function and maintenance. BMC Plant Biol 8: 73.

Wroth J (1998) Variation in pathogenicity among and within *Mycosphaerella pinodes* populations collected from field pea in Australia. Can J Bot 76: 1955–1966.

Wroth J (1999) Evidence suggests that *Mycosphaerella pinodes* infection of *Pisum sativum* is inherited as a quantitative trait. Euphytica 107: 193–204.

Xue AG, Warkentin TD (2001a) Partial resistance to *Mycosphaerella pinodes* in field pea. Can J Plant Pathol 81: 535–540.

Xue AG, Warketin TD (2001b) Reaction of field pea varieties to three isolates of *Uromyces fabae*. Plant Sci 82: 253–255.

Xue AG, Warkentin TD, Gossen BD, Burnett PA, Vandenberg A, Rashid KY (1998). Pathogenic variation of western Canadian isolates of *Mycosphaerella pinodes* on selected *Pisum* genotypes. Can J Plant Pathol 20: 189–193.

Yu J, Gu WK, Provvidenti R, Weeden NF (1995) Identifying and mapping two DNA markers linked to the gene conferring resistance to pea enation mosaic virus. J Am Soc Hort Sci 120: 730–733.

Yu J, Gu WK, Weeden NF and Provvidenti R (1996) Development ASAP marker for resistance to bean yellow mosaic virus in *Pisum sativum*. *Pisum* Genet 28: 31–32.

Zimmermann J, Saalbach I, Jahn D, Giersberg M, Haehnel S, Wedel J, Macek J, Zoufal K, Glunder G, Falkenburg D, Kiprijanov SM (2009) Antibody expressing pea seeds as fodder for prevention of gastrointestinal parasitic infections in chickens. BMC Biotechnol 9: 79.

Zlamal P (1984) Genetics of horizontal resistance to anthracnose in peas. Sb UVTI Genet Slechteni 20: 191–192.

Zohary D, Hopf M (1973) Domestication of pulses in the old world. Science 182: 887–894.

Zong X, Redden RJ, Liu Q Wang S Guan J, Liu J, Xu Y, Liu X, Gu J, Yan L, Ades P, Ford R (2009) Analysis of a diverse global *Pisum* sp. collection and comparison to a Chinese local *P. sativum* collection with microsatellite markers. Theor Appl Genet 118: 193–204.

2

Faba Bean

Ana M. Torres,[1,a,*] *Carmen M. Avila,*[1] *Fred L. Stoddard*[2] and *Jose I. Cubero*[3]

ABSTRACT

Domesticated in ancient times, faba bean (*Vicia faba* L.) provides valuable protein for human and animal consumption. Faba bean seeds are a staple in the diets of many societies in the Middle East and North Africa. Except for the recent increase in production reported in Canada and Australia, the general trend has been towards a slight decrease of the cultivated area around the world, with no appreciable changes in yield and production. Among the reasons for such a decline are the low and unstable yields together with the lack of adapted cultivars resistant to major faba bean diseases. Development of new varieties combining different resistances (mainly to broomrape and *Ascochyta*) continues to be a major task in the crop. Moreover, breeding for quality and abiotic stresses are becoming important challenges in faba bean research. This chapter provides an overview of the genomic resources and molecular markers currently available in faba bean, with putative application in marker-assisted selection of the crop. Molecular breeding for resistance to crenate broomrape, Ascochyta blight, rust and chocolate spot are underway, and promising results have been obtained. Moreover markers to select for growth habit or low tannins, vicine and convicine content have also been reported. Recently, robust markers derived from expressed sequence tags are filling the faba bean genetic maps and genomic tools such as macro- and microarrays are becoming available in the species. In the near future, mapping approaches together with

[1] IFAPA, Centro Alameda del Obispo, Area de Mejora y Biotecnología, Apdo. 3092, E-14080 Córdoba, Spain.
[a] e-mail: *anam.torres.romero@juntadeandalucia.es*
[2] Department of Agricultural Sciences, University of Helsinki, PO Box 27 (Latokartanonkaari 5), FIN-00014 Helsingin Yliopisto, Finland.
[3] Departamento de Genética, Universidad de Córdoba, Campus de Rabanales. Edificio C5, 2nd planta. 14071 Córdoba, Spain.
*Corresponding author

comparative genomics and synteny analyses will reveal candidate genes and selectable markers for use in faba bean crop improvement.

Keywords: abiotic stresses, disease resistance, genome mapping, marker-assisted selection, quantitative trait loci, synteny

2.1 Introduction

Vicia faba L. is a well known crop, particularly in Eurasia and North Africa. In each of the countries where it is grown, it has received many names according to its size and use (vegetable, food legume, feed legume). In English the traditional name, still seen in 19th century dictionaries, was just "bean". The word bean was later applied to other crops whose seeds showed a similar shape, such as the American crop *Phaseolus vulgaris* L., which at present is the most widely known "bean" in the world. *Vicia faba* also gaves its name to other crops, especially to *Phaseolus* and *Vigna* species (for example, in Spanish: *fabes, habichuelas,* etc.). The name "field" bean was widely used in English for a time, and has direct parallels in German "Ackerbohne" and Swedish "bondböna", among others. Since the term "field bean" was often applied to both *V. faba* and *P. vulgaris*, while "bean" is not sufficiently informative, the name "faba bean" was derived by Canadian breeders and spread by ICARDA breeders and agronomists. It is largely used nowadays because of its simplicity and relationship with the scientific name.

2.1.1 Economic Importance

The world cultivated area in 2007 was 2.5 million ha (Mha) with a total production of about 3.74 million tonnes (Mt) and an average yield of some 1.5 t/ha, 93% (about 2.3 Mha and 3.6 Mt) being for dry seeds, the rest for vegetable use (FAOSTAT 2008 and official as well as unofficial estimates). The leading country is China with 1.42 Mt (down from 1.7 Mt in 2001; 0.912 Mha), followed by Ethiopia (0.576 Mt, 0.458 Mha), Egypt (0.302 Mt, 0.089 Mha), France (0.255 Mt, 0.057 Mha), the UK (0.160 Mt, 0.045 Mha), Australia (0.138 Mt, 0.133 Mha) and the Sudan (0.132 Mt, 0.070 Mha). Around 200,000 ha were dedicated to horticultural uses (green pods and seeds), with a total production of almost 1 million tons. Bolivia (34,000 ha) and Algeria (20,000 ha), were countries with the largest area, and the highest production figures were those of China (12,000 t) and Morocco (95,000 t) (FAOSTAT 2008).

Because of its high protein content, dry faba bean is used as a grain legume for feeding animals as well as for human consumption; the latter is seldom found in Europe, except as roasted and salted seeds for a snack, but it is a staple in some regions such as western Asia and northern Africa.

The green pods and seeds are a delicacy in some Mediterranean countries, especially the "baby" seeds (length <12 mm) that can command a high price in the Spanish market; raw semi-matured seeds are also consumed as a snack. Both green seeds and pods are the base of several traditional dishes in Mediterranean countries. Green seeds are also marketed either canned or frozen. Varietal characteristics differ according to their commercial use. At least since classical Roman times, faba bean has also been used as green manure as it is among the most productive annual legumes.

Faba beans were for many centuries, along with other food legumes, especially chickpea (*Cicer arietinum* L.), lentil (*Lens culinaris* Medik.) and pea (*Pisum sativum* L.), permanent components of crop rotations with both wheat and barley. The value as a food crop lies in its high protein and carbohydrate content. Among the amino acids, the lysine content is high and that of the sulfur amino acids low, compensating in both human and animal diets the oposite values of these amino acids in cereals and explaining the wide use of the mixtures of cereals and legumes in diets from the most remote agricultural times (Bressani and Elias 1988). Not negligible, although not indicated above, is the good level of calcium. The widespread use of this crop is therefore not surprising, including in classical times the Celtic peoples (sometimes it was called the "celtic grain") and the Roman legions all over Europe and North Africa.

The typical range of values found in analyses of proximate composition is given in Table 2-1. In exceptional circumstances, protein content may be as low as 20% or as high as 40%. Amylose content is 35–45% of starch, more than in most cereal starches.

Faba bean prefers clay-lime, chalky, well drained and textured soils, with neutral pH, although it adapts to a wide pH range (6 to 9) as well as to sandy-loam soils, especially in humid regions. Exchanges of germplasm between latitudes shows that many faba bean cultivars require long days to flower and mature (photoperiod sensitive) while others appear to be daylength-neutral (Patrick and Stoddard 2010). High temperatures between flowering and early maturity causes loss of both flowers and immature pods and decreases the food quality of the grain.

Like most legumes, faba bean makes nitrogen-fixing symbioses with bacteria of the Rhizobiaceae, in this case *Rhizobium leguminosarum* Frank bv *viciae*. Root nodules formed by other biovars are generally less effective and it is widely recommended that farmers inoculate their seed with the appropriate culture in order to achieve the full potential nitrogen fixation, often in the range of 60–250 kg/ha/year (some reports indicate up to 600 kg/ha/year). The molecular genetics of several *Rhizobium* species has been thoroughly investigated and there is excellent potential to use improved or genetically modified strains to increase crop nitrogen fixation in the near future, but that is outside the range of this review (see Chapter 10).

Table 2-1 Summary of the reported faba bean linkage maps (this table updates the one published by Torres et al. 1996).

Reference	Mapping population	Mapped markers	Linkage groups (Assigned to chromosomes)	Map size (cM)
Van de Ven et al. (1991)	BC: (var. *paucijuga* line 172 x cv. Optica) x var. *paucijuga* line 172	17 markers: (4 morphological, 3 isozymes, 3 RAPDs, 7 RFLPs)	7 (-)	231.4
Torres et al. (1993b)	2 F_2: Vf6 x Vf35 Vf6 x Vf173	51 markers: (7 isozymes, 43 RAPDs, 1 RFLP)	11 (1)	300–350
Ramsay et al. (1995)	BC: (var. *paucijuga* line 172 x cv. Optica) x var. *paucijuga* line 172	23 markers: (8 morphological, 4 isozymes, 4 RAPDs, 7 RFLPs)	7 (-)	~300
Satovic et al. (1996)	7 F_2: Vf6 x Vf2 T5[a] Vf6 x Vf2 T6 Vf6 x Vf33 T3 Vf6 x Vf33 T4 Vf6 x Vf159 T4 Vf6 x Vf159 T5 Vf6 x Vf159 T6	157 markers: (1 morphological, 9 isozymes, 147 RAPDs)	48 (6)	~850
Vaz Patto et al. (1999)	3 F_2: Vf6 (*equina*) x Vf27 (*paucijuga*) T4 Vf6 x Vf27 T6 Vf6 x Vf27	116 markers: (1 morphological, 7 isozymes, 105 RAPDs, 3 seed-protein genes)	13 (7)	~1,200
Román et al. (2002)	F_2: Vf6 x Vf136	121 markers: (2 isozymes, 117 RAPDs, 2 seed-protein genes)	16 (9)	1,445.5

Table 2-1 contd....

Table 2-1 contd....

Reference	Mapping population	Mapped markers	Linkage groups (Assigned to chromosomes)	Map size (cM)
Román et al. (2004)	Composite map of 11 F_2: Satovic et al. (1996) Vaz Patto et al. (1999) Román et al. (2002)	192 markers: (2 morphological, 6 isozymes, 176 RAPDs, 4 SSRs, 4 seed-protein genes)	5 (5)	1,559
Ávila et al. (2004)	F_2: 29H x Vf136	103 markers: (4 isozymes, 94 RAPDs, 2 seed-protein genes, 3 SSRs)	18 (6)	1,308
Ellwood et al. (2008)	RIL: Vf6 x Vf27	151 ITAP markers: (36 MP, 25 ML, 42 MLG, 48 ESTs from GLIP)[b]	12 (0)	~1,685
Diaz-Ruiz et al. (2009a, 2009b, 2010)	RIL: Vf6 x Vf136	277 markers: (238 RAPDs, 4 isozymes, 5 ESTs, 1 SCAR, 6 SSRs, 2 STSs, 21 ITAPs)	21 (9)	~2,856

[a]The number following T refers to the type of F_1 trisomic plant that originated the F2 population (e.g., T5 stands for a F_1 plant trisomic for chromosome 5)
[b]**MP**: primers developed from alignment of *M. truncatula* and *Pisum* spp. database EST sequences; **ML**: primers developed from alignment of *M. truncatula* and *Lupinus* spp.; **MLG**: primers based on alignments between *M. truncatula*, *Lupinus albus*, and *Glycine*

2.1.2 Academic Importance

The faba bean genome is one of the cytogenetically best characterized among plants, with a high DNA content (1C ~13 pg ~11.8 Mpb, Bennett and Leitch 1995) distributed in six chromosomes comprising five acrocentric chromosomes and one metacentric with a satellite. Due to the ease of handling of its large chromosomes from root tip meristems (easily made visible with orcein, fuchsin and other traditional stains), the faba bean has been the species of choice for cytogenetic analysis and many phenomena were observed for the first time in this species. A comprehensive survey of the molecular-cytogenetic studies in faba bean was provided by Fuchs et al. (1998) including nucleus formation during telophase (Heitz 1931), semiconservative replication of chromosomes by using tritium-labeled thymidine (Taylor et al. 1957), fluorescence banding of chromosomes (Caspersson et al. 1969), and heterochromatin extension (Fucik et al. 1970). Faba bean has also been widely used for studies on chromosome breakage and rearrangement induced by either chemicals or radiation (Rieger and Michaelis 1967).

Individual chromosomes may also be distinguished by using different lines of reconstructed genotypes (Schubert et al. 1986; Schubert and Rieger 1991). These lines are excellent tools for cytological studies such as microdissection (Schubert et al. 1993) and flow sorting of chromosomes (Lucretti et al. 1993) for genetic and physical gene mapping, and also immunolocalization of chromosomal proteins (Belyaev et al. 1997, 1998).

The species reproduces by partial allogamy; the outcrossing coefficient, from almost nil to almost total allogamy (85%) depends on both the genotype and the environmental conditions (Stoddard and Bond 1987; Maalouf et al. 2002). Bumblebees, honeybees and large solitary bees such as *Eucera numida* are the main pollen vectors. The success of pollination has a major effect on the potential yield of the crop. Self-fertile cultivars can be obtained, allowing the crop to be grown where the usual pollinators are, for some reason, inadequate in numbers.

The negative correlation between yield and protein content usual in cereals does not exist, or it is very weak in faba bean as in many other grain legumes (Stoddard et al. 1993). This has encouraged breeders to increase the protein content without reducing the yield (but ensuring that the lysine content is not reduced). Important yield components determined by genotype, environment and management system are the number of pods per plant and the number of pods per unit area unit, while those primarily under genetic control are the number of seeds per pod and the seed size (Patrick and Stoddard 2010).

2.1.3 Brief History of the Crop

Probably domesticated in the 6th millennium BC in the Near East, *V. faba* spread quickly (Cubero 1973, 1974) 1) to central Europe and the Eurasiatic plains through Anatolia, the Danube Valley and the Caucasus, 2) to North Africa and the Iberian Peninsula through the Mediterranean coast and islands, 3) from Egypt and Arabia to Ethiopia, 4) from Mesopotamia to India (also from Ethiopia through the Sabean Route, likely active in the second millennium BC), and 5) from the Near East to China probably during the first millenium AD (only land races of *major* type, the latest in being produced, were known in China prior to the 20th century). In the 16th century the crop was introduced to America by European colonizers and since 1980, the crop has been widely grown in Australia.

In historical times in Egypt, faba bean was forbidden as food for priests but appreciated by the common people. Classical Romans believed in a certain connection between the faba bean seed and the spirits of the dead. Besides, in ancient Rome the *Fabaria* was an important religious festival (and the *Fabii* one of the noblest ancient families). The relationship between faba bean and its religious role is not well understood although "favism", an anaemia produced by the consumption of its seeds by persons who are homozygous recessive for a particular allele of glucose 6-phosphate dehydrogenase, could be one of the causes (Marquardt 1982).

The archaeological record shows that the first domesticated materials were characterized by very small and rounded seeds, similar to the *paucijuga* type. "*Minor*" types, also with small rounded seeds, were later obtained, migrating towards the south; all Ethiopian landraces showed *minor* features when collected by Vavilov. Looking for an increased yield produced, by a correlated response, a flattened seed type, similar to the modern *equina* types. Once this new type was obtained, selection proceeded to increase both the flattening and the weight of the seed, reaching 2 g in some vegetable cultivars (Serradilla et al. 1993). A cluster of primitive seed and pod characters similar to those of existing *paucijuga* types is dominant over their modern counterparts, the opposite holding for leaf characters (Martín 1976; Martín and Cubero 1979; Cubero and Martín 1981; Suso and Cubero 1986).

2.1.4 Germplasms and Genepools

The wild ancestor of faba bean has not been found; its closest relative is also unknown. Many attempts to cross it with other *Vicia* species have proved unsuccessful (Cubero 1982, 1984; Maxted et al. 1991). Faba bean DNA content is double to that of most of the other *Vicia* species and all known relatives have a different chromosome composition, comprising seven acrocentric chromosomes. Both facts explain the early abortion of hybrid embryos. Its isolated position within the genus has also been demonstrated by several multivariate studies at the morphological level (Bueno 1976; Cubero 1982). Since *V. faba* does not cross with any other known species, only a primary genepool should be considered according to the Harlan and De Wet system (Harlan and De Wet 1971).

X-ray treatments produced an asynaptic mutant and from this, tetraploid and aneuploid cytotypes were obtained. Tetraploid lines, with no commercial interest, are morphologically very similar to the diploids. Differences are the tetrahedral, not ellipsoidal, pollen grains and undulated margins of leaflets (Poulsen and Martín 1977; Martín et al. 1986). Progenies of the whole set of trisomics has been obtained except for the metacentric chromosome (chromosome I) because of its huge size and genetic content and they have been used in mapping projects to assign different linkage groups (LGs) to specific chromosomes (Martín 1979; Cabrera et al. 1989; Torres et al. 1995). Faba bean chromosomes can now be easily flow-sorted and *in situ* hybridized (Macas et al. 1993a, b; Fuchs et al. 1994), facilitating gene mapping at the sub-chromosomal level and integration of the genetic and physical maps (Macas et al. 1993a; Vaz Patto et al. 1999). This approach has further been used for the development of markers from specific chromosomes and/or regions. Thus, Pozarkova et al. (2002) developed microsatellite markers from chromosome I. Translocation lines have also been obtained, although they have been only moderately used in gene mapping (Sjödin 1971a).

Many mutants affecting all plant organs are known and used in both research and breeding. Most of these mutants are of natural origin but some were also produced by X-rays almost 50 years ago by Sjödin and others, among these the asynaptic line mentioned above. More recently, new sets of mutants have been produced by ethylmethane sulfonate (EMS) treatments. Some mutants of actual or potential agronomical importance are: low content of vicine and convicine (favism agents), low tannin content, determinate growth habit, pod indehiscence, high self-fertility, resistance to key diseases, and high rhizobium nodulation.

Large germplasm collections are maintained at ICARDA (Aleppo, Syria), the N.I. Vavilov Institute (St Petersburg, Russia) and Gatersleben (Germany). Medium term and working collections are maintained in some other places, especially in some European countries (Spain, France, the UK and Germany), Egypt, Australia and China (Sadiki 2006). Faba bean seeds are orthodox from the point of view of conservation, thus are easily stored at low temperature and low humidity. However, the partial allogamy presents a difficulty in the germplasm maintenance, a problem solved by keeping inbred lines as well as the original populations and genepools derived from the breeders work (Witcombe and Erskine 1984; Robertson 1985).

2.1.5 Botanical Description

V. faba belongs to the genus *Vicia*, tribe Fabeae (syn. Vicieae), family Fabaceae (syn. Legumiosae, Papilionaceae), order Fabales. It is a paradox that, although this is the type species for the genus, tribe, family and order, for historical reasons there is no genus named *Faba*. Although the reproductive isolation and cytological differences could support a separation of the species from the genus *Vicia*, in which case the name *Faba bona* would take precedence, the morphological similarities, especially with the *narbonensis* group of species, indicates maintaining the taxonomic status quo.

At the subspecific level, many authors still use Muratova's two subspecies based on the maximum number of leaflets per leaf and related characteristics (Muratova 1931): ssp. *faba* (*eufaba* in Muratova's terminology) and ssp. *paucijuga* with respectively more and fewer than four leaflets per leaf. She also defined three varieties in the former subspecies according to the ratio of seed thickness to seed length (an index strongly correlated with seed size): *minor* (ellipsoidal seeds up to 0.6 g), *equina* (rather flattened seeds between 0.6 and 0.9 g) and *major* (very flattened and large seeds up to 2 g). Later, Hanelt (1972a, b), included *paucijuga* forms in his *minor* subspecies, all the other forms grouped in ssp. *Faba* (Fig. 2-1).

Figure 2-1 Faba bean recombinant inbred lines from a cross between an *equina-minor* and a *paucijuga* type, displaying transgressive segregation for seed color, shape and size. The four types used as descriptors: *paucijuga* (P), *minor* (Mi), *equina* (E) and *major* (Ma), are recovered in the derived population (Courtesy Dr S Cruz).

Color image of this figure appears in the color plate section at the end of the book.

Since all forms cross easily, morphological variation in the key traits is continuous, the only genetic divergence at the isozyme level is related to geographical isolation and not to botanical characteristics, and no wild form has been collected. There are no modern taxonomic grounds for dividing the species into subspecies, but the names "paucijuga", "minor", "equina" and "major" remain very convenient descriptors.

Paucijuga is a rare form collected only in NW India and Pakistan with many peculiar features considered to be primitive (Suso and Cubero 1986): many tiny stems branching from the ground, flowers sessile and only 1–2 per node, small leaflets and very small, almost round seeds of less than 0.20 g. It could be a relic of the wild form, but it was cultivated, and its high self-fertility suggests an introduction in a region lacking pollinators, as happened in historical times in the Sudan.

Major types, characterized by large and flattened seeds ("broad-beans", "fèves", "habas") have been traditionally used for human consumption because of their edible pods and softer and sweeter seeds. For animal feeding, all forms have been used, but particularly *equina* and *minor* types, collectively called "field beans", "féveroles", "habines". The relationships between names and uses transcend language barriers: *equina* (Latin: of horses) types are "horse-bean" in English and "caballar" (related to equines) in Spanish, and *minor* landraces were collectively called "cochineras" (related to pigs) in Spanish.

V. faba is a annual diploid crop with a strong taproot. The plant is more or less erect and branches, if present, usually arise from the crown. Stems are square in cross-section, hollow and glabrous. The compound leaves have no tendrils, and the number (4–8) and shape (oval to lanceolate) of leaflets depend on the variety or landrace, their margins being straight or slightly undulated. Stipules usually show a dark spot with an extrafloral nectary. Where winters are hard, such as central and northern Europe as well as Canada, spring-sown cultivars are used, whereas in zones with milder winters, autumn-sown cultivars are generally preferred and offer higher potential yields than spring-sown.

Its flowers, are papilionate, with a single ovary, usually with 3–5 (in some accessions up to 12) ovules and 10 stamens, of which one is free and filaments of the other nine are united in a sheath around the ovary. Nectaries are present at the bottom of the corolla tube. Racemes are short, the number of flowers varying from 1–2 (especially in *paucijuga*) to 10–12 and even exceptionally more in some *minor* cultivars; common numbers are 4–6.

The pod is a typical legume. The number of pods depends on several genetic and environmental factors. Usually 1–4 pods are produced per node, erect or decumbent; a better fruit/flower ratio is one of the main breeding objectives. Except for some cultivars, especially those used for vegetable purposes, pods are dehiscent. Indehiscent, fleshy pods can be cooked and eaten as those of some varieties of pea (mange tout) and common bean (snap beans). At maturity, pods are brownish and coriaceous, from cylindrical (especially *minor*) to flattened (especially *major*). Pod length can vary from

2–3 cm in *paucijuga* to 30–40 cm, or even 50 cm in some horticultural *major* accessions. Seed number is determined by ovule number, minus a few percent due to abortions.

Its natural growth habit is indeterminate. The determinate habit has been obtained by artificial mutagenesis by X rays; the allele *ti* responsible for this growth habit has being backcrossed into some Spanish *major* cultivars already released for cultivation (Cubero and Nadal 2005) and it has been used in several other countries.

2.2 Classical Genetics and Traditional Breeding

2.2.1 Classical Mapping Efforts

As reported in Torres et al. (2006), prior to 1990 few morphological and isozyme loci had been mapped in the *V. faba* genome (Cabrera et al. 1989) and no extended linkage groups (LGs) had been reported. Nevertheless, it is worth mentioning the early attempts of Erith (1930), Sirks (1931) and Picard (1963) in the assessment of genetic variation and linkage studies, as well as the use of translocation stocks in the assignment of different loci to their respective chromosomes (Sjödin 1971a, b). Comprehensive reviews regarding morphological characters were published by ICARDA (1981, 1986).

Isozyme polymorphisms in faba bean have been widely used in evolutionary and taxonomic studies (Yamamoto 1975a, b; Yamamoto and Plitmann 1980; Polignano and Sonnante 1992), identification of inbred lines (Gates and Boulter 1979, 1980), cultivar identification (Bassiri and Rouhani 1977; Kaser and Steiner 1983), and estimation of outcrossing rates (Peat and Adham 1984; Suso et al. 1995), but details of the inheritance of only few isozyme loci were available (Gates and Boulter 1979; Suso et al. 1993; Peat and Adham 1984; Mancini et al. 1989). Further studies on isozymes (Torres et al. 1995, 1998) identified additional loci and assigned them to specific chromosomes. The advent of the PCR-derived markers facilitated the development of the genetic linkage maps in the species. A summary of the faba bean linkage maps published so far is presented in Section 2.3.

2.2.2 Limitations of Classical Endeavors and Utility of Molecular Mapping

Introgression of desired genes using traditional protocols is lengthy and complicated. Among the multiple uses of molecular markers, the most promising for faba bean is marker-assisted selection (MAS) for disease resistance and quality traits. The approach is based on the selection of markers linked to the desired genes in segregating populations to trace or pyramid favorable alleles in a given genome. This methodology can obviate the need of testing to identify target individuals from early generations, leading to an effective improvement of the breeding procedure.

As in many other crops, MAS has been widely used for major disease resistance genes, selecting alleles with major effects on high-value traits with relatively simple inheritance. By contrast, the approach has been less successful for the improvement of polygenic traits, due to the insufficient precision of quantitative trait loci (QTL) mapping techniques and the unreliable extrapolation of QTL information across multiple populations.

The advantages of MAS compared to conventional phenotypic selection are the simplicity and reliability of the screening process, which can be performed at early seedling stages, thus saving both time and resources. The use of molecular markers and the development of suitable F_2 and advanced inbred populations have allowed significant progress in faba bean mapping to enhance breeding strategies in the species. Sections 2.4, 2.5 and 2.6 provide an overview of the genomic resources currently available in faba bean, with an emphasis on development and application of MAS for genetic improvement of the crop.

2.2.3 *Breeding Objectives*

2.2.3.1 *Yield and Yield Stability*

Faba bean yields are usually very low (0.6–1.0 t/ha worldwide), the traditional care of the crop being very poor. Where modern cultivars and good farming practices are available, yields can be several times that figure. In faba bean "yield stability" is at least as important as yield itself as a breeding objective. Low stability can be attributed partly to the effects of pests and partly to the partial allogamous nature of this species which increases heterogeneity in comparison with fully inbreeding or strictly allogamous species. The indeterminate nature of the crop also allows it to respond to growing conditions.

To improve yield stability, the breeder has several available tools. Modern multivariate statistical methods allow for selecting directly for greater stability. Hybrid cultivars (possible after the pioneering work by Bond et al. 1966a, b) are themselves more stable in yield, but although obtained by French breeders they were not released due to instability of the cytoplasmic male-sterility system. Synthetic cultivars also provide a partial solution for instability (Maalouf et al. 2002; Link 2006). Breeding for wide adaptation has also been performed by crossing central European to Mediterranean materials and testing the progenies in both environments; the populations obtained are an excellent pre-breeding material to widen the genetic base of the crop (Link et al. 1996a, b; Link 2006). Finally, breeding for self-fertile varieties is a well tested solution; the use of insect-proof cages is compulsory, selection being carried simultaneously within the cages and in the open field (Cubero and Moreno 1984; Torres et al. 1993a). Components of self-fertility and indexes to select for it have been proposed (Mondragao-Rodrigues 1994). Self-fertile cultivars also provide some insurance against inadequate numbers or activity of pollinators.

2.2.3.2 Resistance to Diseases and Pests

As for any other crops, disease resistance is one of the main faba bean breeding objectives (Salt 1983; Bond et al. 1994; Sillero et al. 2010). Among the diseases, chocolate spot [caused by *Botrytis fabae* Sard. and sometimes *B. cinerea* (de Bary) Whetzel] and Ascochyta blight [caused by *Ascochyta fabae* Speg.] are important in humid regions as contrasted as the Atlantic Europe and the Nile Valley. Rust [caused by *Uromyces viciae-fabae* (Pers.) J. Schröt.] is widespread but is most important in warm continental climates such as that of New South Wales, Australia, parts of Europe and China, where moderate to substantial yield losses can occur (Rashid and Bernier 1991; Jellis et al. 1998; Stoddard et al. 2010). *Orobanche* (broomrape) parasitic weeds (*O. crenata* Forsk. around the Mediterranean Sea, *O. foetida* Poir in North Africa, and *O. ramosa* L. and *Phelipanche aegyptiaca* (Pers.) Pomel (syn. *O. aegyptiaca* Pers.) in the Near East and humid zones in Egypt) are limiting factors in many Mediterranean countries, affecting yield and even the possibility of growing the crop (Cubero 1983; Cubero et al. 1999). Resistance is known for all these diseases and some resistant cultivars have been released, especially in the case of resistance to *Orobanche* (Pérez-de-Luque et al. 2010; Sillero et al. 2010).

Viruses and bacteria are not considered major pathogens up to now (Bond 1980; Cockbain 1983; Bos et al. 1988) although viruses may become more important in the future. The most frequent at present are broad bean mosaic virus, pea enation mosaic virus, bean yellow mosaic virus, broad bean stain virus and broad bean leaf roll virus. The stain virus can be transmited by seed. Limited research on these diseases, most likely because of their relatively low importance, has not yet produced available resistant genotypes for commercial use. Most are controlled by insecticidal control of the aphid or weevil vectors, for which resistance is also hard to identify and use.

Stem nematode (*Ditylenchus dipsaci* Filipjev), whose symptoms can be mistaken for a chocolate spot infestation, is the most important nematode on faba beans, especially in cold regions. All plant organs are affected, including pods and seeds.

Traditional legume pests are well known on faba beans (Bardner 1983). Among these, *Bruchus rufimanus* Boh. (bruchids), the classical granary pest, is worth mentioning. *Sitona lineatus* (L.) (sitona weevil) adults attack the leaves and larvae the root nodules, eliminating or reducing the nitrogen fixation, which is one of the valuable faba bean agronomical features. *Aphis fabae* Scopoli is a well known polyphagous aphid (some 200 hosts have been identified). No resistant lines have been detected so far, although some work was performed in the past on aphid resistance showing partial resistance in several germplasm accessions (Cammell and Way 1983). The lack of sustained research on this topic can be explained, in the case of insect pests, by available successful treatment with chemicals. The recent limitations by the European Union on active ingredients for agricultural use along with international pressure for reduced agrochemical use can produce either a total crop failure or a new interest in pest resistance as the only solution for farmers (Stoddard et al. 2010).

2.2.3.3 Other Characters

There have been several attempts to select for higher efficiency in nitrogen fixation with positive results, although, as in other cases, there are many constraints to use them in a practical way due to the interaction between the selected bacterial strain and the natural *Rhizobium* population (Roughley et al. 1983; Beringer et al. 1988; Bliss and Miller 1988; Herridge et al. 1994).

The protein content of faba bean, usually 25–30%, is considered excellent and the lysine content is also high. Some bioactive or non-nutritional (formerly called "antinutritional") compounds decrease the biological value of the crop in feed or food. The most important of these components are tannins that combine with the protein in the digestive tract, and vicine and convicine, the two pyrimidine glycosides causing favism. In animals, the effects of vicine-convicine (v-c) are important only in laying hens and on the reproductive performance of sows; no problems have been described for ruminants. Fortunately, there are "low tannin" and "low vicine-convicine" mutations now included in or being transferred to commercial cultivars (Duc 2006).

In the case of tannins, there are two independent recessive genes, *zt-1* and *zt-2*, each one lowering the tannin content (the double heterozygote produces tannins by genic complementation). Both genes produce a white color flower that can be used as an excellent morphological marker for selecting low tannin genotypes. In most European countries white flowered cultivars are available (Picard 1976; Cabrera and Martín 1986; Cubero and Duc 1995). Selecting against v-c content is more complicated as the chemical detection of v-c is costly and difficult, but the discovery of a spontaneous mutant allele named *v-c* that induces a 10–20 fold reduction of v-c content has facilitated the process. New cultivars being registered in Spain have been developed by taking advantage of the linkage (5 cM) of the allele for low v-c content with the colorless hilum of the seed. Molecular markers closely linked to these traits have been also identified (Gutierrez et al. 2006, 2007, 2008), to track the introgression of the desired alleles in the development of new cultivars with improved nutritional value (See Section 2.5).

Mechanical harvesting is a requirement in developed agriculture; faba bean, like all indeterminate crops, has a shortcoming in this sense. Determinate habit was not required in many agricultural systems, as drought did not terminate growth, the crops were swathed (manually harvested before full maturity and left to dry till threshing). For vegetable use, green pods and seeds were gathered as long as they were produced by the plant, but the cost of this operation is not compatible with the requirements of a modern mechanized system. The *ti* gene for determinate habit is therefore now being used (using both traditional and molecular selection) to obtain new cultivars suitable for green products that can be mechanically harvested on a field scale (Nadal et al. 2000).

2.2.4 Limitations of Traditional Breeding and Rationale for Molecular Breeding

Although traditional breeding methods have proved useful, faba bean breeding would greatly benefit from the availability of robust and accurate markers for resistance to biotic stresses such as crenate broomrape and pathogenic fungi, abiotic stresses such as frost and drought tolerance, and traits related to the nutritional value of the seeds.

Genetic and genomic analysis to identify QTLs affecting major diseases (Román et al. 2002, 2003; Avila et al. 2003, 2004) are available together with markers linked to a gene controlling growth habit (Avila et al. 2006, 2007) or to quality breeding traits such as zero tannin and low v-c content (Gutierrez et al. 2006, 2007, 2008) (see Sections 2.5 and 2.6). Nevertheless, enhancing the genetic map saturation to identify tightly linked markers and to determine the accurate position of the corresponding QTL is required. In addition, QTL validation across multiple test locations and independent mapping populations is a prerequisite for a successful application in MAS. Both types of studies are being carried out at present for QTLs controlling broomrape and ascochyta resistance, and preliminary results indicate that saturation and QTL validation efforts are fairly straightforward (Díaz-Ruiz et al. 2009a, b). Moreover, the generation of an updated consensus map will be extremely useful in targeted marker development, a complex task in faba bean where physical maps and bacterial artificial chromosome (BAC) libraries are still not available.

The recent information from the model species *Medicago truncatula* is greatly facilitating the genetic characterization and comparative mapping of traditional legume crops such as faba bean (Choi et al. 2004a, b; Zhu et al. 2005; Ellwood et al. 2008, among many others). Mapping of cross-species markers from *Medicago* is facilitating the establishment of the gene order in other species and the macro- and micro-scale comparison of their genetic maps. In addition, comparative functional genomics may bring some new perspectives to faba bean breeding efforts. Little is known about the functional correspondence of model legume genes and their putative faba bean orthologs but predictions can be made based on the sequence similarities between the relatively few *M. truncatula* and faba bean gene pairs that are available and the high conservation and synteny existing between both species (Ellwood et al. 2008). Thus, identification of QTLs and/or candidate genes involved in stress tolerance and/or quality traits can be use for selection in applied faba bean breeding programs through MAS.

Comparative plant genomics might help to transfer information but transgenesis of target plants will be an unavoidable technology for the functional analysis that can not necessarily be extrapolated from model systems to target crops. In contrast to related species such as *V. narbonensis,* faba bean transformation continues to be a challenging task. Nevertheless, some publications (Böttinger et al. 2001; Hanafy et al. 2005) indicate that also for this more recalcitrant legume, the opportunity for routine genetic modification is available and could be integrated in breeding schemes in the near future (see Chapter 9).

As mentioned above, hybrids could be an excellent option to increasing both the yield and the yield stability. They were obtained by classical methods, by using natural male sterility discovered in the 1960s (Bond et al. 1966a, b). Hybrid cultivar breeding requires use of the limited sources of male sterile cytoplasm and fertility restoring genes already known, narrowing the genotypic content of the hybrids or, requiring long backcrossing programs to transfer these genetic characters to other genotypes. However, the potential shown by the molecular methodology developed in *Brassica napus* L. (oilseed rape or canola) could widen the scope as it can be applied to any set of convenient commercial cultivars.

Finally, as glyphosate is able to control the attack by broomrape, resistance to this herbicide, obtained by molecular methods and nowadays introduced in many transgenic cultivars of maize, cotton and soybean, could be a candidate for use in faba bean molecular breeding.

2.3 Molecular Linkage Maps: Strategies, Resources and Achievements

Plant breeding is based around the identification and utilization of genetic variation. Its success relies upon the selection of the right combination of alleles. The large number of lines needed to produce a single variety and the difficulty or the cost to evaluate some traits, such as yield stability, quality or disease resistance has promoted the development of tools for indirect selection. In this way, Sax (1923) was the first to describe the association between a morphological marker and a yield-related trait, when he reported the association between seed color and seed size in the common bean. Nevertheless, even the management of relatively simple traits usually requires the manipulation of a considerable number of genes. Hence, genetic maps have been evidenced as a useful tool towards the establishment of indirect selection for plant breeding.

As in other crops, the development of linkage maps in faba bean, is parallel to the development and availability of different types of molecular markers. Early genetic studies focused on the inheritance of morphological traits as flower color, plant height and seed morphology (Erith 1930). Subsequently, Sirks et al. (1931) developed detailed analysis on 26 faba bean factors (genes). After calculating the recombination frequency, 19 out of the 26 factors were included in four LGs. Additional morphological markers such as seed coat color were also considered further (Picard 1963). The development of a series of translocation lines represented a valuable tool for locating the first genes or markers in their respective chromosomes (Sjodin 1971a, b). Similarly, the identification and characterization of five of the six primary trisomics of *V. faba* by Martín and Barceló (1984), allowed the assignment of genes and LGs to chromosomes II to VI (Cabrera et al. 1989; Suso et al. 1993, Torres et al. 1995, 1998; Satovic et al. 1996; Vaz Patto et al. 1999).

The advent of molecular markers based on DNA has facilitated the construction of denser genetic linkage maps that provide a useful tool for identification and mapping of QTLs. The first preliminary faba bean map including DNA based markers, three random amplified polymorphic DNAs (RAPDs) and seven restriction fragment length polymorphisms (RFLPs), consisted of 17 markers distributed along seven LGs (van de Ven et al. 1991). Further efforts allowed mapping of five additional loci and the imprecise detection of QTLs related to morphological and biochemical characteristics (Ramsay et al. 1995).

RAPDs have been the most commonly used markers in *V. faba* during the last decade due to the lack of more informative markers. RAPDs and RFLPs were used along with isozyme loci by Torres et al. (1993b) to develop the first meaningful genetic linkage map in the species. Two different F_2 populations sharing the maternal parent, the asynaptic line Vf6, allowed the linkage of 51 markers distributed into 11 LGs. One LG was assigned to chromosome I since it included the 45S ribosomal array (Torres et al. 1993b). Later, 13 F_2 populations derived from trisomic F_1 plants for chromosomes III, IV, V and VI (T3, T4, T5 and T6, respectively) were analyzed for morphological, isozyme and RAPD markers. A composite map was constructed consisting of 157 markers, most of them RAPDs, distributed into 48 LGs, 6 being assigned to specific chromosomes (Satovic et al. 1996). In subsequent investigations on three F_2 populations derived from the crosses Vf6×Vf27, Vf6×Vf27 T4 and Vf6×Vf27 T6, the map expanded to around 1,200 cM and seven out of the 13 LGs identified were assigned to specific chromosomes (Vaz Patto et al. 1999). Vf27 is a *paucijuga* line and Vf6 an *equina*. Several QTLs for seed weight were identified, one of which, on chromosome VI, accounted for around 30% of the phenotypic variation (Vaz Patto et al. 1999).

Although trisomic individuals for chromosome I have been obtained, they have been infertile, precluding their use in genetic mapping experiments. To overcome this limitation, Pozarkova et al. (2002) developed simple sequence repeat (SSR) markers specific for this chromosome using flow sorting methodology.

RAPDs have also proven to be very effective in the development of molecular maps for QTL identification for the main diseases affecting the crop. Crenate broomrape is one of the major constraints for faba bean production in Mediterranean climates, so the identification of genes controlling resistance against this holoparasite is a key subject in faba bean breeding. A set of 196 F_2 individuals were derived from the cross Vf6×Vf136 and a map constructed based on 117 RAPDs, two isozymes and two seed-protein genes. This map was successfully used for QTL detection for broomrape resistance (Román et al. 2002) and Ascochyta blight (Román et al. 2003) as described below. Furthermore, a recombinant inbred line (RIL) population was derived from this cross and a new map consisting of 21 LGs (nine of them assigned to specific chromosomes) and covering 2,856.7 cM has been constructed (Díaz-Ruiz et al. 2009a, b; Díaz et al. 2010). Similarly, an F_2 population derived from the cross Vf29H×Vf136 was used to develop a map useful for QTL mapping against Ascochyta blight, since Vf29H is highly resistant to this pathogen. A total of 103 markers, including 94 RAPDs, three isozymes, two seed-protein genes and three SSRs were mapped into 18 LGs, six of which were unequivocally assigned to specific chromosomes (Avila et al. 2004). An RIL population has recently been developed from this cross and the corresponding genetic map is now being constructed (CM Avila et al. unpubl.data).

With the mapping information available up to 2003, a composite map derived from 11 F_2 populations was constructed (Román et al. 2004). In all, 192 markers, including 176 RAPDs, four SSRs, four seed-protein genes, six isozymes and two morphological traits were mapped into 14 major LGs of which five were assigned to specific chromosomes. This map provided an efficient tool in breeding applications such as QTL analyses and marker-assisted selection (Román et al. 2004). In addition, RAPD markers were mostly used to identify putative QTLs associated with frost tolerance and auxiliary traits in RILs derived from the cross between two frost-tolerant lines, Côte d'Or 1 and Bean Pure Line (BPL) 4628 (Arbaoui et al. 2008b).

The development of genetic markers is complex and costly in species with little fundamental genetic information. Nevertheless, progress has been achieved in the recent years and the utilization of markers cross-amplifying among different species has brought greater efficiency than RAPDs. SSRs derived from the model species *Medicago truncatula* have been successfully amplified in faba bean (Gutierrez et al. 2005) allowing the transfer of knowledge from models to crops. Expressed sequence tag (EST)-SSRs were more transferable (42.6%) than genomic SSRs (24.2%). These markers are now being included in the new composite map under development (AM Torres et al. unpubl.data).

Finally, a RIL population derived from the cross Vf6×Vf27 was used to map orthologous codominant cross-species markers. A map comprising 151 intron-targeted amplified polymorphic (ITAP) markers was constructed. Linkage analysis revealed seven major and five small linkage groups, spanning a total of 1,685.8 cM (Ellwood et al. 2008). This map, anchored with orthologous markers mapped in *M. truncatula*, provides a central reference map for future use of genomic and genetic information in faba bean genetic analysis and breeding. It also provides an excellent tool to transfer information from the model species or related legumes to *V. faba*. Furthermore, the map developed by Ellwood et al. (2008) established macrosyntenic relationships between faba bean, *M. truncatula* and lentil and provide the first step to use the available information for breeding purposes.

2.4 Tagging of Simply Inherited Traits

The development of saturated linkage maps allows the efficient identification of molecular markers for indirect selection of both polygenic and monogenic traits. However, other strategies such as bulked segregant analysis (BSA) or the candidate gene approach constitute cost- and time-effective strategies that have been successfully applied to faba bean breeding.

Important traits such as rust resistance or seed quality parameters including absence of tannins or low content in vicine-convicine (antinutritional factors) are regulated by single genes in faba bean. In these cases, BSA has proved an effective method for the identification of markers tightly linked to the corresponding responsible genes (Avila et al. 2003; Gutierrez et al. 2006, 2007, 2008).

Two different types of incomplete resistance to faba bean rust have been reported. In the first type, resistance is expressed as a reduction of disease severity without any macroscopically visible necrosis (Bernier and Conner 1982; Rashid and Bernier 1986a, b, 1991; Sillero et al. 2000). In the second, hypersensitive resistance to faba bean rust has been described as an incomplete resistance associated with late-acting necrosis of the host tissue resulting in a reduction of the infection type (Sillero et al. 2000). The types of resistance differ only in the presence or absence of macroscopically visible necrosis and in their inheritance (Sillero et al. 2000; Sillero and Rubiales 2002). In this way, hypersensitive resistance is monogenic (Sillero et al. 2000; Avila et al. 2003) and race-specific, which implies that the use of single resistance genes in cultivars would likely result in non-durable resistance in the long term (reviews in Sillero et al. 2006, 2010). Among the strategies proposed to increase the durability of this type of resistance is resistance gene pyramiding (Duvick 1996), the incorporation of different resistance genes into a single cultivar.

The development of MAS programs based on markers tightly linked to resistant genes would greatly facilitate breeding. Accordingly, Avila et al. (2003) used BSA to detect markers tightly linked to the gene causing the hypersensitive resistance to rust (race 1) in the *V. faba* inbred line 2N52. This is the earliest available published report on the use of BSA in faba bean. Three RAPD markers ($OPD13_{736}$, $OPL18_{1032}$ and $OPI20_{900}$) in coupling phase with the resistance gene for race 1 (*Uvf-1*) were mapped in 55 F_2 plants. No recombinants between $OPI20_{900}$ and *Uvf-1* were detected. Two additional markers ($OPP02_{1172}$ and $OPR07_{930}$) were linked to the gene in repulsion phase at a distance of 9.9 and 11.5 cM, respectively (Avila et al. 2003).

Disease resistance in faba bean may be controlled by different genes since several sources of hypersensitive resistance to the same disease have been reported (Sillero et al. 2000). Therefore, the development of adequate molecular markers for MAS would be cost-effective for faba bean breeding programs.

The major bioactive compounds targeted in breeding programs are condensed tannins and vicine-convicine (Section 2.2.3.3) because they decrease digestibility and biological value of the protein (Kumar and Singh 1984; Bartolomé et al. 1994), and useful mutants are available. Two genes, *zt-1* and *zt-2* (Picard 1976), interrupt the tannin biosynthetic pathway, giving plants with white flowers and no tannin content (Bond 1976; Cabrera and Martín 1986). Similarly, vicine-convicine content may decrease up to 20 times due to the presence of the vc- allele that is linked to colorless hilum in the seeds (Duc et al. 1989). BSA has been used to identify molecular markers tightly linked to these genes. Thus, Gutierrez et al. (2007) identified four RAPD loci (OPC_{5551}, OPG_{15600}, OPG_{111171} and $OPAF_{20776}$) tightly linked to *zt-1*. Two sequence characterized amplified regions (SCARs) were developed. One of them, SCC_{5551} was associated to white flowers and zero tannin content, whereas the other, SCG_{111171}, was linked to colored flowers and high tannin content. The simultaneous use of both markers in a collection of 37 faba bean genotypes differing in flower color and tannin content resulted in a 95% selection efficiency of the *zt-1* genotypes (Gutierrez et al. 2007). The same strategy was used to identify a cost-effective marker linked to *zt-2*

(Gutierrez et al. 2008). Five RAPD markers linked in coupling and repulsion phase to the gene were identified and further converted into SCARs, although the initial polymorphism was lost. Hence the authors proceeded to the restriction digestion of the SCAR $SCAD_{16589}$ with *Alu*I ($SCAD_{16\text{-}A}$), Bsp120 ($SCAD_{16\text{-}B}$) and *Hinf* I ($SCAD_{16\text{-}H}$), revealing clear differences due to the amplification of different loci. The consensus sequence of these cleavage amplification polymorphism (CAP) markers allowed discrimination of three bands, from which two new forward SCAR primers were developed, based on the specific sequences from zero tannin and high tannin genotypes. To improve the efficiency of the marker screening, a multiplex PCR was developed that allow the simultaneous amplification of the SCAR with the same advantages as a codominant marker.

Molecular markers linked to the *vc-* allele have been developed using BSA and RAPD markers (Gutierrez et al. 2006). Once again, the development of cleaved amplified polymorphic sequence (CAPS) markers was an efficient tool to overcome the loss of polymorphism detected using SCARs. In this case, the simultaneous use of two CAP markers allowed the correct fingerprinting of low vicine-convicine faba bean plants. The available markers can be used to facilitate the introgression of the *vc-* allele to develop cultivars with low v-c content and improved nutritional value (Fig. 2-2).

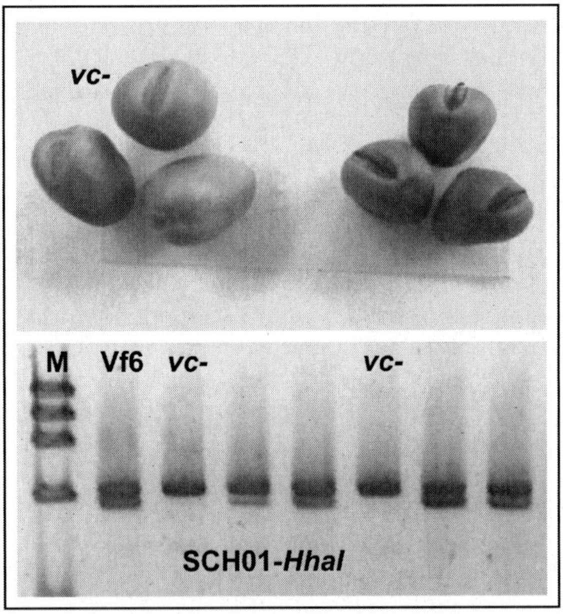

Figure 2-2 Seeds of the mutant *vc-* , with white hilum (above) and CAP marker SCH01-*HhaI* linked to the low vicine and convicine content (below) (Courtesy Dr N Gutierrez).

Color image of this figure appears in the color plate section at the end of the book.

BSA has been extensively and efficiently applied to identify molecular markers linked to single genes in faba bean. However, the efficiency of these markers for the selection of the desired genotypes depends on the distance between the marker and the target gene. Distances higher than 5 cM may lead to a decrease in selection efficiency when transferring the target genes into a new genetic background, therefore, limiting their potential for breeding programs. In contrast, the candidate gene approach allows the development of key markers based on differences between alleles of the gene responsible for the trait under study. These markers are known as diagnostic markers and they have the significant advantage of being completely linked with the selected trait, since no recombination between the marker and the trait is possible. The codominant nature of these markers also facilitates the identification of both homozygous and heterozygous individuals, providing a further advantage. However, this approach is possible only when the target gene is well characterized in the species or in related species carrying orthologous genes.

The strategy was used to develop molecular markers for the selection of determinate growth habit. The traditional faba bean plant has an indeterminate growth habit and may grow to two meters tall. Many cultivars are prone to lodging so are difficult to harvest. Determinate growth habit has been exploited in other species to solve this problem. The transference of this character into horticultural faba bean cultivars would allow obtaining a more synchronous production, suitable for mechanical harvesting. Several mutants of faba bean with determinate growth habit have been described (Sjodin 1971b; Steuckardt et al. 1982; Filippetti 1986), and at present, there are new Spanish cultivars for fresh market production that include the trait. A single recessive gene is responsible for the character (Sjodin 1971b; Filipetti 1986). Genes determining growth habit are well known in various species. First cloned in *Antirrhinum* (gene *CEN*) (Bradley et al. 1996), several orthologs have been found in *Arabidopsis* (*TFL1* (Bradley et al. 1997), tomato, *SELF PRUNING* (*SP*) (Pnueli et al. 1998), tobacco (*CET*) (Amaya et al. 1999), perennial ryegrass (*Lolium perenne*) (*LpTFL1*) (Jensen et al. 2001) and pea (*PsTFL1-a*) (Foucher et al. 2003), and more recently in citrus (Pillitteri et al. 2004), rice (Zhang et al. 2005), apple (Kotoda and Wada 2005) and other fruit species (Esumi et al. 2005). All these genes are extensively recognized as *CEN/TFL1*-homologous or *CEN/TFL1*-like genes.

From this available information, Avila et al. (2006) identified an ortholog of the *CEN/TFL1* responsible of the growth habit in *V. faba*. They then developed a simple PCR-based marker suitable for the selection of determinate growth habit plant in faba bean. However, when tested in a wide collection of germplasm, this marker failed to detect some determinate growth habit genotypes, indicating the existence of different mutations for this trait. Following further investigation, a new diagnostic marker was developed that was able to unambiguously discriminate the gene of interest in a large faba bean germplasm collection (Avila et al. 2007).

2.5 Molecular Mapping of Complex Traits

Progress in breeding faba bean is severely limited by many biotic and abiotic stresses that are under polygenic control. The crop is challenged by numerous pathogens and pests including bacteria, viruses, and fungi. The biotic stresses currently given high priority for molecular mapping include Ascochyta blight (caused by *Ascochyta fabae*), rust (caused by *Uromyces viciae-fabae*) and chocolate spot (caused by *Botrytis fabae*) among the pathogenic fungi and also crenate broomrape (*Orobanche crenata*), a parasitic weed. Drought and frost are the main concerns in breeding for abiotic stress resistance.

The availability of numerous molecular markers, well saturated genetic maps and batteries of ESTs from model species (mainly *Medicago truncatula*) and also from *Pisum sativum* is facilitating QTL analysis and comparative mapping, which eventually will help faba bean breeding using MAS approaches. In order to overcome biotic and abiotic constraints through the use of genetic resistance, a key objective in current breeding programs is the identification of DNA markers linked to key stresses.

As stated in Section 2.2.2, the most promising applications of MAS in faba bean breeding are those associated with major genes (including disease resistance or quality related traits). Nevertheless, the potential of MAS for breeding complex traits controlling resistance to parasitic plants and pathogenic fungi has been convincingly demonstrated (Román et al. 2002, 2003; Avila et al. 2004; Díaz-Ruiz et al. 2009a; Díaz et al. 2010). Moreover, although with relatively minor effects, the first QTLs linked to frost tolerance was reported (Arbaoui et al. 2008b). A summary of the gene(s)/QTLs of agronomic interest reported so far in faba bean and the molecular markers associated was provided by Torres et al. (2010). Molecular mapping of these traits has been performed by analyzing different F_2 and advanced inbred populations and results achieved so far for the principal biotic and abiotic stresses are described below.

2.5.1 Biotic Stresses: Target Traits

2.5.1.1 Ascochyta Blight

Ascochyta blight, caused by *Ascochyta fabae* Speg., is a common and destructive faba bean disease worldwide, being particularly damaging in wet and cool weather. Symptoms, consisting of lesions on leaves, stems, pods and seeds, produce yield losses of 35–40% or as high as 90% in highly susceptible cultivars (Hanounik and Robertson 1989). Infected seeds display dark stains on the coat, reducing their market value and constituting a source of inoculum to the following crop.

Although different disease control methods (crop rotation, chemicals and certified seeds) provide partial crop protection, the most desirable means of control for a long-term strategy is the genetic resistance of the host. Several sources of resistance have been identified and used in breeding programs (Bond and Pope 1980; Tivoli et al. 1988; Hanounik and Robertson 1989; Rashid et al. 1991a, b; Ondrej 1993; Sillero et al. 2001), but little is known about its genetic basis and gene location.

Both polygenic and major gene inheritance of resistance to Ascochyta have been suggested (Rashid et al. 1991b; Hanounik and Robertson 1989; Kohpina et al. 2000; Román et al. 2003; Avila et al. 2004). Furthermore, the complexity of the resistance evaluation, frequently hampered by environmental conditions, and the occurrence of physiological specialization between pathogen isolates and host genotypes (Kharbanda and Bernier 1980; Hanounik and Robertson 1989; Rashid et al. 1991b; Kohpina et al. 1999; Avila et al. 2004) greatly complicates disease scoring. Molecular marker techniques have proved to be an efficient alternative to the tedious task of screening for resistance, allowing the development of saturated maps suitable for locating and characterizing individual QTLs (Lander and Botstein 1989).

Progenies derived from two different crosses were used to detect QTLs affecting Ascochyta blight response (Vf6 x Vf136 and Vf29H x Vf136) where Vf136 is susceptible, Vf6 is partially resistant and Vf29H is resistant. Vf6 was previously used in different mapping projects, facilitating the assignment of linkage groups to specific chromosomes (Torres et al. 2006). First, Román et al. (2003), identified two QTLs (*Af1* and *Af2*) controlling resistance against one isolate of *A. fabae* in a Vf6 x Vf136 F_2 population. A set of 196 F_2-derived F_3 lines was inoculated under controlled growth chamber conditions and evaluated for disease severity (DS) and infection type index (IT). The linkage map was constructed by MAPMAKER V2.0 (Lander et al. 1987) and included 121 markers that were mapped into 16 LGs, nine of which were assigned to specific chromosomes. QTLs were identified using simple interval mapping (SIM) and composite interval mapping (CIM) as implemented in the QTL-CARTOGRAPHER program (Basten et al. 2001). SIM and CIM identified two zones of putative QTL action (*Af1* and *Af2*) in LGs 8 (chromosome III) and 4a (chromosome II), that explained 25.2 and 21.0 % of the total phenotypic variance, respectively. *Af1* displayed pure additive gene action while *Af2* showed considerable dominance effect. In a parallel study, Avila et al. (2004) studied the resistance on leaves and stems in an F_2 population of Vf29H x Vf136, using two pathogenically distinct *Ascochyta* isolates. Linkage mapping software and QTL analysis were as described above. Six QTLs named *Af3* to *Af8* were detected, of which *Af3* and *Af4* were effective against both isolates, *Af5* to *Af8* were effective against only one of them, and *Af3, Af4, Af5* and *Af7* were effective in both leaves and stems while *Af8* was effective only in stems and *Af6* only in leaves. The authors stated that some QTLs detected in one organ may also be active in the other. Kohpina et al. (1999) reported that stem resistance was highly affected by environmental conditions, and this fact might prevent the detection of a real QTL. *Af1* (Román et al. 2003) and *Af3* (Avila et al. 2004) were both assigned to chromosome III, and they might correspond to the same genomic regions.

Prior to the application of MAS approaches, QTL mapping studies should be independently confirmed or verified in different genetic backgrounds or in different testing environments, to demonstrate that DNA markers identified in one population will be useful in different pedigrees (Collard et al. 2005). QTL x environment interactions particularly restrict the usage of identified genomic regions across a wide range of environments, reinforcing the need for multiple field tests. To date, only a small fraction of the QTLs identified for important plant traits have been independently tested for confirmation. The validation may involve replication studies in independent populations (in particular RILs), constructed from the same parental genotypes or with closely related genotypes used in the primary QTL mapping study (Lander and Kruglyak 1995). In faba bean, this endeavor has been recently undertaken by Díaz-Ruiz et al. (2009a) by studying 165 RILs developed by single-seed descent from the previously mentioned cross (Vf6 x Vf136). The protocols for the evaluation and the methodology for map construction and QTL analysis were as described above. The new linkage map constructed consisted of 277 markers (238 RAPDs, four isozymes, one SCAR, six SSRs, two STSs, and 26 ESTs) that mapped into 21 LGs covering 2,856.7 cM. CIM identified two zones of putative QTL action in the RIL population controlling DSL (disease severity on leaves) and DSS (disease severity on stems). *Af1* and *Af2* were detected once more in chromosomes III and II, respectively, jointly explaining 24% of the phenotypic variance of DSL and 16% of DSS. Fifteen common markers between both maps facilitated the comparison of the homologous regions. The higher marker density of the map developed in the advanced generation reduced the confidence interval in both QTLs, thus improving the accuracy of their localization. In both generations and for the two QTLs identified, the resistance alleles were associated with marker alleles derived from the resistant parent (Díaz-Ruiz et al. 2009a). This study allowed for the first time the confirmation of the presence of QTLs for Ascochyta blight resistance in independent populations (F_2 and F_6), re-estimation of their position and genetic effects, and assessment of their stability in different genetic backgrounds. Thus, from the five genomic regions controlling Ascochyta resistance in the F_2 of cross 29H x 136 reported by Avila et al. (2004), *Af3* ascribed to chromosome III seemed to be located in the same region as *Af1* identified and validated in the RIL population from the cross Vf6 x Vf136 (Díaz-Ruiz et al. 2009a). The use of the common marker OPW15_508 tightly linked (<10 cM) to the flanking markers in both maps confirmed the identity of the two QTLs and the assessment of their stability in different pedigrees. Unfortunately, neither of the other QTLs reported can be yet compared due to a lack of common markers between the corresponding linkage groups and the scarce coverage of the target regions.

The prospects of MAS for Ascochyta resistance in faba bean breeding programs will depend on the extent to which QTL results can be extrapolated from one population to another. Mapping additional common anchor markers is a prerequisite to extend this correlation and determine the possible homology of the remaining QTLs or the existence of different resistance loci acting in each of the crosses. This task is being investigated by developing a new map in the RIL population from the cross 29H x Vf136 (CM Avila, unpubl.data). A number of common standard markers (SSRs and ITAPs) are being genotyped, in order to facilitate the confirmation of QTLs across generations (F_2 and RIL) and further to validate additional QTLs across different genetic backgrounds. Moreover, regions bearing stable QTLs are being saturated in both maps as a first step towards the development of reliable molecular markers useful for disease resistance selection.

2.5.1.2 Broomrape

Crenate broomrape is a major faba bean constraint in the Mediterranean basin, North Africa and the Middle East. In addition to *O. crenata*, *P. aegyptiaca* is of importance in the eastern Mediterranean countries, and *O. foetida* in Tunisia. These parasitic weeds produce devastating effects on many crop legumes by obtaining nutrients and water through haustoria, which connect the parasite with the host vascular system (see Chapter 11). As a result, yield losses up to 80% have been reported in faba bean fields with high levels of *O. crenata* infestation (Gressel et al. 2004). Although different control methods have been proposed (Pieterse 1991; Parker and Riches 1993; López-Granados and García-Torres 1998; Joel 2000; Mauromicale et al. 2001; Rubiales et al. 2006), management of *Orobanche* spp. remains a challenge since none of the measures have shown fully satisfactory and economic results (Pérez-de-Luque et al. 2010; Stoddard et al. 2010).

Growing resistant cultivars has been widely recognized as the most appropriate and cost-effective means of disease control. In faba bean, the development of broomrape-resistant cultivars to *O. crenata* is a major objective in many breeding programs, but is a complicated task since resistance against most parasitic weeds is scarce, difficult to evaluate, of complex nature and of low heritability (Cubero 1983; Rubiales 2003). Besides, selection methods rely primarily on field evaluations (highly influenced by environmental conditions), where natural parasite populations occur and homogeneity of seed distribution in the soil is difficult to achieve. Although evaluations in controlled environments could help in avoiding these difficulties, they are extremely expensive and exceptionally difficult to manage for this parasite.

The first study to identify QTLs controlling crenate broomrape resistance in faba bean was conducted by Román et al. (2002) using an F_2-derived population of a cross between the resistant line Vf136 and the susceptible line Vf6. Vf136 derives from a cross between the Spanish cultivar "Alameda" and Vf1071, a resistant line selected from Giza 402 and one of the main sources of broomrape resistance (Nassib et al. 1982; Cubero et al. 1992). The F_2-derived F_3 lines were grown in an alpha lattice design with two replications in field plots heavily infested with crenate broomrape seeds. Broomrape resistance was scored as the number of emerged broomrapes per faba bean individual at plant maturity. To correct for differences in broomrape seed density in the soil between plots, simple regression was carried out and regression corrected values (residuals) calculated. Regression residuals were considered to be the broomrape resistance index of the F_2 population (Román et al. 2002). For mapping purposes, the progeny was analyzed using isozymes, RAPDs, seed protein genes, and microsatellites. As a result, 121 markers were integrated in a genetic map comprising 16 LGs and covering 1,445.5 cM. Nine out of 16 LGs were assigned to specific chromosomes using established markers. The linkage map was constructed by MAPMAKER V 2.0 (Lander et al. 1987) using a LOD score of 3.0 as the threshold for declaring linkage. QTL location was performed using both SIM and CIM in the QTL-CARTOGRAPHER software (Basten et al. 2001). In this study, 105 marker intervals were tested and the experiment-wise type-I error probability was set to $\gamma = 0.05$. Thus, comparison-wise type-I error probability using Bonferroni correction was calculated ($\alpha = 0.000488$) corresponding to an LR threshold of 17.78 (LOD = 3.86) with 3 degrees of freedom. Using SIM, two putative QTLs (*Oc1* and *Oc2*) showed LR scores above the threshold value. The use of the forward-backward stepwise regression revealed one additional QTL (*Oc3*). *Oc1*, *Oc2*, and *Oc3* were identified on chromosome I, chromosome VI and LG 11, respectively, and jointly explained 74% of the total phenotypic variance (37.3%, 11.3% and 25.5%, respectively). All the resistance-enhancing alleles originated from the resistant parent Vf136. *Oc1* and *Oc3* revealed considerable dominant effects, whereas *Oc2* displayed pure additive gene action (Román et al. 2002).

When RILs are used in multi-environment trials, the increased replication of homozygous genotypes results in an increased power for testing differences between genotypic classes and in a greater precision of trait measurement compared with other types of progenies (Austin and Lee 1996). This fact is of special significance in the study of broomrape resistance, a trait highly influenced by the environment, resulting from the infectious ability of the parasite in a specific location together with a range of escape factors and resistance mechanisms acting at different levels of the infection process.

In order to validate both the presence and location of the QTLs associated with broomrape resistance in the F_2 population, 165 RILs (F_6 generation) from the cross Vf6 x Vf136 were further developed (Díaz et al. 2010). The RILs were used to construct a new linkage map by MAPMAKER V2.0 using a LOD score of 4.0 as the threshold for declaring linkage. The map consisted of 277 markers (RAPD, isozymes, seed protein genes, SSRs, SCARs and STSs) grouped in 21 LGs that covered 2,856 cM. Nine of the 21 LGs were assigned to specific chromosomes. To confirm the position and effects of the three QTLs identified in the F_2 population, validation experiments were carried out using the advanced progeny in three different Spanish environments, Córdoba 2003, Córdoba 2004 and Mengíbar 2004. Genotypes from the linkage map and quantitative data for resistance index were input into Windows QTL-CARTOGRAPHER version 2.5 (Wang et al. 2005) with both SIM and CIM. The threshold for the detection of a QTL was fixed at a LOD value of 2 (LR = 9.21). For each LOD peak, the 1-LOD support intervals were determined (Van Ooijen 1992). There were differences in the number of QTLs detected in the F_2 and RIL population. The major QTL, *Oc1*, identified in the F_2 and explaining the highest percentage of the phenotypic variance, was not detected in the RILs in the three environments under study, probably due to the overdominance displayed by *Oc1* in the F_2, as dominance effects are known to be important in this generation but are absent in RILs (Díaz et al. 2010). Moreover, some QTLs detected in early generations of maximum linkage disequilibrium are indeed due to multiple linked genes that may be separated via recombination (Austin and Lee 1996), a fact that could also explain the outcome. Indeed, two additional QTLs (*Oc4* and *Oc5*) with small effects (explaining17 and 9% phenotypic variance), were detected in the same linkage group as *Oc1*, ascribed to chromosome I. Both the QTLs were expressed in only one of the environments assayed. Differences in the level of infestation and/or weather conditions could have contributed to differences in gene expression between environments. In contrast to *Oc1*, both *Oc2* and *Oc3* showed some consistency between F_2 and F_6 generations (Román et al. 2002), as they were detected in at least two of the three environments thereby confirming their environmental stability and pointing to their suitability as targets for MAS (Fig. 2-3).

The total variance explained by all QTLs detected in the F_6 population was 16, 18 and 26% in Córdoba 2003, Mengibar 2004, and Córdoba 2004, respectively. The unexplained variation was attributable to factors such

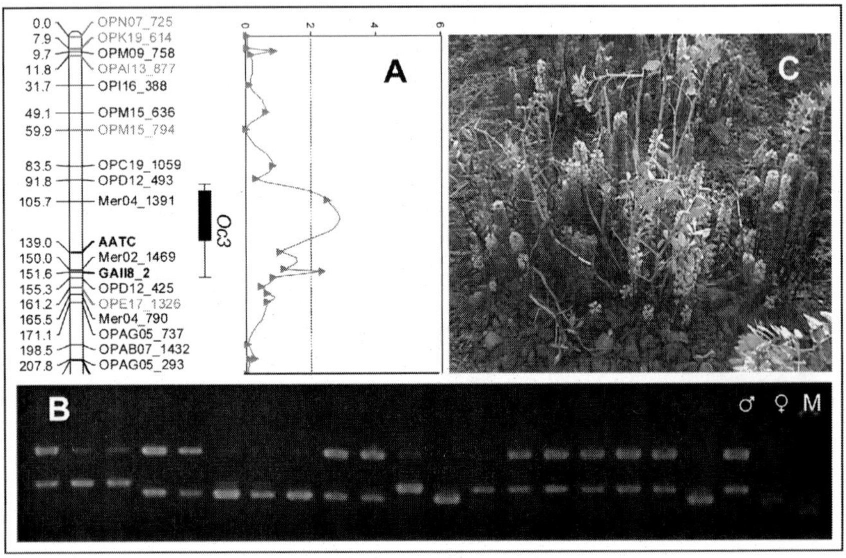

Figure 2-3 QTL *OC3* controlling resistance to broomrape (*Orobanche crenata*) in faba bean (A). Segregation pattern of the marker AATC, present in the QTL interval, in 19 RILs derived from the cross Vf6 (♀) xVf136 (♂) (B). Faba bean plants heavily infected by the parasite (C) (Courtesy Dr R Díaz-Ruíz).

Color image of this figure appears in the color plate section at the end of the book.

as QTLs with minor effect, incomplete map coverage and environmental effects, demonstrating the complexity of the trait that results from a chain of factors such as avoidance of the parasite by root architecture (Ter Borg 1999), low induction of broomrape seed germination (Sillero et al. 1999; Rubiales et al. 2003) and resistance to the parasite penetration or to the establishment of tubercles (Ter Borg 1999; Rubiales et al. 2003; Pérez-de-Luque et al. 2005). Consequently, differences in resistance caused by the diverse mechanisms, which could account for the remaining variation, might be under-represented by the simple scoring of the final number of emerged broomrape shoots.

Although *O. crenata* is the most important parasitic weed attacking legume crops in the Mediterranean region, *O. foetida* has also been the focus of interest since it was found attacking faba bean, vetch and chickpea in Tunisia and Morocco (Kharrat et al. 1992; Rubiales et al. 2005). Taking advantage of the Vf6 x Vf136 RIL population used to identify and locate *O. crenata* QTLs in *V. faba*, Díaz-Ruiz et al. (2009b) used the described map to detect putative QTLs affecting resistance to *O. foetida* in naturally infested fields of Beja (Tunisia) in two seasons (2002–2003 and 2003–2004). Two QTLs for *O. foetida* resistance (*Of1* and *Of2*) were identified in chromosomes I and III explaining 7 and 9% of the trait variation, respectively. Unfortunately neither of the QTLs showed stability across environments. Román et al. (2007) reported molecular variation between *O. foetida* populations growing on different hosts. In this study, the two trials in Beja were set in different experimental areas that could have grown diverse legume crops in previous seasons, so the inconsistency in results could be attributed to molecular dissimilarity of the *O. foetida* populations, and future studies are necessary to test this hypothesis.

The information reported so far, especially that for resistance to crenate broomrape, firmly demonstrates the potential for the use of molecular markers for indirect selection of broomrape resistance in faba bean. Depending on their relative effects and position, some of the QTLs reported so far could be targeted for MAS, following saturation of target regions to refine the QTL position and identify the marker most closely linked to the resistance genes.

2.5.1.3 Other Diseases

Leaf diseases caused by biotrophic pathogens, such as rust, chocolate spot and downy mildew (*Peronospora viciae* Berk.) are other widespread biotic factors limiting faba bean production (Section 2.2.3.2). Rust is more problematic in warm climates whereas chocolate spot and downy mildew are highly destructive diseases in cooler climates of Europe and southern Australia.

The quantitative response to rust and chocolate spot is being analyzed at present in the F_2 population of cross B261 x V274 segregating for resistance against both pathogens (C.M. Avila, unpubl.data). A linkage map has been constructed, QTL analysis is in progress, and the population has been selfed to the F_6 generation that will be used to asses the stability of any QTLs identified in the $F_{2.}$ To our knowledge no marker studies or QTL analyses for downy mildew or for other diseases are available as yet.

2.5.2 Abiotic Stresses

Faba bean crop is exposed to a range of abiotic stresses, including drought, heat, frost, chilling, waterlogging, salinity and mineral toxicities. Responses to these stresses are often measured in field trials, where the severity of the stress is unpredictable and other stresses may also operate, or in controlled-environment trials, where the severity of the stress and its isolation from other stresses can be strictly regulated. In this way we can gather information about the individual stress in isolation as well as its integration in the whole life of the crop.

2.5.2.1 Chilling, Frost Tolerance, and Winter Adaptation

While the faba bean, as a cool-season crop, is generally tolerant of chilling (temperatures between 0 and 10°C), its tolerance to frost is more variable. Most germplasm is tolerant to brief exposure to temperatures down to about –5°C before flowering. Winter and spring faba beans are capable of responding to hardening treatments, with a maximum hardening status achieved after 2–3 weeks of exposure to temperatures of 2–5°C in a 10-hour daylength (Herzog 1988). Hardening confers tolerance to much lower temperatures in winter faba beans than in spring beans, in association with greater increases in membrane linolenic content and decreases in oleic acid content (Arbaoui and Link 2008). This change in cell membrane lipid composition affects membrane flexibility, and is moderately correlated with changes in electrolyte leakage from frozen leaves (Arbaoui et al. 2008a). Proline, an osmoprotectant, accumulates during hardening (Arbaoui et al. 2008a), but the presence of other osmoprotectants such as glycinebetaine, mannitol, raffinose or trehalose has not been demonstrated in faba bean. Similarly, the presence of antifreeze proteins, which inhibit intercellular recrystallization of ice (Griffith and Yaish 2004), has not been tested in faba bean.

Frost tolerance may be assessed in a growth chamber, in which daytime temperatures are maintained at 2°C and night temperatures are lowered progressively by 2–3° each night from –8°C to –21°C or until all plants are dead (Arbaoui and Link 2008). The quantitative measure of frost response was the area under the damage progress curve. Large experiments (36 plants per accession) were required to obtain high heritabilities ($h^2 = 0.89$) and the correlation of frost response with overwintering response was moderate ($r = 0.41$), demonstrating that frost tolerance is a major, but not the only, component of winter hardiness in faba bean (Arbaoui et al. 2008a). QTL analysis of 101 F_6 lines from the cross of Cote d'Or 1 x Bean Pure Line 4628 (both winter-hardy accessions) identified 5 QTLs for frost hardiness (Arbaoui et al. 2008b). Three of these were for unhardened response and the other two for hardened; the favorable allele for one of those for unhardened response was from BPL 4628 and all others were from Cote D'Or 1.

Winter hardiness often requires tolerance to prolonged snow cover. Snow-cover tolerance was associated with short plant height, prostrate growth habit, small petioles and a high content of non-structural carbohydrate in the tissues, in a survey of germplasm (Fukuta and Yukawa 1998), in a review of the development of the hardy cultivar Deneb (Bond and Crofton 1999), and in a comparison of cultivars of several legume species (Annicchiarico and Ianucci 2007).

2.5.2.2 Heat and Drought

Many agricultural scientists consider that drought, an interval of water deficit that leads to a significant reduction in yield, is the most important constraint to crop productivity (Fischer and Turner 1978; Borlaug and Dowswell 2005). In many climates, drought stress and heat stress are coupled, but at high latitudes drought may occur at very moderate temperatures, and at low latitudes, irrigated crops may experience heat stress. Faba bean is considered to be relatively sensitive to water deficit (McDonald and Paulsen 1997) and crop modeling studies have indicated its optimum growth temperature to be around 22–23°C (Boote et al. 2002; Turpin et al. 2002). Temporary water deficit may occur at almost any time in almost any faba bean crop, and in Mediterranean-type climates, terminal water deficit typically occurs during the grain filling period (Wery et al. 1994; Ricciardi et al. 2001).

Crop responses to water deficit have classically been characterized as escape, avoidance or postponement, and tolerance (Levitt 1980; Ludlow and Muchow 1990). Where terminal water deficit predominates, drought is "escaped" through earliness. Dehydration may be "avoided" by reducing water loss through the stomata and maintaining water uptake through a deep and prolific root system. Dehydration may be "tolerated" by physiological mechanisms allowing the plant to withstand low water potential and recover from it (Turner et al. 2001).

Dehydration escape requires a combination of appropriate, early maturing germplasm, with crop management strategies such as early planting that maximize interception and use of the available water. Large, thin leaves that provide a low specific leaf area and rapid ground coverage have been shown to be valuable for minimizing soil evaporation in other crop species (Nautiyal et al. 2002; Rebetzke et al. 2004) and should be sought in faba bean germplasm. The time from sowing to the onset of flowering varies widely among faba bean accessions (Stoddard 1993) and development responds to both temperature and daylength (McDonald et al. 1994; Patrick and Stoddard 2010), so it is possible to identify or select germplasm that flowers at the optimum time for a given environment.

Stomatal closure is one of the first steps in drought avoidance. Stomatal conductance is determined by the density, size, and degree of opening of the stomata. Stomatal size and density vary in faba bean and higher stomatal density was associated with greater yield loss due to dehydration (Ricciardi 1989). Stomatal conductance is much easier to measure than its components and varies at least two-fold among faba bean germplasm (Khan et al. 2007). Stomatal closure inhibits the photosynthetic apparatus from discriminating against ^{13}C, so carbon isotope discrimination is an effective measure for the integrated stomatal conductance through the life of the plant (Condon et al. 2006), but is expensive to measure. Leaf temperature measurement by a non-contact infrared thermometer is a rapid, non-destructive and inexpensive surrogate measure for both stomatal closure and carbon isotope discrimination, as open stomata cool the leaves below atmospheric temperature.

Dehydration may also be postponed when the root traits of the plant are appropriate. Rooting depth is thus important and extensive variation in this trait exists (Manschadi et al. 1998; H Khan and F Stoddard, unpubl. results, EUFABA project). The genetics of this variation is as yet unexplored, although in chickpea (*Cicer arietinum* L.), rooting depth was determined by dominance gene action while the proportion of roots deeper than 60 cm was determined by additive gene action and additive x additive epistasis (Kashiwagi et al. 2008). Root mass is easy to measure and in other species such as chickpea has been shown to be strongly correlated to shoot mass (Kashiwagi et al. 2008), but selection for root mass is often considered to be risky, as it diverts photosynthate from expanding photosynthetic area to heterotrophic root tissue (Nielsen et al. 2001). There is a need for deeper investigation of root traits associated with drought avoidance in faba bean.

Tolerance to dehydration can take many forms, but few of them have been investigated in faba bean. In osmotic adjustment, plants synthesize and accumulate solutes such as soluble sugars, amino acids and amines, that hold water within cells, helping to maintain cellular structure and function (Bartels and Sunkar 2005). To date, osmotic adjustment in faba bean in response to water deficit has not been identified and the closest approach is probably the accumulation of proline in response to hardening, mentioned above. Since osmotic adjustment has been shown in pea (*Pisum sativum* L.) (Sánchez et al. 1998), among other legumes, it is worth seeking it more actively in faba bean germplasm. Closure of stomata decreases CO_2 concentration in the leaves, resulting in the accumulation of superoxide radicals (Cadenas 1989). An antioxidant system such as superoxide dismutase is therefore an important part of dehydration tolerance that has been found in common bean (*Phaseolus vulgaris* L.) (Zlatev et al. 2006) although not yet in faba bean. Variation in the ability to recover from dehydration has been found in soybean (*Glycine max* (L.) Merr.), where it was attributed to variation in the maintenance of leaf area (Lawn and Likoswe 2008), but it has not been identified in faba bean.

2.6 Prospects for Faba Bean Molecular Breeding

There are very few reports of successful single gene transfer by MAS or gene pyramiding in legumes, and these are mostly limited to three crops, namely, soybean, common bean, and pea (Dwivedi 2007). In faba bean, molecular markers have been sought for both simple and complex traits, providing two contrasting sorts of experience with possible application in MAS. Research has focused on developing molecular markers linked to major genes controlling hypersensitive resistance to rust, determinate growth habit (*ti*) or to select against traits affecting the nutritional value of seeds (tannins, vicine and convicine contents). RAPD markers closely linked to these traits have been transformed into CAPs or SCARs to enable the efficient transfer of these genes into selected genotypes (Avila et al. 2003, 2006, 2007; Gutierrez et al. 2006, 2007, 2008). As reviewed above (see Section 2.4), mapping of genetically simple traits has been relatively simple compared with that of genetically complex characters. For major gene traits, phenotyping is relatively straightforward, facilitating the accurate mapping of the gene. In polygenic traits, there are many more factors that influence the feasibility of any MAS-related activity but at the same time, the need for rapid and reliable screening methods is, if anything, even greater than for simple traits. Saturation of the genomic regions associated with target regions, together with QTL validation in multiple environments and genetic backgrounds, are necessary steps in the detection of reliable marker-trait associations, although recent publications on tagging QTLs for important disease resistance traits (Román et al. 2002, 2003; Avila et al. 2004; Díaz-Ruiz et al. 2009a; Díaz et al. 2010) reveal new perspectives on effective MAS schemes in this species.

The availability of well-saturated linkage maps with robust and transferable markers is a key point to accomplish this objective. Up to now, faba bean maps have been mostly built with RAPD markers, a technology particularly useful when there is a need to assay many loci across one of the largest genomes among crops (~13,000 Mb). Although RAPDs are dominant, non-specific and often poorly reproducible in comparison with other markers, and hence limited in their application for accurate MAS breeding approaches, they have proved valuable for finding new markers linked to desirable alleles. Especially for major traits, once such markers are identified, the corresponding bands have been sequenced and used to develop more specific and reliable markers such as CAPs or SCARs that simplify the screening of large progenies.

Faba bean maps are being continually improved, thus, preliminary linkage groups based on RAPDs, morphological traits, isozymes and seed storage protein genes, have been enriched by a number of SSRs (Pozarkova et al. 2002; Zeid et al. 2009). The high levels of polymorphism and the significant association of gene-rich regions with clusters of SSRs (Song et al. 2004) highlight this type of marker as an effective tool for merging maps and saturating regions bearing desired genes or QTLs. A detailed review of faba bean linkage maps was provided by Torres et al. (2006).

The resistance gene analog (RGA) approach, a promising strategy to isolate multiple resistance-gene candidate sequences, has also been assayed in faba bean. RGAs belonging to the nucleotide binding site (NBS) gene families were recently isolated from different faba bean and chickpea genotypes (Palomino et al. 2006). Specific primers for 10 RGA classes have been designed to develop co-dominant CAPS and dCAPS markers able to genotype single nucleotide polymorphisms (SNPs) within these RGA sequences in both species. In chickpea, seven RGAs were successfully genotyped in two progenies. Genomic locations of RGAs were compared with those of known *Cicer* R-genes and previously mapped RGAs and a clear association between RGA05 and genes controlling resistance to Fusarium wilt caused by races 0 and 5 was detected (Palomino et al. 2008). The same approach is being assayed at present in the available faba bean mapping populations segregating for resistance to broomrape, rust and Ascochyta blight to investigate the putative co-segregation of these RGA markers with disease resistance genes (C. Palomino, pers. comm.).

New approaches for a more efficient targeted marker development are also being followed. The availability of whole-genome sequences of many model species and the sequence information generated by ESTs has led to the development of a new generation of markers, gene-targeted markers (also called candidate gene markers) and functional markers (Andersen and Lübberstedt 2003). *Medicago truncatula* and *Lotus japonicus* were proposed and established as model species among legumes, and programs for EST sequencing have led to the creation of more than 300,000 ESTs available in public databases (Cannon et al. 2006; Ané et al. 2008; Sato et al. 2008). Such ESTs constitute a novel source of markers physically associated with coding regions that are being exploited for faba bean gene discovery, and comparative genomics. Thus, within the comparative mapping program of GLIP (http://www.eugrainlegumes.org/) a number of intron-targeted gene-based anchor markers were developed, using *M. truncatula* as a reference genome. A high level of macrosynteny has been found between *Medicago*, lentil, white lupin (*Lupinus albus* L.) and faba bean (Phan et al. 2006; Ellwood et al. 2008). In the faba bean study, 151 markers produced the first exclusively gene-based genetic linkage map of the crop, using an F_6 population developed from Vf6 x Vf27 (see Section 2.3). Linkage analysis revealed seven major and five small linkage groups, spanning a total length of 1,685.8 cM. Markers originally designed from genes on the same *M. truncatula* BACs were found to be grouped together in corresponding syntenic areas in lentil and faba bean, although faba bean and lentil shared a common rearrangement relative to *M. truncatula* (Ellwood et al. 2008). This map is being used to identify genes and QTLs related to yield and other agronomic components (AM Torres et al. unpubl.data). Moreover, the integration of the EST markers of cross Vf6 x Vf27 that might show polymorphisms in other faba bean maps (crosses Vf6 x Vf136 and Vf29H x Vf136) is being carried out, with the aim of developing a composite genetic map in the near future.

The use of conserved genome structure to assist in transferring knowledge among related plant species is well established in grasses, Brassicaceae and Solanaceae where synteny is greatly assisting in gene identification among related species (Bennetzen 2000; Devos and Gale 2000). Further comparative genomic studies between faba bean and other crop legumes can facilitate back-and-forth use of online resources, helping to reduce cost and increase efficiency of genetic research and crop breeding. Importantly, the new composite map anchored with orthologous markers mapped in *M. truncatula*, will provide a central reference map for future use of genomic and genetic information in faba bean comparative mapping. The approach will allow both the identification of candidate genes underlying specific QTLs and the validatation of QTLs in different genetic backgrounds and most likely across species.

The identification of candidate genes and the elucidation of their role can also be facilitated by combining QTL analysis with different sources of information and technological platforms (Pfleiger et al. 2001; Wayne and McIntyre 2002). The progress of the "omic" technologies involving high-throughput profiling of the transcriptome, proteome and metabolome offers the possibility of investigating the concerted response of thousands of genes to key biotic and abiotic stresses. Some of these genomic tools are being used in faba bean crop improvement within the framework of the ERA-PG project LEGRESIST (http://www.genxpro.info/science_and_technologies/Legresist/). SuperSAGE technology (Matsumura et al. 2005) has been employed to resolve the stress transcriptomes of different legume host plants (faba bean, chickpea, pea, lentil and grass pea (*Lathyrus sativus* L.), infected by their corresponding *Ascochyta* pathogens (Kahl et al. 2009). Up to now, more than 7,000 pathogenesis-related genes and around 50 potential resistance genes have been detected that will be further validated by microarray analysis. Assignment of differentially expressed super tags to gene ontology terms has revealed both similar and different responses against this pathogen suggesting overlapping as well as specific defense strategies among crops (Kahl et al. 2009).

Traditional faba bean maps will be greatly enhanced by these new strategies. Most markers used so far in faba bean QTL mapping have been anonymous markers that are unlikely to occur in a gene influencing a quantitative trait. Combining functional, comparative and structural genomics will uncover large numbers of candidate genes that can be excellent targets for QTL studies aimed at identifying markers for use in MAS. Comparative mapping and candidate gene approaches can also utilize such information to search for homologous genes in different legume genomes.

Acknowledgements

The faba bean molecular research has been mostly financed by the projects AGL2005-07497-C02-01/AGR, GEN2006-27798-CO6-5-E/VEG, RTA2007-00030 and AGL2008-02305/AGR, co-financed by FEDER funds and by the EU projects EUFABA (QLK5-CT2002-02307) and GLIP (FP6- FOOD-CT-2004-506223). The authors are grateful to the faba bean coworkers at IFAPA for their excellent contributions.

References

Amaya I, Ratcliffe OJ, Bradley DJ (1999) Expression of *CENTRORADIALIS* (*CEN*) and *CEN*-like genes in tobacco reveals a conserved mechanism controlling phase change in diverse species. Plant Cell 11: 1405–1417.

Andersen JR, Lübberstedt T (2003) Functional markers in plants. Trends Plant Sci 8: 554–560.

Ané JM, Zhu H, Frugoli J (2008) Recent advances in *Medicago truncatula* genomics. Int J Plant Genome, Article ID 256597, doi:10.1155/2008/256597.

Annicchiarico P, Iannucci A (2007) Winter survival of pea, faba bean and white lupin cultivars in contrasting Italian locations and sowing times, and implications for selection. J Agri Sci 145: 611–622.

Arbaoui M, Link W (2008) Effect of hardening on frost tolerance and fatty acid composition of leaves and stems of a set of faba bean (*Vicia faba* L.) genotypes. Euphytica 162: 211–219.

Arbaoui M, Balko C, Link W (2008a) Study of faba bean (*Vicia faba* L.) winterhardiness and development of screening methods. Field Crops Res 106: 60–67.

Arbaoui M, Link W, Satovic Z, Torres AM (2008b) Quantitative trait loci of frost tolerance and physiologically related traits in faba bean (*Vicia faba* L.). Euphytica 164: 93–104.

Austin DF, Lee M (1996) Comparative mapping in F2:3 and F6:7 generations of quantitative trait loci for grain yield components in maize. Theor Appl Genet 92: 817–826.

Avila, CM, Sillero JC, Rubiales D, Moreno MT, Torres AM (2003) Identification of RAPD markers linked to *Uvf-1* gene conferring hypersensitive resistance against rust (*Uromyces viciae-fabae*) in *Vicia faba* L. Theor Appl Genet 107: 353–358.

Avila CM, Satovic Z, Sillero JC, Rubiales D, Moreno MT, Torres AM (2004) Isolate and organ-specific QTLs for ascochyta blight resistance in faba bean. Theor Appl Genet 108: 1071–1078.

Avila CM, Nadal S, Moreno MT, Torres AM (2006) Development of a simple PCR-based marker for the determination of growth habit in *Vicia faba* L. using a candidate gene approach. Mol Breed 17: 185–190.

Avila CM, Atienza SG, Moreno MT, Torres AM (2007) Development of a new diagnostic marker for growth habit selection in faba bean (*Vicia faba* L.) breeding. Theor Appl Genet 115: 1075–1082.

Bardner R (1983) Pests of *Vicia faba* L. other than aphids and nematodes. In: PD Hebblethwaite (ed) The Faba Bean (*Vicia faba* L.). Butterworths, London, UK, pp 371–390.

Bartels D, Sunkar R (2005) Drought and salt tolerance in plants. Crit Rev Plant Sci 24: 23–58.

Bartolomé B, Quesada C, Gómez-Cordibés C, Hernandez T, Estrella I (1994) New contributions to the inhibition study of α-amylase and trypsin by phenolic compounds. In: H Kozlowska, J Fornal, Zdunczyk Z (eds) Bioactive Substances in Food of Plant Origin. Polish Academy of Sciences, Olstyn, Poland, pp 233–238.

Bassiri A, Rouhani I (1977) Identification of broad bean cultivars based on isoenzyme patterns. Euphytica 26: 279–286.

Basten CJ, Weir BS, Zeng ZB (2001) QTL cartographer Version 1.15: a reference manual and tutorial for QTL mapping. North Carolina State University, Raleigh, NC, USA p 162.

Belyaev ND, Houben A, Baranczewski P, Schubert I (1997) Histone H4 acetylation in plant heterochromatin is altered during the cell cycle. Chromosoma 106: 193–197.

Belyaev ND, Houben A, Baranczewski P, Schubert I (1998) The acetylation patterns of histones H3 and H4 along *Vicia faba* chromosomes are different. Chrom Res 6: 59–63.

Bennett MD, Leitch IJ (1995) Nuclear DNA amounts in angiosperms. Ann Bot 76: 113–176.

Bennetzen JL (2000) Comparative sequence analysis of plant nuclear genomes: microcolinearity and its many exceptions. Plant Cell 12: 1021–1030.

Beringer JE, Bisseling, TA, Larue TA (1988) Improving symbiotic nitrogen fixation through the genetic manipulation of *Rhizobium* and legume host plants. In: RJ Summerfield (ed) World Crops: Cool Season Food Legumes. Kluwer Academic Press, Dordrecht, The Netherlands, pp 691–702.

Bernier CC, Conner RL (1982) Breeding for resistance to faba bean rust. In: G Hawtin, C Webb (eds) Faba Bean Improvement, Martinus Nijhoff, Hague, The Netherlands, pp 251–257.

Bliss FA, Miller JC (1988) Selecting and breeding grain legumes for enhanced nitrogen fixation. In: RJ Summerfield (ed) World Crops: Cool Season Food Legumes. Kluwer Academic Press, Dordrecht, The Netherlands, pp 1001–1012.

Bond DA (1976) *In vitro* digestibility of the testa in tannin-free field beans (*Vicia faba* L.). J Agri Sci 86: 561–566.

Bond DA (ed) (1980) *Vicia faba*: Feeding Value, Processing and Viruses. Martinus Nijhoff, Hague, The Netherlands p 422.

Bond DA, Pope M (1980) *Ascochyta fabae* on winter beans (*Vicia faba*): pathogen spread and variation in host resistance. Plant Pathol 29: 59–65: doi: 10.1111/j.1365–3059.1980.tb01181.x.

Bond DA, Crofton GRA (1999) History of winter beans (*Vicia faba*) in the UK. J Roy Agri Soc Engl 160: 200–209.

Bond DA, Jellis GJ, Rowland GG, Le Guen J, Robertson LD, Khalil SA, Li-Juan L (1994) Present status and future strategy in breeding faba beans (*Vicia faba* L.) for resistance to biotic and abiotic stresses. In: FJ Muehlbauer, WJ Kaiser (eds) Expanding the Production and Use of Cool Season Food Legumes. Kluwer Academic Press, Dordrecht, The Netherlands, pp 592–616.

Bond DA, Fyfe JL, Toynbee-Clarke G (1966a) Male sterility in field beans (*Vicia faba* L.) III. Male-sterility with a cytoplasmic type of inheritance. J Agri Sci 66: 359–367.

Bond DA, Fyfe JL, Toynbee-Clarke G (1966b) Male sterility in field beans (*Vicia faba* L.) IV. Use of cytoplasmic male sterility in the production of F_1 hybrids and their performance in trials. J Agri Sci 66: 369–377.

Boote KJ, Mínguez MI, Sau F (2002) Adapting the CROPGRO legume model to simulate growth of faba bean. Agron J 94: 743–756.

Borlaug NE, Dowswell CR (2005) Feeding a world of ten billion people: A 21st century challenge. In: R Tuberosa, RL Phillips, MD Gale (eds) The Wake of the Double Helix: From the Green Revolution to the Gene Revolution, Avenue Media, Bologna, Italy, pp 3–23.

Bos L, Hampton RO, Makkouk KM (1988) Viruses and virus diseases of pea, lentil, faba bean and chickpea. In: RJ Summerfield (ed) World Crops: Cool Season Food Legumes. Kluwer Academic Press, Dordrecht, The Netherlands, pp 591–615.

Böttinger P, Steimetz A, Schieder O, Pickardt T (2001) Agrobacterium-mediated transformation of *Vicia faba*. Mol Breed 8: 243–254.

Bradley D, Carpenter R, Copsey L, Vincent C, Rothstein S, Coen ES (1996) Control of inflorescence architecture in *Antirrhinum*. Nature 379: 791–797.

Bradley D, Ratcliffe O, Vincent C, Carpenter R, Coen E (1997) Inflorescence commitment and architecture in *Arabidopsis*. Science 275: 80–83.

Bressani R, Elias LG (1988) Seed quality and nutritional goals in pea, lentil, faba bean and chickpea breeding. In: RJ Summerfield (ed) World Crops: Cool Season Food Legumes. Kluwer Academic Press, Dordrecht, The Netherlands, pp 381–404.

Bueno MA (1976) Citotaxonomía y cariología en el género *Vicia*. PhD Thesis. Facultad de Ciencias, Universidad Complutense, Madrid, Spain, p 140.

Cabrera A, Martín A (1986) Variation in tannin content in *Vicia faba* L. J Agri Sci 106: 377–382.

Cabrera A, Cubero JI, Martín A (1989) Genetic mapping using trisomics in *Vicia faba* L. FABIS Newsl 23: 5–7.

Cadenas SE (1989) Biochemistry of oxygen toxicity. Annu Rev Biochem 58: 79–110.

Cammell ME, Way MJ (1983) Aphid pests. In: PD Hebblethwaite (ed) The Faba Bean *(Vicia faba* L.). Butterworths, London, UK, pp 315–346.

Cannon SB, Sterck L, Rombauts S, Sato S, Cheung F, Gouzy J, Wang X, Mudge J, Vasdewani J, Schiex T, Spannagl M, Monaghan E, Nicholson C, Humphray SJ, Schoof H, Mayer KF, Rogers J, Quetier F, Oldroyd GED, Debellé F, Cook DR, Retzel EF, Roe BA, Town CD, Tabata S, Van de Peer Y, Young ND (2006) Legume genome evolution viewed through the *Medicago truncatula* and *Lotus japonicus* genomes. Proc Natl Acad Sci USA 103: 14959–14964.

Caspersson T, Zech L, Modest EJ, Foley GE, Wagh U, Simonsson E (1969) Chemical differentiation with fluorescent alkylating agents in *Vicia faba* metaphase chromosomes. Exp Cell Res 58: 128–140.

Choi HK, Kim DJ, Uhm T, Limpens E, Baek JM, Lim H, Kalo P, Penmetsa RV, Seres A, Kulikova O, Bisseling T, Kiss GB, Cook DR (2004a) A sequence-based genetic map of *Medicago truncatula* and comparison of marker co-linearity with *Medicago sativa*. Genetics 166: 1463–1502.

Choi HK, Kim DJ, Zhu H, Mun JH, Beak JM, Roe B, Ellis N, Young ND, Doyle J, Kiss GB, Cook DR (2004b) Estimating genome conservation between crop and model legume species. Proc Natl Acad Sci USA 101: 15289–15294.

Cockbain AJ (1983) Viruses and virus diseases of *Vicia faba* L. In: PD Hebblethwaite (ed) The Faba Bean (*Vicia faba* L.). Butterworths, London, UK, pp 421–462.

Collard BCY, Jahufer MZZ, Brouwer JB, Pang ECK (2005) An introduction to markers, quantitative trait loci (QTL) mapping and marker-assisted selection for crop improvement: The basic concepts. Euphytica 142: 169–196.

Condon AG, Farquhar GD, Rebetzke GJ, Richards RA (2006) The application of carbon isotope discrimination in cereal improvement for water-limited environments. In: JM Ribaut (ed) Drought Adaptation in Cereals. Food Products Press, Binghamton, NY, USA, pp 171–219.

Cubero JI (1973) Evolutionary trends in *Vicia faba* L. Theor Appl Genet 43: 59–65.

Cubero JI (1974) On the evolution of *Vicia faba* L. Theor Appl Genet 45: 47–51.

Cubero JI (1982) Interspecific hybridization in *Vicia*. In: G Hawtin, C Webb (eds) Faba Bean Improvement. Martinus Nijhoff, Hague, The Netherlands, pp 91–108.

Cubero JI (1983) Parasitic diseases in *Vicia faba* L. with special reference to broomrape (*Orobanche crenata* Forsk.). In: PD Hebblethwaite (ed) The Faba Bean (*Vicia faba* L.). Butterworths, London, UK, pp 493–519.

Cubero JI (1984) Utilization of wild relatives of food legumes. In: JR Witcombe, W Erskine (eds) Genetic Resources and Their Exploitation: Chickpeas, Faba Beans and Lentils. ICARDA, Aleppo, Syria, pp 73–84.

Cubero JI, Martín A (1981) Factorial Analysis of yield components. In: R Thompson (ed) *Vicia faba* L.: Physiology and Breeding. Martinus Nijhoff, Hague, The Netherlands, pp 139-151.

Cubero JI, Moreno MT (1984) Breeding for self-fertility. In: PD Hebblethwaite, TCK Dawkins, MC Heath, G Lockwood (eds) *Vicia faba*: Agronomy, Physiology and Breeding. Martinus Nijhoff/Dr. Junk, Hague, The Netherlands, pp 209–217.

Cubero JI, Duc G (1995) To combine zero ANFs with high disease resistance in faba beans. Grain Legumes 10: 11–12.

Cubero JI, Nadal S (2005) Faba bean (*Vicia faba* L.). In: JS Ram, PP Jauhar (eds) Genetic Resources, Chromosome Engineering and Crop Improvement. Series II- Grain Legumes. CRC, Boca Raton, FL, USA, pp 163–186.

Cubero JI, Moreno MT, Hernandez L (1992) A faba bean cultivar resistant to *Orobanche crenata* Forsk. European Association for Grain Legume Research (AEP) (ed) Proc 1st Eur Conf on Grain Legumes. Angers, France, pp 41–42.

Cubero JI, Moreno MT, Rubiales D, Sillero J (eds) (1999) Resistance to Orobanche: The state of the art. Junta de Andalucía, Sevilla, Spain, p 199.

Devos KM, Gale MD (2000) Genome relationships: The grass model in current research. Plant Cell 12: 637–646.

Díaz R, Torres AM, Satovic Z, Gutierrez MV, Cubero JI, Román B (2010) Validation of QTLs for *Orobanche crenata* resistance in faba bean (*Vicia faba* L.) across environments and generations. Theor Appl Genet 120: 909–919.

Díaz-Ruiz R, Satovic Z, Avila CM, Alfaro CM, Gutierrez MV, Torres AM, Román B (2009a) Confirmation of QTLs controlling Ascochyta fabae resistance in different generations of faba bean (*Vicia faba* L.) Crop Pasture Sci 60: 353–361.

Díaz-Ruiz R, Torres A, Gutierrez MV, Rubiales D, Cubero JI, Kharrat M, Satovic Z, Román B (2009b) Mapping of quantitative trait loci controlling *Orobanche foetida* Poir. resistance in faba bean (*Vicia faba* L.). Afr J Biotechnol 8: 2718–2724.

Duc G (2006) New avenues for faba bean: food, feed, industrial uses and seed quality for different markets. In: CM Avila, JI Cubero, MT Moreno, MJ Suso, AM Torres (eds) Faba Bean 2006. Junta de Andalucía, Sevilla, Spain, pp 19–26.

Duc G, Sixdenier, Lila M, Furstoss V (1989) Search of genetic variability for vicine and convicine content in *Vicia faba* L. A first report of a gene which codes for nearly zero-vicine and zero-convicine contents. In: AJM Huisman, ,AFB Van der Poel, IE Liener (eds) Recent Advances of Research in Antinutritionnal Factors in Legume Seeds. Pudoc, Wageningen, The Netherlands, pp 305–313.

Duvick DN (1996) Plant breeding, an evolutionary concept. Crop Sci 36: 539–548

Dwivedi SL, Crouch JH, Mackill DJ, Xu Y (2007) The molecularization of public sector crop breeding: progress, problems, and prospects. Adv Agron 95: 163–318.

Ellwood SR, Phan HTT, Jordan M, Torres AM, Avila CM, Cruz-Izquierdo S, Oliver RP (2008) Construction of a comparative genetic map in faba bean (*Vicia faba* L.); conservation of genome structure with *Lens culinaris*. BMC Genom 9: 380.

Erith AG (1930) The inheritance of colour, size, form of seeds and of flower colour in *Vicia faba* L. Genetica 13: 477–510.

Esumi T, Tao R, Yonemori K (2005) Isolation of *LEAFY* and *TERMINAL FLOWER 1* homologues from six fruit tree species in the subfamily Maloideae of the Rosaceae. Sex Plant Reprod 17: 277–287.

FAOSTAT (2008): http://www.fao.org.

Filippetti A (1986) Inheritance of determinate growth habit induced in *Vicia faba major* by ethyl methane sulphonate (EMS). FABIS Newsl 15: 12–14.

Fischer RA, Turner NC (1978) Plant productivity in the arid and semiarid zone. Annu Rev Plant Physiol 29: 277–317.

Foucher F, Morin J, Courtiade J, Cadioux S, Ellis N, Banfield MJ, Rameau C (2003) *DETERMINATE* and *LATE FLOWERING* are two *TERMINAL FLOWER 1/CENTRORADIALIS* homologs that control two distinct phases of flowering initiation and development in pea. Plant Cell 15: 2742–2754.

Fuchs J, Pich U, Meister A, Schubert I (1994) Differentiation of field bean heterochromatin by *in situ* hybridization with a repeated *Fok*I sequence. Chrom Res 2: 25–28

Fuchs J, Strehl S, Brandes A, Schweizer D, Schubert I (1998) Molecular-cytogenetic characterization of the *Vicia faba* genome—heterochromatin differentiation, replication patterns and sequence localization. Chrom Res 6: 219–230.

Fucik V, Michaelis A, Rieger R (1970) On the induction of segment extension and chromatid structural changes in *Vicia faba* chromosome after treatment with 5–azacytidine and 5-azadeoxycytidine. Mutat Res 9: 599–606.

Fukuta N, Yukawa T (1998) Varietal differences in snow tolerance and growth characteristics of bread bean (*Vicia faba* L.). Jpn J Crop Sci 67: 505–509.

Gates P, Boulter D (1979) The use of seed isoenzymes as an aid to the breeding of field beans (*Vicia faba* L.). New Phytol 83: 783–791.

Gates P, Boulter D (1980) The use of pollen isoenzymes as an aid to the breeding of field beans (*Vicia faba* L.). New Phytol 84: 501–504.

Gressel J, Hanafi A, Head G, Marasas W, Obilana AB, Ochanda J, Souissi T, Tzotzos G (2004) Major heretofore intractable biotic constraints to African food security that may be amenable to novel biotechnological solutions. Crop Protec 23: 661–689.

Griffith M, Yaish MWF (2004) Antifreeze proteins in overwintering plants: a tale of two activities. Trends Plant Sci 9: 399–405.

Gutierrez MV, Vaz Patto MC, Huguet T, Cubero JI, Moreno MT, Torres AM (2005) Cross-species amplification of *Medicago truncatula* microsatellites across three major pulse crops. Theor Appl Genet 110: 1210–1217.

Gutierrez N, Avila CM, Duc G, Marget P, Suso MJ, Moreno MT, Torres AM (2006) CAPs markers to assist selection for low vicine and convicine content in faba bean (*Vicia faba* L.). Theor Appl Genet 114: 59–66.

Gutierrez N, Avila CM, Rodriguez-Suarez C, Moreno MT, Torres AM (2007) Development of SCAR markers linked to a gene controlling absence of tannins in faba bean. Mol Breed 19: 305–314.

Gutierrez N, Avila CM, Moreno MT, Torres AM (2008) Development of SCAR markers linked to *zt-2*, one of the genes controlling absence of tannins in faba bean. Aust J Agri Res 59: 62–68.

Hanafy M, Pickardt T, Kiesecker H, Jacobsen HJ (2005) *Agrobacterium*-mediated transformation of faba bean (*Vicia faba* L.) using embryo axes. Euphytica 142: 227–236.

Hanelt P (1972a) Die infraspezifische Variabilität von *Vicia faba* und ihre Gliederung. Kulturpflanze 20: 75–128.

Hanelt P (1972b) Zur Geschichte des Anbaues von *Vicia faba* L. und ihrer verschiedenen Formen. Kulturpflanze 20: 209–223.

Hanounik SB, Robertson LD (1989) Resistance in *Vicia faba* germplasm to blight caused by *Ascochyta fabae*. Plant Dis 73: 202–205.

Harlan JR, De Wet JMJ (1971) Towards a rational classification of cultivated plants, Taxon 20: 509–517.

Heitz E (1931) Nukleolen und Chromosomen in der Gattung *Vicia*. Planta Med (Berlin) 15: 495.

Herridge DF, Rupela OP, Serraj R, Beck DP (1994) Screening techniques and improved biological nitrogen fixation in cool season food legumes. In: FJ Muehlbauer, WJ Kaiser (eds) Expanding the Production and Use of Cool Season Food Legumes. Kluwer Academic Press, Dordrecht, The Netherlands, pp 472–492.

Herzog H (1988) Freezing resistance in faba bean: a limitation to winter growing in the Mediterranean and subtropical highlands? In: JP Srivastava, MC Saxena, S Varma, M Tahir (eds) Winter Cereals and Food Legumes in Mountainous Areas, ICARDA, Aleppo, Syria, pp 235–243.

ICARDA, International Center for Agricultural Research in the Dry Areas (1981) Genetic Variation within *Vicia faba*. ICARDA, Aleppo, Syria, p 12.

ICARDA, International Center for Agricultural Research in the Dry Areas (1986) Third Conspectus of Genetic Variation within *Vicia faba*. ICARDA, Aleppo, Syria, p 54.

Jensen CS, Salchert K, Nielsen KK (2001) A terminal flower1-Like gene from perennial ryegrass involved in floral transition and axillary meristem identity. Plant Physiol 125: 1517–1528.

Jellis GJ, Bond DA, Boulton RE (1998) Diseases of faba bean. In: DJ Allen, JM Lenné (eds), The Pathology of Food and Pasture Legumes. CAB International, Wallingford, UK, pp 371–422.

Joel DM (2000) The long-term approach to parasitic weeds control: Manipulation of specific developmental mechanisms of the parasite. Crop Protec 19: 753–758.

Kahl G, Winter P, Horres R, Rotter B, Jüngling R and the LEGRESIST Consortium (2009) Pathogenesis-related genes and genetic variation in potential resistance genes of major European legumes: the LEGRESIST project. In: Ascochyta 2009, Proc 2nd Int Ascochyta Workshop. Pullman, Washington, USA, p 47.

Kaser HR, Steiner AM (1983) Subspecific classification of *Vicia faba* L. by protein and isozyme patterns. FABIS Newsl 7: 19–20.

Kashiwagi J, Krishnamurthy L, Gaur PM, Chandra S, Upadhyaya HD (2008) Estimation of gene effects of the drought avoidance root characteristics in chickpea (*C. arietinum* L.). Field Crops Res 105: 64–69.

Khan HR, Link W, Hocking TJH, Stoddard FL (2007) Evaluation of physiological traits for improving drought tolerance in faba bean (*Vicia faba* L.). Plant Soil 292: 205–217.

Kharbanda PD, Bernier CC (1980) Cultural and pathogenic variability among isolates of *Ascochyta fabae*. Can J Plant Pathol 2: 139–142.

Kharrat M, Halila MH, Linke KH, Haddar T (1992) First report of *Orobanche foetida* Poiret on faba bean in Tunisia. FABIS Newsl 30: 46–47.

Kohpina S, Knight R, Stoddard FL (1999) Variability of *Ascochyta fabae* in South Australia. Aust J Agri Res 50: 1475–1481.

Kohpina S, Knight R, Stoddard FL (2000) Genetics of resistance to ascochyta blight in two populations of faba bean. Euphytica 112: 101–107.

Kotoda N, Wada M (2005) *MdTFL1*, a *TFL1*-like gene of apple, retards the transition from the vegetative to reproductive phase in transgenic *Arabidopsis*. Plant Sci 168: 95–104.

Kumar R, Singh M (1984) Tannins: their adverse role in ruminant nutrition. J Agri Food Chem 32: 447–453.

Lander E and Botstein D (1989) Mapping Mendelian factors underlying quantitative traits using RFLP linkage maps. Genetics 121: 185–199.

Lander ES, Kruglyak L (1995) Genetic dissection of complex traits: guidelines for interpreting and reporting linkage results. Nat Genet 11: 241–247, doi:10.1038/ng1195–241.

Lander ES, Green P, Abrahamson J, Barlow A, Daly MJ, Lincoln SE, Newburg L (1987) MAPMAKER: an interactive computer program for constructing genetic linkage maps of experimental and natural populations. Genomics 1: 174–181, doi:10.1016/0888-7543(87)90010-3.

Lawn RJ, Likoswe AA (2008) Genotypic differences in leaf area maintenance contribute to differences in recovery from water stress in soybean. Aust J Agri Res 59: 1075–1085.

Levitt J (1980) Responses of Plants to Environmental Stresses, vol 2: Water, Radiation, Salt and Other Stresses. Academic Press, New York, USA, p 607.

Link W (2006) Methods and objectives in faba bean breeding. In: CM Avila, JI Cubero, MT Moreno, MJ Suso, AM Torres AM (eds) Faba Bean 2006. Junta de Andalucía, Sevilla, Spain, pp 35–40.

Link W, Schill B, Kittlitz E von (1996a) Breeding for wide adaptation in faba bean. Euphytica 92: 185–190.

Link W, Schill B, Barbera AC, Cubero JI, Filippetti A, Stringi L, Kittlitz E von, Melchinger AE (1996b) Comparison on infra and inter-pool crosses in faba beans (*Vicia faba* L.) I. Hybrid performance and heterosis in Mediterranean and German environments. Plant Breed 115: 352–360.

Lopez-Granados F, Garcia-Torres L (1998) Short and long term economic implications of controlling crenate broomrape (*Orobanche crenata*) in broad bean under various management strategies. Crop Protec 17: 139–143.

Lucretti S, Dolezel J, Schubert I, Fuchs J (1993) Flow karyotyping and sorting of *Vicia faba* chromosomes. Theor Appl Genet 85: 665–672.

Ludlow MM, Muchow RC (1990) A critical evaluation of traits for improving crop yields in water-limited environments. Adv Agron 43: 107–153.

Maalouf FS, Suso MJ, Moreno MT (2002) Comparative performance of faba bean synthetics developed from different parental numbers. J Genet Breed 56: 251–258.

Macas J, Dolezel J, Lucretti S, Pich U, Meister A, Fuchs J, Schubert I (1993a) Localization of seed genes on flow-sorted field bean chromosomes. Chrom Res 1: 107–115.

Macas J, Weschke W, Baumlein H, Pich U, Houben A, Wobus U, Schubert I (1993b) Localization of vicilin genes via polymerase chain reaction on microisolated field bean chromosomes. Plant J 3: 883–886.

Mancini R, De Pace C, Scarscia Mugnozza GT, Deire V, Vittori D (1989) Isoenzyme gene markers in *Vicia faba* L. Theor Appl Genet 77: 657–667.

Manschadi AM, Sauerborn J, Stützel H, Göbel W, Saxena MC (1998) Simulation of faba bean (*Vicia faba* L.) root system development under Mediterranean conditions. Eur J Agron 9: 259–272.

Marquardt RR (1982) Favism. In: G Hawtin, C Webb (eds) Faba Bean Improvement. Martinus Nijhoff, Hague, The Netherlands, pp 343–353.

Martín A (1976) Genética de los componentes del rendimiento en *Vicia faba*. PhD Thesis. ETSIA, Universidad de Córdoba, Córdoba, Spain, p 383.

Martín A (1979) Aneuploidy and tetraploidy in *Vicia faba*, FABIS Newsl 1: 17

Martín A, Cubero JI (1979) Inheritance of quantitative characters in *Vicia faba*. In: DA Bond, GT Scarascia-Mugnozza, MH Poulsen (eds) Some Current Research on *Vicia faba* in Western Europe. Commission of the European Communities, Luxemburg, pp 383–395.

Martín A, Barceló P (1984) The cytology and morphology of *Vicia faba* trisomics. In: GP Chapman, SA Tarawai (eds) Systems for Cytogenetic Analysis in *Vicia faba* L. Martinus Nijhoff, Hague, The Netherlands, pp 63–76.

Martín A, Gonzalez-García JA, Serradilla JM (1986) Progress in the diploidisation of a *Vicia faba* tetraploid. Biol Zentralbl 105: 85–94.

Matsumura H, Ito A, Saitoh H, Winter P, Kahl G, Reuter M, Kruger DH, Terauchi R (2005) SuperSAGE. Cell Microbiol 7: 11–18.

Mauromicale G, Restuccia G, Marchese M (2001) Soil solarization, a nonchemical technique for controlling *Orobanche crenata* and improving yield of faba bean. Agronomie 21: 757–765.

Maxted N, Khatabb AMA, Bisbi FA (1991) The newly discovered relatives of *Vicia faba* L. do little to resolve the enigma of its origin. Bot Chron 10: 435–465.

McDonald GK, Paulsen GM (1997). High temperature effects on photosynthesis and water relations of grain legumes. Plant Soil 196: 47–58.

McDonald GK, Adisarwanto T, Knight R (1994) Effect of time of sowing on flowering in faba bean (*Vicia faba*). Aust J Exp Agri 34: 395–400.

Mondragao-Rodrigues F (1994) Estudio del sistema reproductivo de *Vicia faba* L: autofertilidad y alogamia. ETSIAM, Universidad de Córdoba, Córdoba, Spain, p 253.

Muratova V (1931) Commons beans (*Vicia faba* L.). Bull Appl Bot Genet Plant Breed, Suppl 50, Leningrad, USSR, p 298.

Nadal S, Moreno MT, Cubero JI (2000) Nuevas variedades de habas de crecimiento determinado. Agricultura 812: 108–109.

Nassib AM, Ibrahim AA, Khalil SA (1982) Breeding for resistance to *Orobanche*. In: G Hawtin, C Webb (eds) Faba Bean Improvement. Martinus Nijhoff, Hague, The Netherlands, pp 199–206.

Nautiyal PC, Rachaputi NR, Joshi YC (2002) Moisture-deficit-induced changes in leaf-water content, leaf carbon exchange rate and biomass production in groundnut cultivars differing in specific leaf area. Field Crops Res 74: 67–79.

Nielsen KL, Eshel A, Lynch JP (2001) The effect of phosphorus availability on the carbon economy of contrasting common bean (*Phaseolus vulgaris* L.) genotypes. J Exp Bot 52: 329–339.

Ondrej M (1993) Response of resistant lines of horse bean to pathogenous fungus *Ascochyta fabae* Speg. Plant Genet Resour 2: 45–48.

Palomino C, Satovic Z, Cubero JI, Torres AM (2006) Identification and characterization of NBS-LRR class resistance gene analogs in faba bean (*Vicia faba* L.) and chickpea (*Cicer arietinum* L.). Genome 49: 1227–1237.

Palomino C, Fernandez-Romero MD, Rubio J, Torres AM, Moreno MT, Millán T (2008) Integration of new CAPS and dCAPS-RGA markers into a composite chickpea genetic map and their association with disease resistance. Theor Appl Genet 118: 671–682.

Parker C, Riches CR (1993) Parasitic Weeds of the World: Biology and Control. CAB International, Wallingford, UK, p 332.

Patrick JW, Stoddard FL (2010). Physiology of flowering and grain filling in faba bean. Field Crops Res 115: 234–242.

Peat WE, Adham JY (1984). The use of isozyme genes as markers in the population genetics of *Vicia faba* L. In: GP Chapman, SA Tarawai (eds) Systems for Cytogenetic Analysis in *Vicia faba* L. Martinus Nijhoff/Dr W. Junk Publishers, Hague, The Neherlands, pp 109–117.

Pérez-de-Luque A, Rubiales D, Cubero JI, Press MC, Scholes J, Yoneyama K, Takeuchi Y, Plakhine D, Joel DM (2005) Interaction between *Orobanche crenata* and its host legumes: Unsuccessful haustorial penetration and necrosis of the developing parasite. Ann Bot 95: 935–942.

Pérez-de-Luque A, Eizenberg H, Grenz JH, Sillero JC, Avila C, Sauerborn J, Rubiales D (2010) Broomrape management in faba bean. Field Crops Res 115: 319–328.

Pflieger S, Lefebvre V and Causse M (2001) The candidate gene approach in plant genetics: a review. Mol Breed 7: 275–291.

Phan HTT, Ellwood SR, Ford R, Thomas S, Oliver R (2006) Differences in syntenic complexity between *Medicago truncatula* with *Lens culinaris* and *Lupinus albus*. Funct Plant Biol 33: 775–782.

Picard J (1963) La coloration des téguments du grain chez la féverole (*Vicia faba* L.). Étude de l'hérédité des différentes colorations. Ann Amélior Plant 13: 97–117.

Picard J (1976) Aperçu sur l'hérédite du caractère absence de tanins dans les graines de féverole (*Vicia faba* L.). Ann Amelior Plant 26: 101–106.

Pieterse AH (1991) The effect of nitrogen fertilizers on the germination of seeds of *Striga hermonthica* and *Orobanche crenata*. In: K Wegmann, LJ Musselman (eds) Proc Int Workshop in *Orobanche* Research, Eberhard-Karls-Universität, Tübingen, Germany, pp 115–124.

Pillitteri LJ, Lovatt, CJ, Walling LL (2004) Isolation and characterization of a *TERMINAL FLOWER* homolog and its correlation with juvenility in citrus. Plant Physiol 135: 1540–1551.

Pnueli L, Carmel-Goren L, Hareven D, Gutfinger T, Alvarez J, Ganal M, Zamir D, Lifschitz E (1998). The *SELF-PRUNING* gene of tomato regulates vegetative to reproductive switching of sympodial meristems and is the ortholog of *CEN* and *TFL1*. Development 125: 1979–1989.

Polignano GB, Sonnante G (1992) Characterization of faba bean populations by isoenzyme patterns of GOT and PGI. FABIS Newsl 31: 12–16.

Poulsen MH, Martín A (1977) A reproductive tetraploid *Vicia faba* L. Hereditas 87: 123–126.

Pozarkova D, Koblizkova A, Román B, Torres AM, Lucretti S , Lysak M, Dolezel J, Macas J (2002) Development and characterization of microsatellite markers from chromosome 1-specific DNA libraries of *Vicia faba*. Biol Plant 45: 337–345.

Ramsay G, van de Ven W, Waugh R, Griffiths DW, Powel W (1995) Mapping quantitative trait loci in faba beans. In: AEP (ed) 2nd Eur Conf on Grain Legume. Copenhagen, AEP, Paris, France, pp 444–445.

Rashid KY, Bernier CC (1986a) The genetics of resistance in *Vicia faba* to two races of *Uromyces viciae-fabae* from Manitoba. Can J Plant Pathol 8: 317–322.

Rashid KY, Bernier CC (1986b) Selection for slow rusting in faba bean (*Vicia faba* L.) to *Uromyces viciae-fabae*. Crop Protec 5: 218–224.

Rashid KY, Bernier CC (1991) The effect of rust on yield of faba bean cultivars and slow-rusting populations. Can J Plant Sci 71: 967–972.

Rashid KY, Bernier CC, Conner RL (1991a) Evaluation of faba bean for resistance to *Ascochyta fabae* and development of host differential for race identification. Plant Dis 75: 852–855.

Rashid KY, Bernier CC, Conner RL (1991b) Genetic of resistance in faba bean inbred lines to five isolates of *Ascochyta fabae*. Can J Plant Pathol 13: 218–225.

Rebetzke GJ, Botwright TL, Moore CS, Richards RA, Condon AG (2004) Genotypic variation in specific leaf area for genetic improvement of early vigour in wheat. Field Crops Res 88: 179–189.

Ricciardi L (1989) Plant breeding for resistance to drought. I: Stomatal traits in genotypes of *Vicia faba* L. Agri Medit 119: 297–308.

Ricciardi L, Polignano GB De Giovanni C (2001) Genotypic response of faba bean to water stress. Euphytica 118: 39–46.

Rieger R, Michaelis A (1967) Die chromosomen aberrationen. In: H Stubbe (ed) Genetik-Grundlagen. Ergebnisse und Problemes in Einzeldarstellungen. Beitrag 6 Jena VEB Gustav Fischer, p 433.

Robertson LD (1985) Faba bean germplasm collection, maintenance, evaluation and use. In: MC Saxena, S Varma (eds) Faba Beans, Kabuli Chickpeas and Lentils in the 1980's. ICARDA, Aleppo, Syria, pp 15–21.

Román B, Torres AM, Rubiales D, Cubero JI, Satovic Z (2002) Mapping of quantitative trait loci controlling broomrape (*Orobanche crenata* Forsk.) resistance in faba bean (*Vicia faba* L.). Genome 45: 1057–1063.

Román B, Satovic Z, Avila, CM, Rubiales, D, Moreno, MT, Torres, AM (2003) Locating genes associated with *Ascochyta fabae* resistance in *Vicia faba* L. Aust J Agri Res 54: 85–90.

Román B, Satovic Z, Pozarkova D, Macas J, Dolezel J, Cubero JI, Torres AM (2004) Development of a composite map in *Vicia faba* L. breeding applications and future prospects. Theor Appl Genet 108: 1079–1088.

Román B, Satovic Z, Alfaro C, Moreno MT, Kharrat M, Pérez-de-Luque A, Rubiales D (2007) Host differentiation in *Orobanche foetida* Poir. Flora 202: 201–208.

Roughley RJ, Sprent JI, Day JM (1983) Nitrogen fixation. In: PD Hebblethwaite (ed) The Faba Bean (*Vicia faba* L.). Butterworths, London, UK, pp 232–269.

Rubiales D (2003) Parasitic plants, wild relatives and the nature of resistance. New Phytol 160: 459–461.

Rubiales D, Pérez-de-Luque A, Cubero JI, Sillero JC (2003) Crenate broomrape (*Orobanche crenata*) infection in field pea cultivars. Crop Protec 22: 865–872.

Rubiales D, Sadiki M, Román B (2005) First report of *Orobanche foetida* on common vetch (*Vicia sativa*) in Morocco. Plant Dis 89: 528.

Rubiales D, Perez-de-Luque A, Fernandez-Aparicio M, Sillero JC, Román B, Kharrat M, Khalil S, Joel DM, Riches C (2006) Screening techniques and sources of resistance against parasitic weeds in grain legumes. Euphytica 147: 187–199.

Sadiki M (2006) Faba bean genetic resources status and prospects. In: CM Avila, JI Cubero, MT Moreno, MJ Suso AM Torres (eds) Faba Bean 2006. Junta de Andalucía, Sevilla, Spain, pp 185–187.

Salt GA (1983) Root diseases of *Vicia faba* L. In: PD Hebblethwaite (ed) The Faba Bean (*Vicia faba* L.). Butterworths, London, UK, pp 393–420.

Sánchez FJ, Manzanares M, Andrés EF de, Tenorio JL, Ayerbe L (1998) Turgor maintenance, osmotic adjustment and soluble sugar and proline accumulation in 49 pea cultivars in response to water stress. Field Crops Res 59: 225–235.

Sato S, Nakamura Y, Kaneko T, Asamizu E, Kato T, Nakao M, Sasamoto S, Watanabe A, Ono A, Kawashima K, Fujishiro T, Katoh M, Kohara M, Kishida Y, Minami C, Nakayama S, Nakazaki N, Shimizu Y, Shinpo S, Takahashi C, Wada T, Yamada M, Ohmido N, Hayashi M, Fukui K, Baba T, Nakamichi T, Mori H, Tabata S (2008) Genome structure of the legume *Lotus japonicus*. DNA Res 15: 227–239.

Satovic Z, Torres AM, Cubero JI (1996) Genetic mapping of new morphological, isozyme and RAPD markers in *Vicia faba* L. using trisomics. Theor Appl Genet 93: 1130–1138.
Sax K (1923) The association of size differences with seed coat pattern and pigmentation in *Phaseolus vulgaris*. Genetics 8: 552–560.
Schubert I, Rieger R (1991) Catalogue of chromosomal and morphological mutants of faba bean in the Gatersleben Collection, 1991. FABIS Neswl 28/19: 14–22.
Schubert I, Rieger R, Michaelis A (1986) Structural and numerical manipulation of the *Vicia faba* karyotype: results and perspectives. Biol Zbl 105: 9–17.
Schubert I, Dolezel J, Houben A, Scherthan H, Wanner G (1993) Refined examination of plant metaphase chromosome structure at different levels made feasible by new isolation methods. Chromosoma 102: 96–101.
Serradilla JM, De Mora, T, Moreno MT (1993) Geographic dispersion and varietal diversity in *Vicia faba* L. Genet Resour Crop Evol 40: 143–151.
Sillero JC, Rubiales D (2002) Histological characterization of the resistance of faba bean to faba bean rust. Phytopathology 92: 294–299.
Sillero JC, Rubiales D, Moreno MT (1999) New sources of resistance to broomrape (*Orobanche crenata*) in a collection of *Vicia* species. In: JI Cubero, MT Moreno, D Rubiales, J Sillero (eds) Resistance to *Orobanche*: The State of the Art. Junta de Andalucía, Sevilla, Spain, pp 25–41.
Sillero JC, Moreno MT, Rubiales D (2000) Characterization of new sources of resistance to *Uromyces viciae-fabae* in a germplasm collection of *Vicia faba*. Plant Pathol 49: 389–395.
Sillero JC, Avila CM, Moreno MT, Rubiales D (2001) Identification of resistance to *Ascochyta fabae* in *Vicia faba* germplasm. Plant Breed 120: 529–531: doi:10.1046/j.1439-0523.2001.00653.x.
Sillero JC, Fondevilla S, Davidson J, Vaz Patto MC, Warkentin TD, Thomas J, Rubiales D (2006) Screening techniques and sources of resistance to rusts and mildews in grain legumes. Euphytica 147: 255–272.
Sillero JC, Villegas-Fernández AM, Thomas J, Rojas-Molina MM, Emeran AA, Fernández-Aparicio M, Rubiales D (2010) Faba bean breeding for disease resistance. Field Crops Res 115: 297–307.
Sirks HJ (1931) Beiträge zur einer genotypischen Analyse der Ackerbohne, *Vicia faba* L. Genetica 13: 209–631.
Sjödin J (1971a) Induced translocation in *Vicia faba* L. Hereditas. 68: 1–34.
Sjödin J (1971b) Induced morphological variation in *Vicia faba* L. Hereditas 67: 155–180.
Song QJ, Marek LF, Shoemaker RC, Lark KG, Concibido VC, Delannay X, Specht JE, Cregan PB (2004) A new integrated genetic linkage map of the soybean. Theor Appl Genet 109: 122–128.
Steuckardt R, Dietrich M, Griem H (1982) Ergebnisse von Kreuzungsanalysen mit terminalinfloreszenten *(ti)* Mutanten und daraus entwickelte Zuchtstamme bei *Vicia faba* L.. Arch Zuchtungsfors 12: 33–42.
Stoddard FL (1993) Limits to retention of fertilized flowers in faba beans (*Vicia faba* L.). J Agron Crop Sci 171: 251–259.
Stoddard FL, Bond DA (1987) The pollination requirements of the faba bean (*Vicia faba* L.). Bee World 68: 144–152.
Stoddard FL, Marshall DR, Ali SM (1993) Variability in grain protein in peas and lentils in Australia. Aust J Agri Res 44: 1415–1419.
Stoddard FL, Nicholas AH, Rubiales D, Thomas J, Villegas-Fernández AM (2010) Integrated pest management in faba bean. Field Crops Res 115: 308–318.
Suso MJ, Cubero JI (1986) Primitive and advanced forms of Vicia faba. Differences concerning the quantitative inheritance of characters related to yield. Genet Agr 40: 265–282.
Suso MJ, Moreno MT, Cubero JI (1993) New isozyme markers in *Vicia faba*: inheritance and linkage. Plant Breed 111: 170–172.
Suso MJ, Cubero JI, Moreno MT (1995) Outcrossing in two faba bean cultivars under dryland conditions in Spain. Can J Plant Sci 75: 441–443.

Taylor JH, Woods PS, Hughes WL (1957) The organization and duplication of chromosomes as revealed by autoradiographic studies using tritium-labeled thymidine. Proc Natl Acad Sci USA 43: 122–127.

Ter Borg S (1999) Broomrape resistance in faba bean: what do we know? In: JI Cubero, MT Moreno, D Rubiales, J Sillero (eds) Resistance to *Orobanche*: The State of the Art. Junta de Andalucía, Sevilla, Spain, pp 25–41.

Tivoli B, Berthelem P, LeGuen J, Onfroy C (1988) A study of the performance of certain faba bean genotypes in relation to *Botrytis fabae* and *Ascochyta fabae* in France. FABIS Newsl 21: 36–39.

Torres AM, Moreno MT, Cubero JI (1993a) Genetics of six components of auto-fertility in *Vicia faba*. Plant Breed 110: 220–228.

Torres AM, Weeden NF, Martín A (1993b) Linkage among isozyme, RFLP and RAPD markers in *Vicia faba*. Theor Appl Genet 85: 937–945.

Torres AM, Satovic Z, Cánovas J, Cobos S, Cubero JI (1995) Genetics and mapping of new isozyme loci in *Vicia faba* L. using trisomics. Theor Appl Genet 91: 783–789.

Torres AM, Vaz Patto MC, Satovic Z, Cubero JI (1998) New isozyme loci in faba bean (*Vicia faba* L.): Genetic analysis and mapping using trisomics. J Hered 89: 271–274.

Torres AM, Román B, Avila CM, Satovic Z, Rubiales D, Sillero JC, Cubero JI, Moreno MT (2006) Faba bean breeding for resistance against biotic stresses: towards application of marker technology. Euphytica 147: 67–80.

Torres AM, Avila CM, Gutierrez N, Palomino C, Moreno MT, Cubero JI (2010) Marker-assisted selection in faba bean (*Vicia faba* L.). Field Crops Res 115: 243–252: doi:10.1016/j.fcr.2008.12.002.

Turner NC, Wright GC, Siddique KHM (2001) Adaptation of grain legumes (pulses) to water-limited environments. Adv Agron 71: 193–223.

Turpin JE, Robertson MJ, Hillcoat NS, Herridge DF (2002) Faba bean (*Vicia faba*) in Australia's northern grains belt: canopy development, biomass, and nitrogen accumulation and partitioning. Aust J Agri Res 53: 227–237.

Van de Ven WTG, Waugh R, Duncan N, Ramsay G, Dow N, Powell W (1991). Development of a genetic linkage map in *Vicia faba* using molecular and biochemical techniques. Asp Appl Biol 27: 49–54.

Van Ooijen JW (1992) Accuracy of mapping quantitative trait loci in autogamous species. Theor Appl Genet 84: 803–811.

Vaz Patto MC, Torres AM, Koblizkova A, Macas J Cubero JI (1999) Development of a genetic composite map of *Vicia faba* using F_2 populations derived from trisomic plants. Theor Appl Genet 98: 736–743.

Wang S, Basten CJ, Gaffney P, Zeng ZB (2005) Windows QTL Cartographer 2.5. North Carolina State University, Bioinformatics Research Center: http://statgen.ncsu.edu/qtlcart/WQTLCart.htm.

Wayne ML and McIntyre L (2002) Combining mapping and arraying: an approach to candidate gene identification. Proc Natl Acad Sci USA 99: 14903–14906.

Wery J, Silim SN, Knights EJ, Malhotra RS, Cousin S (1994) Screening techniques and sources and tolerance to extremes of moisture and air temperature in cool season food legumes. Euphytica 73: 73–83.

Witcombe JR, Erskine W (eds) (1984) Genetic Resources and Their Exploitation: Chickpeas, Faba Beans and Lentils. ICARDA, Aleppo, Syria, p 256.

Yamamoto K (1975a) Estimation of genetic homogeneity by isozymes from interspecific hybrids of *Vicia*. I. Amylase isozyme patterns in the hybrid progenies between *Vicia pilosa* and *Vicia macrocarpa*. Jpn J Breed 25: 60–64.

Yamamoto K (1975b) Estimation of genetic homogeneity by isozymes from interspecific hybrids of *Vicia*. II. Amylase isozyme patterns in the hybrid progenies between *Vicia amphicarpa* and *V. macrocarpa*. Jpn J Breed 29: 59–65.

Yamamoto K, Plitmann U (1980) Isozyme polymorphism in species of the genus *Vicia* (*Leguminosae*). Jpn J Genet 55: 151–164.

Zeid M, Mitchell S, Link W, Carter M, Nawar A, Fulton T, Kresovich S (2009) Simple sequence repeats (SSRs) in faba bean: new loci from *Orobanche*-resistant cultivar Giza 402. Plant Breed 128: 149–155.

Zhang SH, Hu WJ, Wang LP, Lin C, Cong B, Sun CR, Luo D (2005) *TFL1/CEN*-like genes control intercalary meristem activity and phase transition in rice. Plant Sci 168: 1393–1408.

Zhu H, Choi H, Cook DR, Shoemaker RC (2005) Bridging model and crop legumes through comparative genomics. Plant Physiol 137: 1189–1196.

Zlatev ZS, Lidon FC, Ramalho JC, Yordanov IT (2006) Comparison of resistance to drought of three common bean cultivars. Biol Plant 50: 389–394.

3

Lentil

Marcelino Pérez de la Vega,[1,a,*] *Richard M. Fratini*[1]
and *Fred J. Muehlbauer*[2]

ABSTRACT

Lentil (*Len culinaris* Medik.) is one of the first domesticated plant species in the Fertile Crescent and has been cultivated for 10 thousand years in the most difficult agricultural environments. Lentils are an important cool season food legume valued as healthy food, although they are globally a minor crop. Lentils are traditionally considered as a source of energy, proteins and iron in human nutrition. In addition they are an important dietary source of fiber, minerals vitamins and antioxidants, although they also contribute with non-nutritional components such as protease inhibitors, tannins or phytic acid. Different processing and cooking conditions are employed to improve lentil organoleptic properties and reducing the levels and or effects of non-nutritional components. The domestication process most likely determined the low amount of genetic variation detected in the cultivated species. This scarcity in genetic variability represents a limitation for lentil breeding programs. Other *Lens* species are important sources of genetic variation for breeding new lentil varieties adaptable to new environments and resistant to biotic and abiotic stresses. Lentil breeding programs throughout the world have similar objectives with larger and more stable seed yield being the most important. Increasing yield often entails breeding for adaptation to environmental stress factors such as heat or cold, drought, and diseases caused by fungi, bacteria or viruses. The implementation of genetic markers for routine use in lentil breeding programs is still very limited and much of the focus has been on developing molecular markers for selecting resistance to a few

[1]Area de Genética, Universidad de León, 24071 León, Spain.
[a]e-mail: *m.perez.delavega@unileon.es*
[2]US Department of Agriculture, Agriculture Research Service, 303 Johnson Hall, Washington State University, Pullman, WA 99164-6434, USA.
*Corresponding author

major diseases; although, the number of molecular markers available for genetic mapping and breeding purposes in *Lens* is increasing dramatically. Genetic maps in lentil are still unsaturated and a consensus map is lacking. However significant advances in identifying molecular markers associated to simply inherited trait loci related to both abiotic and biotic stress resistance have been achieved. Genetic improvement through genetic transformation has not yet been achieved in lentil. This chapter summarizes the state of the art of the most important aspects on lentil genetics and breeding.

Keywords: abiotic stresses, biotic stresses, gene mapping, marker-assisted selection, marker development, taxonomy, healthy food

3.1 Basic Information on Lentil

The scientific name of lentil, *Lens culinaris*, was given in 1787 by Medikus, a German botanist and physician (Cubero et al. 2009) and is the scientific name currently accepted. However, the name of this cultivated grain legume was a matter of some confusion since *Lens esculenta* Moench was widely used and can be still found in many publications. Other less common synonymous names are *Cicer lens*, *Vicia lens* (L.) Coss. & Germ., *Ervum lens* L. or *Lentilla lens*. There was also controversy concerning the author to whom the genus name *Lens* had to be credited. Even in recent times, well known Floras have credited it sometimes to Adanson, Moench, and Tournefort and also to Grenier and Godron. Finally, the Congress on Botanical Nomenclature decided in 1966 to conserve Miller as the author. *Lens* Miller is thus a *nomen conservatum* and has priority over older names (Gunn 1969; Cubero et al. 2009).

Lentil is a predominantly self-pollinated diploid ($2n = 14$) annual grain legume species adapted to growth in dry-temperate climates, traditionally as a rain fed crop. According to FAOSTAT (http://faostat.fao.org), lentils were produced in 52 countries in 2007 on around 3.8 million hectares. The distribution includes all continents, except Antarctica, from cold-winter to subtropical areas and high altitudes in tropical regions. Lentil (*Lens culinaris* ssp. *culinaris*) is a bushy annual herb with erect, semi-erect or spreading growth habit ranging from 25 to 30 cm in height for the majority of genotypes, but may vary from 15 to 75 cm, depending on the genotype and environmental conditions. Stems (square and ribbed) and branches are slender while the root system is formed by a slender tap root with fibrous lateral roots. Roots in the surface layer of the soil have numerous and indeterminate nodules that vary in shape from round to elongate. Leaves are described as pinnate or imparipinnate and comprise as many as 14 sessile, ovate or elliptic, or obovate or lanceolated leaflets that vary in length from 1 to 3 cm. Each leaf is subtended by two small stipules and may or may not terminate in a tendril. The flower, complete and typical papilionaceous and 4-8 mm long, are borne singly or in multiples on peduncles that originate from upper plant nodes. Flower color is white, pale purple or purple blue. The legume fruits (referred to as pods) are oblong, laterally compressed, 6 to 20 mm long and 3.5 to 11 mm wide, and usually contain one, two or rarely three seeds. They are lens shaped and weigh between 20 and 80 mg and are a rich source of protein and dietary fiber. Seed diameter is the main characteristic of Barulina's classification of lentil genotypes into the large seeded macrosperma type (6 to 9 mm) or small and medium sized microsperma (2 to 6 mm) (see taxonomic status in Section 3.3.3) (Muehlbauer et al. 1995).

3.1.1 Economic Importance and Production

Lentils are valued as a healthy food. They were usually rated highly and consumed where they were grown and they were and still are a globally minor crop. Currently, global production averages 3.5 Mt yearly with approximately 1.3 Mt traded internationally (FAOSTAT 2006).

Different lentil varieties are preferred in different countries. For example, while red cotyledon dehulled and split lentils are preferred in the Indian subcontinent. In Europe green or yellow-creamy cotyledon types are favored and lentils are consumed whole. These preferences indicate that lentil production in the principal exporting countries is strongly based on market preferences. Currently, most of the Canadian and Australian productions are red lentils destined for markets in the Indian subcontinent; while a large share of the production in the US Pacific Northwest is comprised of small sized, brown yellow cotyledon Pardina lentils destined for markets in Spain. However, in countries where lentils have been consumed for centuries, preferences are often shared between different lentils types. Therefore, international trade greatly determines the lentil types grown in countries in which export predominates over internal consumption; while in traditional consuming countries, local demand greatly determines the lentil types cultivated.

Lentil yield and production varies widely within and between producing countries, and even within wider geographic areas on a yearly basis. This fact also widely affects export-import trade between countries. Thus, to buffer these oscillations, all data referred in this section are the average of the last five years in which statistics in FAOSTAT are available. The average world production for the period 2003–2007 was approximately 3.59 Mt, ranging from 2.98 Mt in 2003 to 4.04 Mt in 2005. The principal producer was India with an average of 1.05 Mt, followed by Canada (773 kt) and Turkey (571 kt). During these years world lentil production increased approximately 3.82-times (on the basis of the five-year average, from 940,259 t to 3.59 Mt) from the start of FAOSTAT statistics to 2007, while the area of production increased 2.28 times during these years (from 1.67 Mha to 3.82 Mha), the average yield (kg ha^{-1}) only increased 1.67-times (from 562 kg ha^{-1} to 937 kg ha^{-1}).

The principal lentil exporting countries (average of years 2002–2006) are Canada, Australia, India, Turkey, the United States, and Syria, while the principal importers are the European Union (EU-27 countries), Bangladesh, Sri Lanka, Pakistan, Colombia, and Algeria (Table 3-1). Spain is the largest importer within the EU with an average of nearly 50,000 t annually. Examples of wide variations in the lentil trade in this period are Australia with a range between 242,049 and 84,595 export t, and Bangladesh with a range between 147,179 and 63,329 import t.

Table 3-1 Principal countries importing and exporting lentils. Average of the years 2002 to 2006 in ton.

Country	Import	Export	Net
Algeria	**62,070.2**	88.2	–61,982.0
Australia	1,751.8	**151,667.0**	149,915,2
Bangladesh	**110,886.2**	174.4	–110,711,8
Canada	8,783.2	**471,084.2**	462,301.0
Colombia	**63,110.8**	849.4	–61,707.4
Egypt	**87,041.2**	1,403.4	–85,637.8
European Union (27)	**203,882.0**	28,313.0	–175,569.0
Spain	**49,569.0**	3,624.8	–45,944.2
India	45,309.6	**141,731.2**	96,421,6
Pakistan	**63,112.0**	2,176.2	–60,935.8
Sri Lanka	**90,537.6**	102.0	–90,435,6
Syria	287.6	**58,288.0**	58,000.4
United States	14,260.4	**116,227.8**	101,967.4
Turkey	35,665.0	**139,248.6**	103,583.6

3.1.2 Nutritional Composition

Lentils are traditionally valued as a source of energy, proteins and iron in human nutrition. In addition they are an important dietary source of fiber, minerals vitamins and antioxidants, although they also contribute with non-nutritional components such as protease inhibitors, tannins, α-galactoside oligosaccharides and phytic acid. The amounts of these components vary among varieties, thus the ranges for different components per 100 g of raw lentil dry matter are: energy 1483–2010 kJ, protein 20.6–31.4 g, fat 0.7–4.3 g, carbohydrates 43.4–69.9 g, fiber 5.0–26.9 g, ash 2.2–4.2 g (Urbano et al. 2007). Narrower ranges were found in Australian grown lentils for protein content (19.8–28.0 g, in 853 accessions) and for fiber (2.9–6.0 g) (Petterson et al. 1997). Variation in the food composition of these components are due in part to the two main ways in which lentils are consumed, as complete seeds or as dehulled split seeds. Although seed coats represent 8 to 20% of the lentil seed weight, dehulling causes significant alterations in food composition (Cuadrado et al. 2002; Urbano et al. 2007). Ghavidel and Prakash (2007) showed the effects of dehulling on lentil dietary components (Table 3-2). Almeida-Costa et al. (2007) showed the effects of cooking on lentil components in comparison with raw uncooked lentils: proteins, carbohydrates and insoluble dietary fiber increased (20.6 to 23.4, 56.4 to 61.8, and 19.0 to 21.4, respectively, in g/100 g) while soluble dietary fiber decreased (1.44 to 1.37). Similar results were obtained in a combined analysis of cooking and dehulling on the nutritional composition of lentil varieties (Wang et al. 2009).

Table 3-2 Effect of germination and dehulling[a] in some dietary components of lentil seeds (data on dry weigth/100g) (Summarized from Ghavidel and Prakash 2007).

	Raw	Germinated	Dehulled		Raw	Germinated	Dehulled
Fat (g)	0.9	0.8	1.2	Thiamin (mg)	0.5	0.7	0.8
Protein (g)	26.5	28.5	29.6	Phytic acid (g)	0.2	0.1	0.1
Starch (g)	38.2	34.3	37.3	Tannin (g)	0.8	0.6	0.4
Soluble dietary fiber (g)	0.9	1.5	0.3	Iron (mg)	8.5	6.7	5.5
Insoluble dietary fiber (g)	15.6	15.5	11.4	Calcium (mg)	77.0	63.0	50.3

[a]Seeds were soaked for 12 hours and allowed to germinate for 24 additional hours, then they were dried and a part of the germinated sample was dehulled. Components were compared with ungerminated raw seeds.

Lentil protein concentration is relatively high; however, the proteins are relatively low in sulfur-containing amino acids and tryptophan. On the other hand, the protein is highly digestible and heating does not cause losses of quality. Digestive utilization of lentil protein determined in vivo using animal models ranges from 62.3 to 92.0%, while using *in vitro* techniques it varies from 37.6–49.6% to 82.3–92.5% depending upon the experimental conditions used (Urbano et al. 2007). The fat content of lentils is low (0.7–4.3%) and unsaturated fatty acids roughly represent the 80% of the total fatty acid concentration (linoleic > oleic > linolenic), with low amounts of saturated fatty acids, mainly palmitic acid (Petterson et al. 1997; Urbano et al. 2007). The non-germinated lentil seeds contain some non-protein amino acids with known or proposed physiological effects such as γ-hydroxyarginine and homoarginine affecting the biosynthesis of nitric oxide. The isoxazolinones, α-aminoadipic acid, GABA and lathyrine [β-(2-amino-pyrimidine-4-yl)-alanine], a heterocyclic neurotoxic non-protein amino acid previously found in some *Lathyrus* species were only found after germination. Seedlings contain secondary metabolites that are absent from the seeds and on which there is insufficient nutritional, toxicological or pharmacological information (Rozan et al. 2001). In that study *L. orientalis* was most different from the other four species studied (*L. culinaris, L. ervoides, L. nigricans*, and *L. odemensis*) by the absence of homoarginine, lathyrine and isoxazolinone derivatives. That contrasting result in relation to the *Lens* taxonomy (see Section 3.3.3) could be due to the particular accessions used, and suggests that additional analyses with more accessions should be carried out.

An analysis of the nutritional components and the *in vitro* digestibility of lentil (two cultivars) in comparison with pea (three cultivars) and chickpea (three cultivars), grown under identical environmental conditions, revealed that there is an overlapping of the starch content among the samples of the three grain legumes. Lentil flour contained the lowest amount of rapidly digestible starch and it was hydrolyzed more slowly and to a lesser extent than pea and chickpea flours. The expected glycemic index of lentil flour was the lowest among the pulse flours (Chung et al. 2008).

Lentils are an important source of iron and other dietary essential minerals (e.g., 100 g serving of dry lentils is within the range of the recommended Fe and Zn daily intake) and water soluble vitamins. They generally have a low content of phytate (0.1–0.8%) and tannins (0.5–10.5%), which is important since phytic acids form complexes with protein and minerals and tannins with proteins reducing the availability of these food components. Mineral micronutrient bioavailabilty can be increased by using natural low-phytic acid lentil varieties, preferably decorticated before cooking, which reduced total phytic acid by more than 50%, likewise it can be increased by breeding (Thavarajah et al. 2009a, b). In addition to phytic acid and tannins, lentils seeds also have other antinutritive compounds such as α-amylase, trypsin and chymotrypsin inhibitors, lectins and saponins. The level of lectins is low and trypsin inhibitors and saponins are lower than in other legumes except lupins. Fortunately, much of their activity is lost by heat treatment or cooking. Lentil lectin is a homodimeric protein similar to pea lectin, which agglutinates erythrocytes independent of the ABH blood and has a reported specificity toward glucose and mannose (Iskratsch et al. 2009).

Among the enzyme inhibitors of pulses, trypsin and chymotrypsin inhibitors elicited the largest number of studies in the 1970s and 1980s, due to their impact on animal nutrition. Fortunately they can be inactivated by hydrothermic treatments. Trypsin inhibitors might deplete the sulfur-containing amino acids that are already scarce in most grain legumes. In a comparative study of the inhibitory activity of the seeds of 18 legumes, lentils were found to inhibit human chymotrypsin much more than bovine chymotrypsin, human trypsin and bovine trypsin. Further, four isoinhibitors were described in 38 lentil samples tested. The four lentil inhibitors belonged to the Bowman-Birk inhibitor family (see Champ 2002; Guillamón et al. 2008 for further references). The analysis of four Spanish lentil cultivars and other grain legumes, including among others faba bean, pea, chickpea, and grass pea, revealed a range of variation in lentil trypsin inhibitor units (TIU) between 3 and 8 TIU mg^{-1}. These TIU values were similar to faba bean and only higher than *Lupinus* species, and agreed with the range activities previously reported (Guillamón et al. 2008). Protease inhibitors are known to defend the plant against predators. Consequently, removal of these products by breeding could have deleterious effects for the plant.

Lentils are also valued as fodder in particular for small farm animals. Lentil straw is also a preferred alternative to other fodder. It is interesting to note that the nutritive value of lentil straw is greater than vetch straw and wheat straw and close to alfalfa hay for replacement Awassi ewe lambs (Haddad and Husein 2001).

3.1.2.1 New Products and Alternative Uses

Some new uses for less useful lentil parts and alternative uses for the traditional ones have been suggested. Thus, it has been proposed that lentil husk is a suitable source for the production of extracellular alkaline α-amylase by solid state fermentation with strains of *Bacillus* sp. (Baysal et al. 2008). Besides, locally available and useless lentil shells could be a low-cost and promising adsorbents for the removal of heavy metals such as Cu (II) from aqueous solutions (Aydin et al. 2008). Red lentil flour and bakers yeast cells could be an inexpensive nitrogen source (replacing peptone and yeast extract) for the production of L-(+) lactic acid by anaerobic fermentation by *Lactobacillus amylophilus* GV6 (Altaf et al. 2007). In addition to the nutritional values previously mentioned, some lentil components have been associated with a reduction in the incidence of various cancers, thus, for instance, lentil Bowman-Birk protease inhibitor is able to inhibit the growth of colon adenocarcinoma cells (Caccialupi et al. 2010; Roy et al. 2010).

3.1.3 Academic Importance: Use in Genetics, Cytogenetics, Breeding and Genomics Works

Lentil is a relatively minor crop when compared to common bean, pea and chickpea and as a result genomic information is limited. Furthermore, its large genome size and the scarce information on lentil gene sequences are major obstacles to genomic studies in this species. Perhaps one of the most relevant academic contributions of lentil was in the formulation of the

Vavilov's Law of Homologous Series in Variation . Vavilov (1922) used *Lens, Vicia, Pisum* and other legume species to support his Homologous Series theory: "Genetically closely related species and genera are characterized by similar series of hereditary variation with such regularity that, knowing the series of forms within one species, one can predict the existence of parallel forms in other species and genera. The genetically closer the genera and species within the general system are, the greater is the similarity in their variation series."

Genetic improvement thorough genetic transformation has not yet been achieved in lentil, mainly because efficient and reproducible gene transfer systems are not available. Although Sarker and Erskine (2006) reported that the development of transgenic lentil with appropriate *Bt* toxin gene for sitonia weevil control and with herbicide resistance to broomrape control was underway at ICARDA, publications on lentil transformation are scarce and no direct breeding results have been published to date (see an updated review in Chapter 9).

No bacterial artificial chromosome (BAC) library, BAC-end sequence nor a physical map have been described in lentil (Varshney et al. 2009). Table 3-3 shows the results of two queries in the NCBI data sets (http://www.ncbi.nlm.nih.gov/Taxonomy/) carried out within a year (July 2009–July 2010). In comparison with other crop species the number of *Lens* data indexed is scarce, although this situation is rapidly changing, in particular for nucleotide expressed sequence tags (ESTs) (Table 3-3). Moreover, recent reports on the characterization of the nucleotide binding site-leucine rich repeat (NBS-LRR) disease resistance genes from several legume species, including lentils, and on transcriptome sequencing of cool season food legumes using 454 GS-FLX Titanium technology (Penmetsa et al. 2010; Kaur et al. 2010) also reveal this tendency change. The GEO Dataset, one of the largest gene expression resources, registers the "transcriptional profiling of Ascochyta blight resistance in lentil" as well as other legume transcriptional profilings (including lentils). The genome project is indexed as *Lens culinaris*

Table 3-3 Genomic and transcriptomic *Lens* data indexed at the NCBI data sets.

Database name	Number of records	
Year	2009	2010
Nucleotide	218	254
Nucleotide EST	0	9,513
Nucleotide GSS[a]	485	485
Proteins	138	153
Genome projects	1	1
Gene-expression datasets	9	9

[a]Genome Survey Sequences, sequences similar to EST with the exception that most of the sequences are genomic in origin, rather than cDNA.

(Project ID: 36533) at the International Center for Agricultural Research in the Dry Areas (ICARDA), although no genomic results are linked to it. The first report on the use of the microarray technology to study gene expression in lentil was recently published (Mustafa et al. 2009). Highly resistant (ILL7537) and highly susceptible (ILL6002) lentil germplasm accessions were inoculated with *Ascochyta lentis*. Ninety genes were differentially expressed in ILL7537 and 95 genes were differentially expressed in ILL6002. The expression profiles of the two accessions showed substantial difference in the type of genes and the times of expression in response to the pathogen. The resistant variety showed early up-regulation of PR4 and 10 proteins and other defense-related genes. The susceptible genotype showed early down-regulation of defense-related genes. Real-time PCR (RT-PCR) was used to verify microarray expression ratios. The resistant and susceptible lentil accessions differed not only in the type of genes expressed but also in the time and level of expression in response to *A. lentis* inoculation. SuperSAGE expression profiling in response to some biotic and abiotic stresses have been carried out within the LEGRESIST and other projects and microarrays for lentil genomic analysis are being developed and will be ready soon (P. Winter, pers. comm.). This technology has been recently used to analyze the allele-specific differential expression of transcripts potentially involved in frost-tolerance, using bulked segregant analysis among 90 F_9 RILs derived from the Precoz x WA8649041 lentil cross (Barrios et al. 2010).

Lentil proteomics is also underdeveloped. Two-dimensional electrophoretic maps of proteins from mature lentil seed have been recently described (Scippa et al. 2008, 2010). Among the hundreds of protein species, 122 were further identified by MALDI-TOF PMF and/or LC-ESI-LIT-MS/MS. Map comparisons revealed that 103 protein spots were differentially expressed within and between populations (Scippa et al. 2010).

3.1.4 Brief History of the Lentil Crop

Lentil is one of the first domesticated plant species in the Fertile Crescent. The origin of the cultivated form is in the wild *L. culinaris* ssp. *orientalis* (syn. *L. orientalis*). According to archaeological data, the distribution of wild species and the existence of remains of both wild and cultivated lentils in the same region, the Near East is the obvious candidate. Ladizinsky (1999) found the cultivated lentil to be monomorphic for several characters, but considerable polymorphism was found to exist in the wild accessions of subsp. *orientalis*. However, three accessions from eastern Turkey and northern Syria were shown to share the above characters with the cultigen and, therefore, can be regarded to be members of the genetic stock from which the cultivated lentil was domesticated. Zohary (1999), based on founder effects revealed by chromosome and DNA polymorphisms, as well as on evidence from domestication traits and species diversity, came to the conclusion that lentil had been domesticated once or only a few times. Lev-Yadun et al. (2000) suggested a place close to or overlapping the area where einkorn and emmer wheats were domesticated in the Fertile Crescent. The other possible center of origin of the cultivated lentil was proposed by Barulina (1930): the region between Afghanistan, India and Turkistan (i.e., the Himalaya-Hindu Kush junction). This center of origin of the cultivated forms was proposed because it showed the highest proportion of endemic varieties (a pure Vavilovian criterion to define centers of origin). The high degree of endemisms that exists in this area is better explained, as in other species, by an intense genetic drift, typical of highly diversified environments, coupled with artificial selection carried out by very diverse human populations, with drastic genetic fixation and losses providing secondary centers of diversity (Cubero et al. 2009). Available data point to the region of southern Turkey–northern Syria as the most likely place of lentil domestication and where some populations of *L. orientalis* were unconsciously subjected to automatic selection leading to *L. culinaris* as a new crop.

The expansion of cultivated lentil most likely occurred simultaneously with other of the Fertile Crescent first domesticated crops as the agriculture expanded to Europe to the west, to central Asia and South Asia to the east, and to North Africa and the Arabic Peninsula to the south. Either according to the demic model or the cultural model of the diffusion of agriculture, this diffusion of lentils occurred during the earliest period of agriculture diffusion since lentil remains have been recovered in several archaeological sites corresponding to the earliest agricultural sites in each of these geographical areas. For instance, there are lentil archaeological remains in the Iberian Peninsula since the early Neolithic, the oldest ones corresponding to the eastern part of the Peninsula. In particular in the cave "de les Cendres", there are remains of several crops (*Triticum monococcum, T. dicoccum, T. aestivum*, barley, pea, grass pea, lentil and faba bean, i.e. a typical Near East crop complex) dated by ^{14}C to 7540 ± 140 BP. According to the seed size, the archaeological lentil findings in the Iberian Peninsula fall within the range of the microsperma sizes (Buxó 1997).

The use of lentil was probably popular in prehistoric times around the Mediterranean. For instance, lentils were best represented among pulses in some Iron Age archaeological sites in the eastern part of the Iberian Peninsula (Alonso 2008). Lentils were well-known to the ancestors of most of the modern European nations from time immemorial. The majority of the words denoting lentil in modern Indo-European languages of Europe originate from very few Proto-Indo-European roots, with a subsequent diversification within each of its branches. The Proto-Indo-European root responsible for the words denoting lentil in the largest number of its both extinct and modern descendants is **lent-*, with the same unchanged meaning (Mikic 2010).

The arrival of lentils to the Nile delta had to be much earlier than the archaeological remains suggest due to its proximity, both geographical and cultural, to the Fertile Crescent; but the delta environment is not helpful in preserving organic materials. Ethiopia was probably reached from the Arabian coast rather than by the Nile. Lentils did not reach India before 4000 BP, probably carried by the Indo-European invasion through Afghanistan (De Candolle 1882), but again there is the problem of availability of archaeological findings. That primeval introduction was probably performed by very small samples of a common origin as the variability found in the Indian subcontinent in the local landraces is very limited despite being the largest lentil growing region in the world (Cubero et al. 2009). The asynchrony in flowering of the local pilosae ecotype is probably a consequence of a long reproductive isolation period now being broken by plant breeding methods designed to broaden the genetic base (Erskine et al. 1998). The crop was introduced to Spanish ruled America during the XVIth century.

According to Cubero et al. (2009), if lentils have been maintained by farmers through ages, it is most likely because they grow in poor soils, rough climates, and harsh conditions for humans, animals and crops. In many cases, they may be the only source of protein available to them.

3.1.5 Germplasm and GenePools

Managed germplasm collections are available for many orphan legume species, and characterization of genetic diversity within these collections is a necessary prelude to their efficient use (Varshney et al. 2009). There are at least 43 lentil germplasm collections available for breeding and research purposes. Muehlbauer et al. (1995) reviewed the most important institutions maintaining lentil collections. Prominent among them are the International Center for Agricultural Research in the Dry Areas (ICARDA; http://www.icarda.org), the Indian Agricultural Research Institute (http://www.iari.res.in), the N I Vavilov Research Institute of Plant Industry (http://www.vir.nw.ru), and the USDA collection at the Regional Plant Introduction Station, Pullman, Washington (http://www.ars.usda.gov); all of them conserving more than one thousand accessions. Another national Germplasm Bank holding an important collection of lentils is the Spanish Plant Genetic Resource Center (http://www.inia.es) with more than 600 accessions. A complete picture of the genetic resources available can be obtained in the Germplasm Collection Directory web page of Biodiversity International (http://www.bioversityinternational.org/Information_Sources/Germplasm_Databases/Germplasm_Collection_Directory/), which integrates information from most of the world germplasm collections. For instance, the query on *Lens culinaris* landraces and traditional cultivars generate a list of 43 GeneBanks and Institutions holding a total of 12,097 accessions. A query on wild or weedy forms of *Lens* yields a list of six Institutions and 296 accessions (Table 3-4). The URL address http://gisweb.ciat.cgiar.org/GapAnalysis/?p=271 describes gaps in the collected *Lens* genepool and provides recommendations on new collecting priorities, and particularly *L. tomentosus*, which is given high priority.

Table 3-4 Summary of the *Lens culinaris* materials registered in the database of Bioversity International.

Type of material	Number of Institutions	Number of accessions
Advanced cultivars	20	2,772
Breeding and inbred lines	12	1,819
Cultivated	1	2
Genetic stocks	0	0
Introgressed forms	0	0
Landrace or traditional cultivar	43	12,097
Mutants	0	0
Wild\Weedy species	6	296

It is remarkable that no genetic stocks, introgressed forms and mutants are listed among the available materials. Nevertheless, obtaining new useful mutants by artificial mutagenesis has been attended in lentils. Chemical mutagens and ionizing radiation have been used to produce mutant lentil released for commercial production (review by Toker et al. 2007), and gamma rays induced mutations have been used to obtain and release new varieties of outstanding performance ("NIA- MASOOR-05") in Pakistan (Ali and Shaikh 2007). Moreover, lentil core collections have been developed from the ICARDA and USDA-ARS collections (Varshney et al. 2009).

Hybridization barriers support the idea of six differentiated *Lens* species. *L. culinaris* ssp. *orientalis* obviously belongs to the primary genepool and *L. odemensis* rather to the second. Although the success in such crosses (needing embryo rescue or not) depends on the particular accessions used in the study. *L. nigricans* and *L. ervoides* belong to the tertiary genepool, but can become part of the secondary genepool by means of embryo rescue. Further hybridization studies are needed to establish whether *L. tomentosus* and *L. lamottei* belong to the secondary or tertiary genepool (Cubero et al. 2009). Wild *Lens* species are a source of donor genes. Resistance to Ascochyta blight was found in an ICARDA wild germplasm collection: 24 out of 86 accessions of *L. culinaris* ssp. *orientalis* were resistant, as were 12 out of 35 accessions of *L. odemensis*, 3 out of 35 accessions of *L. nigricans*, and 36 out of 89 accessions of *L. ervoides* (Bayaa et al. 1994). Ye et al. (2002) listed a series of cultivated and wild lentil materials as sources of partial resistance to Ascochyta blight and Muehlbauer and Chen (2007) reported that the most notable of the partially resistant germplasm accessions are lentil accessions PI 339283, PI 374118, ILL5588, ILL5684, PR86-360, and ILL7537. Wild materials have also been used as a source of genes for winter hardiness. An evaluation of 245 accessions of wild lentils and 10 cultivated lentils in Turkey showed that, on average, *L. orientalis* exhibited the highest level of winter hardiness; whereas, accessions of *L. ervoides* were the most susceptible (Hamdi et al. 1996). Cultivated material with large seeds was found in West Asia to be less susceptible to winter cold than that with small seeds, and the seed yield of the large-seeded group was better than the small seeds in average temperatures below 10°C from January to April, with the converse true at higher temperatures (Erskine 1996).

Resistance to fungal diseases (wilt, powdery mildew, and rust), as well as to other agronomic traits, was evaluated in wild taxa, finding that *L. nigricans* had the maximum number of resistant accessions for biotic and tolerance to abiotic stresses (Gupta and Sharma 2006). Tullu et al. (2006) tested the six wild *Lens* species for resistance to *Colletotrichum truncatum*. Accessions of *L. ervoides* revealed the highest level of resistance (3–5) to race Ct1 and Ct0 followed by *L. lamottei* in both field and greenhouse assays. *L. orientalis, L. odemensis, L. nigricans* and *L. tomentosus* were highly susceptible (8–9) to race Ct0 in the greenhouse. The highest levels of resistance were found among *L. ervoides* accessions originating from Syria and Turkey.

Recently, resistance against the parasitic plant *Orobanche crenata* was found in wild *Lens* materials (Fernández-Aparicio et al. 2009). The screening of 23 wild accessions for resistance to *O. crenata* under field conditions showed a wide range of responses from complete resistance to high susceptibility. The higher levels of resistance were observed in accessions of *L. ervoides, L. odemensis* and *L. orientalis*. Likewise resistance against *Sitona crinitus* Herbst, a major insect-pest limiting lentil productivity mainly in the countries of West Asia and North African region, was found in wild species (*L. ervoides, L. odemensis, L. nigricans* and *L. c.* ssp. *orientalis*). Eight accessions were identified as resistant, with nodule damage lower than 10%, compared to more than 56% damage recorded on the cultivated lentil (El-Bouhssini et al. 2008).

Ladizinsky et al. (1990) described tertiary trisomic lines of lentils with *L. orientalis* and *L. ervoides*. Trisomic plants were weaker than euploid plants and many did not reach maturity. There are no further records of its use in genetic analyses or breeding. Colchicine-induced lentil polyploids have been reported (Singh et al. 1992). Polyploidy results in gigantism for most determinate plant parts but causes reduced fertility and yield, although fertility and yield increased dramatically after selection from C_1

to C_4 generations. However, no commercial tetraploid lentil varieties have been described. Dixit and Dubey (1986) described the effect of alkylating agents generating chromosome mutations, mainly translocations, after seed soaking treatments with the agents and observing plant meiosis.

3.1.6 Botanical Descriptions: Taxonomy, Karyotype, Ploidy Levels, Genome Size

The genus *Lens* Miller is included in the Tribe Vicieae DC. (syn. Fabeae Rchb.) (Fabaceae, Papilionoideae) together with three other genera, namely: *Lathyrus* L., *Pisum* L., and *Vicia* L. Other genera, such as *Cicer* L. or *Vavilovia*, have been sometimes included (Gunn and Kluve 1976), but recent phylogenetic analyses excluded *Cicer* from this tribe, and the inclusion of *Vavilovia* within it is also discussed. The Tribe Vicieae is included in the inverted repeat-lacking clade (IRLC) (Wojciechowski et al. 2000), so-called because it is uniquely marked by the loss of one copy of the large inverted repeat (approx. 25 kb) in the chloroplast genome. The clade includes all members of tribes Cicereae, Hedysareae, Trifolieae, and Vicieae, as well as at least three other genera (Wojciechowski 2006; Jansen et al. 2008). Thus interesting crop legume genera such as *Trifolium, Medicago*, and *Cicer* are included in close phylogenetic groups to lentil and other Vicieae species.

The C (unreplicated haploid) genome size of lentil was determined by flow cytometry in an amount of 4.41 pg equivalent to 4,063 Mbp (Arumuganathan and Earle 1991). This amount agrees with a previous evaluation of 4.6 pg using Feulgen microdensitometry (Bennet and Smith 1976), and is similar to the genome size of closely related species such as pea (3,947–4,397 Mpb). Neither the mitochondrial nor the chloroplast genomes of lentil have been sequenced to date according to available data in the NCBI Eukaryote Organelles and the Organelle Genome Megasequencing Program web sites. Although a comparative restriction fragment length polymorphism (RFLP) analysis of the chloroplast of the *Lens* species has been carried out (Van Oss et al. 1997).

All lentil species have the same chromosome number, $2n = 14$, and share more or less the same karyotype (Ladizinsky 1993). According to this author the lentil karyotype consists of three pairs of metacentric and submetacentric chromosomes, three pairs of acrocentric chromosomes, and a pair of metacentric chromosomes with the nucleolar organizing region (NOR) secondary constriction proximal to the centromere location. Balyan et al. (2002) described the karyotype of *L. culinaris* as constituted by two metacentric chromosomes, one of them with the proximal NOR, two submetacentric chromosomes and three acrocentric chromosomes. This karyotype agrees with the general *Lens* karyotype described by Ladizinsky (1993); however Balyan et al. (2002) found in other *Lens* species karyotypes with variations in relation to this general one. In addition to the standard ssp. *culinaris* karyotype, Ladizinsky (1993) described other karyotypes in the ssp. *orientalis*: a single accession (No 113) showed a karyotype in which about three quarters of the satellite was translocated to another chromosome and the metacentric-satellited chromosome became acrocentric. Accessions from Iran differed by chromosomal rearrangement involving two metacentric chromosomes, and accessions from Turkey differed by a paracentric inversion and two or even three chromosome rearrangements. Accordingly, ssp. *orientalis* appears to be the most chromosomally variable *Lens* taxon (Ladizinsky 1993). Balyan et al. (2002) using fluorescence *in situ* hybridization (FISH) identified chromosome 3 as the NOR chromosome, chromosome 2 as the submetacentric chromosome with a 5S locus distally located on the long arm (the major site of hybridization) and chromosome 6 as the acrocentric chromosome with the other 5S locus located proximal on the short arm (minor site). Further data based on *Lens* homologous probes, which differentially hybridize with the 5S non-transcribed spacer (NTS), allowed better differentiation between the two 5S loci (Fernández et al. 2005a). Locus 5S on chromosome 2 corresponds to the longer NTS family sequence while the other on chromosome 6 corresponds to the shorter NTS family. Further, the FISH data indicated that in some wild ssp. *orientalis* accessions (e.g., ILWL-7), 3/4th of the satellite were transferred to another chromosome, the metacentric-satellited chromosome became acrocentric and one of the submetacentric chromosomes was lengthened. In this *orientalis* accession a single 5S locus was detected using the longer NTS probe on chromosome 6, thus it is possible that some ssp. *orientalis* karyotypes lack one of the 5S loci (Fernández et al. 2005a).

3.2 Classical Genetics and Traditional Breeding

Mating system is a major determinant of the genetic structure of plant populations and of the breeding methods in crops. Lentil is a highly predominant self-pollinating species in which outcossing rates are usually low. The outcrossing rate of lentils ranged from 0.06 to 5.12% among three varieties as measured in central Europe using a dominant orange cotyledon allele as a genetic marker. The outcrossing rate of individual plants ranged from 0 to 22.2%. Results were strongly influenced by cultivar, year and location, thus, for instance, the outcrossing rates of the cultivar Pisarecka Perla in the same locality were 0.89, 5.12 and 3.01% in the years 1999, 2000 and 2001, respectively. These outcrossing rates agree with previously published estimates ranging from 0.01 to 5.66% (Horneburg 2006).

3.2.1 *Classical Genetic Mapping*

The inheritance of several morphological and agronomical traits in lentil was first described at the end of 70's and early 80's (Haddad et al. 1978; Ladizinsky 1979a; Muehlbauer and Slinkard 1981). Several traits such as seed coat color, epicotyl and flower color, and pod dehiscence, were found to be monogenic and thus appropriate to be used as morphological markers. Linkage analysis in lentil started in the 80's and was initially based on morphological and isozyme markers. The first genetic linkage analysis in lentil was reported by Zamir and Ladizinsky (1984). They mapped one morphological and eight isozyme markers in two linkage groups (LGs) using segregations obtained from intersubspecific crosses between *L. c. culinaris* and *L. c. orientalis*. Later on, in an interspecific cross between *L. c. culinaris* and *L. ervoides*, Tadmor et al. (1987) determined five LGs in which they mapped two morphological, nine isozyme markers, four seed protein loci and a translocation breakpoint. Muehlbauer et al. (1989) described the allozyme polymorphisms for 18 loci linkage, demonstrating their monogenic inheritance, and the linkage relationships among them and with four genes controlling morphological traits. The linkage analysis was carried out with seven F_2 mainly from intersubspecific crosses with *L. c. orientalis* and one interspecific cross with *L. odemensis*, and resulted in six small LGs, which contained 14 of the loci analyzed. This work also contributed to the first evidence of shared synteny between *Lens* and *Pisum* since several of the LGs were conserved between both species. Tahir and Muehlbauer (1994) were the first to use recombinant inbred lines (RILs) for mapping lentil markers. They identified six LGs encompassing a total of 17 isozyme loci and four morphological trait loci.

3.2.2 *Limitations of Classical Endeavors and Utility of Molecular Mapping*

The low amount of variation detected in the cultivated species was a limitation in using intraspecific crosses for linkage mapping purposes, thus genetic maps were, and to some extend still are, based on intersubspecific or interspecific segregating populations with a consequently large proportion of segregation distortion among markers. Furthermore, some of the morphological markers and other traits initially described as monogenic seem now to be of oligogenic nature. For instance, pod dehiscence, which in the segregating populations shows several degrees between total dehiscent and non-dehiscent, has been classified in IBPGR/ICARDA (1985) lentil descriptor into four classes on a scale from 0 to 9 when scored one week after maturity.

However, classical mapping efforts using intraspecific segregating populations of lentil have continued to establish linkage relationships among morphological markers. In this respect, Sarker et al. (1999) found linkage between loci controlling pod pubescence (*Pep*), seed coat pattern (*Scp*) and flowering time (*Sn*), while Hoque et al. (2002) described linkage between leaf color (*Dgl*) and plant pubescence (*Pub*). More recently, Kumar et al. (2004a, 2005a, b) have described two LGs, the first containing the loci for stem pigmentation (*Gs*), leaf pigmentation (*Bl*), pod pigmentation (*Rdp*) and erect growth habit (*Ert*), while the second LG included leaf color (*Gl*), plant pubescence (*Pub*), leaflets per leaf (*Hl*) and plant height (*Ph*).

3.2.3 Breeding Objectives

Lentil breeding programs throughout the world have similar objectives with larger and more stable seed yield being the most important. Increasing yield often entails breeding for adaptation to environmental stress factors such as heat and drought. However, breeding for resistance to diseases and pests is a common breeding objective. Priorities and breeding goals usually differ between regions depending on specific problems and special considerations related to the needs of farmers and consumer demands. While increasing yield per hectare can be the first priority in some highly import-dependent countries, in some developed countries obtaining stable yields while maintaining the visual and quality characteristic of traditional and highly valued varieties is a usual priority. Although mechanical harvesting is common in countries with large-farm lentil production, breeding suitable varieties for mechanical harvesting in smaller fields is still a major goal in some developing countries. Lentil straw is valued as animal feed or as an aid in soil erosion control in some countries, while in others lentil straw has minimal or no value. On the other hand, increased seed yields, improved disease resistance, and improved seed quality are key breeding goals in the major exporting countries (Muehlbauer et al. 1995).

Muehlbauer et al. (2006) pointed out that several abiotic stresses (such as drought, heat, salt susceptibility, iron deficiency, among others) are important constraints in lentil production worldwide. The severity of these stresses is unpredictable in field experiments, so field trials are increasingly supplemented with controlled environment testing and physiological screening. Lentils are mainly grown as a rain fed crop, and often are seeded in marginal areas where limited rainfall and deficient soil moisture are encountered. Thus, lentil breeding for improved drought resistance is a major goal in most regions where lentils are grown and particularly in areas with limited and erratic rainfall such as the Middle East and areas of the Mediterranean region. Drought-tolerant cultivars are required to stabilize lentil production and also to extend cultivation to those dry areas.

Muehlbauer et al. (1995) indicated that screening for drought tolerance was difficult because of the lack of knowledge of what to screen for and the absence of efficient screening techniques.

Screening for drought tolerance has improved with controlled environment testing and physiological screening. In a recent review on screening techniques and sources of resistance to abiotic stresses in cool-season food legumes (Stoddard et al. 2006), it was stated that over the past decade, there has been considerable progress at increasing the adaptation of lentil to autumn-sowing and there is a large body of literature on drought responses, particularly in chickpea and faba bean. The available literature on salinity tolerance and waterlogging is minimal. The review of Stoddard et al. (2006) also reviewed the information on the identified lentil sources of resistance or tolerance to drought and heat stress [e.g., MI-30, a mutant line showing high yields at limited water supply identified by Salam and Islam (1994)], to chilling or freezing stress (Hamdi et al. 1996; Ali et al. 1999; Kahraman et al. 2004a), to salinity stress (Mamo et al.1996; Ashraf and Zafar 1997; Rai and Singh 1999), to waterlogging (Ashraf and Chishti 1993), and to boron concentration (Hobson et al. 2006). Another important component in drought resistance is the plant root system. Sarker et al. (2005) analyzed the variation in shoot and root characteristics and their association with drought tolerance in several lentil landraces. Stem length, taproot length and lateral root number were highly correlated, both amongst themselves and with yield. High heritability estimates provided reliability in screening based on these traits. Among a total of 40 genotypes, only one line (ILL 6002) was strikingly different from all the other tested genotypes. Another approach to drought resistance screening is to select for drought avoidance through early maturity, particularly when the major moisture stress occurs at the end of the growing season. Early planting can contribute to this avoidance since lentil plants usually develop and initiate flowering before serious drought conditions are reached. This is the case of fall-seeding versus spring seeding lentil in cold areas. On the other hand, lentils seem to be among the least tolerant cool season legumes to transient waterlogging in greenhouse experiments (Solaiman et al. 2007).

A factor for increasing the production in the Mediterranean region and other areas with cool winters is earlier sowing dates and often sowing in fall rather than at the end of winter. Available winter hardy lentil germplasm has prompted interest in the development and use of cultivars that can be fall planted in cold highland areas. This change in production of lentils from normally spring sowing to fall sowing increases yield potential (Kahraman et al. 2004a). Early sowing has proved to increase lentil production in several countries, for instance, by shifting sowing from spring to early spring or to autumn-sowing in areas such as the highlands of Turkey, Iran, Afghanistan, and Balochistan in Pakistan. Early fall-seeded of lentil increased yield from

480 to 590 kg ha^{-1} when compared to early spring sowing of non-winter hardy varieties in the US Pacific Northwest and northern Great Plains of Canada. Fall sowing in Australia allowed a longer period for vegetative and reproductive growth, rapid canopy development, greater absorption of photosynthetically active radiation, and better water use resulting in greater water use efficiency, increased dry matter and improved seed production when compared with later sowing. Early-sown lentils began flowering and forming seeds earlier in the growing season, at a time when vapor pressure deficits and air temperatures were lower, and used more water in the post-flowering period when compared to those treatments where sowing was delayed (Siddique et al. 1998; Chen et al. 2006). Delaying spring seeding in the Northern Great Plains of Canada by 4 weeks negatively affected crop development parameters due to the exacerbation of summer drought stress. Seed yields decreased at the majority of site-years and the reduction averaged 38%. Seeding delays of 2 and 4 weeks decreased seed size averaging 10% (Miller et al. 2006). However, early sowing increased the incidence of some fungal and bacterial diseases (Davidson and Kimber 2007).

According to Muehlbauer et al. (1995), diseases of lentil are in general less damaging than those of most other food legume crops. However, there are some important and potentially devastating diseases (Taylor et al. 2007). Several reviews on screening techniques and sources of resistance to root and foliar diseases in cool season grain legumes have been recently published (Infantino et al. 2006; Sillero et al. 2006; Tivoli et al. 2006).

The root rot/wilt complex is probably the most important disease problems of lentil worldwide (Muehlbauer et al. 1995). This set of diseases is caused by several fungi of the genera *Fusarium, Pythium, Rhizoctonia* and *Sclerotinia*, in particular *F. oxysporium* f. sp. *lentis, R. solani* and *S. sclerotiorum*. Another fungal species, which can be included in this complex, is *Aphanomyces euteiches*, a soil-borne pathogen causing Aphanomyces root rot of several legumes, including alfalfa, bean, lentil, and pea. Relatively little is known about the population biology of this legume pathogen (Grunwald and Hoheisel 2006). Resistance to Fusarium wilt has been described in germplasm from India (Kamboj et al. 1990); and Bayaa et al. (1997) screened a core collection of 577 germplasm accessions from 33 countries and re-screened a subset of the 88 mostly resistant accessions. The inheritance of resistance to Fusarium vascular wilt was investigated by Eujayl et al. (1998) who found several random amplified polymorphic DNA (RAPD) markers linked to the *Fw* locus. Wang et al. (2006) found small but consistent differences among 14 lentil cultivars for their reaction to Rhizoctonia seedling blight and root rot caused by *Rhizoctonia solani* in greenhouse trials, and in field experiments conducted in Canada. *Fusarium redolens* has been also described as a causal agent of Fusarium wilt in lentil (Riccioni et al. 2008).

Rust caused by *Uromyces viciae-fabae* Pers. is a serious problem in areas where mild temperatures and humid conditions favor the development of this fungus (Muehlbauer et al. 1995). Incomplete (partial) resistance to *Uromyces* has been described in lentil cultivars (Negussie et al. 2005). These authors analyzed the response to four components of resistance, namely: latent period, infection efficiency, pustule size and spore production, and concluded that, since there was an interdependence of the components, selection based on more than one component should help in obtaining lines with higher levels of partial resistance. *Uromyces* resistant cultivars, such as Laird or Calpun-INIA derived from Laird, have been described (Penazola et al. 2007).

Ascochyta blight, a seed-borne disease described in at least 16 countries in five continents, is probably one of the most widely distributed and devastating disease. The fungus, *Ascochyta lentis* (syn. *A. fabae* f.sp. *lentis;* teleomorph: *Didymella lentis*), attacks leaves, stems and pods causing necrotic spots. In some cases kills the plants before harvest or infects seeds to such a degree that lentils are unmarketable. Ascospores of *Didymella lentis* can be found on lentil stubble under natural conditions, and the presence of the teleomorph has implications in the long-distance dispersal of *A. lentis* (Galloway et al. 2004). Species of the genus *Ascochyta* cause most of the major diseases of many cool season grain legumes and a considerable amount of information on the pathogens, plant hosts and disease resistance responses is available. For example, the European Journal of Plant Pathology (2007) published a monographic issue on *Ascochyta* (Ascochyta Blights of Grain Legumes).

The mating type (*MAT*) locus of *Ascochyta lentis* was cloned and characterized, and a multiplex PCR assay for mating type was developed based on *MAT* idiomorph and flanking sequences (Cherif et al. 2006). Molecular diagnostic techniques have been developed to differentiate the *Ascochyta* pathogens that infect cool season food and feed legumes, as well as to improve the sensitivity of detecting latent infection in plant tissues (Taylor and Ford 2007). Recent multilocus phylogenetic analyses of a worldwide sample of *Ascochyta* fungi causing Ascochyta blights of chickpea (*Cicer arietinum*), faba bean (*Vicia faba*), lentil (*Lens culinaris*), and pea (*Pisum sativum*) have revealed that the fungi causing disease on each host formed a monophyletic group. Host inoculations of these fungi demonstrated that they were host-specific, causing disease only on the host species from which they were isolated (Hernández-Bello et al. 2006; Peever 2007). Tivoli et al. (2006) provided a thorough review on the sources of resistance and screening techniques for Ascochyta blight of lentil and Ye et al. (2002) gave an account of breeding techniques for selection of lentils with resistance to Ascochyta blight. Partial resistance to the disease is available in the germplasm. The most notable lentil accessions among the partially resistant germplasm are PI 339283, PI 374118, ILL5588, ILL5684, PR86-360 and ILL7537.

Anthracnose is another important disease in lentil and is caused by the fungus *Colletotrichum truncatum* (Schwein) Andrus and Moore [teleomorph *Glomerella truncate*, described by Armstrong-Cho and Banniza (2006)]. A total of 1,771 lentil accessions from the US lentil collection (Pullman, WA) and the Institut fur Pflanzengenetik und Kulturpflanzenforschung (Gatersleben, Germany) were screened for resistance to *C. truncatum*. About 95% of the accessions were susceptible when inoculated with a single isolate in the field. Resistance was found in 16 accessions. The presence of two distinct pathogen races (Ct0 and Ct1) was demonstrated (Buchwaldt et al. 2004). Tullu et al. (2006) searched for sources of resistance to anthracnose in cultivated and wild materials and found resistant cultivars (Indianhead and PI 320937) to race Ct1 but none to race Ct0. Some accessions of wild *Lens* species are potential sources of genes for resistance to both the races. Transfer of resistance to both races, Ct1 and Ct0, from *L. ervoides* accession L-01-827A to the cultivar Eston by single-seed descent from F_2 individuals to $F_{7:8}$ RILs, has been recently described (Fiala et al. 2009).

Powdery mildew is a minor disease in lentil, although it can be severe on certain lentil cultivars and in some parts of the world. *Erysiphe pisi* is the powdery mildew pathogen that has been reported in several countries, but *E. trifolii* has been recently added as a causal agent of this disease in the Pacific Northwest of the United States (Attanayake et al. 2009).

Multiple resistant materials exist among both breeding and landraces. For instance, Laird is a macrosperma multi resistant cultivar and Assano is a recently released cultivar highly resistant to rust caused by *Uromyces viciae-fabae* with moderate resistance to Ascochyta blight and to wilt-root rot complex (*Fusarium* sp. and *Rhizoctonia* sp.) (Fikru et al. 2007). Moreover, Pakistan accession 66013-6 was found to have a high level of resistance to blight, rust and pea seed-borne mosaic virus (Hussain et al. 2008).

Lentils have proved to be susceptible to more than 25 viruses as listed in the Plant Virus Online (http://www.agls.uidaho.edu/ebi/vdie/). In a field experiment 12 of 16 lentil genotypes were ranked as highly susceptible and four as susceptible to the pea seed-borne mosaic virus (PSbMV). Once infected, plant sensitivities (symptom severities) were low in most lentil genotypes. Seed lots harvested from PSbMV-infected plants of lentil were usually less affected than other cool season grain legumes. Seed transmission of PSbMV (6%) was detected in seed from the infected lentil plants (Coutts et al. 2008). In a study to assess the variation among lentil germplasm for their reaction to natural virus infection and to identify sources of resistance against viruses, lentil accession and breeding lines were infected by pea seed-borne mosaic virus (PSbMV), cucumber mosaic virus (CMV), bean yellow mosaic virus (BYMV), broad bean stain virus (BBSV) and faba bean necrotic yellows virus (FBNYV). Out of 108 lentil germplasm accessions/ breeding lines tested, none of the genotypes was found as highly resistant. However, 30 (64%) and nine (35%) genotypes were found as resistant and moderately resistant respectively to viral disease infection. All other lines were susceptible to highly susceptible (Bashir et al. 2005). Some differences in the sequence of the *eIF4E* lentil gene have been found (Saenz de Miera, pers. comm.); this gene is related to virus resistance in pea and other plants species (Robaglia and Caranta 2006).

Another major threat for lentil cultivation in the Mediterranean area and western Asia is crenata broomrape (*Orobanche crenata*). This weedy root parasite constitutes a serious threat to lentils and other legumes. Control of broomrape is difficult because of the large number of wind-blown seeds which can be produced each year and which may remain dormant in soils for several years thereafter. Control strategy has centered on agronomic practices and the use of herbicides. Resistance breeding is hampered by scarcity of proper sources of resistance and of a reliable and practical screening procedure. A Spanish germplasm collection of 234 accessions of lentils was screened for resistance to crenate broomrape under field conditions. A wide range of responses was observed, however, complete resistance was not detected. Some accessions exhibited a substantial reduction in broomrape infection (Fernández-Aparicio et al. 2007). Other *Orobanche* species can also parasitize lentil plants (Muehlbauer et al. 1995).

Lentil is damaged by many types of insects and other pests. Among insects, the major pests in the field are aphids (*Aphis craccivora*, *Acyrthosiphon pisum*), leaf weevil (*Sitona* spp.), lygus bugs (*Lygus* spp.) and the cutworm (*Agrotis ipsilon*). Another major pest problem causing great seed losses are seed insect species: *Bruchus ervi* and *B. lentis* with *Callosobruchus chinensis* and *C. maculatus* (Stevenson et al. 2007). Little progress in breeding for insect resistance was indicated by Muehlbauer et al. (1995) and, in comparison with breeding for diseases, the situation is similar nowadays. There was no source of resistance to *Sitona* and ICARDA materials had shown to be only effective against the adult insect but not to larvae (Stevenson et al. 2007). However, resistance to *S. crinitus* Herbst was further described in wild materials (El-Bouhssini et al. 2008). An extensive search on the bruchid species infecting lentil was carried out for 10 years at the National Bureau of Plant Genetic Resources, New Delhi, where 2,517 imported lentil samples from 40 countries were X-ray screened for insects hidden in seeds. About 30% of the samples were infected and the species found were *Bruchus ervi* Froelich, *B. lentis* Froelich, *B. tristiculurs* Fahraeus, *Callosobruchus analis* (Fabr.), *C. chinensis* (L.) and *C. maculatus* (Fabr.) (Bhalla et al. 2004).

Genotypes with resistance to various biotic stresses, particularly resistance to vascular wilt, rust and Ascochyta blight have been identified, and directly exploited or used in breeding programs (Sarker and Erskine 2006).

Recently a clonal propagation protocol to evaluate disease data on infections has been proposed in lentil by Vail and Vandenberg (2010). They tested the feasibility of using clonal propagation of individual lentil plants to generate replicated ratings to evaluate the response to anthracnose (caused by *Colletotricum truncatum*), and proposed that this method may be more dependable than using single-plant evaluation.

Root characteristics are important in nutrient uptake, and this is particularly relevant in lentil culture in developing countries where often they are cultivated in poor soil with little or no fertilization. Differences in root traits exist in lentil and uptake of most nutrients differ significantly among genotypes, but root length was, in general, weakly correlated to the uptake of most nutrients (Gahoonia et al. 2006).

In the semiarid regions available water is the most limiting factor influencing crop production for rainfed crops, even more than the availability of nitrogen. Lentils have been traditionally planted in rotation with cereals (wheat, barley, etc.) not only in traditional subsistence farming but also under modern farming systems. The advantage of lentils in these rotations is not only from the nitrogen afforded to the soil by symbiotic nitrogen fixation but also in the efficiency of water use (Campbell et al. 2007; Pala et al. 2007).

The consumption of organic farming products is increasing in some countries that create new breeding opportunities for improving lentil yield and quality under organic culture conditions. Vlachostergios and Roupakias (2008) found lentil genotypes with higher yielding ability under the organic management in a collection of 36 varieties from several countries, opening new opportunities for breeding.

There has been increasing interest in sustainable disease, weed, and pest management in low-input farming systems. In an integrated approach, the development of cropping systems such as appropriate spatial arrangement and efficient tillage will help crops themselves to compete with weeds. Another issue which can contribute to more sustainable agriculture is the use of biopesticides, biofertilizers and the biological control of diseases. These trends are opening new lines in breeding goals in many crops including lentils. Living and heat-killed bacterial cells of *Rhizobium leguminosarum* totally protected lentil plants against infection by *Fusarium oxysporum* MR 84 (Essalmani and Lahlou 2003). Seed treatments with formulations of *Bacillus subtilis* on glucose, talc and peat decreased disease severity by reducing plant colonization by *F. oxysporum*, promoting lentil plant growth and increasing the dry weight (El-Hassan and Gowen 2006). Several rhizobacteria isolated in western Canada promoted the growth of lentil or suppressed the growth of pathogens such as *Pythium* sp., *Fusarium avenaceum* or *Rhizoctonia solani* (Hynes et al. 2008).

The root-knot nematode *Meloidogyne javanica* can be biocontrolled on lentil using four plant growth-promoting rhizobacteria (*Pseudomonas putida, P. alcaligenes, Paenibacillus polymyxa* and *Bacillus pumilus*) and root nodule *Rhizobium* species (Siddiqui et al. 2007). Huang and Erickson (2007) reported that seed treatment with *R. leguminosarum* bv. *viceae* is effective in control of *Pythium* damping-off of pea and lentil and that the efficacy of control is strain specific, thus, among the strains tested, R12 was useful for the control of this disease in lentil.

Co-inoculation with arbuscular mycorrhizal fungi and some rhizobial strains varying in phosphorus-solubilizing ability affects plant growth and nutrient uptake in a calcareous soil with low amounts of available phosphorus and nitrogen (Zarei et al. 2006). The results showed that the effects of arbuscular mycorrhizal fungi, rhizobial strains and phosphate fertilizers were highly significant ($p < 0.01$) for all the characteristics studied. Likewise, growing lentil cv. Titore with compatible arbuscular mycorrhizal inoculum (*G. mosseae*) can minimize arsenic toxicity (a major constraint due to contamination of irrigation water in some countries such as Bangladesh) and increase growth and phosphorus uptake (Ahmed et al. 2006). The beneficial effects of nodulation can be enhanced when rhizobial inoculation is combined with plant-growth-promoting bacteria. The strain *Bacillus thuringiensis*-KR1 was found to promote plant growth of field pea (*Pisum sativum* L.) and lentil (*Lens culinaris* L.) under several conditions when co-inoculated with *Rhizobium leguminosarum*-PR1. The enhancement in nodulation due to co-inoculation was 73.3% in lentil compared to *R. leguminosarum*-PR1 treatment alone (Mishra et al. 2009).

Defensins are a class of host defense peptides which participate in the host defense systems of plants, invertebrates and mammals. Plant defensins are small (45–54 amino acid residues), basic, cysteine-rich peptides and have comparable three-dimensional structures (Broekaert et al. 1995). Most of them were characterized in seeds despite being expressed in various plant tissues. Their amino acid sequences are characterized by the presence of eight structurally important and conservative Cys residues forming four disulfide bridges. Like many other pathogenesis-related (PR) proteins, plant defensins are encoded by multigenic family, and their expression is induced by environmental stress factors (Broekaert et al. 1995; García-Olmedo et al. 1998). Plant defensins are characterized by a broad spectrum of biological activities including inhibition effects on pathogens (bacteria and fungi), digestive enzymes (trypsin and α-amylases), HIV-1 reverse transcriptase and protein translation. The defensin VrD1 from mung bean (VrCRP) displays insecticidal activity (Finkina et al. 2008). A promising effect of a sunflower defensin Ha-DEF1 against *Orobanche* parasitic plants was described (Zelicourt et al. 2007). Recently, a defensin named Lc-def was isolated from lentil germinated seeds shown to inhibit *Aspergillus niger* growth but did not affect trypsin or chymotrypsin activities. The peptide belongs to a new, yet unclassified group of plant defensins (Finkina et al. 2008). The diversity of biological activities of plant defensins points to their medical and agricultural potential value (Finkina et al. 2008). Likewise a peptide, with a molecular mass of 11 kDa, was isolated from dry seeds of red lentils which inhibited mycelial growth in *Mycosphaerella arachidicola* and also exhibited antifungal activity against *Fusarium oxysporum* (Wang and Ng 2007).

3.2.4 Achievements and Limitations of Traditional Breeding: Molecular Breeding

In comparison with old classical early-domesticated crops, including some other legume species, systematic lentil breeding research started later (Sarker and Erskine 2006). This review article stressed that since 1970s there have been significant national and international lentil breeding programs and that research progress has been made in various aspects of the crop. The main objective was to obtain phonologically adapted, stress resistant and high-yielding cultivars. Traditional breeding in lentil has increased and/or stabilized yield by development of cultivars that are able to cope with major biotic or abiotic constraints. Noticeable achievements have been reached, mainly in relation to resistance-tolerance to diseases and some abiotic factors, as it has been pointed out in the previous section and in recent reviews (Muehlbauer et al. 2006; Sarker and Erskine 2006; Materne and McNeil 2007). As examples of achievement it is worth mentioning that ICARDA alone has released 91 improved cultivars during the past three decades. Improved cultivars together with new production technologies have increased productivity from 611 kg ha^{-1} to 966 kg ha^{-1} (Sarker and Erskine 2006). However traditional lentil breeding is still limited by: 1) the apparent minimal genetic variability and relatively few described Mendelian genes; 2) the difficulties in realizing crosses by hand emasculation and pollination (that means, for instance, that gene introgression using the backcross method of breeding is difficult); 3) the lack of information on genetic, physiological, and other aspects of the crop; and 4) the lack of screening methods under controlled conditions. Traditional breeding has often been directed to address major production constraints, which limit consideration of other possible breeding goals.

Molecular markers may be a means of overcoming these breeding limitations and rapidly improve the relatively neglected lentil crop. In general, molecular markers and gene tagging has provided plant breeders with a means to accelerate breeding programs and represent an efficient tool for indirect selection of traits that would otherwise be difficult to select using more conventional procedures. Many of the PCR-based markers are available and much more will be available in the future. High efficient analysis techniques such as microarrays will be available soon (see Section 3.1.2). Genetic transformation useful for breeding purposes seems to be still far away. TILLING and transposon-tagging will probably be available in a shorter period than transformation (e.g., lentil transposons are being systematically analyzed at the University of Leon, Spain). All of them hold promise for improving our knowledge of the lentil genome and accelerating progress through breeding.

The implementation of markers for routine use in lentil breeding programs is currently very limited. To date, much of the focus has been on developing molecular markers for selecting resistance to a few major diseases, such as Ascochyta blight, but in the near future, the potential exists to develop markers for many other highly sought traits that are difficult to breed conventionally. The key to both research and implementation of markers for lentil lies in the integration of the markers within the breeding program to ensure that cost effective utilization of the technology is achieved. Globally, trait selection that would greatly benefit from the availability of robust and accurate markers includes: drought, frost, boron and salinity tolerance, and resistances to Ascochyta blight, anthracnose, rust, Fusarium and other diseases (Muehlbauer et al. 2006).

3.3 Diversity Analysis

3.3.1 Phenotype-based Diversity Analysis

The first detailed and complete study of the cultivated lentil was made by Barulina (1930), a disciple of Vavilov. She considered two subspecies, *microsperma* and *macrosperma*, according to the size of the seed, the main objective of human selection. Barulina also considered the geographical distributions of cluster of characters, defining regional groups or *greges*. The main characters used to define subspecies were pods, seeds and distinct differences in the length of flowers. Secondary characters included size of leaflets, length of vegetation and height of plants. Important characters with which to define the groups within the subspecies included dehiscence, length of calyx teeth, and number of flowers per peduncle (Cubero et al. 2009). A study of the quantitative agro morphological variation in germplasm from 13 major lentil producing countries identified three major regional groups (Erskine et al. 1989). The close resemblance between germplasm from adjacent climatically similar countries indicates that environmental conditions are the major evolutionary force in the cultivated lentil.

Furthermore, it appeared that phenological adaptation to the environment is central to the evolution of the crop. Landrace accessions from India and Ethiopia were together in one such group, strikingly similar for the nine quantitative traits surveyed. The group was characterized by early flowering and maturity, low biological yield, short stature and lowest pod height and small seeds. This similarity for quantitative traits was previously noted by Barulina (1930), but she indicated that germplasm from these countries contrast markedly for qualitative characters with the Ethiopian material characterized by the endemic elongated pod apex and the Indian material characterized by *pilosae* traits. The regional grouping in lentil has a practical breeding use in the selection of parental stocks from germplasm collections (Erskine et al. 1989).

South Asia is the largest lentil growing region in the world, however, indigenous lentils (the specific ecotype is *pilosae*) show a marked lack of variability. According to Erskine et al. (1998) this lack of variability is the result of a genetic bottleneck from its introduction from Afghanistan around 2000 BC (most likely a founder effect). Further evidence of the paucity of variation within Indian germplasm comes from a study of the sensitivity of a world lentil collection to temperature and photoperiod (Erskine et al. 1990). In the agro-morphological study (Erskine et al. 1989), germplasm from Afghanistan was found markedly different to the Indo-Ethiopian group and had greater similarity to germplasm from Iran and Turkey. Afghan accessions were among the latest to mature of all the accessions tested. Germplasm from Pakistan was of the *pilosae* group for qualitative characters, but intermediate between Afghan and Indian material for quantitative traits. Strikingly, countries with much smaller area dedicated to lentil (e.g., Lebanon and Chile) than South Asian countries (e.g., India and Pakistan) showed a greater mean coefficient of variation over nine morpho-agronomic characters (Erskine et al. 1998).

3.3.2 Genotype-based Diversity Analysis, Molecular Markers Applied

The implementation of markers for routine use in lentil breeding programs is currently very limited and much of the focus has been on developing molecular markers for selecting resistance to a few major diseases, such as Ascochyta blight (Muehlbauer et al. 2006). The introduction of biochemical and molecular markers in the analysis of diversity in lentil represents a high step forward in this purpose since it allows in overcoming the limitations caused by the relative low number of morphological characteristics used previously, which further were poorly understood in relation to their genetic control. Isozymes and storage proteins were first used in the 1980s to study their genetic control, difference between *Lens* species and between lentil germplasm collections and geographical distribution (Skibinski and Warren 1984; Zamir and Ladizinsky 1984; Pinkas et al. 1985; Hoffman et al. 1986; Muehlbauer et al. 1989; Rosa and Jouve 1992). Nevertheless, their use rapidly declined when DNA markers were introduced. Restriction fragment length polymorphism (RFLP) markers were the first type of molecular marker used in lentil genetic and linkage analyses (Havey and Muehlbauer 1989). Subsequently, random amplified polymorphic DNA (RAPD) was used, among other purposes, to study diversity, phylogeny and taxonomy of *Lens* (Ford et al. 1997; Ferguson et al. 1998a, b, c). Arbitrarily produced amplified fragment length polymorphism (AFLP) markers have also been used to study genetic diversity (Sharma et al. 1996) and to differentiate cultivars (Závodná et al. 2000). More recently, simple sequence repeat (SSR or microsatellite) and inter-simple sequence repeat (ISSR) markers have been included in the list of markers used in lentils. Sets of microsatellite markers have been developed by Závodná et al. (2000) and Hamwieh et al. (2005, 2009) at ICARDA, others being currently produced at the Universidad de León (R de la Puente, unpubl. data). A whole transcriptome list of EST-SSRs has been recently published (Kaur et al. 2011). Recently, cross-genera SSRs have been used in a diversity analysis of *Lens* species (Reddy et al. 2010). The use of molecular markers has been recently reviewed (Muehlbauer et al. 2006). Recent papers on lentil diversity analysis, many of them analyzing local or regional germplasm collections, have been published. Thus, Italian and Spanish materials have been analyzed using RAPDs and ISSRs (Sonnante and Pignone 2001, 2007; Duran and Pérez de la Vega 2004) or AFLPs (Fiocchetti et al. 2009). Besides, ISSR and two-dimensional polyacrylamide gel electrophoresis (PAGE), in addition to morphological characters, were the basis to distinguish between two appreciated (good qualaity) Italian landraces (Scippa et al. 2008). Seed proteins were used to cluster Turkish cultivars (Yuzbasioglu et al. 2008) and SSRs were used to analyze a central Asia and Caucasian collection (Babayeva et al. 2009) and a core collection from the ICARDA (Hamwieh et al. 2009). ISSRs and AFLPs were also used for the analysis of landraces from Southeast Turkey (Toklu et al. 2009). These authors pointed out that Turkish landraces exhibited high genetic variation ($H_e = 0.180$) and found that one landrace from Karacadag/Diyarbakir region was significantly different from the rest of the germplasm analyzed. The Karacadag mountain range, in Southeast Turkey, is located within the core area of plant domestication in the Fertile Crescent. Lev-Yadun et al. (2000) suggested this region as the lentil domestication center. Furthermore, two centers of diversity for *L. culinaris* ssp. *orientalis* have been described: 1) Southeast Turkey and Northwest Syria, and 2) West and North Jordan and southern Syria (Ferguson et al. 1998a). Recently, Inder et al. (2008) introduced lentil-specific sequenced tagged microsatellite site (STMS) markers for genotyping elite lentil materials in Australia. The extensive use of additional molecular markers types such as single nucleotide polymorphisms (SNPs) will further improve the usefulness of markers on lentil for basic genetic studies and breeding.

A systematic comparative study on the extent of genetic variation in *Lens* is lacking, but it is generally considered that the wild species have more genetic variation when compared to cultivated material. Also, when compared to other diploid sexual-reproduced cultivated crop species, lentils would have less variation but similar to that of other cool season grain legumes. A recent paper (Alo et al. 2011) assess diversity in the genus *Lens*.

3.3.3 *Relationship with Wild Lens Relatives*

The taxonomic status of *Lens culinaris* and the phylogenetic relationships between *Lens* species has been a subject of scientific discussion. The classical taxonomic structure of the genus based on morphological analyses and/or biochemical markers included four species, namely, *L. culinaris, L. orientalis, L. nigricans* and *L. ervoides* (Ladizinsky 1979b). However, Ladizinsky et al. (1984) proposed only two species: *L. culinaris* and *L. nigricans* with the former having subspecies *culinaris* [the cultigen], *orientalis* and *odemensis,* while *L. nigricans* comprised ssp. *nigricans* and ssp. *ervoides.* Two new species, *L. tomentosus* (ex *L. orientalis;* Ladizinsky 1997a) and *L. lamottei* (ex *L. nigricans;* Van Oss et al. 1997), were added to the genus *Lens* in 1997. Ferguson et al. (2000) using morphological as well as molecular markers, considered both *odemensis* and *tomentosus* as subspecies of *L. culinaris*. The following classification of the genus *Lens* was proposed: *L. culinaris*, with four subspecies, namely, *culinaris, orientalis, tomentosus* and *odemensis, L. ervoides, L. nigricans* and *L. lamottei.* This nomenclature is currently being used by the National Center for Biotechnology Information (NCBI). However, the current more accepted number of species in the genus *Lens* is of six, namely, *L. culinaris* [ssp. *culinaris,* the cultivated lentil and ssp. *orientalis* (Boiss.) Ponert, the wild ancestor of the cultivated forms], *L. odemensis* (Godr.) Ladiz., *L. tomentosus* Ladiz., *L. nigricans* (Bieb.) Godr., *L. ervoides* (Bring.) Grande. and *L. lamottei* Czfr.

In addition to the subspecific status of *L. orientalis*, the confused taxonomy of *Lens* has another lingering doubt on the phylogenetic position of *L. nigricans* within the genus (i.e. see the placement of some *L. nigricans* accessions within *Lens* dendrogram in a recent study with cross-genera SSR markers by Reddy et al. 2010). As *L. orientalis* is clearly the wild ancestor of the cultigen, its status has to be considered as subspecific (i.e. *L. culinaris* ssp. *orientalis*) according to the nomenclature for cultivated plants proposed by Harlan and De Wet (1971). Traditional *Lens* taxonomy was based on morphology. Nevertheless, morphological similarities or dissimilarities do not necessarily reflect phylogenetic relationships, since sometimes they are a consequence of evolutionary convergence. Studies are very much dependent on both the sets of accessions and the characteristics evaluated. Thus, even with the same *Lens* accessions, different clusters have been obtained depending on whether quantitative or qualitative characters were scored (Ahmad et al. 1997). On the contrary, the use of molecular data has helped to reach a more precise *Lens* phylogeny. Recent phylogenetic analyses based upon ribosomal DNA spacer sequences, molecular markers, and convicilin sequences strongly suggest:

1) A divergent position of *nigricans* from the remaining *Lens* species. For instance a probe with the ribosomal intergenic spacer (IGS) sequence of *L. culinaris* hybridized *in situ* on nucleolar organizer region (NOR) chromosomes of *L. odemensis* but did not on NOR chromosomes of *L. nigricans*. Moreover, some sequence characteristics of the *nigricans* IGS are different from that in other *Lens* species except *L. tomentosus*. Likewise, the nontrascribed ribosomal 5S spacer (NTS) suggests that *L. nigricans* and *L. tomentosus* are less related to the other *Lens* species (Fernández et al. 2005a, b); and the phylogenetic analyses of the ribosomal DNA internal transcribed spacer (ITS) pointed out the divergent status of *L. nigricans* (Mayer and Bagga 2002; Sonnante et al. 2003). Furthermore, the comparison of the deduced amino acid sequence of the convicilin N-terminal extension, a highly variable protein domain within the Vicieae, showed that all *Lens* species differed from the consensus sequence by a single amino acid change (an amino acid deletion in the case of *L. tomentosus*, all other were amino acid substitutions) except for *L. nigricans*, which differed by 12 amino acid substitutions and two deletions, of 10 and 15 amino acids, respectively (Saenz de Miera et al. 2008). Likewise, *L. nigricans* showed the most diverging sequences for Bowman–Birk protease inhibitors the nucleotide and the amino acid levels (Sonnante et al. 2005). This divergent situation of *L. nigricans* agrees with genetic distance analyses based on allozyme data (Pinkas et al. 1985).
2) The close relationship between *culinaris* and *orientalis* supports their subspecific status.
3) The relationships between the other taxa are less clear cut; however a) *tomentosus* is the sister taxon to the *culinaris-orientalis* group, b) *lamottei* is probably the closest relative to *ervoides*, and c) *odemensis* is generally joined to the *culinaris–orientalis–tomentosus* cluster (Mayer and Bagga 2002; Sonnante et al. 2003; Durán and Pérez de la Vega 2004).

In addition, Galasso (2003) by means of *in situ* hybridization of highly repeated DNA sequences found a typical fluorescent *in situ* hybridization (FISH) karyotype for each species with very little variation within species (except in *odemensis*) and a great similarity between ssp. *culinaris* and ssp. *orientalis*. Despite their morphological similarity, results favored the separation of *tomentosus* and *culinaris* into different species, as well as that of *lamottei* and *nigricans* (Van Oss et al. 1997) (see Cubero et al. 2009 for further data).

3.3.4 Relationship with Geographical Distribution and Extent of Genetic Diversity

The most extensive analysis of the genetic variation in cultivated lentil was carried out by Ferguson et al. (1998c) who analyzed isozyme and RAPD variation in accessions from 16 different countries. The germplasm from Afghanistan clustered with that from the South Asian countries. Accessions from these two areas were strikingly different to that from other countries studied and showed low variability. Another noticeable result was that the classification into *macrosperma* and *microsperma* types did not reflect overall country relationships, suggesting that variation in seed size has little relation to between-country genetic similarities. Similar results were found in relation to seed size by Duran and Pérez de la Vega (2004). Using RAPD and ISSR markers they found that Continental Spanish *macrospermas* and *microspermas* were more similar between them than with *microspermas* from different origins. Most of the studies on genetic variability after 1998 have been limited to local or regional germplasm samples (Sonnante and Pignone 2001, 2007; Duran and Pérez de la Vega 2004; Babayeva et al. 2009; Fiocchetti et al. 2009; Toklu et al. 2009). The general overview is that genetic variability in lentil is higher in or near the area where the domestication of lentil occurred in the Fertile Crescent.

In relation to wild *Lens* species the analysis of molecular variance with isozyme and RAPD data revealed that between 78 and 99% of the variation was attributable to between-population differences (Ferguson et al. 1998b). Areas of high diversity and unique diversity were located for four wild species, and indicate regions where further germplasm collection is likely to yield novel genetic variation (Ferguson et al. 1998a).

3.4 Molecular Linkage Maps: Strategies, Resources and Achievements

3.4.1 Classical Mapping Efforts

After an initial use of isoenzymes and morphological markers to obtain linkage genetic maps in lentil (see Section 3.2.1), the first map using DNA-based markers was described by Havey and Muehlbauer (1989) and was based on an F_2 mapping population derived from a cross between *L .c. culinaris* and *L. c. orientalis*. Nine linkage groups (LGs) comprising 333 cM were identified based on segregation of six morphological, eight isozyme and 20 RFLP markers. From a study of F_3 families from an interspecific cross with *L. ervoides*, Weeden et al. (1992) obtained 11 LGs covering a total distance of of 560 cM. Their map used information provided by 64 markers (three morphological, 37 isozymes, and 26 RFLPs). A comparison of this lentil map with *Pisum* maps confirmed previous results on shared synteny obtained by Muehlbauer et al. (1989) and identified eight syntenic regions between *Lens* and *Pisum*. The syntenic marker groups constituted 250 cM or approximately 40% of the then known genetic map of *Lens*.

Subsequent lentil genetic maps were also obtained from inter-subspecific or interspecific segregating populations (Tahir et al. 1993; Vaillancourt and Slinkard 1993; Tahir and Muehlbauer 1994). Tahir et al. (1993) unified the data from different studies and proposed 10 LGs based on the segregation of seven morphological, 25 isozyme, 38 RFLP, four seed protein markers, plus one chloroplast and one ribosomal marker. Vaillancourt and Slinkard (1993) used F_2 data to construct four LGs based on 17 isozyme and 11 morphological markers. Tahir and Muehlbauer (1994) identified six LGs encompassing a total of 17 isozyme loci and four morphological trait loci in lentil recombinant inbred lines (RILs). More recently Kumar et al. (2005a, b), using classical linkage analysis, have mapped eight morphological markers in two LGs (Table 3-5). Monogenic and dominant genes for plant height, leaf color, plant pubescence and leaflets per leaf were found to be linked and arranged in the order of *Ph-Gl-Pub-Hl*, respectively, with map distances of 21.1, 28.9, and 17.5 cM between them (Kumar et al. 2005b). This short sequence of four linked genes spanning over 37.2 cM was designated as LG 2 of lentil. The second LG (Kumar et al. 2005a) included simply inherited dominant genes for stem, leaf and pod pigmentation, and erect growth habit, linked in the order *Gs-Bl-Rdp-Ert*, respectively, with map distances of 12.3, 25.5, and 20.8 cM between them and spanning over 33.9 cM.

Genes for simply inherited traits mapped in lentil some of which may be oligogenic instead of monogenic are listed in Table 3-5. Current data indicate that cotyledon color is controlled by three genes (Emami and Sharma 1996a; Sharma et al. 2004). A total of 46 traits of monogenic inheritance have been described in different studies; however, only 28 have been included in linkage studies and just 22 have been mapped. The morphological markers most extensively used have been pod indehiscence (*Pi*) and anthocyanin production in the stem (*Gs*), seed coat spotting (*Scp*) and cotyledon color (*Yc*) being also used in several studies. Several of these markers have not been yet assigned to linkage groups. Thus, the lack of common markers used in different studies and the location of morphological markers in different linkage groups, have prevented the development of a consensus map in the species.

Table 3-5 List of gene symbols of lentil traits with simple monogenic[a] inheritance and linkage studies in which they have been used.

Symbol	Trait	References on inheritance	References[c] on linked markers (BSA)	References on linkage and mapping[b, c]
AbR$_1$	Resistance to ascochyta blight	Ford et al. (1999)	12, 16	
B	Brown Cotyledon	Emami and Sharma (1996a)		
Bl	Leaf pigmentation	Emami and Sharma (1996b)		19, 21
Blf	Leaflet size	Kumar et al. (2004b)		
Blsc	Black seed coat	Vaillncourt and Slinkard (1992)		**6**
Blt	Black testa	Emami and Sharma (2000)		
Chl	Chlorina chlorophyll mutant	Vandenberg and Slinkard (1989)		
Df$_1$, *Df*$_2$	Dwarfing genes	Ladizinsky (1997b)		
Dg	Dark green cotyledon	Sharma and Emami (2002)		
Dgl	Leaf color	Hoque et al. (2002)		15
Erf	Erect growth habit	Kumar et al. (2004a)		19, 21
Fe	Resistance to iron deficiency	Ali et al. (1997)		
Fn	Number of flowers per inflorescence	Gill and Malhorta (1980)		
Frt (*Rft*)	Radiation frost tolerance	Eujayl et al. (1999)	11	22
Fw	Fusarium vascular wilt resistance	Eujayl et al. (1998b)	10	22
Ggc	Grey seed coat ground color	Vandenberg and Slinkard (1990)		**6**, 18
Gh	Plant growth habit	Ladizinsky (1979a)		3, **4**, 7
Gl	Leaf color	Kumar et al. (2004a)		19, 20
Glp	Glabrous pod	Vandenberg and Slinkard (1989)		
Grp	Green pod color	Vandenberg and Slinkard (1989)		**6**
Gs	Epicotyl color	Ladizinsky (1979a)		1, 2, 3, 4, 5, 6, 7, 8, 13, 18, 19, 21
Hl	Leaflets per leaf	Kumar et al. (2004a)		19, 20
Hsc	Hard seed coat	Ladizinsky (1985)		**6**
I	Cotyledon color	Slinkard (1978)		
LCt-2	Resistance to anthracnose	Tullu et al. (2003)	16, 17	

Mot	Seed coat pattern	Vandenberg and Slinkard (1990)		
O	Cotyledon color	Singh (1978)		
P	Flower color	Lal and Srivastava (1975)		
PdP	Pod Pigmentation	Havey and Muehlbauer (1989)		4
Pep	Pubescent Peduncle	Sarker et al. (1999)		13
Ph	Plant height	Tahir et al. (1994)		19, 20
Pi	Pod indehiscence	Ladizinsky (1979a)		2, 3, **<u>4</u>**, 5, 6, 7, 9, **<u>18</u>**, 22
Pub	Plant pubescence	Hoque et al. (2002)		15, 19, 20
ral_2	Resistance to ascochyta blight	Chowdhury et al. (2001)	14, 16	
Rdp	Red pod	Kumar et al. (2004a)		19, 21
Sbv	Resistance to pea seedborne virus	Haddad et al. (1978)		
Scp	Seed coat spotting	Ladizinsky (1979a)		3, 5, **<u>6</u>**, 9, 13, 18, 22
Sn	Early flowering	Sarker et al. (1999)		13
Tgc	Tan seed coat ground color	Vandenberg and Slinkard (1990)		**<u>3</u>**, 6
Tnl	Tenderilled leaf	Vandenberg and Slinkard (1989)		**<u>6</u>**, 13
Urf_1, Urf_2, Urf_3	Rust resistance	Kumar et al. (2001)		
V	Flower color	Lal and Srivastava (1975)		
W	Flower color	Wilson and Hudson (1978)		9, 22
Xan	Xantha chlorophyll mutant	Vandenberg and Slinkard (1987)		
Y	Yellow cotyledon	Emami and Sharma (1996a)		
Yc	Cotyledon color	Slinkard (1978)		4, 6, 7, 18

[a]Some traits are oligogenic instead of being strictly monogenic.

[b]Numbers in bold and underlined indicate that the trait was not assigned to any linkage group in that reference.

[c]List of references: 1, Zamir and Ladizinsky (1984); 2, Tadmor et al. (1987); 3, Havey and Muehlbauer (1989); 4, Muehlbauer et al. (1989); 5 Weeden et al. (1992); 6, Vaillancourt and Slinkard (1993); 7, Tahir and Muehlbauer (1994); 8, Eujayl et al. (1997); 9; Eujayl et al. (1998a); 10, Eujayl et al. (1998b); 11, Eujayl et al. (1999); 12, Ford et al. (1999); 13, Sarker et al. (1999); 14, Chowdhury et al. (2001); 15, Hoque et al. (2002), 16, Tar'an et al. (2003); 17, Tullu et al. (2003); 18, Durán et al. (2004); 19, Kumar et al. (2004a); 20, Kumar et al. (2005a); 21, Kumar et al. (2005b); 22, Hamwieh et al. (2005).

3.4.2 Evolution of Marker Types and Mapping Efforts

With the introduction of PCR-based technologies, the number of molecular markers mapped across the genome of *Lens* increased dramatically. Thus, RAPDs were evaluated for genetic mapping and tested for segregation distortion in F_2 populations (Eujayl et al. 1997). This study analyzed 76 segregating loci at an LOD score of 3.0 for linkage using MAPMAKER V1.9 and assigned 28 RAPD, one RFLP, one morphological and three oligonucleotide markers to nine LGs spanning over 206 cM. Then, the first extensive lentil linkage genetic map (Eujayl et al. 1998a) was constructed using a total of 177 markers (89 RAPDs, 79 AFLPs, six RFLPs and three morphological traits), which were scored in 86 RILs ($F_{6:8}$) derived from an intersubspecific cross with *L. c. orientalis*. Marker segregation analyzed with MAPMAKER/EXP V3.0 using a LOD score established seven LGs covering 1,073 cM with an average distance of 6 cM between adjacent markers.

Rubeena et al. (2003) described the first intraspecific lentil linkage map. It was based on the analysis of an F_2 population developed from a cross of the lines ILL5588 and ILL7537 differing in resistance to Ascochyta blight. The segregation information provided by 114 markers (100 RAPD, 11 ISSR, and three resistance gene analogs (RGAs) analyzed by using MAPMAKER/EXP V3.0 (LOD score of 4.0 and maximum recombination fraction of 0.25) generated nine linkage groups, which included from six to 18 markers spanning over 784.1 cM. A total of 16 out of the 114 mapped markers (14%) showed segregation distortion.

PCR-based markers of codominant nature and molecular markers originated from known gene sequences has recently started to enable development of transferable and function-associated lentil genetic maps. The most representative example of codominant markers is the SSRs. They are of particular interest as they are unilocus and multi-allelic. In addition, some SSRs that can be sourced from within known gene sequences of interest can be useful for future functional association studies.

The first lentil linkage genetic map to include any SSR marker was described by Durán et al. (2004) using an intersubspecific F_2 population of 113 individuals, obtained from the cross of a Spanish cultivar and *L.c. orientalis*. By means of MAPMAKER using a LOD score of 3.0 and a maximum distance of 30 cM, 161 out of 200 markers screened (71 RAPDs, 39 ISSRs, 83 AFLPs, two SSRs and five morphological traits) were found to be linked in 10 LGs covering a total of 2,172.4 cM with an average distance between markers of 15.9 cM. Next, Hamwieh et al. (2005) used 86 recombinant intersubspecific inbred lines previously studied by Eujayl et al. (1998a) and added 39 SSRs and 49 AFLPs (using JOINMAP, LOD scores between 3.00–6.00 and a maximum recombination fraction of 25 cM). Combined data were used to construct a comprehensive *Lens* map containing 283 markers distributed in 14 LGs and covering 751 cM with an average marker interval of 2.6 cM. The most recently published molecular linkage map was developed in an RIL population (Precoz x WA8949041) of 94 lines with 166 markers (AFLPs, ISSRs, RAPDs and two morphological). The resulting linkage map consists of 11 LGs covering 1,396.3 cM with an average map distance between framework markers of 8.4 cM (Tanyolac et al. 2010). Finally, the first predominantly gene-based genetic linkage map of lentil has been constructed (Phan et al. 2007) using an F_5 intraspecific population of RILs. Seventy-nine intron-targeted amplified polymorphic (ITAP) gene-based markers and 18 SSR markers were included using the MULTIPOINT program with a maximum recombination fraction of 0.32 and LOD value of 9.0. The analysis revealed seven LGs containing from five to 25 markers and varying in length from 80.2 to 274.6 cM. This genetic map spanned over a total length of 928.4 cM and facilitated clear evidence of simple and direct macrosyntenic relationships between *Lens* and *Medicago truncatula*.

To date, marker density in the published genetic maps is rather limited. In general, the lack of codominant common markers prevents comparisons of the available genetic maps and the efficient identification of markers closely linked to genes of interest for breeding purposes. However, the situation is changing rapidly as more gene-based markers are being developed as more lentil genomic sequences are available thanks to the information from legume model species, i.e., *Medicago truncatula* and *Lotus japonicus*, and other crop related legume species.

Table 3-6 summarizes the above mentioned data of genetic mapping in lentil.

3.4.3 Molecular-Cytogenetic Maps

The chromosomes of the cultivated lentil were identified by physical mapping of repetitive sequences by Galasso et al. (2001). FISH was carried out with two short-repeated DNA sequences (pLc30 and pLc7), isolated from lentil genomic DNA. The pLc30 sequence hybridizes on six out of seven lentil chromosome pairs, while pLc7 hybridizes on only one. The distribution of the nine different hybridization sites of pLc30 allows the discrimination of all seven chromosome pairs and the construction of a karyotype of *L. culinaris* ssp. *culinaris*. Additionally, the combination of simultaneous and successive FISH with pLc7, 5S rRNA, 18S-5.8S-25S rRNA genes, and a telomeric sequence allowed the assembly of a physical map based on lentil karyotype. Other cytogenetical *Lens* maps were also described by Galasso (2003). The mitotic chromosomes maps of seven *Lens* taxa were obtained by multiple-target FISH using two highly repetitive sequences (pLc30 and pLc7) and the multigenic families for the rDNA (pTa71 and pTa794) from wheat as probes.

Table 3-6 List of the main genetic mapping studies in lentil.

Reference	Population[a]	Number and type of markers	Analysis method[b]	LOD (Maximum diatance[3])	Number of linkage groups	Average Distance between markers[c]	Total coverage[c]
Eujayl et al. 1997	inter., F_2	**33** (28 RAPD; 1 RFLP; 3 Oligonucleot.; 1 Morph.)	MM	3 (27)	9	6.24	206
Eujayl et al. 1998a	inter., RILs	**177** (89 RAPD; 79 AFLP; 6 RFLP; 3 Morph.)	MM	4 (25)	7	6.06	1073
Rubeena et al. 2003	intra., F_2	**114** (100 RAPD; 11 ISSR; 3 RGA)	MM	4 (25)	9	6.87	784
Durán et al. 2004	inter., F_2	**161** (62 RAPD; 65 AFLP; 29 ISSR; 1 SSR; 4 Morph.)	MM	3 (30)	10	15.87	2172
Kahraman et al. 2004a	intra., RILs	**130** (27 RAPD; 48 AFLP; 55 ISSR)	MM	3 (30)	9	9.17	1192
Hamwieh et al. 2005	inter., RILs	**283** (110 RAPD; 129 AFLP; 39 SSR; 5 Morph.)	JM	3–6 (25)	14	2.65	751
Rubeena et al. 2006	intra., F_2	**72** (38 RAPD; 30 AFLP; 3 ISSR; 1 Morph.)	MM	4 (25)	11	5.72	412
Phan et al. 2007	intra., RILs	**97** (18 SSR; 79 ITAP)	MP	9 (32)	7	9.57	928
Tullu et al. 2008	intra., RILs	**207** (54 RAPD; 144 AFLP; 9 SSR)	MM	3 (25)	12	9.02	1868
Tanyolac et al. 2010	Intra., RILs	**166** (75 RAPD; 64 AFLP; 23 ISSR; 4 Morph.)	MM	3 (30)	11	8.40	1396

[a]Inter., intersubspecific (*L. c.* ssp. *culinaris* x *L. c.* spp. *orientalis*); Intra., intraspecific.
[b]MM, MAPMAKER; JM, JOINTMAP; MP, MULTIPOINT.
[c]Distance in cM.

3.4.4 Comparative Mapping

As it has been mentioned, comparative legume mapping can be traced back to the Law of Homologous Series in Variation proposed by Vavilov (1922). Earlier pioneering studies using small numbers of morphological, isozyme and RFLP markers suggested a simple genetic relationship between pea and lentil (Weeden et al. 1992). However, precise comparative mapping requires a relatively high number of genetic markers with demonstrated orthology. When such molecular markers are available, it is then possible to conduct meaningful comparative mapping among legume species. Comparative mapping was able to show extensive macrosynteny between *Medicago truncatula* and *L. culinaris* (Phan et al. 2007). These authors generated the first gene-based genetic linkage map of lentil (*L. culinaris* ssp. *culinaris*) using an F_5 population developed from a cross between the cultivars Digger (ILL5722) and Northfield (ILL5588). They used 79 ITAP and 18 genomic SSR markers (see Section 3.4.2). The evidence of macrosynteny between lentil and *M. truncatula* was based on the fact that 66 out of the 71 gene-based markers, which were previously assigned to *M. truncatula* genetic and physical maps, were found in regions syntenic between the *L. c.* ssp. *culinaris* and *M. truncatula* genomes. However, there was evidence of moderate chromosomal rearrangements, which may account for the difference in chromosome numbers between these two legume species. Comparative mapping showed a direct and simple relationship between the *M. truncatula* and *L. culinaris* ssp. *culinaris* chromosomes, with complete homology evident between LGs. Thus lentil LG 1, 2, 3, 4, 5, 6, and 7 were proposed to be syntenic to *M. truncatula* LG 4+7, 8, 1+6, 2, 5, 3 and 3, respectively; and then, by considering lentil or pea LGs syntenic to the same *M. truncatula* or *M. sativa* LGs to be syntenic to each other, lentil LG 1, 2, 3, 4, 5, 6, and 7 were proposed to be syntenic to pea LG VII + V, IV, II + III + VI, III + IV, I, III and III, respectively.

Eighteen common SSR markers were also used by Phan et al. (2007) to connect their map with another comprehensive map of lentil (Hamwieh et al. 2005), providing the syntenic context of four important domestication traits. Ellwood et al. (2008) utilized 151 markers out of 796 ITAP markers screened, to construct a comparative genetic map between *L. culinaris* and *Vicia faba*. A simple and direct macrosyntenic relationship between faba bean and *M. truncatula* was evident, while faba bean and lentil shared a common rearrangement relative to *M. truncatula*. This study also provides a preliminary indication for finer scale macrosynteny between *M. truncatula*, lentil and faba bean. Markers originally designed from genes on the same *M. truncatula* bacterial artificial chromosomes (BACs) were found to be grouped together in corresponding syntenic areas in lentil and faba bean. Other research has also demonstrated marker and/or gene order conservation between species closely related to lentil such as pea: *Pisum sativum* and *Medicago sativa* (Kalo et al. 2004), *P. sativum* and *M. truncatula* (Aubert et al. 2006), *P. sativum, M. truncatula, M. sativa, Vigna radiata, Glycine max* and *Phaseolus vulgaris* (Choi et al. 2004) (see also Chapter 7).

3.5 Mapping and Tagging of Simply Inherited Traits

Several studies based on the use of bulked segregant analysis (BSA) have identified molecular markers associated to simply inherited trait loci related to both abiotic and biotic stress resistance. All BSA studies related to biotic stresses have used MAPMAKER to associate markers to the simple traits. Thus, Eujayl et al. (1998b) identified an RAPD marker linked at 10.8 cM to the locus *Fw* conferring resistance to Fusarium wilt. Ford et al. (1999) identified RAPD markers flanking at 6 cM and 14 cM apart a major dominant locus for Ascochyta blight resistance (AbR_1). However, in the lentil cultivar Indian head, resistance to Ascochyta blight is determined by a single recessive gene (ral_2). Chowdhury et al. (2001) identified two flanking RAPD markers linked to ral_2 at 12 and 16 cM apart and one of the RAPD markers was converted into a sequence-characterized amplified region (SCAR) marker. With regard to anthracnose, Tullu et al. (2003) documented a major resistance gene *LCt-2* linked to flanking RAPD markers at 6.4 and 10.5 cM apart; likewise, three AFLP markers were also found to be linked to this locus (Tables 3-5 and 3-6).

Tar'an et al. (2003) analyzed 156 RILs to identify markers linked to the Ascochyta blight resistance loci AbR_1 (Ford et al. 1999) and ral_2 (Chowdhury et al. 2001), and to the anthracnose resistance locus *LCt-2* (Tullu et al. 2003). They found that 11 RILs retained all three resistance genes, in a practical demonstration of marker-assisted selection (MAS) to pyramid genes for resistance.

With regard to abiotic stresses only one study has been published (Eujayl et al. 1999). These authors identified an RAPD marker linked at 9.1 cM distance to the radiation-frost tolerance (*Frt*) locus, although resistance to low temperatures is usually analyzed as polygenic characters (see next section).

The most recent report on molecular markers associated to a resistance locus is the study of Hamwieh et al. (2005) who included data related to both biotic an abiotic stresses. From the data of Eujayl et al. (1998a) on the genetic map from 86 RILs, and of Eujayl et al. (1999) on the genotypic data for radiation-frost (*Frt*) tolerance, Hamwieh et al. (2005) added 39 microsatellites and 49 AFLP markers to the existing genetic map and screened the 86 RILs for Fusarium vascular wilt resistance. Using the JOINMAP program, the locus *Fw* was localized on LG 6 flanked by a microsatellite marker at 8.0 cM apart and an AFLP marker at 3.5 cM apart. Therefore, of the five traits of simple inheritance for which associated markers have been identified, only the radiation-frost tolerance (*Frt* or *Rft*) and the Fusarium wilt resistance (*Fw*) loci have been mapped to date (Tables 3-5 and 3-6).

3.6 Molecular Mapping of Complex Traits

Association of quantitative trait loci (QTL) to markers was first reported in lentil by Tahir et al. (1994). Using isozymes and eight sets of F_6 RILs derived from an intersubspecific cross with *L. c. orientalis*, seven quantitative traits (days to flower, days to maturity, plant height, biomass production, seed yield, harvest index and seed weight) were associated by single marker analysis to 14 isozyme markers. The study detected five QTLs for seed yield and weight, four QTLs for days to flower, plant height and biomass production, three QTLs for days to maturity and two QTLs for harvest index (Tahir et al. 1994). Similarly, using isozymes and single marker analysis and scoring the same characteristics in 10 sets of backcross inbred lines developed from three different inter-subspecific crosses with *L. c. orientalis*, Tahir and Muehlbauer (1995) detected one locus affecting variation in days to flower, days to maturity, and plant height, whereas two to three loci were detected for biomass production, seed yield, harvest index and seed weight. In both cases, the tentative position of the QTLs on the lentil genetic map was determined by the known position of the isozyme loci in various LGs of previous studies.

A controlled environment test was developed for frost tolerance of lentil, involving hardening seedlings at 10°C day/0°C night for 1–6 weeks, then bringing them to –15°C for 3 hour (Ali et al. 1999). Under these conditions, narrow-sense heritability was very high and useful transgressive segregants were identified (Ali and Johnson 2000). Similarly, heritability estimates for winter hardiness ranged up to 0.907 in a range of populations of RILs exposed to winterkill conditions in several environments (Kahraman et al. 2004a). Heritability analysis for cold tolerance in lentil was conducted using parental, F_2 and F_3 populations at two locations in Balochistan, Pakistan, and one controlled environment in Fort Collins, Colorado, USA (Ali and Johnson 2000). Estimates of narrow-sense heritability ranged from 0.32 ± 0.06 to 0.71 ± 0.16 under field conditions. Under controlled conditions, the estimated heritability was maximized at 1.00 ± 0.17 using 6- to 8-week-old lentils. Significant transgressive segregants were found in five of the six populations in the F_3 under controlled conditions but were not observed in field environments. The mode of inheritance of winter hardiness has been analyzed by growing RILs at Haymana, Turkey, and Pullman, WA, USA, between 1997 and 2001 (Kahraman et al. 2004a). General combining ability (GCA) and specific combining ability (SCA) effects were significant at all locations. Heritability estimates among the 10 RIL populations ranged from 15.9 to 90.7%. Inheritance patterns of winter hardiness appeared to be quantitative on the basis of frequency distributions and the lack of discretely segregating classes. For QTL analysis of winter survival, three QTLs were detected at Haymana in 1997–1998, one QTL was detected at Pullman in 1998-1999, and three QTLs were identified at Haymana in 1999–2000. Only one of the QTLs was common to all environments. For winter injury scores at Pullman in 1999, four QTLs were identified that influenced winter survival. Overall results indicated that winter hardiness is influenced by several genes and the cumulative effects of winter stress (Kahraman et al. 2004b). In a screening for winter hardiness sources Hamdi et al. (1996) described that winter hardiness was concentrated among accessions originating from high elevation areas and among the wild materials *L. orientalis* exhibited, on average, the highest level of winter hardiness.

Molecular markers (56 RAPDs, 106 ISSRs and 94 AFLPs) were used by Kahraman et al. (2004b) to locate QTLs related to winter hardiness. The MAPMAKER software was used to construct the genetic map and QTLs were detected by means of QGENE3 using simple interval mapping (SIM) and multiple regression. Three QTLs for winter survival in two locations and one QTL at the remaining location were located. These QTLs explained 23% of the total phenotypic variance for winter survival, however, only one QTL was common to all locations but the effect and position differed across environments. For winter injury scores undertaken at only one location, four QTLs were identified explaining 43% of the phenotypic variation (Kahraman et al. 2004b).

Rubeena et al. (2006) used linkage genetic maps developed with MAPMAKER from two intraspecific F_2 populations of 150 individuals, a population previously mapped (Rubeena et al. 2003) and the second one containing 72 markers (38 RAPD, 30 AFLP, three ISSR and one morphological trait), to locate QTLs related to Ascochyta blight resistance. Five QTLs were identified in one population and three in the other by means of QTL-CARTOGRAPHER and composite interval mapping (CIM). Two of these QTLs coincided with regions identified using multiple interval mapping (MIM) and were shown to be conditioned by dominant and partial dominant gene action. Both QTLs altogether accounted for about 50% of the phenotypic variance of disease severity. Fratini et al. (2007) identified with MAPMAKER/QTL and QTLCartographer using SIM and CIM a total of 23 QTLs for nine quantitative traits. The analysis was based on a previous linkage map (Duran et al. 2004) constructed with MAPMAKER according to the segregation of 161 markers (62 RAPDs, 29 ISSRs, 65 AFLPs, 1 SSR and four morphological). Four QTLs were localized for number of branches at the first node, three QTLs for each of the following traits: plant height, flowering time, dehiscence, seed weight and seed diameter, and two QTLs for the height of the first node and two for total number of branches. The multiple QTL model explained more than 80% of the observed phenotypic variance for three of the quantitative traits analyzed (branches at first node, flowering time, and dehiscence). Finally, an intraspecific linkage map was constructed by Tullu et al. (2008) using MAPMAKER; the QTL analysis was carried out by QTL-CARTOGRAPHER and CIM for earliness and plant height at two locations. Two hundred and seven markers were used to genotype 94 RILs, the genetic map included a previously mapped resistance gene to anthracnose (Tullu et al. 2003) and a QTL for Ascochyta blight resistance (Rubeena et al. 2006). Tullu et al. (2008) identified two and five QTLs affecting earliness and plant height, respectively, at two locations, explaining 37 to 46% and 31 to 40% of phenotypic variance, respectively.

3.7 Molecular Breeding

Although many different types of molecular markers have been used, for instance, in the estimation of genetic variability and genetic mapping in lentil, their use in molecular breeding is still scarce. Molecular markers have been used in the assessment of distinctness of cultivars or lines (see Section 3.3.2), their use in marker-assisted selection is limited only to a few examples (see Section 3.5) and examples of marker-assisted introgression has not been described in detail.

3.7.1 Gene Pyramiding

Tar'an et al. (2003) were able to pyramid the two genes for resistance to Ascochyta blight, caused by *A. lentis,* and the gene for resistance to anthracnose, caused by *Colletotrichum truncatum,* into lentil breeding lines, using 156 RILs. Genetic analysis of disease reactions demonstrated that the observed segregation ratios of resistant versus susceptible fit the two gene model for resistance to Ascochyta blight and a single gene model for resistance to anthracnose. Using markers linked to the two genes conferring resistance to *A. lentis* (UBC 2271290) to *ral*1, and (RB18680) to *AbR1*, and to the major gene for resistance to anthracnose (OPO61250), respectively, they found that 11 RILs retained all the three resistance genes. Therefore, selecting materials with both markers for resistance to Ascochyta blight and a marker for resistance to anthracnose can clearly facilitate progress toward pyramiding resistance in the population.

3.7.2 Limitations and Prospects of Marker-Assisted Selection and Breeding

Although several genetic linkage maps have been proposed based on different molecular markers, it is essential to convert arbitrary markers to sequence-specific markers that are reproducible and robustly transferable among genetic backgrounds. It is also required to incorporate more morphological markers into genetic maps to establish homologies between studies and to allow indirect selection in the field, in addition, it is also necessary to involve more quantitative traits of economic importance for their tagging or establishing tight linkage for practical use in MAS. The function-associated molecular maps that are being developed in lentil will enable direct selection of the candidate genes (Phan et al. 2006, 2007). Together with information of the genomic regions that quantitatively account for genetic portions of determined phenotypes (Rubeena et al. 2006), these maps will allow precise selection of quantitative characteristics in lentil, for future trait/gene pyramiding (Tar'an et al. 2003) and for the adaptation to multiple environments (Ford et al. 2007).

References

Ahmad M, McNeill D, Seedcole J (1997) Phylogenetic relationships in *Lens* species and their interspecific hybrids as measured by morphological characters. Euphytica 94: 101–111.

Ahmed FRS, Killham K, Alexander I (2006) Influences of arbuscular mycorrhizal fungus *Glomus mosseae* on growth and nutrition of lentil irrigated with arsenic contaminated water. Plant Soil 283: 33–41.

Ali A, Johnson DL (2000) Heritability estimates for winter hardiness in lentil under natural and controlled conditions. Plant Breed 119: 283–285.

Ali A, Riaz-ul-Haque M, Bhatti MS (1997) Inheritance of resistance to iron-deficiency chlorosis in lentil. LENS 24: 28–29.

Ali A, Johnson DL, Stushnoff C (1999) Screening lentil (*Lens culinaris*) for cold hardiness under controlled conditions. J Agri Sci 133: 313–319.

Ali JAF, Shaikh NA (2007) Genetic exploitation of lentil through induced mutations. Pak J Bot 39: 2379–2388.

Almeida-Costa GE, Queiroz-Monici KS, Machado Reis SMP, Costa de Oliveira A (2007) Chemical composition, dietary fibre and resistant starch contents of raw and cooked pea, common bean, chickpea and lentil legumes. Food Chem 94: 327–330.

Alo F, Furman BJ, Akhunov E, Dvorack J, Gepts P (2011) Leveraging genomic resources of model species for the assessment of diversity and phylogeny in wild and domesticated lentil. J Hered 102: 315–329.

Alonso N (2008) Crops and agriculture during the Iron Age and late antiquity in Cerdanyola del Valles (Catalonia, Spain). Veg Hist Archaeobot 17: 75–84.

Altaf M, Venkateshwar M, Srijana M, Reddy G (2007) An economic approach for L-(+) lactic acid fermentation by *Lactobacillus amylophilus* GV6 using inexpensive carbon and nitrogen sources. J Appl Microbiol 103: 372–380.

Armstrong-Cho CL, Banniza S (2006) *Glomerella truncata* sp nov, the teleomorph of *Colletotrichum truncatum*. Mycol Res 110: 951–956.

Arumuganathan K, Earle ED (1991) Nuclear DNA content of some important plant species. Plant Mol Biol Rep 9: 208–218.

Ashraf M, Chishti SN (1993) Waterlogging tolerance of some accessions of lentil (*Lens culinaris* Medic.). Trop Agri 70: 60–67.

Ashraf M, Zafar ZU (1997) Effect of potassium on growth and some biochemical characteristics in two lines of lentil (*Lens culinaris* Medic.). Acta Physiol Plant 19: 9–15.

Attanayake RN, Glawe DA, Dugan FM, Chen W (2009) *Erysiphe trifolii* causing powdery mildew of lentil (*Lens culinaris*). Plant Dis 93: 797–803.

Aubert G, Morin J, Jacquin F, Loridon K, Quillet M, Petit A, Rameau C, Lejeune-Henaut I, Huguet T, Burstin J (2006) Functional mapping in pea, as an aid to the candidate gene selection and for investigating synteny with the model legume *Medicago truncatula*. Theor Appl Genet 112: 1024–1041.

Aydin H, Buluta Y, Yerlikaya C (2008) Removal of copper (II) from aqueous solution by adsorption onto low-cost adsorbents. J Environ Manag 87: 37–45.

Babayeva S, Akparov Z, Abbasov M, Mammadov A, Zaifizadeh M, Street K (2009) Diversity analysis of Central Asia and Caucasian lentil (*Lens culinaris* Medik.) germplasm using SSR fingerprinting. Genet Resour Crop Evol 56: 293–298.

Balyan HS, Houben A, Ahne R (2002) Karyotype analysis and physical mapping of 18S–5.8S–25S and a 5S ribosomal RNA loci in species of genus *Lens* Miller (Fabaceae). Caryologia 55: 121–128.

Barrios A, Martin-Sanz A, Ramos S, Rotter B, Horres R, Plotner A, Kretzdorn N, Constantino Saldaña C, Perez de la Vega M, Winter P (2010) Allele-specific differential expression of transcripts potentially involved in cold tolerance QTL in lentil revealed by bulked extremes SuperTag digital gene expression (BE-STDGE) profiling and qRT-PCR. In: 5th Int Food Legumes Res Conf & 7th Eur Conf on Grain Legumes, April 26–30, Antalya, p 41: http://www.iflrc-ecgl.org/BookofAbstracts.pdf (Cited 21 July 2010)

Barulina H (1930) Lentils of the USSR and other countries. Bull Appl Bot Genet Plant Breed 40: 265–304 USSR [in Russian with English summary].

Bashir M, Azeem T, Ali A, Ijaz M (2005) Variation in lentil germplasm for reaction to virus infection. Pak J Bot 37: 53–60.

Bayaa B, Erskine W, Hamdi A (1994) Response of wild lentil to *Ascochyta fabae* f.sp. *lentis* from Syria. Genet Resour Crop Evol 41: 61–65.

Bayaa B, Erskine W, Singh M (1997) Screening lentil for resistance to fusarium wilt: methodology and sources of resistance. Euphytica 98: 69–74.

Bhalla S, Kapur ML, Singh C, Gupta K, Kumar N, Lal B (2004) Interception of bruchids in imported lentil (*Lens* spp) germplasm. Indian J Agri Sci 74: 332–333.

Baysal Z, Uyar F, Dogru M, Alkan H (2008) Production of extracellular alkaline alpha-amylase by solid state fermentation with a newly isolated *Bacillus* sp. Prep Biochem Biotechnol 38: 191–197.

Bennet MD, Smith JB (1976) Nuclear DNA amounts in angiosperms. Phil Trans Roy Soc Lond B 274: 227–274.

Broekaert W, Terras FR, Cammue BP, Osborn RW (1995) Plant defensins: novel antimicrobial peptides as components of the host defense system. Plant Physiol 108: 1353–1358.

Buchwaldt L, Anderson KL, Morrall RAA, Gossen BD, Bernier CC (2004) Identification of lentil germ plasm resistant to *Colletotrichum truncatum* and characterization of two pathogen races. Phytopathology 94: 236–243.

Buxó R (1997) Arqueología de las Plantas. Crítica, Barcelona, Spain Caccialupi P, Ceci LR, Siciliano RA, Pignone D, Clemente A, Sonnante G (2010)

Bowman-Birk inhibitors in lentil: Heterologous expression, functional characterisation and anti-proliferative properties in human colon cancer cells. Food Chem 120: 1058–1066.

Campbell CA, Zentner RP, Basnyat P, Wang H, Selles F, McConkey BG, Gan YT, Cutforth HW (2007) Water use efficiency and water and nitrate distribution in soil in the semiarid prairie: Effect of crop type over 21 years. Can J Plant Sci 87: 815–827.

Champ MM-J (2002) Non-nutrient bioactive substances of pulses. Brit J Nutr 88: S307–S319.

Chen CC, Miller P, Muehlbauer F, Neill K, Wichman D, McPhee K (2006) Winter pea and lentil response to seeding date and micro- and macro-environments. Agron J 98: 1655–1663.

Cherif M, Chilvers MI, Akamatsu H, Peever TL, Kaiser WJ (2006) Cloning of the mating type locus from *Ascochyta lentis* (teleomorph: *Didymella lentis*) and development of a multiplex PCR mating assay for *Ascochyta* species. Curr Genet 50: 203–215.

Choi HK, Mun JH, Kim DJ, Zhu H, Baek JM, Mudge J, Roe B, Ellis N, Doyle J, Kiss GB, Young ND, Cook DR (2004) Estimating genome conservation between crop and model legume species. Proc Natl Acad Sci USA 101: 15289–15294.

Chowdhury MA, Andrahennadi CP, Slinkard AE, Vandenberg A (2001) RAPD and SCAR markers for resistance to Ascochyta blight in lentil. Euphytica 118: 331–337.

Chung H-J, Liu Q, Hoover R, Warkentin TD, Vandenberg B (2008) *In vitro* starch digestibility, expected glycemic index, and thermal and pasting properties of flours from pea, lentil and chickpea cultivars. Food Chem 111: 316–321.

Coutts BA, Prince RT, Jones RAC (2008) Further studies on Pea seed-borne mosaic virus in cool-season crop legumes: responses to infection and seed quality defects. Aust J Agri Res 59: 1130–1145.

Cuadrado C, Grant G, Rubio, LA, Muzquiz M, Bardocz S, Pusztai A (2002) Nutritional utilization by the rat of diet based on lentil (*Lens culinaris*) seed meal or its fractions. J Agri Food Chem 50: 4371–4376.

Cubero JI, Pérez de la Vega M, Fratini R (2009) Origin, Phylogeny, Domestication and Spread. In: W Erskine, FJ Muehlbauer, A Sarker, B Sharma (eds) The Lentil: Botany, Production and Uses. CABI Press, Wallingford, UK, pp 13–33.

Davidson JA, Kimber RBE (2007) Integrated disease management of ascochyta blight in pulse crops. Eur J Plant Pathol 119: 99–110.

De Candolle AP (1882 reprinted 1967) Origins of cultivated species. Hafner, London, UK, p 468.

Dixit P, Dubey DK (1986) Alkylating agent induced chromosomal aberrations in lentil (*Lens culinaris* Med. Var T 36). Cytologia 51: 325–331.

Durán Y, Pérez de la Vega M (2004) Assessment of genetic variation and species relationships in a collection of *Lens* using RAPD and ISSR. Span J Agri Res 4: 538–544.

Durán Y, Fratini R, García P, Pérez de la Vega M (2004) An intersubspecific genetic map of *Lens*. Theor Appl Genet 108: 1265–1273.

El-Bouhssini M, Sarker A, Erskine W, Joubi A (2008) First sources of resistance to Sitona weevil (*Sitona crinitus* Herbst) in wild *Lens* species. Genet Resour Crop Evol 55: 1–4.

El-Hassan SA, Gowen SR (2006) Formulation and delivery of the bacterial antagonist *Bacillus subtilis* for management of lentil vascular wilt caused by *Fusarium oxysporum* f. sp *lentis*. J Phytopathol 154: 148–155.

Ellwood SR, Phan HTT, Jordan M, Hane J, Torres AM, Avila CM, Cruz-Izquierdo S, Oliver RP (2008) Construction of a comparative genetic map in faba bean (*Vicia faba* L.); conservation of genome structure with *Lens culinaris*. BMC Genom 9: 380.

Emami MK, Sharma B (1996a) Digenic control of cotyledon color in lentil. Indian J Genet 56: 357–361.

Emami MK, Sharma B (1996b) Inheritance of brown leaf pigmentation in lentil. Indian J Genet 56: 362–365.

Emami MK, Sharma B (2000) Inheritance of black testa colour in lentil (*Lens culinaris* Medik.). Euphytica 115: 43–47.

Erskine W (1996) Seed-size effects on lentil (*Lens culinaris*) yield potential and adaptation to temperature and rainfall in West Asia. J Agri Sci 126: 335–341.

Erskine W, Adham Y, Holly L (1989) Geographic distribution of variation in quantitative traits in a world lentil collection. Euphytica 43: 97–103.

Erskine W, Ellis RH, Summerfield RJ, Roberts EH, Hussain A (1990). Characterization of responses to temperature and photoperiod for time to flowering in a world lentil collection. Theor Appl Genet 80: 193–199.

Erskine W, Chandra S, Chaudary M, Malik IA, Sarker A, Sharma B, Tufail M, Tyagi MC (1998) A bottleneck in lentil: widening the genetic base in South Asia. Euphytica 101: 207–211.

Essalmani H, Lahlou H (2003) Mécanismes de bioprotection des plantes de lentille par *Rhizobium leguminosarum* contre *Fusarium oxysporum* f. sp. *Lentis*. Com Rend Biol 326: 1163–1173.

Eujayl I, Baum M, Erskine W, Pehu E, Muehlbauer FJ (1997) The use of RAPD markers for lentil genetic mapping and the evaluation of distorted F_2 segregation. Euphytica 96: 405–412.

Eujayl I, Baum M, Powell W, Erskine W, Pehu E (1998a) A genetic linkage map of lentil (*Lens* sp.) based on RAPD and AFLP markers using recombinant inbred lines. Theor Appl Genet 97: 83–89.

Eujayl I, Erskine W, Bayaa B, Baum M, Pehu E (1998b) Fusarium vascular wilt in lentil: Inheritance and identification of DNA markers for resistance. Plant Breed 117: 497–499.

Eujayl I, Erskine W, Baum M, Pehu E (1999) Inheritance and linkage analysis of frost injury in a lentil population of recombinant inbred lines. Crop Sci 39: 639–642.

European Journal of Plant Pathology (2007) Special issue: Ascochyta blights of grain legumes. Eur J Plant Pathol 119: 1–141.

Ferguson ME, Ford-Lloyd BV, Robertson LD, Maxted N, Newbury HJ (1998a) Mapping the geographical distribution of genetic variation in the genus *Lens* for the enhanced conservation of plant genetic diversity. Mol Ecol 7: 1743–1755.

Ferguson ME, Newbury HJ, Maxted N, Ford-Lloyd BV, Robertson LD (1998b) Population genetic structure in *Lens* taxa revealed by isozyme and RAPD analysis. Genet Resour Crop Evol 45: 549–559.

Ferguson ME, Robertson LD, Ford-Lloyd BV, Newbury HJ, Maxted N (1998c) Constrasting genetic variation amongst lentil landraces from different geographical origins. Euphytica 102: 265–273.

Ferguson ME, Maxted N, Van Slageren M, Robertson LD (2000) A reassessment of the taxonomy of *Lens* Miller (*Leguminosae, Papilionoideae, Vicieae*). Bot J Linn Soc 133: 41–59.

Fernández-Aparicio M, Sillero JC, Pérez-de-Luque A, Rubiales D (2007) Identification of sources of resistance to crenate broomrape (*Orobanche crenata*) in Spanish lentil (*Lens culinaris*) germplasm. Weed Res 48: 85–94.

Fernández-Aparicio M, Sillero JC, Rubiales D (2009) Resistance to broomrape in wild lentils (*Lens* spp.). Plant Breed 128: 266–270.

Fernández M, Ruiz ML, Linares C, Fominaya A, Pérez de la Vega M (2005a) 5S rDNA genome regions of *Lens* species. Genome 48: 937–942.

Fernández M, Ruiz M, Pérez de la Vega M (2005b) Length polymorphism in the ribosomal DNA intergenic spacers of *Lens* species. Span J Agri Res 3: 410–417.

Fiala JV, Tullu A, Banniza S, Seguin-Swartz G, Vandenberg A (2009) Interspecies transfer of resistance to anthracnose in Lentil (*Lens culinaris* Medic.). Crop Sci 49: 825–830.

Fikru E, Sarker A, Fikre A, Ahmed A, Ali K, Erskine W (2007) Registration of Assano lentil. J Plant Regist 1: 41.

Finkina EI, Shramova EI, Tagaev AA, Ovchinnikova TV (2008) A novel defensin from the lentil *Lens culinaris* seeds. Biochem Bioph Res Com 371: 860–865.

Fiocchetti F, Laddomada B, Roselli M, Crino P, Lucretti S (2009) Fingerprinting of three typical macrosperma Italian lentil (*Lens culinaris* Medik.) landraces using fluorescence-based AFLP markers. Sci Hort 121: 383–387.

Ford R, Pang ECK, Taylor PWJ (1997) Diversity analysis and species identification in *Lens* using PCR generated markers. Euphytica 96: 247–255.

Ford R, Pang ECK, Taylor PWJ (1999) Genetics of resistance to Ascochyta blight (*Ascochyta lentis*) of lentil and the identification of closely linked RAPD markers. Theor Appl Genet 98: 93–98.

Ford R, Mustafa B, Inder P, Rubeena S, Materne M, Taylor P (2007) Lensomics: advances in genomics and molecular techniques for lentil breeding and management. In: SS Yadav, D McNeil, PC Stevenson (eds) Lentil. An Ancient Crop for Modern Times. Springer, Dordrecht, The Netherlands, pp 275–289.

Fratini R, Durán Y, García P, Pérez de la Vega M (2007) Identification of quantitative trait loci (QTL) for plant structure, growth habit and yield in lentil. Span J Agri Res 5: 348–356.

Gahoonia TS, Ali O, Sarker A, Nielsen NE, Rahman MM (2006) Genetic variation in root traits and nutrient acquisition of lentil genotypes. J Plant Nutr 29: 643–655.

Galasso I (2003) Distribution of highly repeated DNA sequences in species of the genus *Lens* Miller. Genome 46: 1118–1124.

Galasso I, Schmidt T, Pignone D (2001) Identification of *Lens culinaris* ssp. *culinaris* chromosomes by physical mapping of repetitive DNA sequences. Chrom Res 9: 199–209.

Galloway J, MacLeod WJ, Lindbeck KD (2004) Formation of *Didymella lentis*, the teleomorph of *Ascochyta lentis*, on lentil stubble in the field in Victoria and Western Australia. Australas Plant Pathol 33: 449–450.

García-Olmedo F, Molina A, Alamillo JM, Rodriguez-Palenzuella P (1998) Plant defense peptides. Biopolymers 47: 479–491.

Ghavidel R, Prakash J (2007) The impact of germination and dehulling on nutrients, antinutrients, *in vitro* iron and calcium bioavailability and *in vitro* starch and protein digestibility of some legume seeds. Food Sci Technol 40: 1292–1299.

Gill AS, Malhorta RS (1980) Inheritance of flower color and flower number per inflorescence in lentils. LENS 7: 15–19.

Grunwald NJ, Hoheisel GA (2006) Hierarchical analysis of diversity, selfing, and genetic differentiation in populations of the oomycete *Aphanomyces euteiches*. Phytophatology 96: 1134–1141.

Guillamón E, Pedrosa MM, Burbano C, Cuadrado C, de Cortes Sánchez M, Muzquiz M (2008) The trypsin inhibitors present in seed of different grain legume species and cultivar. Food Chem 107: 68–74.

Gunn CR (1969) Genera, types and lectotypes in the tribe Vicieae (Fabaceae). Taxon 18: 725–733.

Gunn CR, Kluve J (1976) Androecium and pistil characters for Tribe Vicieae (Fabaceae). Taxon 25: 563–575.

Gupta D, Sharma SK (2006) Evaluation of wild *Lens* taxa for agro-morphological traits, fungal diseases and moisture stress in North Western Indian hills. Genet Resour Crop Evol 53: 1233–1241.

Haddad N, Muehlbauer FJ, Hampton RO (1978) Inheritance of resistance to pea seed-borne mosaic virus in lentils. Crop Sci 18: 613–615.

Haddad SG, Husein MQ (2001) Nutritive value of lentil and vetch straws as compared with alfalfa hay and wheat straw for replacement ewe lambs. Small Ruminant Res 40: 255–260.

Hamdi A, Küsmenoglu I, Erskine W (1996) Sources of winter hardiness in wild lentil. Genet Resour Crop Evol 43: 63–67.

Hamwieh A, Udapa SM, Choumane W, Sarker A, Dreyer F, Jung C, Baum M (2005) A genetic linkage map of lentil based on microsatellite and AFLP markers and localization of fusarium vascular wilt resistance. Theor Appl Genet 110: 669–677.

Hamwieh A, Udupa SM, Sarker A, Jung C, Baum M (2009) Development of new microsatellite markers and their application in the analysis of genetic diversity in lentils. Breed Sci 59: 77–86.

Harlan JR, De Wet JMJ (1971) Towards a rational classification of cultivated plants. Taxon 20: 509–517.

Havey MJ, Muehlbauer FJ (1989) Linkages between restriction fragment length, isozyme, and morphological markers in lentil. Theor Appl Genet 77: 395–401.

Hernández-Bello MA, Chilvers MI, Akamatsu H, Peever TL (2006) Host specificity of Ascochyta spp. infecting legumes of the Viciae and Cicerae tribes and pathogenicity of an interspecific hybrid. Phytopathology 96: 1148–1156.

Hobson K, Armstrong R, Nicolas M, Connor D, Materne M (2006) Response of lentil (Lens culinaris) germplasm to high concentrations of soil boron. Euphytica 151: 371–382.

Hoffman DL, Soltis DE, Muehlbauer FJ, Ladizinsky G (1986) Isozyme polymorphism in *Lens* (Leguminosae). Syst Bot 11: 392–402.

Hoque ME, Mishra SK, Kumar Y, Kumar R, Tomar SMS, Sharma B (2002) Inheritance and linkage of leaf colour and plant pubescence in lentil (*Lens culinaris* Medik.). Indian J Genet 62: 140–142.

Horneburg B (2006) Outcrossing in lentil (*Lens culinaris*) depends on cultivar, location and year, and varies within cultivars. Plant Breed 125: 638–640.

Huang HC, Erickson RS (2007) Effect of seed treatment with *Rhizobium leguminosarum* on *Pythium* damping-off, seedling height, root nodulation, root biomass, shoot biomass, and seed yield of pea and lentil. J Phytopathol 155: 31–37.

Hussain S, Ali A, Zahid MA, Iqbal SM (2008) Multiple disease resistance in lentil germplasm from Pakistan. Pak J Bot 40: 897–903.

Hynes RK, Leung GCY, Hirkala DLM, Nelson LM (2008) Isolation, selection, and characterization of beneficial rhizobacteria from pea, lentil, and chickpea grown in western Canada. Can J Microbiol 54: 248–258.

Infantino A, Kharrat M, Riccioni L, Coyne CJ, McPhee KE, Grunwald NJ (2006) Screening techniques and sources of resistance to root diseases in cool season food legumes. Euphytica 147: 201–221.

Inder P, Materne M, Taylor PWJ, Ford R (2008) Genotyping elite genotypes within the Australian lentil breeding program with lentil-specific sequenced tagged microsatellite site (STMS) markers. Aust J Agri Res 59: 222–225.

Iskratsch T, Braun A, Paschinger K, Wilson IBH (2009) Specificity analysis of lectins and antibodies using remodelled glycoproteins. Anal Biochem 386: 133–146.

Jansen RK, Wojciechowski MF, Sanniyasi E, Lee S-B, Daniell H (2008) Complete plastid genome sequence of the chickpea (*Cicer arietinum*) and the phylogenetic distribution of *rps12* and *clpP* intron losses among legumes (Leguminosae). Mol Phylogenet Evol 48: 1204–1217.

Kahraman A, Kusmenoglu I, Aydin N, Aydogun A, Erskine W, Muehlbauer FJ (2004a) Genetics of winter hardiness in 10 lentil recombinant inbred line populations. Crop Sci 44: 5–12.

Kahraman A, Kusmenoglu I, Aydin N, Aydogan A, Erskine W, Muehlbauer FJ (2004b) QTL mapping of winter hardiness genes in lentil. Crop Sci 44: 13–22.

Kalo P, Seres A, Taylor S, Jakab J, Kevei Z, Kereszt A, Endre G, Ellis T, Kiss G (2004) Comparative mapping between *Medicago sativa* and *Pisum sativum*. Mol Genet Genom 272: 235–246.

Kamboj RK, Pandey MP, Chaube HS (1990) Inheritance of resistance to Fusarium wilt in India lentil germplasm. Euphytica 50: 113–117.

Kaur S, Pembleton L, Cogan N, Hobson K, Leonforte T, Paull J, Forster J, Materne M (2010) Transcriptome sequencing from cool season food legumes (lentil, chickpea, pea and faba bean) using 454 GS-FLX Titanium technology. In: 5th Int Food Legumes Research Conf & 7th Eur Conf on Grain Legumes, 26–30April, Antalya, p 40: http://www.iflrc-ecgl.org/BookofAbstracts.pdf (Cited 21 July 2010).

Kaur S, Cogan NO, Pembleton LW, Shinozuka M, Savin KW, Materne M, Forster JW (2011) Transcriptome sequencing of lentil based on second-generation technology permits large-scale unigene assembly and SSR marker discovery. BMC Genomics 12: 265.

Kumar R, Mishra SK, Sharma B (2001) Genetics of rust resitance in lentil (*Lens culinaris* Medik.). Indian J Genet 61: 238–241.

Kumar Y, Mishra SK, Tyagi MC, Sharma B (2004a) Detection of two linkage groups in lentil (*Lens culinaris* Medik.). Indian J Genet 64: 306–309.

Kumar Y, Sharma B, Mishra SK, Tyagi MC, Singh SP (2004b) Inheritance of leaflet size in lentil (*Lens culinaris* Medik.). J Lentil Res 1: 26–29.

Kumar Y, Mishra SK, Tyagi MC, Singh SP, Sharma B (2005a) Inheritance of genes for thre pigmentation traits in lentil (*Lens culinaris* Medik.). J Genet Breed 59: 107–111.

Kumar Y, Mishra SK, Tyagi MC, Singh SP, Sharma B (2005b) Linkage between genes for leaf colour, plant pubescence, number of leaflets and plant height in lentil (*Lens culinaris* Medik.). Euphytica 145: 41–48.

Ladizinsky G (1979a) The genetics of several morphological markers in lentil. J Hered 70: 135–137.

Ladizinsky G (1979b) The origin of lentil and its wild gene pool. Euphytica 28: 179–187.

Ladizinsky G (1985) The genetics of hard seed coat in the genus *Lens*. Euphytica 34: 539–543.

Ladizinsky G (1993) Wild lentils. Crit Rev Plant Sci 12: 169–184.

Ladizinsky G (1997a) A new species of *Lens* from south-east Turkey. Bot J Linn Soc 123: 257–260.

Ladizinsky G (1997b). Dawrfing genes in the genus *Lens* Mill. Theor Appl Genet 95: 1270–1273.

Ladizinsky G (1999) Identification of the lentil wild genetic stock. Genet Resour Crop Evol 46: 115–118.

Ladizinsky G, Braun D, Goshen D, Muehlbauer FJ (1984) The biological species of the genus *Lens*. Bot Gaz 145: 253–261.

Ladizinsky G, Weeden NF, Muehlbauer FJ (1990) Tertiary trisomic in lentil. Euphytica 51: 179–184.

Lal S, Srivastava RS (1975).Inheritance of flower color in lentils. Indian J Genet 35: 29–30.

Lev-Yadun S, Gopher A, Abbo S (2000) The cradle of agriculture. Science 288: 1602–1603.

Mamo T, Richter C, Heiligtag B (1996) Salinity effects on the growth and ion contents of some chickpea (*Cicer arietinum* L.) and lentil (*Lens culinaris* Medic.) varieties. J Agron Crop Sci 176: 235–247.

Materne M, McNeil Dl (2007) Breeding methods and achievements. In: SS Yadav, D McNeil, PC Stevenson (eds) Lentil. An Ancient Crop for Modern Times. Springer, Dordrecht, The Netherlands, pp 241–253.

Mayer MS, Bagga SK (2002) The phylogeny of *Lens* (Leguminosae): new insight from ITS sequence analysis. Plant Syst Evol 232: 145–154.

Mikic A (2010) Words denoting lentil (*Lens culinaris*) in European languages. J Lentil Res. (in press).

Miller PR, Brandt SA, McDonald CL, Waddington J (2006) Chickpea, lentil, and pea response to delayed spring seeding on the Northern Great Plains. Can J Plant Sci 86: 1059–1070.

Mishra PK, Mishra S, Selvakumar G, Bisht JK, Kundu S, Gupta HS (2009) Coinoculation of *Bacillus thuringeinsis*-KR1 with *Rhizobium leguminosarum* enhances plant growth and nodulation of pea (*Pisum sativum* L.) and lentil (*Lens culinaris* L.). World J Microbiol Biotechnol 25: 753–761.

Muehlbauer FJ, Slinkard AE (1981) Genetic and breeding methodology. In: C Webb, GC Hawtin (eds) Lentils. CAB, Farham Royal, Slough, UK, pp 69–90.

Muehlbauer FJ, Chen W (2007) Resistance to ascochyta blights of cool season food legumes. Eur J Plant Pathol 119: 135–141.

Muehlbauer FJ, Weeden NF, Hoffman DL (1989) Inheritance and linkage relationships of morphological and isozyme loci in lentil (*Lens* Miller). J Hered 80: 298–303.

Muehlbauer FJ, Kaiser WJ, Clement SL, Summerfield RJ (1995) Production and breeding of lentil. Adv Agron 54: 283–332.

Muehlbauer FJ, Cho S, Sarker A, McPhee KE, Coyne CJ, Rajesh PN, Ford R (2006) Application of biotechnology in breeding lentil for resistance to biotic and abiotic stress. Euphytica 147: 149–165.

Mustafa BM, Coram TE, Pang ECK, Taylor PWJ, Ford R (2009) A cDNA microarray approach to decipher lentil (*Lens culinaris*) responses to *Ascochyta lentis*. Australas Plant Pathol 38: 617–631.

NegussieT, Pretorius ZA, Bender CM (2005) Components of rust resistance in lentil. Euphytica 142: 55–64.

Pala M, Ryan J, Zhang H, Singh M, Harris HC, (2007) Water-use efficiency of wheat-based rotation systems in a Mediterranean environment. Agri Water Manag 93: 136–144.

Peever TL (2007) Role of host specificity in the speciation of *Ascochyta* pathogens of cool season food legumes. Eur J Plant Pathol 119: 119–126.

Penazola E, Tay J, France A (2007) Calpun-INIA, a large seed size and rust-resistant lentil (*Lens culinaris* Medik.) cultivar. Agri Téc 67: 68–71.

Penmetsa RV, Carrasquilla-Garcia N, Rosen B, Gao J, Sarma BK, Datta S, Vail SL, Garzon Luz, Bett K, Vandenberg B, Woodward J, Blair M, Bertioli D, May GD, He G, Bruening GE, Varshney RK, Cook DR (2010) Studies of genome evolution in Papilionoid legumes. In: 5th Int Food Legumes Res Conf & 7th Eur Conf on Grain Legumes, 26–30 April, Antalya, p 41: http://www.iflrc-ecgl.org/BookofAbstracts.pdf (Cited 21 July 2010).

Petterson DS, Sipsas S, Mackintosh JB (1997) The Chemical Composition and Nutritive Value of Australian Pulses. 2nd edn. Grain Research & Development Corporation, Camberra, Australia, p 65.

Phan HTT, Ellwood SR, Ford R, Thomas S, Oliver RP (2006) Differences in syntenic complexity between *Medicago truncatula* with *Lens culinaris* and *Lupinus albus*. Funct Plant Biol 33: 775–782.

Phan HTT, Ellwood SR, Hane JK, Ford R, Materne M, Oliver RP (2007) Extensive macrosynteny between *Medicago truncatula* and *Lens culinaris* ssp. *culinaris*. Theor Appl Genet 114: 549–558.

Pinkas R, Zamir D, Ladizinsky G (1985) Alozyme divergence and evolution in the genus *Lens*. Plant Syst Evol 151: 131–140.

Reddy MRK, Rathour R, Kumar N, Katoch P, Sharma TR (2010) Cross-genera legume SSR markers for analysis of genetic diversity in *Lens* species. Plant Breed 129: 514–518.

Rai R, Singh RP (1999) Effect of salt stress on interaction between lentil (*Lens culinaris*) genotypes and *Rhizobium* sp. Strains: symbiotic *N*-2 fixation in normal and sodic soils. Biol Fert Soils 29: 187–195.

Riccioni L, Haegi A, Valvassori M (2008) First report of vascular wilt caused by *Fusarium redolens* on lentil in Italy. Plant Dis 92: 1132–1132.

Robaglia C, Caranta C (2006) Translation initiation factors: a weak link in plant RNA virus infection. Trends Plant Sci 11: 40–45.

Rosa L de la, Jouve N (1992) Genetic variation for isozyme and proteins in Spanish primitive cultivars and wild species of *Lens*. Euphytica 59: 181–187.

Roy F, Boye JI; Simpson BK (2010) Bioactive proteins and peptides in pulse crops: Pea, chickpea and lentil. Food Res Int 43: 432–442.

Rozan P, Kuo Y-H, Lambein F (2001) Amino acids in seeds and seedlings of the genus *Lens*. Phytochemistry 58: 281–289.

Rubeena, Taylor PWJ, Ford R (2003) Molecular mapping the lentil (*Lens culinaris* ssp. *culinaris*) genome. Theor Appl Genet 107: 910–916.

Rubeena, Taylor PW, Ades PK, Ford R (2006) QTL mapping of resistance in lentil (*Lens culinaris* ssp. *culinaris*) to Ascochyta blight (*Ascochyta lentis*). Plant Breed 125: 506–512.

Ryland D, Vaisey-Genser M, Arntfield SD, Malcolmson LJ (2010) Development of a nutritious acceptable snack bar using micronized flaked lentils. Food Res Int 43: 642–649.

Saenz de Miera LE, Ramos J, Pérez de la Vega M (2008) A comparative study of convicilin storage protein gene sequences in species of the tribe Vicieae. Genome 51: 511–523.

Salam MA, Islam MT (1994) Growth, yield and leaf-water attributes of some advanced mutant lentil lines under different moisture regimes. Lens Newsl 21: 32–35.

Sarker A, Erskine W (2006) Recent progress in the ancient lentil. J Agri Sci 144: 19–29.

Sarker A, Erskine W, Sharma B, Tyagi MC (1999) Inheritance and linkage relationships of days to flowering and morphological loci in lentil (*Lens culinaris* Medikus subsp. *culinaris*). J Hered 90: 270–275.

Sarker A, Erskine W, Singh M (2005) Variation in shoot and root characteristics and their association with drought tolerance in lentil landraces. Genet Resour Crop Evol 52: 87–95.

Scippa GS, Trupiano D, Rocco M, Viscosi V, Di Micheli M, D'Andrea A, Chiatante D (2008) An integrated approach to the characterization of two autochthonous lentil (*Lens culinaris*) landraces of Molise (south-central Italy). Heredity 101: 136–144.

Scippa GS, Rocco M, Ialicicco M, Trupiano D, Viscosi V, Di Michele M , Arena S, Chiatante D, Scaloni A (2010) The proteome of lentil (*Lens culinaris* Medik.) seeds: Discriminating between landraces. Electrophoresis 31: 497–506.

Sharma B, Emami MK (2002) Discovery of new gene causing dark green cotyledons and pathway of pigment synthesis in lentil (*Lens culinaris* Medik.). Euphytica 124: 349–353.

Sharma B, Tyagi MC, Mishra SK, Kumar Y (2004) Three gene control of cotyledon color in lentil (*Lens culinaris* Medik.). J Lentil Res 1: 1–10.

Sharma SK, Knox MR, Ellis THN (1996) AFLP analysis of the diversity and phylogeny of *Lens* and its comparison with RAPD analysis. Theor Appl Genet 93: 751–758.
Siddique KHM, Loss SP, Pritchard DL, Regan KL, Tennant D, Jettner RL, Wilkinson D (1998) Adaptation of lentil (*Lens culinaris* Medik.) to Mediterranean-type environments: effect of time of sowing on growth, yield, and water use. Aust J Agri Res 49: 613–626.
Siddiqui ZA, Baghel G, Akhtar MS (2007) Biocontrol of *Meloidogyne javanica* by *Rhizobium* and plant growth-promoting rhizobacteria on lentil. World J Microbiol Biotechnol 23: 435–441.
Sillero JC, Fondevilla S, Davidson J, Patto MCV, Warkentin TD, Thomas J, Rubiales D (2006) Screening techniques and sources of resistance to rusts and mildews in grain legumes. Euphytica 147: 255–272.
Singh RK, Raghuvanski SS, Prakash D (1992) Genetic improvement in advanced generations of colchitetraploid lentil (*Lens culinaris* Medikus). J Genet Breed 46: 151–154.
Singh TP (1978) Inheritance of cotyledon color in lentil. Indian J Genet 38: 11–12.
Skibinski DOF, Warren ME (1984) Variation at soluble enzyme loci in four *Lens* species. Lens Newsl 11: 10–13.
Slinkard AE (1978) Inheritance of cotyledon color in lentil. J Hered 69: 139–140.
Solaiman Z, Colmer TD, Loss SP, Thomson BD, Siddique KHM (2007) Growth responses of cool-season grain legumes to transient waterlogging. Aust J Agri Res 58: 406–412.
Sonnante G, Pignone D (2001) Assessment of genetic variation in a collection of lentil using molecular tools. Euphytica 120: 301–307.
Sonnante G, Pignone D (2007) The major Italian landraces of lentil (*Lens culinaris* Medik.): Their molecular diversity and possible origin. Genet Resour Crop Evol 54: 1023–1031.
Sonnante G, Galasso I, Pignone D (2003) ITS sequence analysis and phylogenetic inference in the genus *Lens* Mill. Ann Bot 91: 49–54.
Sonnante G, De Paolis A, Pignone D (2005) Bowman-Birk inhibitors in *Lens*: identification and characterization of two paralogous gene classes in cultivated lentil and wild relatives. Theor Appl Genet 110: 596–604.
Stevenson PC, Dhillon MK, Sharma HC, El-Bouhssini M (2007) Insect pest of lentil and their management. In: SS Yadav, D McNeil, PC Stevenson (eds) Lentil. An Ancient Crop for Modern Times. Springer, Dordrecht, The Netherlands, pp 331–348.
Stoddard FL, Balko C, Erskine W, Khan HR, Link W, Sarker A (2006) Screening techniques and sources of resistance to abiotic stresses in cool-season food legumes. Euphytica 147: 167–186.
Tadmor Y, Zamir D, Ladizinsky G (1987) Genetic mapping of an ancient translocation in the genus *Lens*. Theor Appl Genet 73: 883–892.
Tahir M, Muehlbauer FJ (1994) Gene mapping in lentil with recombinant inbred lines. J Hered 85: 306–310.
Tahir M, Muehlbauer FJ (1995) Association of quantitative trait loci with isozyme markers in lentil (*Lens culinaris* L.). J Genet Breed 49: 145–150.
Tahir M, Simon CJ, Muehlbauer FJ (1993) Gene map of lentil: A review. Lens Newsl 20: 3–10.
Tahir M, Muehlbauer FJ, Spaeth SC (1994) Association of isozyme markers with quantitative trait loci in random single seed derived lines of lentil (*Lens culinaris* Medik.). Euphytica 75: 111–119.
Tanyolac B, Ozatay S, Kahraman A, Muehlbauer F (2010) Linkage mapping of lentil (*Lens culinaris* L.) genome using recombinant inbred lines revealed by AFLP, ISSR, RAPD and some morphologic markers. J Agri Biotechnol Sust Develop 2: 1–6.
Tar'an B, Buchwaldt L, Tullu A, Banniza S, Warkentin TD, Vandenberg A (2003) Using molecular markers to pyramid genes for resistance to ascochyta blight and anthracnose in lentil (*Lens culinaris* Medik). Euphytica 134: 223–230.
Taylor PWJ, Ford R (2007) Diagnostics, genetic diversity and pathogenic variation of ascochyta blight of cool season food and feed legumes. Eur J Plant Pathol 119: 127–133.

Taylor P, Lindbeck K, Chen W, Ford R (2007) Lentil diseases. In: SS Yadav, D McNeil, PC Stevenson (eds) Lentil. An Ancient Crop for Modern Times. Springer, Dordrecht, The Netherlands, pp 291–313.

Tivoli B, Baranger A, Avila CM, Banniza S, Barbetti M, Chen WD, Davidson J, Lindeck K, Kharrat M, Rubiales D, Sadiki M, Sillero JC, Sweetingham M, Muehlbauer FJ (2006) Screening techniques and sources of resistance to foliar diseases caused by major necrotrophic fungi in grain legumes. Euphytica 147: 223–253.

Toker C, Yadav SS, Solanki IS (2007) Mutation breeding. In: SS Yadav, D McNeil, PC Stevenson (eds) Lentil. An Ancient Crop for Modern Times. Springer, Dordrecht, The Netherlands, pp 209–224.

Toklu F, Karakoy T, Hakli E, Bicer T, Brandolini A, Kilian B, Özkan H (2009) Genetic variation among lentil (*Lens culinaris* Medik) landraces from Southeast Turkey. Plant Breed 128: 178–186.

Thavarajah D, Thavarajah P, Sarker A, Vandenberg A (2009a) Lentils (*Lens culinaris* Medikus Subspecies *culinaris*): A whole food for increased iron and zinc intake. J Agri Food Chem 57: 5413–5419.

Thavarajah P, Thavarajah D, Vandenberg A (2009b) Low phytic acid lentils (*Lens culinaris*): A potential solution for increased micronutrient bioavailability. J Agri Food Chem 57: 9044–9049.

Tullu A, Buchwaldt L, Warketin T, Taran B, Vandenberg A (2003) Genetics of resistance to anthracnose and identification of AFLP and RAPD markers linked to the resistance gene in PI 320937 germplasm of lentil (*Lens culinaris* Medikus). Theor Appl Genet 106: 428–434.

Tullu A, Buchwaldt L, Lulsdorf M, Banniza S, Barlow B, Slinkard AE, Sarker A, Tar'an B, Warkentin T, Vandenberg A (2006) Sources of resistance to anthracnose (*Colletotrichum truncatum*) in wild Lens species. Genet Resour Crop Evol 53: 111–119.

Tullu A, Tar'an B, Warkentin T, Vandenberg A (2008) Construction of an intraspecific linkage map and QTL analysis for earliness and plant height in lentil. Crop Sci 48: 2254–2264.

Urbano G, Porres, JM, Frías J, Vidal-Valverde C (2007) Nutricional value. In: SS Yadav, D McNeil, PC Stevenson (eds) Lentil. An Ancient Crop for Modern Times. Springer, Dordrecht, The Netherlands, pp 47–93.

Vail S, Vandenberg A (2010) Evaluation of a clonal propagation protocol to obtain replicated disease data on infection by *Colletotrichum truncatum* in *Lens culinaris*. Crop Sci 50: 926–932.

Vaillancourt R, Slinkard AE (1992). Inheritance of new genetic markers in lentil. Euphytica 64: 227–236.

Vaillancourt RE, Slinkard AE (1993) Linkage of morphological and isozyme loci in lentil, *Lens culinaris*. Can J Plant Sci 73: 917–926.

Vandenberg A, Slinkard AE (1987) Inheritance of a xantha chlorophyll deficiency in lentil. J Hered 78: 130.

Vandenberg A, Slinkard AE (1989) Inheritance of four new quantitative genes in lentil. J Hered 80: 320–322.

Vandenberg A, Slinkard AE (1990) Genetics of seed coat color and pattern in lentil. J Hered 81: 484–488.

Van Oss H, Aron Y, Ladizinsky G (1997) Chloroplast DNA variation and evolution in the genus *Lens* Mill. Theor Appl Genet 94: 452–457.

Varshney RK, Close TJ, Singh NK, Hoisington DA Cook DR (2009) Orphan legume crops enter the genomics era! Curr Opin Plant Biol 12: 202–210.

Vavilov NI (1922) The law of homologous series in variation. J Genet 12: 47–88.

Vlachostergios DN, Roupakias DG (2008) Response to conventional and organic environment of thirty-six lentil (*Lens culinaris* Medik.) varieties. Euphytica 163: 449–457.

Wang HX, Ng TB (2007) An antifungal peptide from red lentil seeds. Peptides 28: 547–552.

Wang H, Chang KF, Hwang SF, Gossen BD, Turnbull GD, Howard RJ, Strelkov SE (2006) Response of lentil cultivars to rhizoctonia seedling diseases in Canada. J Plant Dis Protec 113: 219–223.

Wang N, Hatcher DW, Toews R, Gawalko EJ (2009) Influence of cooking and dehulling on nutritional composition of several varieties of lentils (*Lens culinaris*). LWT-Foof Sci Technol 42: 842–848.

Weeden NF, Muehlbauer FJ, Ladizinsky G (1992) Extensive conservation of linkage relationships between pea and lentil genetic maps. J Hered 83: 123–129.

Wilson VE, Hudson LW (1978) Inheritance of lentil flower color. J Hered 69: 139–140.

Wojciechowski MF (2006) IRLC (Inverted Repeat-lacking clade). Version 11 July 2006: http://tolweb.org/IRLC_%28Inverted_Repeat-lacking_clade%29/60358/2006.07.11 in The Tree of Life Web Project, http://tolweb.org/.

Wojciechowski MF, Sanderson MJ, Steele KP, Liston A (2000) Molecular phylogeny of the "temperate herbaceous tribes" of papilionoid legumes: a supertree approach. In: P Herendeen, A Bruneau (eds) Advances in Legume Systematics, part 9. Royal Botanic Garden, Kew,UK, pp 277–298.

Ye G, McNeil DL, Hill GD (2002) Breeding for resistance to lentil Ascochyta blight. Plant Breed 121: 185–191.

Yuzbasioglu E, Acik L, Ozcan S (2008) Seed protein diversity among lentil cultivars. Biol Plant 52: 126–128.

Zamir D, Ladizinsky G (1984) Genetics of allozyme variants and linkage groups in lentil. Euphytica 33: 329–336.

Zarei M, Saleh-Rastin N, Alikhani HA, Aliasgharzadeh N (2006) Responses of lentil to co-inoculation with phosphate-solubilizing rhizobial strains and arbuscular mycorrhizal fungi. J Plant Nut 29: 1509–1522.

Závodná M, Kraic J, Paglia G, Gregova E, Morgante M (2000) Differentiation between closely related lentil (*Lens culinaris* Medik.) cultivars using DNA markers. Seed Sci Technol 28: 217–219.

Zelicourt A de, Letousey P, Thoiron S, Campion C, Simoneau P, Elmorjani K, Marion D, Simier P, Delavault P (2007) Ha-DEF1, a sunflower defensin, induces cell death in *Orobanche* parasitic plants. Planta 226: 591–600.

Zohary D (1999) Monophyletic vs. polyphyletic origin of crops on which agriculture founded in the Near East. Genet Resour Crop Evol 46: 133–142.

4

Grass Pea

Maria Carlota Vaz Patto,[1,*] *Colin D. Hanbury,*[2]
Marijke Van Moorhem,[3,4] *Fernand Lambein,*[4]
Sergio Ochatt[5] and *Diego Rubiales*[6]

ABSTRACT

Grass pea (*Lathyrus sativus* L.) is an annual legume crop with high protein content and remarkable resistance to extreme environmental conditions, including flooding, drought, salinity and low soil fertility, and a significant degree of resistance to biotic stress agents. It is rightly considered one of the most promising sources of calories and protein for the vast and expanding populations of drought-prone and marginal areas of Asia and Africa and as an interesting alternative for cropping systems diversification in marginal lands in Europe, Australia and America. It is a dual purpose crop with great agronomic potential as a grain and forage legume. Nevertheless, as a result of the little breeding effort invested in it compared to other legumes, grass pea cultivation has shown a regressive pattern in many areas in recent decades. The presence of a neurotoxin proposed to be responsible for lathyrism in humans and animals has also contributed to its abandonment. Recently, there has been renewed interest in the improvement of this high-potential legume species, with the integration of new breeding technology that anticipates great achievements in the development of new varieties.

[1]Instituto de Tecnologia Química e Biológica (ITQB), Universidade Nova de Lisboa (UNL), Apartado 127, 2781-901 Oeiras, Portugal; e-mail: *cpatto@itqb.unl.pt*

[2]Department of Agriculture and Food, Western Australia, 3 Baron Hay Court, South Perth 6151, Australia.

[3]Department of Basic Medical Sciences, Faculty of Medicine and Health Sciences, Ghent University, K.L. Ledeganckstraat 35, 9000 Ghent, Belgium.

[4]Institute of Plant Biotechnology for Developing Countries (IPBO), Ghent University, K.L. Ledeganckstraat 35, 9000 Ghent, Belgium.

[5]INRA, Centre de Recherches de Dijon, UMRLEG, BP 86510, 21065 Dijon Cedex, France.

[6]Instituto de Agricultura Sostenible, CSIC, Apartado 4084, 14080, Córdoba, Spain.

*Corresponging author

In this chapter we review the present economic importance, advantages and constraints, available genetic resources, major breeding achievements and we report on recent research approaches to be used in the improvement of grass pea.

Keywords: biotic and abiotic stresses, breeding, diversity analysis, expression analysis, genetic transformation, genome mapping, lathyrism, ODAP, resistance

4.1 Introduction

Only a limited number of crops produce the bulk of food consumed in the world. Many neglected and underutilized species are extremely important for food production in low-income food deficit countries (LIFDC) (Ochatt et al. 2007). They are locally well adapted to marginal lands and constitute an important part of the local diet, providing valuable nutritional elements (e.g., protein, vitamins, and minerals) and spices, often lacking in staple crops. Neglected and underutilized crops also represent an important source of revenue for local economies and are part of the rich cultural and traditional heritage of communities around the world. In addition, these crops are important sources of resistance genes for biotic and abiotic stress breeding that can be utilized also for the genetic improvement of commodity crops. Compared to major crops, they require relatively low inputs and, therefore, contribute to sustainable agricultural production. So far there has been little research and development of these species due to lack of government funding and interest of the private sector. These traditional crops are often low yielding, because they are often grown on marginal land and cannot compete with major crops, even though many have the potential to be economically viable.

The annual legume, *Lathyrus sativus* L., also known as grass pea (*khesari* in Bangladesh, *san lee dow* in China, *guaya* in Ethiopia, *pois carré* in France, *khesari dal* or *theora* in India, *gharas* in Pakistan, *zaailathyrus* in Holland, *almorta* in Spain, *chícharo* in Portugal), is one such neglected species being widely grown as a pulse crop for human food but also for animal feed and as forage in many of the harshest agro-environments of the world (Hanbury 2000). Nowadays, for over 200 million people in drought-prone areas of Asia and Africa, grass pea is a traditional popular crop because of its easy cultivation, its resistance to drought, flood and insect attacks, and its good yield of protein-rich and tasty seeds with relatively little labor expenditure (Khan et al. 1994; Lambein et al. 1994; Getahun et al. 2002b). It can be produced without any fertilizers and hence its cost of production is very low compared to other crops (Getahun et al. 2002a). *L. sativus* is also important in crop rotation systems because of its efficient nitrogen fixation, improving the fertility of the soil and increasing the yield of the subsequent (mostly cereal) crop. Its extreme resistance to drought is a characteristic likely to become even more important in many regions as a result of climate change.

Lathyrus sativus is a diploid species ($2n = 2x = 14$) (Smartt et al. 1984) that belongs to the family Leguminosae (= Fabaceae), subfamily Papilionoideae, tribe Vicieae (= Fabaea). The composition of *L. sativus* seeds is similar to those of other commonly used feed grain legumes, with a high protein content of 29% (Hanbury et al. 1999). Grass pea is also characterized by a low content of the essential amino acid methionine (Lambein et al. 2001) and high levels of antinutritional factors (Wang et al. 1998), particularly the neurotoxic β-N-oxalyl-L-α,β-diaminopropionic acid (β-ODAP, hereafter in this chapter designated as ODAP). In years when conditions are particularly harsh, human consumption of the crop may increase through lack of any suitable alternative, especially for the poorest rural people, to a level at which there is a severe risk of the consumer succumbing to lathyrism, a neurological disorder caused by the presence of ODAP.

Statistics on grass pea area, production and yield are not readily accessible; however, the information that is available indicates that India is the largest producer of grass pea grain (Skiba et al. 2007) and in this country the crop is the third most important pulse crop after chickpea and pigeon pea (Kuo et al. 2000). Between 1980/81 and 1994/95 the area sown to grass pea in India was 0.9–1.4 million ha, with an average yield of 0.4–0.6 t/ha (Gautam et al. 1998). Bangladesh and Ethiopia are the second and third largest producers of grass pea grain, with 0.24 million ha (average yield 0.7 t/ha) in Bangladesh (as cited in Campbell 1997), where grass pea is the most important pulse crop (Kuo et al. 2000), and 0.14 million ha (average yield 0.7 t/ha) in Ethiopia (as cited in Tadesse and Bekele 2003b) where it ranks fourth, in both cultivated area and production, after faba bean, field pea and chickpea (Kuo et al. 2000). It is also cultivated to a lesser extent in central, southern and eastern Europe, West Asia and North Africa (Campbell 1997). Additionally, it is being investigated as a potential crop for other parts of the world, including the southern Australian Mediterranean-type environments (Siddique et al. 1996; Hanbury et al. 1999). In Europe, where diversification of cropping systems and more sustainable and environmentally safe agriculture are the main concerns, grass pea has potential to be reintroduced in marginal areas providing an alternative to imported soya or the hazardous use of protein of animal origin in animal feed (Vaz Patto et al. 2006b).

Although there are relatively few efforts throughout the world to genetically improve grass pea, some important programs exist that aim to improve its yield, resistance to biotic and abiotic stresses and, most importantly, to reduce the percentage, or ideally eliminate, the neurotoxin from the seed. All these breeding efforts require access to suitable genetic resources. However, local landraces and cultivars are being lost as farmers switch to alternative crops, potentially limiting the progress that can be made through genetic enhancement. Due to its importance as a survival food for some of the world's poorest people, yet recognizing the dangers that excessive consumption can cause, in 1991 grass pea was listed among the crops included in the multilateral system of access and benefit sharing under the International Treaty on Plant Genetic Resources for Food and Agriculture (ITPGRFA) (Crop Trust 2007).

Some significant collections of cultivated and wild *Lathyrus* species have already been assembled and are maintained *ex situ* in a number of different institutes throughout the world. The largest collections are maintained by the University of Pau (IBEAS) in France (4,477 accessions) and by ICARDA in Syria (3,239 accessions), with the Indian (NBPGR), Bangladeshi (GRC Bangladesh Agricultural Research Institute) and Russian (VIR) collections coming next with 2,619, 2,432 and 1,835 accessions, respectively (Crop Trust 2007). Nearly all (98%) of the material held by NBPGR in India and 40% of that in Russia are reportedly of national origin. Many other collections also report high levels of indigenous material including Ethiopia (98% of 588 accessions), Nepal (100% of 164 accessions), Portugal (100% of 256 accessions), Spain (89% of 377 accessions) and Turkey (100% of 363 accessions). The two largest collections (France and ICARDA) both comprise about 50% *L. sativus*. In terms of species, the collections in the main grass-pea-producing countries all have high percentages of *L. sativus*: those in Bangladesh, Ethiopia, India and Nepal all comprise at least 70% *L. sativus* (Crop Trust 2007).

Lathyrus sativus is described as being a much-branched suberect, straggling or climbing herbaceous winter annual. Although considered self-pollinating, the percentage of cross-pollination in *L. sativus*, which is primarily due to bees, appears to be within the range 4–28% (Kaul et al. 1986; Rahman et al. 1995).

It was previously believed that the natural distribution of grass pea had been obscured by cultivation, even within one of its presumed centers of origin located in south-west or central Asia (Smartt 1984). After combining archaeobotanical and phytogeographical evidences, it has been found that cultivation of grass pea began in the Balkan Peninsula, during the early Neolithic period, around the beginning of the 6th millennium BC (Kislev 1989) as a result of the expansion of Near Eastern agriculture of annuals into the region.

According to the genepool concept developed by Harlan and de Wet (1971) the cultivated and wild races of *L. sativus* are included in the grass pea primary genepool. The secondary genepool includes the other species that will cross with some difficulty with *L. sativus*. Therefore, here

Heywood et al. (2007) included: *L. chrysanthus, L. gorgoni, L. marmoratus* and *L. pseudocicera,* with which *L. sativus* can cross and produce ovules, and possible more remotely *L. amphicarpos, L. blepharicarpus, L. chloranthus, L. cicera, L. hierosolymitanus* and *L. hirsutus,* with which *L. sativus* can cross and with which pods are formed. Nevertheless, based on crossability and cytological evidence between grass pea and 15 wild species in the section *Lathyrus,* Yunus and Jackson (1991) placed only *L. cicera* and *L. amphicarpos* in the grass pea secondary genepool. The tertiary genepool include species that can cross with the original crop species only with use of specialized techniques such as embryo rescue and culture or the use of bridging species, therefore the remaining species of the genus can be considered members of the tertiary genepool (Heywood et al. 2007).

4.2 Lathyrism

The term "lathyrism" was introduced by Cantani in 1873 for a crippling condition occurring after ingestion of large amounts of "certain peas" and known in Greek and Indian literature for about 2,500 years. Later, the term "neurolathyrism" was coined by Selye (1957) to differentiate between the spastic paraparesis of the legs occurring in humans and certain animals after a period of dependency on grass pea or related species, and osteolathyrism indicating skeletal changes induced in various experimental animals by β-aminopropionitrile (BAPN) present in bound form in the seeds of sweet pea (*L. odoratus*) or by some synthetic chemicals. The latter condition is also called experimental lathyrism or angiolathyrism. While neurolathyrism is never lethal, osteo-lathyrism can cause aortic rupture (angiolathyrism). Only rare cases of osteolathyrism among sufferers of neurolathyrism have been reported (Haque et al. 1997).

The discovery of a neuro-active compound in seeds of grass pea, identified as β-oxalyl-α,β-diaminopropionic acid (ODAP) in 1964 by two independent research groups has triggered intensive studies on the precise action of this ODAP in animals and, recently, on cultured neuronal cells. It was considered that neurolathyrism was the only upper motor neuron disease for which the specific cause was established, and neurolathyrism was seen as a model for other motor neuron diseases. Plant breeders have performed many and often long-term experiments trying to achieve grass pea varieties with low- or zero-level ODAP. These studies did not result in stable ODAP-free varieties of grass pea. A possible explanation for this may be that enzymes of primary metabolism are involved in the biosynthesis of ODAP (Kuo et al. 1998) and their elimination would be lethal for the plant.

The drought tolerance and easy cultivation of grass pea has made it a survivor since the Neolithic era and its seeds have saved uncounted numbers of human lives during drought-triggered famines in Ethiopia and the Indian subcontinent where it remains the cheapest source of dietary protein. But the double stigma of being the food of the very poor and the cause of the irreversible neurolathyrism (in up to 6% of the population during severe epidemics) has made grass pea a neglected crop with a reputation as being a toxic plant. However, this toxic reputation has to be taken *cum granum salum*, as it would take at least 50 kg of grass pea seed ingested as principal dietary component over an uninterrupted period to have a 6% chance of becoming crippled, while 500 g of kitchen salt (NaCl) or one meal of unprocessed bitter cassava roots can be lethal.

4.2.1 History

Grass pea has been an important crop both for human consumption and for animal feed or fodder since ancient times. The origin of grass pea cultivation is around 6000 BC (Kuo et al. 2000). Neurolathyrism was known to ancient Hindus (the Indian physician Charak associated the consumption of grass pea with a disease state characterized by trembling and limping of legs and loosening of joints), to Hippocrates (460–377 BC), Pliny the Elder (23–79 AD), Pedanius Dioscorides (50 AD) and Galen (130–210 AD). Centuries later, in 1671, Duke George of Württenberg banned the use of grass pea in bread, principally because of its ability to "paralyze" legs. It was Cantani of Naples (1873), who coined the name "lathyrismus" to describe the disease. During the 18th, 19th and 20th centuries, outbreaks of neurolathyrism occurred throughout Europe, northern Africa, the Middle East, Afghanistan, Russia, and India (see Section 4.4.2) (Spencer 1995; Getahun et al. 2002a).

The most remarkable account comes from the description of the epidemic in the German prisoners of war camp in Vapniarca, a town in Ukraine occupied by Germany during the Second World War. There, the highest incidence of neurolathyrism on record developed between December 1942 and the end of February 1943. One of the prisoners, Dr. A. Kessler, noted many details on the diet and neurological symptoms of the prisoners and the neurolathyrism victims. The prisoners received daily rations of 400 g boiled grass pea and 200 g barley bread, and 60% of the inmates developed various stages of neurolathyrism, the highest incidence ever reported in the literature. From Kessler's report (Kessler 1947) it has been deduced that a threshold level of ODAP intake may be between 500 mg and 1 g per day per adult person under the conditions of the forced labor camp (Lambein et al. 2001).

Surprisingly, extracts from grass pea have long been used for homeopathic treatment: the substance prepared from the ripe seeds was recommended in spastic spinal paralysis, multiple sclerosis, myelitis, weakness and paralysis of the lower extremities and rheumatic paralysis (Getahun et al. 2002a).

4.2.2 Epidemiology

Although neurolathyrism can occur sporadically (Haque et al. 1994), it is supposed to be epidemic in nature. Affected families usually have only a single case of neurolathyrism (Hugon et al. 1988; Haimanot et al. 1990, 2005).

Since antiquity, neurolathyrism has been reported in many parts of the world (Fig. 4-1): in the Indian subcontinent (Bangladesh, India, Nepal and Pakistan), Afghanistan, China, Ethiopia and in parts of Europe (Greece, Spain and Ukraine) (Haimanot et al. 1990, 2005; Haque et al. 1994; Spencer 1995; Getahun et al. 2005).

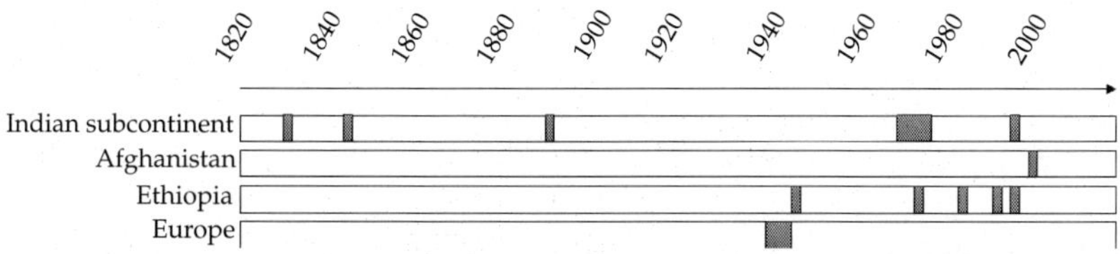

Figure 4-1 Reported neurolathyrism epidemics in the world. In the Indian subcontinent neurolathyrism was reported in 1831 (Bengal), 1844 (Bengal), 1892 (Bengal), 1970–1976 (Bangladesh), 1998 (Nepal). In Afghanistan it was reported in 2001, whereas in Ethiopia in 1947, 1976–1977, 1984–1985, 1994 and 1997–1998. In parts of Europe neurolathyrism occurred during World War II. In Spain it occurred during the Civil War.

The prevalence is higher in Ethiopia (0.6–3%; Getahun et al. 1999, 2002a) than in Bangladesh (0.029–1.88%; Haque et al. 1996). The reported male:female ratio varies between 2:1 and 10:1 (Haque et al. 1996). Not only female hormones, but also cultural habits have been considered to explain this sex ratio. For example in Bangladesh males traditionally eat first, so during famine, relatively less food is left for the female family members (Haque et al. 1996).

4.2.3 Risk Factors

The variability in the susceptibility to neurolathyrism is such that no threshold of toxicity for ODAP content in the seed, grass pea intake or ODAP intake could be scientifically established. Several factors have been reported to affect the susceptibility to neurolathyrism. The major risk factors are famine and chronic food shortage (Getahun et al. 2003), but also young age and male sex (Haimanot et al. 1990; Haque et al. 1996), heavy physical labor (Haimanot et al. 1990), a diarrheal episode, a history of acute illness, zinc deficiency (Khan et al. 1993; Lambein et al. 1994), illiteracy, large family size, exclusive use of clay utensils (Getahun et al. 2005) and blood group O (Getahun et al. 2002b) have been reported as statistically significant risk factors.

A reduced risk of paralysis was associated with soaking grass pea seeds in water before preparation (Haimanot et al. 1993). In addition, the type of grass pea seed preparation is a critical factor: consumption of grass pea as roasted, boiled and raw unripe seeds has been associated with an increased risk of neurolathyrism, whereas no raised risk has been noted for the fermented pancake, unleavened bread or gravy preparations (Getahun et al. 1999, 2002a, 2003). Soaking grass pea in water before cooking, fermentation, mixing the food with gravy that contains condiments with antioxidant activity or mixing with cereals rich in sulfur amino acids is shown to be protective (Getahun et al. 2005). There are also some beliefs, often described as popular misconceptions, that steam from boiling, smoke from roasting, the dust from harvesting grass pea seeds or walking through a flowering field or lying on the straw and the stalks of grass pea are responsible for the disease. The pancake and raw forms are considered to be more poisonous when consumed with milk (Haimanot et al. 1990; Getahun et al. 2002a).

Studies to optimize the extraction of ODAP with solutions of various culinary products suggested that ODAP could best be removed from seeds when they are treated with limewater overnight and then cooked for 25 minutes. This method can serve as a simple procedure to detoxify the grass pea seeds in real household conditions (Jahan and Ahmad 1993). Risk factors are discussed below (see Section 4.2.6).

4.2.4 Clinical Aspects

Most frequently, a latency period of 2–3 months is observed after continuous daily intake of at least 400 g grass pea seeds, although some report consumption of less than a month before contracting the disease (Haimanot et al. 1990). Three modes of initial presentation of neurolathyrism have been described: most common is a sudden onset in which the affected individual complains of heavy, weakened legs after falling down or awakening from sleep, often following physical exhaustion. Some subjects report a subacute onset of walking disability, whereas others experience a weakness and progression of spastic paraparesis up to three months after onset. Some patients report a significant improvement of symptoms early in the course of the disease when discontinuing consumption of grass pea. After these initial changes disappear, the irreversible spastic paraparesis develops gradually. This results in toe-walking, scissoring and unsteady gait. This clinical picture is consistent with the neuropathological evidence of degeneration of Betz cells and of anterior and dorsolateral corticospinal tracts in thoracic, lumbar and sacral regions of the spinal cord. The age of onset is between 2 and 70 years, although the disorder most commonly affects youngsters (mean age of onset is 16.4 years) (Ludolph et al. 1987; Haque et al. 1994; Spencer 1995, 1999; Ludolph and Spencer 1996).

The diagnosis relies on three factors: a history of heavy ingestion of *L. sativus* seeds (at least 2 weeks prior to and at the time of onset of walking difficulty), symmetrical leg weakness with exaggerated thigh adductor, patellar, and ankle reflexes and normal sensory examination, intact mentation, cerebellar function and cranial nerve function (Spencer 1995).

In India, the degree of neurological impairment has been crudely but usefully classified according to the extent of walking difficulty and the need for physical support (Hugon et al. 1993). The majority is mildly affected having difficulty running, but without aid ("no-stick stage"). The more severely affected walk with a characteristic scissoring gait and require the support of one ("1-stick stage") or two crutches ("2-stick stage"). The most severe cases are forced to crawl or use a wheelchair ("crawler stage") (Hugon et al. 1988). Most neurolathyrism patients reside in one of the first two stages (Haimanot et al. 1990).

With the exception of the most severely affected individuals, the arms are clinically uninvolved, speech, swallowing, eye movements, cognitive state, sensation, and proprioception appear normal in all cases and life expectancy is not affected (Hugon et al. 1988, 1993; Spencer 1995, 1999; Paleacu et al. 1999). However, some people do report cognitive deficits, mood disturbances and vivid dreams (Kunig et al. 1994).

4.2.5 Treatment

Most health workers and neurolathyrism patients know that neurolathyrism is incurable and preventable. However, some patients seek help from a medical doctor, from holy water, massage with different oils and lubricants, an Islamic service or a traditional healer (Haimanot et al. 1990; Haque et al. 1996; Getahun et al. 2002a).

A clinical study has been performed with tolperisone HCl, an inhibitor of multisynaptic reflex activity thereby reducing strychnine-induced hyperreflection. When administered to neurolathyrism patients for a period of three months (150 mg/d), tolperisone HCl had a beneficial effect on several parameters both in Bangladesh and in Ethiopia (Haque et al. 1994; Melka et al. 1997). Dorsal rhizotomy or incomplete surgical transsection of the thigh adductor muscles has been tried with some occasional success. Little relief was offered by chemical spasmolytics, and continued treatment of impoverished subjects in remote rural settings is impractical (Spencer 1995). However, such treatments are also not possible for financial or ethical reasons.

In India, attempts to ban the cultivation of grass pea did not succeed, but the intervention of market forces had some effect. The price of grass pea became higher than that of wheat, making this crop well beyond the reach of the poor. In Bangladesh the price of grass pea is now higher than the price of rice, and no new cases of lathyrism are reported. During the last epidemic (1970–1976) grass pea was the cheapest food available in Bangladesh.

4.2.6 Grass pea and ODAP

The non-protein amino acid present in grass pea seeds, ODAP, was suggested to be the causal agent of human neurolathyrism (Spencer et al. 1986). ODAP was discovered by two research groups in 1964 (Murti et al. 1964; Rao et al. 1964). Before the discovery of β-ODAP, it was pointed out that the seeds of grass pea itself were harmless, but that the danger of the disease neurolathyrism laid probably in the contamination of the latter with other seeds, for instance *Vicia sativa* (common vetch) containing β-cyano-L-alanine (Ressler 1962). It was also noted that methionine excretion falls in the urine of neurolathyrism patients (Rudra and Chaudhury 1952). Since the discovery of ODAP, all attention was directed towards this neuro-excitant compound, recently described as a multifunctional plant metabolite of toxic reputation (Lambein et al. 2007).

It can be questioned, whether this focusing on ODAP might have blinded researchers to other neurotoxic effects of grass pea, together with ODAP. The ODAP content in grass pea seeds ranges from 0.3 to 22 g/kg, making the estimated averaged daily intake of ODAP 15–150 mg/kg/d (Spencer 1995). If ODAP would be the only responsible agent for developing neurolathyrism, the prevalence would be higher than up to 6%, so other factors must play an additional role to make some people more vulnerable for this neurodegenerative disease compared to the majority. For example, environmental factors such as drought, zinc deficiency, iron or manganese oversupply in the soil can not only considerably increase the level of ODAP in the seeds, but these same environmental factors can also influence the patients' susceptibility to neurolathyrism. For instance zinc deficiency, which is even aggravated by ODAP being a carrier for zinc-ions (Lambein et al. 1994), has been shown to induce oxidative stress and apoptotic death in neurons (Mackenzie et al. 2007). Very recently, it has been described that manganese, which is abundantly present in Ethiopian soils, may potentiate the neurotoxicity of grass pea seeds or ODAP by altering the blood–brain barrier permeability (Mishra et al. 2009). Manganese overexposure is also found to correlate with increased susceptibility to other neurodegenerative diseases such as Parkinson's disease (Roth and Garrick 2003). Next to environmental factors, the characteristic methionine and cysteine deficiency of grass pea seeds, the lowest among commercial legumes, can also contribute to neurodegeneration in neurolathyrism patients by its effect on oxidative stress. That this deficiency in the seeds has consequences in humans was already reported before the discovery of ODAP. Apparently, the methionine excretion of neurolathyrism patients falls (Rudra et al. 1952), indicating a deficiency of this sulfur-containing amino acid as induced by the consumption of grass pea. This was later confirmed by Nunn et al. (1994), who found a significant decrease of methionine in the serum of volunteers after ingesting a single meal of grass pea. This amino acid is a crucial precursor for glutathione, responsible for the protection of cells (including motor neurons) against oxidative stress.

However, neurotoxicity of ODAP is the best studied aspect in neurolathyrism research, though its mechanism has still not been conclusively elucidated. Excitotoxicity mediated through activation of glutamate receptors (Ross et al. 1989), more specifically α-amino-3-hydroxy-5-methyl-4-isoxazolepropionic acid (AMPA) receptors (Kusama-Eguchi et al, 1996), mitochondrial (Sriram et al. 1998) as well as enzymatic dysfunctions (Pai and Ravindranath 1993) have been proposed to explain at least part of its neurotoxicity (Getahun et al. 2005). For all these proposed mechanisms, entry of ODAP into the central nervous system is crucial, and evidence for that is hard to find. Entry of this toxin into the blood circulation has been suggested by studies of Nunn et al. (1994), indicating that when subjects consumed a single meal of 200 g of cooked grass pea seeds, which contained about 30 mmol/kg of ODAP, it produced plasma concentrations of 2.6–5.25 μM after 150–200 min (Nunn et al. 1994; Warren et al. 2004). When amino acid levels of cerebrospinal fluid (CSF) were studied in 50 patients with neurolathyrism and 12 healthy volunteers, ODAP was not detected. However, the levels of excitatory amino acids glutamate and aspartate were 281 and 71%, respectively, of control values. The concentration of inhibitory amino acids glycine and taurine were 277 and 198%, respectively, of the levels in CSF from control individuals. There was a significant correlation between the level of glycine and the duration of the disease. Also increased levels of threonine, serine and alanine were found. In contrast to reports on other motor neuron diseases where an increase of isoleucine was observed, a significant decrease of isoleucine was found. So these results suggest a disturbance of amino acid metabolism due to excitotoxic damages caused by ODAP (Khan et al. 1995).

4.2.7 Conclusions

Because of its rich history, its high traditional and nutritive value and easy cultivation, the consumption of grass pea can not just be abandoned to protect people from developing neurolathyrism, the problem is much more complicated than that. All epidemiological studies on neurolathyrism report overconsumption of grass pea during extended periods of time prior to the sudden onset of neurolathyrism symptoms, and the poverty and often illiteracy of the victims. Apparently, neurolathyrism only occurs when grass pea is the cheapest food available to the poor. Banning the sale of grass pea without control of the consumption and production can only increase the risk for neurolathyrism for the poor farmers producing it. Also the toxic reputation of grass pea, ignored by millions of consumers, must affect the market value of grass pea, which is lower than other food legumes.

Since the discovery of ODAP 45 years ago, attempts to remove the neurotoxin from the otherwise nutritive (with a protein content of more than 25%) and environment-resistant (against drought, flood, insect attack and selection) crop have been made worldwide by many different methods (Kuo et al. 2000), unfortunately still without any stable or reliable result. Although several attempts were made to genetically transform the plant, this technology is clearly underexploited to develop grass pea varieties with better nutritional qualities. Only one success has been published (Barik et al. 2005), while this may be the only sure way to develop stable varieties with better balanced composition of essential amino acids and low or zero-level ODAP. The semantic discussions on the benefit of ODAP for the survival of the plant under harsh environments can only be settled when varieties without ODAP can be studied.

ODAP remains a useful neuro-active molecule that has been instrumental in the elucidation of important molecular mechanisms. It affects various neurological and enzymatic activities and its effects are more complex than originally proposed or envisaged. It is essential to develop an animal model of neurolathyrism (Kusama-Eguchi et al. 2005) that can be important for the elucidation of more general molecular mechanisms of neurodegeneration.

The best approach to protect people from neurolathyrism may be to inform them about the safe consumption of grass pea. Therefore, more scientific evidence is needed to prove some hypotheses and to define safe nutritional standards.

4.3 Resistance to Biotic and Abiotic Factors

4.3.1 Resistance to Fungal Diseases

4.3.1.1 Powdery Mildew (Erysiphe pisi)

Powdery mildew is among the major diseases that affect *L. sativus* (Campbell et al. 1994) and *L. odoratus* crops (Cook and Fox 1992). *Erysiphe pisi* is a biotrophic, ascomycete fungus, characterized by its gray to white colonies formed on leaves, stems and pods of infected plants. It also causes severe damage to other cultivated members of the Leguminosae family (Dixon 1978; Poulter et al. 2003). Existence of specialized forms and races is still unclear, but different ability to infect different plant species has been reported. Cook and Fox (1992) reported that a strain of *E. pisi* collected on *L. odoratus* was able to infect faba bean (*Vicia faba*) but not pea, whereas a different strain collected on *L. latifolius* was able to infect pea and faba bean.

Little is known about the availability of resistance against powdery mildew in the *Lathyrus* genus and its different responsible mechanisms. Qualitative resistance to *E. pisi*, characterized by a collapse of sporelings shortly after germination, associated with contacted epidermal host cell collapse was reported on *L. belinensis* and its hybrids with *L. odoratus* (Poulter et al. 2003). *Lathyrus odoratus* × *L. belinensis* hybrid plants and those produced by backcrossing to *L. odoratus* were resistant to *E. pisi* (Poulter et al. 2003). Continued backcrossing resulted in introgressed plants that closely resembled the *L. odoratus* parent, but segregated for complete resistance/susceptibility to *E. pisi*, with a ratio of 2.5:1 resistant to susceptible plants, suggesting the presence of a single dominant gene that confers resistance. Although spores germinated on the leaves of both resistant and susceptible plants, spores present on resistant plants collapsed shortly after germination.

Lathyrus sativus lines showing moderate resistance to powdery mildew have been reported in India and Syria (Campbell et al. 1994; Robertson and El-Moneim 1996; Asthana and Dixit 1998), but the disease reactions were not critically studied. Resistance has been only recently characterized in *L. sativus* and *L. cicera* germplasm (Vaz Patto et al. 2006a, 2007). Under growth chamber conditions, *L. sativus* showed in most cases a compatible reaction of high infection type (ITy) with no macroscopically visible hypersensitivity. However, disease severity (DS) varied significantly among accessions. Accessions with reduced disease severity of a high-infection-type powdery mildew have been identified (Vaz Patto et al. 2006a). The combination of different resistance genes and or resistance mechanisms in new cultivars should provide longer durability of resistance than each resistance gene by itself. Some of the *L. sativus* accessions segregated for resistance reaction with plants with no symptoms and plants with well-formed colonies with no associated macroscopically visible chlorosis or necrosis. This could be due to a certain level of heterogeneity affecting the polygenetic partial resistance to powdery mildew. If this is the case, it will be possible for breeders to select plants with higher levels of resistance from within these accessions. Otherwise, this could be understood as a case of low infectibility. From these accessions, selected resistant plants were self-pollinated and crossed with susceptible accessions. The obtained progeny will allow the study of the inheritance of this resistance.

4.3.1.2 *Downy Mildew* (Peronospora *spp.*)

Downy mildew is a serious grass pea disease in South Asia, where the climatic conditions are very favorable for endemic outbreak of this pathogen (Campbell 1997). Downy mildew resistance has been reported in *L. sativus* (Asthana and Dixit 1998) and *L. aphaca* (Campbell 1997). However, more detailed studies are needed to discern if this is due to true resistance or just to escape mechanisms (Campbell 1997).

4.3.1.3 *Rusts* (Uromyces *spp.*)

Several rust species can infect the major grain legumes, most of them belonging to the genus *Uromyces* (Barilli et al. 2011). Rusts are important diseases of grass pea in northwestern Ethiopia (Campbell 1997). In the *Lathyrus* genus infection by different rust species has been reported, although on *L. sativus* and *L. cicera*, data only refer to *U. pisi* and *U. viciae-fabae* (Farr et al. 2008, on line Fungal database). *Uromyces viciae-fabae* is commonly known as the rust of faba beans but is a species complex infecting several different plant genera (*Lathyrus, Lens, Phaseolus, Pisum* and *Vicia*) (Farr et al. 2008). Moreover, recent studies have shown high specialization within *U. viciae-fabae*, with each morphologically distinct isolate exclusively infecting cultivars of the species from which it was collected, allowing subdivisions such as *U. viciae-fabae* ex. *V. faba*, ex. *L. culinaris* or ex. *V. sativa* (Emeran et al. 2005). Results from cross-species inoculations (Emeran 2003) agreed with the above, where *L. sativus* was only infected by *U. pisi*, and slightly by *U. ciceri-arietini* and *U. viciae-fabae* ex. *V. sativa* but not by *U. viciae-fabae* ex. *V. faba* or by other *Uromyces* species (Emeran 2003).

A wide range of disease reaction severity has been found in *L. sativus* (Vaz Patto and Rubiales 2009) and *L. cicera* (Vaz Patto et al. 2009) germplasm against these rust species. A compatible reaction (high infection type) was the rule with *U. pisi*, although in this case the level of resistance varied in terms of disease severity among accessions. *L. sativus* accessions were, in general, more resistant (lower DS) to *U. pisi* than accessions of *L. cicera* both under field and growth chamber conditions. However, low infection type was only found on some *L. cicera* accessions. Relatively low infection in spite of high ITy was detected in 17% of *L. cicera* and in almost all *L. sativus* accessions, indicating levels of partial resistance. In some *L. cicera* accessions, a phenotype mixture was detected, with plants with hypersensitivity reaction and plants with high ITy. This is an indication of a certain level of variability in these accessions. *L. sativus* accessions were also highly resistant against *U. ciceris-arietini* and *U. viciae-fabae* ex. *V. sativa*.

Differences among the selected *Lathyrus* accessions became evident once the stomata were penetrated by the infection structures. Resistance to

U. pisi was only detected microscopically after stomata penetration and both pre-haustorial and post-haustorial resistance mechanisms were identified. In the partially resistant *L. cicera* and *L. sativus* accessions, resistance was due to a restriction of haustorium formation with a high level of early abortion of the colonies, a reduction in the number of haustoria per colony and, in some cases, also with a reduction in the intercellular growth of infection hyphae, representing a clear example of pre-haustorial resistance with no associated necrosis. In some *L. cicera* accessions, post-haustorial resistance with necrosis of the host cells associated with infection hyphae and a reduced growth of the colony were identified. Hence, the resistance reaction to *U. pisi* shows not only typical host resistance reactions such as hypersensitivity, but also pre-haustorial resistance based both on hampered haustoria formation and low spore germination and bad germ tube orientation with no associated necrosis (partial resistance sensu Parlevliet 1979). Such partial resistance against rusts not associated with host cell necrosis is common in major grain legumes, but hypersensitive reaction against rust fungus is also found on certain legume species (reviewed in Sillero et al. 2006). Macroscopically visible hypersensitive resistance was detected not only in some *L. cicera* accessions to *U. pisi*, but also in *L. sativus* to *U. ciceris-arietini* and in *L. cicera* to *U. viciae-fabae*. At microscopic level, host cell death was also detected on *L. cicera* against *U. ciceris-arietini* and in *L. sativus* against *U. viciae-fabae* which resulted in reduced colony size. However, in these host–pathogen interactions the importance of a potential pre-haustorial resistance mechanism with no associated necrosis cannot be excluded and requires a more detailed evaluation in the future.

4.3.1.4 *Ascochyta* Blight

Lathyrus spp. are known to be resistant to *Mycosphaerella pinodes*, the causal agent of pea Ascochyta blight. Gurung et al. (2002) showed that *L. sativus*, *L. ochrus* and *L. clymenum* accessions were significantly more resistant to stem infection than field pea cultivars. Resistance of *L. sativus* accessions was confirmed by Skiba et al. (2004), but sufficient variation was identified in the level of resistance to justify a genetic study that suggested that resistance in *L. sativus* may be controlled by two independently segregating genes, operating in a complementary epistatic manner. Resistance to *Ascochyta pisi* has also been reported (Robertson and Abd El-Moneim 1996).

4.3.1.5 *Fusarium* Wilt

Good levels of resistance to *Fusarium oxysporum* have been reported in Slovakian grass pea germplasm (Benková and Záková 2001). *L. sativus* and *L. tingitanus* are very susceptible to *F. solani* allowing more hyphal proliferation than pea (Tewari et al. 2003).

4.3.1.6 Cercospora

Mishra et al. (1986) have reported variable levels of resistance to *Cercospora pisi-sativae* f. sp. *lathyrisativae* among grass pea lines, also mentioning the existence of a few resistant or immune lines.

4.3.2 Resistance to Parasitic Weeds

Orobanche is a genus of holoparasitic plants (see Chapter 11), which is completely dependent on the host due to its lack of chlorophyll. Once attached on the host roots, *Orobanche* is able to survive and develop, sucking carbohydrates from the phloem and water and minerals from the xylem of the host by means of a bridge to the host vascular tissues. *Orobanche crenata* has been known to threaten legume crops since antiquity. It is widely distributed in the Mediterranean region and West Asia (Rubiales et al. 2006) where it is a major constraint to the production of grain and forage legumes, including *L. sativus* (Robertson and Abd El Moneim 1996).

Several strategies have been developed for the control of broomrape, from cultural practices to chemical control (Rubiales et al. 2009a, b; Pérez-de-Luque et al. 2010), but none with unequivocal success. So far, a combination of sowing *Orobanche*-free crop seeds and delaying the sowing date or use of early maturing cultivars to escape *O. crenata* damage has been the only available strategy for controlling the broomrape infection in many cool-season legume crops (Rubiales and Fernández-Aparicio 2011).

Breeding for resistance is the most economic, feasible and environmentally friendly method of control. Resistance to *O. crenata* is scarce and of complex nature, but in spite of these difficulties, significant success has been achieved in some legume crops (Rubiales et al. 2006; Pérez-de-Luque et al. 2009). However, no sources of resistance are available in *L. sativus* (Linke et al. 1993; Robertson and Abd El Moneim 1996). Recent results show varying levels of infection under field conditions in *L. cicera* accessions (Fernández-Aparicio et al. 2009). Reduced level of infection was identified as the result of a combination of different avoidance and resistance mechanisms, each acting at different phases of the infection process as is the case in other crops (Rubiales 2003). Avoidance due to early flowering plays a major role in the reduction of broomrape infection in *L. cicera* and *L. sativus* (Fernández-Aparicio et al. 2007, 2009). Calculation of the deviations from the polynomic regression allowed an improved approximation of the true genetic resistance levels present in the germplasm, here defined as those components of resistance that do not depend on the precocity, although it also includes environmental error (Fernández-Aparicio et al. 2009). Likewise, besides early maturing germplasm avoiding infection, valuable *L. cicera* germplasm was reported with additional interesting levels of genetic resistance that might be further exploited in breeding programs. In addition to escape due to precocity, vigor or root biomass, different mechanisms of resistance and avoidance to *O. crenata* are operative in *L. cicera* germplasm (Fernández-Aparicio 2005, Fernández-Aparicio and Rubiales 2010).

High levels of resistance to *O. crenata* have been reported in the species *L. ochrus* and *L. clymenum* (Sillero et al. 2005). In both species the main mechanism of resistance seems to be an early barrier to the establishment of the broomrape, as none or few broomrape tubercles were recorded under field conditions. This low establishment is due to the low induction of germination of the broomrape seeds. Additional mechanisms of resistance preventing development of tubercles were suggested in *L. chloranthus* by the fact that tubercle formation was allowed, but they did not emerge.

4.3.3 Resistance to Viruses

Lathyrus sativus is generally moderately resistant to the aphid-borne virus bean yellow mosaic virus (BYMV). When aphids transmit the virus and infection occurs, symptoms include initial vein clearing of leaves followed by leaf mottle, leaf deformation and plant stunting. Aphids normally spread BYMV to grass pea crops from infected pastures dominated by subterranean clover. However, the virus is seed borne at a low level (about 0.2%) in *L. sativus*, therefore sowing contaminated seed can also introduce the virus to a crop. BYMV is most likely to damage grass pea crops in high rainfall areas. Infection with BYMV is minimized by sowing healthy seed distant from legume pastures dominated by subterranean clover. Sowing at high seeding rates to rapidly generate a dense plant canopy may also shade over early-infected plants thereby reducing the virus source for further virus spread. Insecticide application to control aphids is unlikely to stop the spread of the virus.

Alfalfa mosaic virus (AMV) and pea seed-borne mosaic virus (PSbMV) also infect *L. sativus* and are seed-borne in this species (1–5% seed transmission rates recorded). They are most likely to cause problems when virus-infected grass pea seed is sown. Symptoms of AMV include leaf mottle, chlorotic spotting of leaves, tip death and severe plant stunting. Symptoms of PSbMV are leaf pallor, leaf mottle, leaf deformation, tip necrosis and plant stunting. Infection with AMV and PSbMV is minimized by sowing healthy seed.

Field resistance to pea enation mosaic virus (PEMV) has been reported in *L. odoratus*, but not to common pea mosaic virus (CPMV) (Hull and Selman 1965).

4.3.4 Resistance to Pests

Thrips can be a serious pest of grass pea, and there appears to be little known resistance to this insect problem except in *L. aphaca* (Pandey et al. 1996). Only moderate resistance has been reported in Indian accessions JRL6, JLR41 (Gautam et al. 1998), RLK1, RLK26, RLK281, RLK617 (Asthana and Dixit 1998), RLK273-1 and RLK273-3 (Pandey et al. 1998).

Root knot (*Meloidogyne artiella*) and cyst nematodes (*Heterodera ciceri*) attacks cause some damage on grass pea but, luckily, resistance has been identified in *L. sativus* (Campbell 1997). Reaction to the northern root-knot nematode, *Meloidogyne hapla* was examined in accessions representing 16 *Lathyrus* species. There were differences in the percentage of root tissue galled and in the nematode reproductive index among species and among accessions within species. Root biomass of infected plants was significantly less than that of uninfected plants of the same accession. Accessions of *L. ochrus* and *L. tingitanus* were susceptible to *M. hapla*, whereas *L. latifolius*, *L. sylvestris* and *L. hirsutus* were resistant. *Lathyrus sativus* accessions ranged from moderately susceptible to susceptible.

4.3.5 Resistance/Tolerance to Abiotic Stresses

Lathyrus sativus is adapted to grow under drought stress conditions. High proline accumulation (Tyagi et al. 1999) and induction of lipoxygenase expression has been observed in *L. sativus* under water stress (Tyagi et al. 1995). Xiong et al. (2006) studied interrelationships among abscisic acid (ABA) content, accumulation of free polyamines and biosynthesis of β-N-oxalyl-L-α,β-diaminopropionic acid (ODAP) in *L. sativus* seedlings under drought stress induced by 10% polyethylene glycol. Increase of ABA content occurred prior to that of ODAP and polyamine contents, and there was a significant positive correlation between ABA and ODAP contents. Addition of exogenous ABA increased the ODAP content in leaves. On the other hand, pretreatment with α-difluoromethylarginine (DFMA), a polyamine biosynthesis inhibitor, significantly suppressed the accumulation of free putrescine, free spermidine and free spermine, which in turn inhibited biosynthesis of ODAP in well-watered leaves. Meanwhile, addition of exogenous putrescine alleviated DFMA-induced inhibition on the biosynthesis of putrescine and spermidine, but did not affect the biosynthesis of spermine and ODAP in well-watered leaves. Similar results were also found in drought-stressed leaves.

Despite its tolerance to drought, grass pea is also tolerant to excessive rainfall and can be grown on land subject to flooding (Kaul et al. 1986; Campbell et al. 1994). This is exploited in South Asia where its seeds are scattered into the water of standing rice crop (Abd El Moneim et al. 2000).

Also its capacity to withstand moderate salinity has been recognized (Campbell et al. 1994; Hoque et al. 1996). It has a very hardy and penetrating root system and therefore can grow on a wide range of soil types, including very poor soil and heavy clays (Campbell 1997). A high biological nitrogen fixation rate, which assists in maintaining soil fertility, requiring few production inputs, allows this crop to be an important component in sustainable farming systems (Campbell et al. 1994; Peoples et al. 1995).

Lathyrus sativus has been reported to be tolerant to a deficiency in essential nutrients and able to store large amounts of lead in its root tissues (Brunet et al. 2008). Over three quarters of the lead was not mobile. Compared with control plants, lead-exposed plants showed a six-fold, two-fold and three and a half-fold reduction in their root calcium, zinc and copper contents, respectively. Therefore, *L. sativus* could be used for the development of new rhizofiltration systems.

Lathyrus sativus accessions are generally very susceptible to cold, whereas resistance is frequently found in *L. cicera* accessions (Robertson and Abd El-Moneim 1996). Tolerance to cold has also been reported in a Portuguese accession of *L. ochrus* (Abd El-Moneim and Cocks 1993).

4.4 Conventional Breeding

Grass pea has been a crop since Neolithic times (Zohary and Hopf 2000) and the center of origin has often been thought long obscured by cultivation. In this long period of cultivation, farmer selection has developed many landraces adapted to varied environments in the Indian subcontinent, the Middle East, North Africa, Ethiopia and throughout Europe.

There are evident differences in grass pea morphology between regions. Seed size tends to be greater in Mediterranean locations (Hanbury et al. 1999) where selection has often been for forage, in comparison to the Indian subcontinent where small seeds were more desirable (Chowdhury and Slinkard 2000). A larger seed size is desirable in some specific European markets (Mera et al. 2003b). Larger seeds are a product of human selection (Chowdhury and Slinkard 2000), similar to other Old World pulses such as lentil and pea; and small-seeded grass pea tend to be hard-seeded and generally more primitive. In grass pea the small-seeded types are associated with darker seeds and blue flowers. Blue flowers are dominant over white (Quader et al. 1986), with simple inheritance of possibly two genes. Seed color is closely related to flower color.

As a crop plant, *L. sativus* is comparatively unimproved. This is most likely due to the presence of ODAP and two consequences: grass pea has become a "neglected" crop compared to other crops due to the presence of ODAP and the health problems for humans and animals, and modern grass pea breeding has concentrated on reducing seed ODAP as a priority over improvements in other characters.

There is considerable genetic diversity within *L. sativus*, indicating much potential for crop improvement by conventional means using this primary genepool. The greatest variation found by Chowdhury and Slinkard (2000) was in accessions from North Africa and the Near East, presumably close to the center of origin.

Most variety development of grass pea has been via selection from germplasm collections that include landraces. Mutation breeding has also been employed on several occasions (Rybinski 2003; Kumar and Tripathi 2007) and molecular techniques are increasing in use. Different forms of male sterility have been reported (Quader 1987; Tiwari and Campbell 1997) but not exploited in grass pea breeding.

There have been successful interspecific crosses between grass pea and other *Lathyrus* spp., particularly *L. pseudocicera*. Embryo rescue has increased the range of species in successful interspecific crosses (Addis and Narayan 2000). Interspecific crosses have not been used for grass pea improvement to date. Yunus and Jackson (1991) considered *L. cicera* and *L. amphicarpos* to be in the secondary genepool of *L. sativus* and so may provide an easily accessed source of genes. However, considering the largely unused potential of the primary genepool, intraspecific crosses are not likely to be attractive at this stage of grass pea development.

There have been conventional breeding programs in several countries including Australia (Hanbury et al. 1995), Bangladesh (Malek 1998), Canada (Campbell and Briggs 1987), China (Yang and Zhang 2005), Chile (Mera et al. 2003a), Ethiopia (Tadesse 1997), India (Lal et al. 1986; Pandey et al. 1996), Nepal (Yadav 1996) and Syria (Abd El-Moneim et al. 2000). Some of these breeding programs are still active and most are small in comparison to other legume crops.

4.4.1 Outcrossing

Although grass pea is primarily self-pollinated it has a significant outcrossing rate; 2% according to Chowdhury and Slinkard (1997), and 4–16% according to Kaul et al. (1986). The extent of outcrossing was 28% in Bangladesh (Rahman et al. 1995) and Gutierrez-Marcos et al. (2006) calculated outcrossing to be 36%. It is highly likely that bee activity influences outcrossing rates (Chowdhury and Slinkard 1997), and possibly other insects also have an effect.

Outcrossing has led to a major problem in establishing cultivars of low ODAP (and probably of other characters, but these have generally been of secondary importance) that in areas of established grass pea cultivation, newer and improved varieties can be genetically "swamped" by outcrossing with the larger population of older landraces. This is particularly the case in countries like Bangladesh, where grass pea is widely cultivated and so there is a background of high ODAP levels.

4.4.2 Breeding for ODAP Content

For some time the emphasis on grass pea breeding has been reduction in ODAP levels. Since ODAP was not recognized as a problem until it was identified in the 1960s (Murti et al. 1964; Rao et al. 1964) the breeding for reduced levels has had at most 45 years to progress. Most of the initial progress in cultivars low in ODAP was by direct selection from landraces and lines. There has been consistent lowering of ODAP levels in seed over time.

Unimproved grass pea types have ODAP levels of up to 1.5% (Hanbury et al. 2000) and typically around 0.5% (Campbell et al. 1994). Various attempts to link readily observable characters with ODAP for ease of selection have not been successful (Kaul et al. 1982; Campbell et al. 1994) and ODAP has been shown to be quantitatively inherited (Tiwari and Campbell 1996a).

The Indian variety Pusa 24 was selected from a field crop in 1966 (Mehra et al. 1995) and was the first line known to have reduced seed ODAP levels; it was released in 1973–1974 with 0.2% ODAP (0.26% as determined by Campbell and Briggs 1987). Pusa 24 has been the basis of a number of other low ODAP varieties in India and elsewhere.

In Canada, LS82046 was selected from Pusa 24, with seed ODAP levels as low as 0.03% (Campbell and Briggs 1987). This value may vary, since LS82046 was found to have 0.18% ODAP and Pusa 24 to have 0.21% at one site in western Australia (CD Hanbury, unpubl. data). Environment does affect ODAP levels, although genotype has a much stronger effect (Hanbury et al. 1999). Thus, ODAP levels should be assessed in a range of environments to establish a genotype mean level.

In Bangladesh, two varieties were released in 1995 and 1996 with high yield potential and lower seed ODAP content: Barikhesari-1 and Barikhesari-2 with 0.29 and 0.27% seed ODAP, respectively (Rahman et al. 2001).

A small conventional breeding program in Australia produced cv. Ceora with low ODAP levels of 0.04–0.09% (Siddique et al. 2006). Ceora was a product of a conventional cross between K33 (from Pakistan; ODAP of 0.43%) as the female parent and 8604 (from Bangladesh; ODAP 0.05%) as the male (Hanbury et al. 1995). Under Western Australian conditions K33 grew vigorously and had a larger seed size, and 8604 grew poorly and had smaller seeds. Ceora has been tested in Bangladesh since 2007 (C Johannsen, pers. comm.).

ODAP has been shown to be independent of such characters as seed protein content and seed size (Kaul et al. 1982; Bisignano et al. 2003) and so it should be relatively easy to breed in conjunction with a number of other desirable plant characteristics. Quader et al. (1988) reported a positive correlation between seed protein and ODAP levels, but this seems to be an exception and referred to a small number of accessions only from the Indian sub-continent.

There has been little assessment of forage and general plant ODAP levels and no breeding for this purpose. Considering that there are reports of lathyrism in animals on forage (Hanbury et al. 2000) this is an important neglected area.

4.4.3 Other Characters

Over the long period of farmer selection many of the traits of landraces have been improved, such as the loss by pod shattering, although pod shattering can still limit usefulness in regions such as Poland (Milczak et al. 2001).

Due to the dual use of grass pea as fodder and grain in different locations, selection has not been in just the one direction. In some environments there has been selection for increased seed yield and for increased plant size in others. Selection from lines in Canada has recently produced cv. AC Greenfix for green manuring purposes, suited for much of North America (Krause and Krause 2003) and requiring high biomass and high nitrogen fixation.

There has been some characterization of cultivars with greater number of branches, pods per plant, seeds per pod and seeds per plant, but in some cases also a lower weight of seeds per pod and per plant (Rybinski et al. 2008). There is still little application of improvement in such agronomic characteristics in many regions.

Seed weight generally has high heritability in grass pea (Milczak et al. 2001; Mera et al. 2003a; Turk et al. 2007) and can be positively correlated with seed yield (Tiwari and Campbell 1996b), but this may be environment specific (e.g., under higher rainfall conditions). Turk et al. (2007) also found that biological yield was highly heritable. Under Indian conditions seed size and biomass were both positively associated with seed yield (Kumari and Mehra 1991); however, it should be noted that Indian genotypes are predominantly small-seeded. Pods per plant has also been associated with higher yields (Quader et al. 1988; Islam et al. 1989). For yield and yield components, the single seed descent method seemed superior to pedigree and random bulk methods under Indian conditions (Mitra and Mehra 2001).

Characteristics such as prostrate plant habit, indeterminate growth, late maturity and pod shattering are important factors limiting the broader adoption in Poland. Other than ODAP, there can be high levels of antinutritionals in seed of grass pea (Hanbury et al. 2000), and beyond assessment of some accessions there has been little effort in improvement in these qualities.

Considering the range of environments in which grass pea is grown, and the purposes it is put to, there are a number of attributes that require future breeding attention including:

- Better ability for seed harvest mechanization, as the prostrate nature is a problem.
- Less prostrate character/more rigid stems.
- High biomass as fodder or green manure.
- Lower levels of antinutritionals (e.g., tannins).
- Higher levels of protein, lysine, methionine and cysteine in both seed and forage.
- Ability to reshoot following grazing where seeds are also required.
- Increased tolerance to waterlogging where relay sowing after rice occurs.

Many of such characters have not been widely screened for in grass pea. Considering the previously noted wide variation available in the *L. sativus* primary genepool alone, it is highly likely that significant improvements could rapidly be made given the resources

4.5 Diversity Analysis

An International Board for Plant Genetic Resources (IBPGR) forage task force drew attention to the severe genetic erosion of *Lathyrus* throughout the Mediterranean and suggested certain species (*L. ochrus, L. gorgoni* and *L. cicera*) to be given high priority for *ex situ* conservation (IBPGR 1985). However, to achieve effective conservation and enhance the use of the germplasm collections there is a need for detailed characterization of the existing diversity. The evaluation of the genetic diversity is a preliminary step that ensures the conservation and development of genetic resources. Accordingly, it is important to know which species are the most related to *L. sativus*, since these will be the most promising sources of traits. The genus *Lathyrus* is composed of about 160 different species organized into 13 sections (Kupicha 1983). Kupicha's (1983) morphology-based monograph represents the only worldwide treatment of the genus. She attempted to reflect natural groups and to provide a functional framework for horticultural, agronomic, and biogeographic discussion. *Lathyrus sativus* is included in section *Lathyrus* together with species such as *L. cicera, L. amphicarpus, L. gorgoni, L. latifolius* and *L. sylvestris*.

4.5.1 *Interspecific Analysis*

Morphological and molecular data have been used in several interspecific diversity and phylogenetic studies of the *Lathyrus* genus, allowing identification of the species most closely related to *L. sativus*. Electrophoretic patterns of total seed protein, seed globulin A or albumin data (El-Shanshoury 1997; Przybylska et al. 1999; Emre 2009) reflected the profound interspecific hybridization barriers in the genus *Lathyrus*, with all the examined species distantly related to *L. sativus*. However, morphological, interspecific hybridization, some molecular studies (isozymes, RAPD and RFLP markers) and seed protein analysis (globulin B, glutelin and convicilin), have detected a close phylogenetic proximity between *L. sativus* and *L. cicera* (Jackson and Yunus 1984; Yunus and Jackson 1991; Croft et al. 1999; Chtourou-Ghorbel 2001, 2002; Ben Brahim et al. 2002; Sáenz de Miera et al. 2008; Emre 2009), but even so the compliance with the classification of Kupicha (1983) is not complete. Different levels of diversity have been detected in the different species, reflecting their different perenniality and breeding systems (Ben Brahim et al. 2002). In this way, *L. latifolius* and *L. sylvestris* maintain high levels of isozyme diversity, probably associated with their perenniality and the predominant outcrossing reproduction system. *Lathyrus sativus*, being an annual species and with high levels of autogamy (Chowdhury and Slinkard 1997), presented low levels of genetic diversity and a low level of gene flow when compared to the previously reported *Lathyrus* species.

To test the monophyly and the relationships of the infrageneric classification of the *Lathyrus* genus proposed by Kupicha (1983), chloroplast DNA restriction fragment length polymorphisms (RFLP) were analyzed (Asmussen and Liston 1998) and suggested that Kupicha's *Lathyrus* section, where *L. sativus* is included, should be combined with the *Orobon* and *Orobastrum* sections to render monophyletic groups interpreted as sections. Based on amplification fragment length polymorphism (AFLP) data, Badr et al. (2002) confirmed the monophyly of the section *Lathyrus*, but only for the species sampled. Unfortunately, in this study the sections *Orobon* and *Orobastrum* were not sampled. Additional DNA sequence data (nuclear and chloroplastic), from the study of Kenicer et al. (2005), ratify as well the monophyly of the *Lathyrus* section, providing support for a circumscription of section *Lathyrus* in Asmussen and Liston's broader sense (merging at least the *Orobastrum* section) and showed also that *L. cicera* and *L. sativus* have a close relationship and clustered together.

4.5.2 Intraspecific Analysis

The germplasm resources profile of *L. sativus* shows a tertiary genepool that is most extensive, but with a restricted secondary genepool (*sensu* Yunus and Jackson 1991), with which transfer of genes will be difficult. It is clear that exploitation of germplasm resources for the improvement of grass pea must concentrate on the primary genepool, including wild and landrace materials.

High improvement potential exists within this material since high variability has been found at the primary genepool within *L. sativus* accessions in several published studies.

4.5.2.1 Morphological and Agronomical Studies

Through the application of multivariate methods, Jackson and Yunus (1984) analyzed and described the morphological variation in *L. sativus*. By studying a collection of 49 different accessions to cover the known worldwide geographical distribution of the species, these authors showed that grass pea is differentiated into several distinct forms, primarily on the basis of flower color, seed size and size of leaves. In this way, they identified a clear distinction between the blue-flowered forms from Southwest Asia, Ethiopia and the Indian subcontinent, and the white and blue flowered forms with white seeds which have a more western distribution (from the Canary Islands to the western ex-republics of the Soviet Union). This array of variation is undoubtedly the result of geographical separation as well as selection by man.

Later on, Hanbury et al. (1999) grouped a worldwide collection of *L. sativus* accessions, evaluated under Mediterranean-type environments, into two clusters based on the phenology, seed yield and ODAP content matching their geographical origin: Indian subcontinent and Mediterranean/European. Those lines of Mediterranean/European origin were consistently higher yielding, with much larger seeds and later phenology.

On a more restricted germplasm study, Tavoletti et al. (2005), by field evaluation of a collection of 16 potential grass pea landraces from the Marche region (Italy) detected a high level of phenotypic variability (morphological and agronomic traits) between commercial, cultivated by market-oriented farms, and household populations (not market-oriented farms). Moreover, a clear differentiation was found among accessions collected in different locations of the Marche region, justifying their classification as real landraces.

Tadesse and Bekele (2003a, b) studying the phenotypic diversity of Ethiopian grass pea germplasm (50 populations), observed a large amount of morphological variation both within altitude and within regions and suggested that character variability in grass pea is not uniform in all regions. To tackle genetic erosion they identified the highest diversity regions where collection and *in situ* conservation should be primarily focused. Diversity among and within populations has been detected for several of the studied traits indicative of high breeding potential in these materials and that these traits can be improved further through a strong selection program. Similar studies on field evaluation of grass pea landraces, where high variability in plant, pod and seed traits was detected, were also performed with Indian (Kumari 2001), Spanish (De la Rosa and Martín 2001), Slovak (Benková and Záková 2001) or Chilean germplasm (Tay et al. 2000).

In a more widespread germplasm study, but more related with quality traits, Granati et al. (2003) characterized a large collection of Italian and foreign grass pea germplasm. The observed similarity trend enabled identification of five groups without any well-defined differences in the pattern distribution of origins, species, cultivars and ecotypes for the traits considered. This provided interesting information for the establishment of a subset of grass pea collection upon which future work should be concentrated. Variability was also detected for other important agronomic traits such as ODAP content or resistance to several fungal diseases (as reported above in this chapter), suggesting that the primary genepool is a valuable source of genetic material for plant improvement of these particular important traits.

4.5.2.2 Biochemical and Molecular Studies

The high agronomical and morphological diversity within grass pea germplasm is also found at the biochemical and molecular level. Considerable genetic diversity, as revealed by isozyme and random amplified polymorphic DNA (RAPD) polymorphisms, exists in grass pea throughout the world. Chowdhury and Slinkard (2000), using a wide grass pea germplasm collection, managed to associate different levels of genetic diversity, measured by isozymes, with the different geographical origins. These authors identified the Near East and North Africa regions as the ones with the most isozyme variability (and also the most closely related), suggesting that the center of diversity (center of origin) for grass pea was this general area. The lowest variability was found in accessions from South America (also the most differentiated one from the rest), followed by those from Sudan–Ethiopia.

In a more restricted study, mainly considering Ethiopian grass pea populations selected based on morphological diversity (Tadesse and Bekele 2001), isozyme diversity was not correlated with the previously studied morphological diversity. This absence of correlation had already been observed also on a wider isozyme grass pea germplasm collection study (Yunus et al. 1991). In this study the isozyme data showed considerable polymorphism although there was no apparent correlation with morphology. The frequency of putative heterozygous loci was greater than we would have expected given the knowledge of the predominantly autogamous breeding system of *L. sativus* (Chowdhury and Slinkard 1997). Likewise, Gutierrez-Marcos et al. (2006) found more heterozygosity than expected analyzing isozyme variation in a wide grass pea collection; these authors found that a greater genetic variability exited at the intra-regional level than at the inter-regional level. Also in a more restricted germplasm study, Alba et al. (2001) analyzed a selected collection of Italian grass pea populations detecting also a wide polymorphism for the enzyme systems tested. No association between the isozyme and the morphological diversity was investigated in this study.

Seed proteins have also been used to study diversity among grass pea populations. Electrophoretic analysis of seed albumins (Przybylska et al. 1998) and seed globulins (Przybylska et al. 2000) was performed on a wide collection of germplasm. The presence of considerable intraspecific variation was revealed in both studies. The electrophoretic data obtained for both seed albumins and globulins tended to separate two groups of *L. sativus* accessions: white-seeded with large seeds, originating mainly from Europe and North Africa, and colored-seeded with relatively small seeds, originating mainly from Asia and Ethiopia. Apparently, the electrophoretic seed protein data reflect changes in the genetic structure of populations that take place on developing large-seeded cultivars from rather primitive forms. Still, no correlation was found between occurrence of either the seed albumin patterns or the major albumin bands and geographical origin of the forms under study. The lack of correlation between genetic diversity and the region of origin, support the idea that the natural distribution of *L. sativus* has been completely obscured by cultivation.

Since the study of Chtourou-Ghorbel et al. (2002), PCR-based molecular markers have proven to be efficient in distinguishing between different grass pea accessions and to assess the within-species genetic variability. More recently, Barik et al. (2007) analyzed a collection of Indian grass pea populations using also RAPD markers and concluded that though the grass pea genotypes considered were of apparently similar morphology, polymorphism exists at molecular level, which can be exploited in breeding programs aimed at crop improvement. Belaid et al. (2006) using inter-simple sequence repeat (ISSR) markers provided evidence of high molecular polymorphism at intraspecific level, showing that both wild and cultivated grass pea forms constitute important pools of diversity and also showing the effectiveness of the ISSR procedure for assessing the genetic diversity from grass pea populations. Also in this study, the high intrapopulation variation seems to have no relation with the geographical origin. It is probable that the rates of outcrossing in the species do not restrict intra-population diversity.

More recently, AFLP molecular markers have also been used to study the genetic diversity of grass pea germplasm. Tavoletti and Iommarini (2007) evaluated a collection of grass pea populations collected in central Italy. Two main clusters were found: one included large-seeded populations from farms that were not market-oriented (named household populations) and the second, small-seeded populations, cultivated in market-oriented farms (named commercial populations). Sardinha et al. (2007) and Almeida et al. (2009) studied the genetic diversity of a collection of Iberian grass pea germplasm as a first step towards the selection of appropriate parental lines for the establishment of a disease-resistance crossing breeding scheme also using AFLP markers. The detected genetic diversity was compared with previous morphological (passport data) and disease resistance data. No clear clustering was detected either based on morphological or molecular data. Furthermore the correlation between the molecular and the morphological distance matrices was not significant; and no significant genetic differentiation was found among the disease resistance classes. These findings support the joint use of molecular and morphological traits to properly exploit the natural diversity within this collection.

4.6 Genome Mapping

Apart from the use of DNA markers in assessing the level of genetic diversity in phylogenetic studies, these markers have numerous applications in plant breeding. They provide the necessary tools to determine the number, position and individual effects of genes/quantitative trait loci (QTLs) controlling interesting traits, such as genes for disease or insect resistance, reduced ODAP concentration, increased protein content, and other characters of agronomic importance (Campbell et al. 1994), via, for instance, genetic mapping and QTL analysis. Subsequently, the use of molecular markers may accelerate the generation of new varieties, by providing a powerful tool to plant breeders that allows the early selection of desirable individuals based on genotypic configuration rather than phenotype (Tanksley et al. 1989). They also allow the association of phenotypic characters with the genomic loci responsible for them, which could facilitate gene transfer to suitable agronomic types.

4.6.1 Linkage Maps and Mapping Studies

There is limited literature on the development of molecular linkage maps for grass pea. The first linkage map constructed for *L. sativus* was based on morphological traits (Sen and Ghosh 1962). However, more recently two *L. sativus* linkage maps have been developed using molecular markers to identify regions in the *Lathyrus* genome linked to desirable traits (Chowdhury and Slinkard 1999; Skiba et al. 2004).

The first linkage map incorporated 65 RAPDs, three isozymes and one morphological marker (flower color), and was based on an F_2 population of 100 individuals derived from a cross between a blue-flowered and a white-flowered parent (Chowdhury and Slinkard 1999). The map comprised mainly dominant markers and consisted of 14 linkage groups, covering 898 cM, with an average distance between markers of 17.2 cM. This map was used to locate a flower color marker. These authors expected that since ODAP concentration was associated with flower color, the genes controlling ODAP would map near the flower color trait. Nevertheless, until now there is no report in the literature pointing in that direction.

The second molecular linkage map available for *L. sativus* incorporated 47 RAPD, 7 sequence-tagged microsatellite sites (STMS) and 13 sequence-tagged sites-cleaved amplified polymorphic sequence (STS-CAPS) markers and was constructed using 92 backcross individuals derived from the cross between an accession resistant to Ascochyta blight (*Mycosphaerella pinodes*) and a susceptible accession (Skiba et al. 2004). A total of 64 markers were arranged into nine linkage groups and covered 803.1 cM with an average spacing between markers of 15.8 cM. One consistent putative QTL for Ascochyta blight resistance was identified, located on linkage group 1, after combining the data from simple and composite interval mapping and single-point analysis. A second potential QTL was detected only by single-point analysis in linkage group 2. QTL1 explained 12% of the phenotypic variance in the backcross population. The second QTL, QTL2, accounted for 9% of the trait variation. No significant interactions were detected between QTL1 and QTL 2. Two of the three markers located near QTL1 were generated from RAPD primers. The third marker mapped in the vicinity of QTL1 was generated from an STS primer pair designed from a *L. sativus* cDNA clone (Skiba et al. 2003) that showed sequence similarity to the *Cf-9* resistance gene of tomato. This *Cf-9* gene confers resistance to *Cladosporium fulvum* in tomato (Parniske et al. 1997). This STS primer pair may have amplified a gene in *L. sativus* similar to the *Cf-9* resistance gene; therefore, the marker *Cf-9* could represent an Ascochyta blight resistance gene in *L. sativus*. However, to accurately determine whether this marker represents a *Cf-9*-like gene in the *L. sativus* genome, the amplicon for this marker needs to be cloned, sequenced and subjected to database searches (Skiba et al. 2004).

Both linkage maps were not adequately saturated with markers, presenting numerous gaps, short linkage groups and unlinked markers. Due to the lack of common markers these two maps could not be aligned (Vaz Patto et al. 2006b).

The construction of a linkage map is not necessarily a prerequisite for identifying markers closely linked to a particular trait of interest, if single inherited traits are considered. Using bulked segregant analysis (BSA), Hanada and Hirai (2003) identified RAPD markers linked to the gene responsible for the tendril trait of sweet pea (*L. odoratus*), without having to generate a whole linkage map of the species' genome that was later on transformed into a more specific sequence characterized amplified region (SCAR) marker.

4.6.2 Cross Species Amplification

Future linkage mapping studies in *L. sativus* should aim at incorporating common markers in order to allow map alignment and comparison. STS markers generated by primers designed from defense-related expressed sequence tag (EST) sequences as the ones obtained from a *L. sativus* cDNA library (Skiba et al. 2003) are a good option. These markers are derived from coding DNA (representing genes of known function), which generally have a high degree of sequence conservation and are possibly codominantly inherited, being more likely transferable across species boundaries compared with markers derived from non-expressed sequences such as AFLPs or RAPDs. Therefore, the polymorphic *L. sativus* STSs may be useful as markers in the more closely related species (*L. cicera*) and to serve as anchors for future comparative mapping. Also the incorporation of codominant markers would be very important for a correct estimation of genetic distances among markers in repulsion phase.

A recently initiated work, with the objective to obtain codominant markers to apply in the development of a grass pea linkage map, explored the potential use of the already developed molecular markers (EST-SSR and gSSR) from the closely related pea and lentils in *Lathyrus* spp. (Almeida et al. 2008). This study reports that, despite the low transferability, the cross-amplification of pea SSR in *Lathyrus* spp. is a good alternative to the development of specific grass pea microsatellites (Almeida et al. 2008). These authors tested 249 SSRs developed from *Pisum sativum* (Loridon et al. 2005), and 30 primer SSR pairs developed from *Lens culinaris* (Hamwieh et al. 2005) with two *L. sativus* accessions. This study confirmed that the pea microsatellite motif was maintained, but in most cases with different repeat numbers. Thus, from the 249 pea markers tested, 183 did not amplify in *L. sativus*, but went well in the controls. For the 66 *P. sativum* SSR that amplified, 24 produced one single fragment and the other 42 produced unspecific amplification. The single amplicons were sequenced to confirm the microsatellite motif. Of the 24 SSRs, only 18 had data available in the NCBI allowing a comparison between pea and grass pea SSR motifs. For two SSRs the number of repeats was maintained. For the others, in some cases there were more than one mutation event in each microsatellite, with deletions and transitions being the most common mutations, and the least frequent transversions and insertions. In the lentil SSRs, 28 did not yield any amplification product and the others amplified a non-specific fragment. This approach reduces the cost of development of new SSR and allows comparative mapping between species, since the donor species (pea) has already several linkage maps developed (Loridon et al. 2005; Fondevilla et al. 2008).

4.6.3 Expression Analysis

Understanding the molecular mechanisms underlying host-pathogen interactions is of primary importance for devising strategies to control diseases. The molecular mechanism of quantitative resistance to *Ascochyta*, powdery mildew or rust fungi of *Lathyrus* has not yet been clarified. Knowledge of the underlying resistance genes (candidate genes) is mandatory before transferring these quantitative resistances into new lines. The conventional approach for candidate gene identification uses genetic dissection of a population at a certain resolution. This approach becomes impractical due to the prohibitively large number of individuals required to achieve satisfactory resolution. Alternative, more efficient approaches, that allow estimating gene content for any given QTL and obtaining some information about functionality of the genes underlying it, are needed. For this purpose, gene expression analysis could be extremely helpful.

The potential defense-related ESTs, besides being useful as molecular markers for the construction of high-density genetic linkage maps, are also very interesting for expression analysis studies in order to identify and assess the function of putative genes thought to be involved in plant stress responses. Expression analysis studies were performed on *L. sativus* inoculated with *Mycosphaerella pinodes* using 29 ESTs representing genes coding for enzymes and proteins involved in different levels of defense (Skiba et al. 2005). These included the recognition events between plant and pathogen, signaling transduction and regulation, oxidative stress and hypersensitive response, and specific defense-related/pathogenesis-related proteins. One of the ESTs expressed earlier or at greater levels in the stem tissue of the resistant *L. sativus* accession tested when compared with the susceptible accession was the one similar to the *Cf-9* resistance gene previously located near a resistance QTL on the *L. sativus* genome (Skiba et al. 2004). This class of R-gene resides in complex loci carrying multiple genes and encodes membrane-bound proteins. They contain leucine-rich-repeats, which may be important for pathogenic recognition, through the binding of ligands secreted by the pathogen, encoded by the pathogen's *avr* gene (Kobe and Deisenhofer 1995). If present and active in *L. sativus*, then the interaction between this type of gene and the *M. pinodes avr* gene would presumably activate a cascade of defense responses (Skiba et al. 2005).

The availability of this EST set allied with the microarray technology will provide a powerful way to identify further defense-related genes and pathways involved in *L. sativus* disease resistance, and also in biotic or abiotic resistance in other related pulses such as chickpea (Coram and Pang 2006; Mantri et al. 2007).

Despite considerable investment and progress in the use of molecular tools for resistance breeding, any application of marker technologies for durable resistance is still hampered by insufficient knowledge of allelic diversity in resistance genes and our limited insight into plant–pathogen interactions. Moreover, cost-effective modern tools for advanced resistance breeding are missing (Kahl et al. 2009). The advent of next generation sequencing technologies has boosted genome-wide quantitative gene expression profiling by SuperSAGE (Matsumura et al. 2003). SuperSAGE can decipher the transcriptome of two interacting eukaryotic organisms simultaneously (i.e., detect and quantify all transcripts present in the interaction space between a host and a pathogen). This approach has been recently applied to the interaction *L. sativus*/*Ascochyta lathyri* within the frame of an ERA-PG European funded project (LEGRESIST) (Kahl et al. 2009). This technique is also particularly adapted to organisms for which no extended DNA database is available (Matsumura et al. 2003; Coemans et al. 2005; Molina et al. 2008) as is the case for *L. sativus*. With this approach it was possible to identify the differential expression in *L. sativus* of more than 3000 ($P < 0.05$) or 900 ($P < 0.005$) transcripts, respectively, between infected and control tissue (2010 Winter, pers. comm.).

In the future, through quantitative real time PCR (qRT-PCR) assays that will be available from this European consortium, it will be possible to perform eQTL (expression QTL) mapping in the *L. sativus/Ascochyta* interaction. The transcript abundance (also referred as gene expression level) can be analyzed as any higher order trait in terms of QTL detection. Variation in gene expression can be mapped in segregating populations using "genetical genomics" approaches (Jansen and Nap 2001) by expression profiling and marker-based fingerprinting of each individual of a segregating population.

This same European project is trying to understand the allelic diversity of resistance genes on the level of single nucleotide polymorphisms (SNPs) for genetic mapping of all expressed resistance genes in major crop legumes. At this moment, the 3´-RACE and the 5´-RACE sequences of the expressed potential resistance genes (EPRG) are being screened for SNPs and small indels with the objective of delivering a SNP-array for rapid and cost efficient mapping in *L. sativus* (Kahl et al. 2009) providing potential links to the structural genes themselves. EPRGs were isolated from cDNA using resistance gene specific primers.

Co-localization of EPRGs, expression QTLs (eQTLs), and QTL for conventional phenotypic traits on the same genetic map will provide a powerful route to identify candidate genes and study resistance gene networks in *L. sativus*.

4.7 Genetic Transformation, Interspecific Hybridization and *In Vitro* Selection

The grass pea, with its large eco-physiological plasticity and very low production costs, is a cheap source of dietary protein available for subsistence farmers in low-income food deficit countries, where large areas of this crop are grown (Ochatt et al. 2007). However, there are some limits to grass pea consumption deriving from the sometimes relatively high seed content of the toxin held responsible for neurolathyrism (ODAP). As already mentioned in this chapter (see Section 4.2) and elsewhere (Yadav and Mehta 1995; Vaz Patto et al. 2006b; Ochatt et al. 2007), it would be interesting to develop cultivars with decreased levels of this toxin to ensure a wider and safer use of this environmentally adaptable, yet largely underutilized crop (Campbell 1997; Ochatt et al. 2007).

Added to this, the neglected *Lathyrus* species are members of the Vicieae (syn. Fabae) tribe, the same as *Pisum*, *Lens* and *Vicia*, and could share genes, physiological processes and defense mechanisms with those more widely cultivated crops (Campbell 1997; Vaz Patto et al. 2006a and b, 2009). In this way grass pea could also be a useful genetic resource for resistance breeding of field pea (Campbell 1997; McCutchan et al. 1999; Ochatt et al. 2001, 2002, 2007; Vaz Patto et al. 2006b). Unfortunately, grass pea has a long generation cycle, is cross-incompatible with most other cultivated seed legumes and insufficient research input has been devoted to it in the past, so that little is known of its floral biology, cytogenetics and breeding in general.

In this context, biotechnology and *in vitro* approaches in general would permit to produce novel grass pea genotypes that, while retaining their rusticity, are better adapted for human consumption (Ochatt et al. 2001, 2007; Vaz Patto et al. 2006b) and also to introduce interesting traits from *Lathyrus* into the cross-incompatible major legume crops, including pea, by somatic hybridization (Durieu and Ochatt 2000).

4.7.1 *Lathyrus* In Vitro *culture and Plant Regeneration*

In line with the scanty research background available for all neglected crops (Ochatt and Jain 2007), few groups have developed *in vitro* approaches with *Lathyrus* species and, as with other legumes, the main bottleneck has been the regeneration of whole plants from cultured tissues.

Pecket and Selim (1965) authored the first formal article on the *in vitro* culture of *L. sativus* explants, where they reported culture of embryos from which they obtained calluses. Twenty years later, Delgado-Montero and Moreno (1985) attempted the culture of explants from various tissue sources, but also obtained only callus. Other authors, using juvenile tissues comprising meristems at the time of explanting, reported on the subsequent regeneration of buds and shoots from callus (Mukhopadhyay and Bhojwani 1978; Gharyal and Maheshwari 1980; Sinha et al. 1982, 1983; van Dorrestein et al. 1998). Rooted plants were difficult to produce and rarely obtained, coupled with the fact that, often, shoots only grew on the callus surfaces and probably developed from meristems pre-existing in the initial explants employed for culture. Only twice was de novo regeneration obtained from callus (Gharyal and Maheshwari 1980; Roy et al. 1992), while organogenesis was reported from leaf discs (Roy et al. 1991) and entire seeds (Malik et al. 1993). Fratini and Ruiz (2003) described a highly efficient rooting procedure for *in vitro* regenerated shoots of several cool season grain legumes including *L. sativus*.

Only recently were fertile plants efficiently and prolifically regenerated both from meristematic tissues (Zambre et al. 2002) and from hypocotyl segments (Ochatt et al. 2002). This latter report had an additional interest as, for the first time information on the fundamental genetic mechanism(s) underlying the occurrence of hyperhydricity was provided, by analyzing the cultured tissues and regenerated organs and plants with flow cytometry (Ochatt et al. 2002). Hyperhydricity, formerly known as vitrification (Debergh et al. 1992), has been observed *in vitro* in cultures of many species. Thus, the morphology of cultures had been described and the influence of various physiological and ecological parameters extensively reviewed (Gaspar et al. 1987), but it was not until this work with grass pea (Ochatt et al. 2002) that proof was shown that hyperhydric tissues consistently display an abnormal flow cytometry profile, with supernumerary peaks, which indicate disorders in the nuclear DNA content of nuclei in such tissues (Ochatt 2006, 2008).

4.7.2 *Protoplast Technology and Interspecific Hybridization*

Many researchers have attempted interspecific hybridization in the genus *Lathyrus*, but most attempts have failed (Campbell 1997; Vaz Patto et al.

2006b) and, in the few where viable F_1 hybrids were successfully produced, they presented low fertility (Yunus and Jackson 1991). In such cases, the rate of success can be increased through utilization of embryo rescue techniques after fertilization (Campbell 1997; Vaz Patto et al. 2006b; Ochatt et al. 2007). At the intergeneric level, incompatibility problems are even more serious (McCutchan et al. 1999; Ochatt et al. 2004). Thus, from the information available in the literature on crossing, fertility and chromosome behavior of hybrids, it is apparent that breeding strategies involving alien genetic transfer for the improvement of grass pea are only possible through the readily crossable species *L. cicera* and *L. amphicarpos* (Vaz Patto et al. 2006b), while crosses involving other species will have to be assisted by biotechnology tools (Ochatt et al. 2007).

One such biotechnology tool, which would help for interspecific hybridization would be protoplast technology, including the isolation and culture of protoplasts and their subsequent fusion with those of other species, and even other genera, for the recovery of somatic hybrid plants. Unfortunately, advances in this area have been quite limited for *Lathyrus*. Razdan et al. (1980) produced leaf mesophyll protoplast-derived callus of the ornamental sweet pea, *L. odoratus*. In this respect, despite its economic importance, information on biotechnology approaches applied to the ornamental sweet pea, *L. odoratus*, is surprisingly scanty (Razdan et al. 1980; Ochatt et al. 2010a, b). More recently, McCutchan et al. (1999) successfully isolated leaf mesophyll protoplasts and established suspension cultures of *L. sativus* from which they also isolated viable protoplasts, but they both failed to undergo sustained division. In parallel, Durieu and Ochatt (2000) reported on the isolation and sustained division of protoplasts isolated from leaves of a white-seeded grass pea genotype. It then appeared that methods employed for protoplast isolation from grass pea tissues by others had been too harsh (particularly in terms of the cellulase used for digestion), provoking a reduction in viability that would not be compensated for by the slightly higher yield of isolated protoplasts obtained, as it also impaired their subsequent competence for division. Different from previous data for sweet pea (Razdan et al. 1980), a high initial plating density and a high-saline-strength medium were required to foster sustained division of leaf mesophyll protoplasts of grass pea (Durieu and Ochatt 2000). Such proliferation-competent grass pea protoplasts were fused with pea mesophyll protoplasts, with a heterokaryon production efficiency of around 20% when polyethyleneglycol (30%, w/v, PEG6000) or electrofusion (three successive pulses at 1,500 V/cm, 75 µF, at 10 s intervals) were used as fusogens. Also interesting in this report was the use of a differential staining of the fusion partners. Thus FDA (fluorescein diacetate), which emits a yellow-green fluorescence under UV light was used for *Lathyrus*, while *Pisum* protoplasts were stained with Rhodamine B isothiocyanate that emits a red fluorescence under UV light. Subsequently, the heterokaryons could be unequivocally identified and scored by their dual fluorescence under UV light excitation, as compared to the product of homofusions, which would exhibit only one fluorochrome under UV. Such double fluorescence was retained during the first weeks of culture and thus permitted monitoring of the initial division and proliferation of heterokaryon-derived cell colonies (Durieu and Ochatt 2000), with the ensuing production of somatic hybrid microcalluses (Ochatt et al. 2001). To date, calluses produced were confirmed as being of somatic hybrid origin but have not been able to regenerate plants.

These could be the first steps into the introduction of Ascochyta blight resistance from grass pea into pea (McCutchan et al. 1999; Ochatt et al. 2001, 2004, 2007), of resistance to rust fungi from *L. cicera* (Vaz Patto et al. 2009), of broomrape resistance from *L. ochrus* (Sillero et al. 2005) or from *L. cicera* (Fernández-Aparicio et al. 2009), or yet of powdery mildew resistance from sweet pea into grass pea (Poulter et al. 2003; Ochatt et al. 2010a).

4.7.3 Haplo-Diploidization

Another useful biotechnology breeding approach is haplo-diploidization (Maluszynski et al. 2003). Haploids have the same chromosome complement as the gametes and, following chromosome doubling, doubled haploids (DHs) become homozygous in all loci, whereby they can be used as parental genotypes for breeding and also for independent study (Grewal et al. 2009; Ochatt et al. 2009). This approach, which is not new (Guha and Maheshwari 1964), has been attempted by culturing female (gynogenesis) or male (androgenesis) organs or reduced gametophytes from many species, but haploid plants have not been obtained for all (Maluszynski et al. 2003). As with other biotechnology tools, legumes have been described as extremely unresponsive to DH research (Croser et al. 2006; Skrzypek et al. 2008), and until quite recently (Ochatt et al. 2009; Grewal et al. 2009) there were no successful reports for cool season food legumes in the literature.

Ochatt et al. (2009) cultured isolated microspores of a number of legume species, including 10 genotypes of *Lathyrus* belonging to four different species (*L. sativus, L. cicera, L. ochrus* and *L. clymenum*), whereby initial microspore division was observed for most genotypes albeit at different frequencies, and resulted in the formation of microcalluses of around 50 cells each. With various *Lathyrus* genotypes, growth was arrested very early on (at the eight-celled colony stage), but microspores of four landraces of *L. sativus* proliferated first to microcalluses then to calluses and, ultimately, a reduced number of plants was recovered and confirmed as haploid by flow cytometry for two genotypes of grass pea (LB and L3). Unfortunately, they were very enfeebled and did not survive to greenhouse transfer (Ochatt et al. 2009). The ultimate goal of this work was to efficiently produce DHs from isolated microspores, which has not been achieved to date in legumes. Further efforts are under way designed to improve these responses and also to extend the strategies developed to other genotypes and to intact anthers, as was recently done with chickpea (Grewal et al. 2009).

4.7.4 *Shortening of Generation Cycles* In Vitro *for Faster Breeding*

Associated with studies on the development and exploitation of biotechnology for breeding of legumes, including grass pea, a series of experiments were devised to shorten the duration of generation cycles (Ochatt et al. 2004), comparing a greenhouse strategy used as control with an *in vitro* strategy. In all cases, shoots (1 cm tall and comprising two internodes) that had either been germinated *in vitro* from excised embryo axes on B5 modified medium (Ochatt et al. 2000), or regenerated *in vitro* from hypocotyl explants (Ochatt et al. 2001, 2002) were transferred to a simple, hormone-free MS medium where they elongated, flowered and ultimately set seed, or they were transferred onto a rooting medium prior to their eventual flowering and fruiting. Flowering and seed set were consistently obtained *in vitro*, and there was no prerequisite for rooting of shoots preceding it. On the contrary, an *in vitro* rooting phase both lengthened each generation cycle and also affected the flowering competence of shoots. Likewise, optimum flowering responses were obtained on a hormone-free medium, whereas the addition of growth regulators consistently reduced flowering, and a halving of salt strength provoked a reduced seed set, and was coupled with a subsequently lower germination competence of the seeds produced. Use of this strategy with three different *L. sativus* genotypes and over 10 successive generation cycles showed a shortening of each cycle from the standard 150–180 days in the field to 100–120 days *in vitro* (Ochatt et al. 2004). With this procedure, a constant supply of flower buds can be ensured, e.g., for haplo-diploidization experiments (see above), and a more efficient exploitation of those approaches for breeding where seed supply is short (such as for single seed descent-SSD) or where ethical or political issues may render field testing difficult, e.g. with GMO material (see Section 4.7.6), can be envisaged.

4.7.5 *Mutagenesis*

Mutation breeding can be a valuable supplement to conventional plant breeding methods, aimed at creating additional genetic variability that may be utilized by the plant breeder in the development of cultivars for specific purposes or with specific adaptabilities (Biswas 2007). Most mutation studies in this genus concerned *L. sativus*, and a wide spectrum of morphological mutations has been found, affecting plant habit, maturity, branching, stem shape, leaf size, stipule shape, flower color and structure, pod size, and seed size and color (Nerkar 1976; Rybinski 2003; Biswas 2007). Interestingly, variations in the neurotoxic protein compound ODAP have been induced by treatments with both physical and chemical mutagens in four different cultivars of *L. sativus* (Nerkar 1972), and the results obtained revealed the characteristic of a monogenic F_2 segregation for this trait, whereby ODAP content may exhibit a simple Mendelian inheritance. In addition to this, mutagenesis also induced an enhancement in the contents of methionine and lysine following X-ray irradiation (Chaudhury and Das 1988).

In addition to induced mutagenesis, extensive variation concerning a range of traits (from plant growth habit to ODAP content) appeared spontaneously in the course of tissue culture studies (somaclonal variation), and was found among regenerated plants of grass pea (Roy et al. 1993; Misra et al. 1994; Mehta 1997; Patil et al. 1997; Ochatt et al. 2002; Zambre et al. 2002; Lambein and Kuo 2004). This type of genetic variability may also be successfully exploited to produce agronomic ally desirable types, providing a faster means of improvement than that allowed by conventional crossing and backcrossing methods.

4.7.6 Gene Transfer

Insertional mutagenesis, by genetic transformation using *Agrobacterium* vectors or biolistic techniques has also been tried repeatedly and with many species, but only twice has this been reported for grass pea so far, and only Barik et al. (2005) obtained stable transformed plants. Thus, in the first of such attempts, Barna and Mehta (1995) co-cultured leaflets and nodal segments with *Agrobacterium*, and 8–13% of transformation frequency of calluses or embryos was obtained, but no transgenic plants were regenerated. These authors did not provide any information on the parameters influencing transformation nor on the molecular characterization of the transformed tissues. In the second report, Barik et al. (2005) developed a reproducible procedure for gene transfer in grass pea through co-cultivation of epicotyl segments with two disarmed *A. tumefaciens* strains, EHA 105 and LBA 4404, both carrying the binary plasmid p35SGUSINT with the neomycin phosphotransferase II (*nptII*) gene and the β-glucuronidase (GUS)-intron, as vector systems. The latter was found to have a higher transforming ability, and several key factors modifying the transformation rate were optimized, whereby up to 36% transient expression based on the GUS histochemical assay was achieved. Southern hybridization of genomic DNA of the kanamycin-resistant GUS-expressive shoots to a *gus* intron probe confirmed the integration of the transgene. The transformed shoots were rooted, acclimatized and established in the field, where germ-line transformation was evident through progeny analysis. In this respect, kanamycin-resistant and -sensitive plants segregated in a ratio close to 3:1 among T_1 seedlings of most transgenic plant lines. With the recent availability of such a strategy, *Agrobacterium*-mediated genetic transformation aimed at reducing the level of ODAP in the grain may be at hand-reach, as the coding sequence for an ODAP-degrading gene (approx. 3.3 kb) has been characterized and cloned from a soil microbe, *Pseudomonas stutzeri* (Sachdev et al. 1995). However, given that regeneration protocols for grass pea are often genotype specific (Ochatt et al. 2007), it may be necessary to either develop more generally applicable protocols or to adapt the protocol after transformation.

4.7.7 Cytological, Biochemical, Molecular and Cytogenetic Characterization

Whatever the biotechnological approach assessed and regardless of its efficiency, it is of outmost importance to reliably characterize the regenerants, haploids, somatic hybrids, mutants or primary transformants eventually obtained, as compared to the maternal plants of a given genotype. Moreover, wild type genotypes also need to be characterized, as this information is not always available in the literature for *Lathyrus* genotypes and landraces. In recent years, studies were undertaken with this aim, using biochemical and, sometimes, molecular approaches too (see Section 4.5.2.2). However, data at the cytogenetic level are scantier (Yamamoto et al. 1984; Ali et al. 2000), and when it comes to nuclear DNA content they are practically nonexistent (Narayan 1982; Nandini et al. 1997; Ali et al. 2000). Flow cytometry is a tool of choice to palliate this lack.

More than 35 years have elapsed since the first formal report on the use of flow cytometry with plant cells (Heller 1973), but it is in the last decade that the usefulness of this method was clearly demonstrated, for understanding of several complex fundamental mechanisms underlying plant cytogenetics, and for breeding of agronomic traits, as well as for the identification of subcellular fractions and the content of a number of chemicals in cells and tissues (Ochatt 2008). The most frequent use of flow cytometry with plant cells to date has certainly been for the determination of the ploidy level of tissues and plants following their manipulation *in vitro* or their treatment with mitogenic substances. Nevertheless, with the advent of new molecular tools flow cytometry with plant cells has left its infancy (Dolezel et al. 2007; Ochatt 2008).

The range of variation in genome size in the plant kingdom spans nearly 2,000-fold (Leitch and Bennett 2007), and this area of research has been rather controversial over the years because of significant differences in the nuclear DNA content of a same species analyzed over a several-year interval, perhaps due to the more refined practices used in the more recent studies (Dolezel et al. 2007; Ochatt 2008).

For flow cytometry analysis of biotechnology-derived material, the DNA content and ploidy level of tissues studied (either calluses or leaves harvested from regenerated plants) is compared with leaf controls from seed-derived plants of the respective species and landraces. Nuclei are mechanically isolated by chopping the tissues with a razor blade in a two-step method (nuclei extraction buffer plus staining buffer) or directly in a one-step procedure (Partec GmbH, Germany), and the suspension is filtered through a fine mesh prior to readings (Elmaghrabi and Ochatt 2006; Ochatt 2006, 2008; Koné et al. 2007). A range of stains of differing specificity, but which all bind stoicheometrically to the DNA bases can be used and, therefore, permit to determine the relative nuclear DNA content when compared to an internal or, eventually, external standard of known DNA content. The in-built program of the flow cytometer then transforms the signal of the epifluorescence emitted into numerical data which are plotted as graphs where the abscissa shows the intensity of epifluorescence and the ordinate the count of objects emitting at each intensity. Typically, a flow cytometry profile will show two peaks where the first one corresponds to nuclei in the G0/G1 phase of the cell cycle (2C DNA) while the second peak includes nuclei in the G2/M phase (4C DNA) (Dolezel et al. 2007; Ochatt 2008).

In a recent work, Ochatt et al. (2010b) studied in terms of *in vitro* approaches a number of genotypes of the five species *L. sativus, L. cicera, L. ochrus, L. clymenum* and *L. odoratus* that were also characterized by flow cytometry. They analyzed the cell cycle and calculated the relative DNA content per nucleus and the AT/GC ratio of all these *Lathyrus* genotypes. Leaf tissues of the model legume *Medicago truncatula* were used as the internal standard (i.e., run simultaneously with the *Lathyrus* nuclei suspension), since the DNA content measured is of more magnitude than an accurate content. For assessments of the AT/GC ratio (Barow and Meister 2002), two fluorochromes were used: DAPI (4',6-diamidino-2-phenylindole), which forms a fluorescent complex by attaching in the minor grove of A-T rich sequences of DNA (Kapuscinski 1995), and propidium iodide (PI), which although usually believed to be non-base pair specific exhibits nevertheless a clear GC-preference (Vinogradov 1994). Other fluorochromes that are specific for G-C such as chromomycine A3 (Barow and Meister 2002; Dolezel et al. 2007; Ochatt 2008) were also tested.

The fact that no intermediate callus phase intervened in the regeneration strategy from explants used permitted to envisage that true-to-typeness would be commonplace among the regenerated plants, but further analyses were nevertheless undertaken, using flow cytometry. All regenerants obtained were normal, true-to-type in terms of phenotype and euploid, except for one regenerant of *L. sativus* L156, which appeared to be a tetraploid (Ochatt et al. 2010b). In addition to this, the mother plants of all the genotypes tested *in vitro* were also analyzed, and data for each species fitted well with the few available in the literature, where other methods of measurement of DNA had been used, but they were not exactly the same, particularly because here the samples had been contrasted with an internal standard which, interestingly, had never been done before. This stresses how different ecotypes, genotypes and even clones within a given species may have a different relative DNA content per nucleus. This is particularly true when dealing with neglected and/or underutilized crops where little information on the cytogenetics of the species and even less on its landraces is available (Ochatt and Jain 2007).

The use of nuclear genome size as a taxonomical tool has been fostered by the knowledge that the amount of nuclear DNA remains remarkably constant within most species, while it varies considerably among species, even when closely related (Leitch and Bennett 2007). In previous flow cytometry studies with vertebrates (Vinogradov 1994), the AT-specific fluorochrome Hoechst 33258 (HO) and the GC-specific fluorochrome olivomycin (OM) were used to measure base pair specific DNA content in 20 different species, and data found were in good correlation with the biochemical literature on base pair frequencies ($r = 0.972$, $P < 1 \times 10^{-8}$), thus allowing to determine the percent of GC/AT-pairs and genome size from the flow cytometric data. On the other hand, different plant species vary in terms of the ratio of nucleotide base pairs of genomic DNA (Barow and Meister 2002), and a correlation between genome size and base pair ratio has been claimed (Dolezel et al. 2007). Base composition can be analyzed by base-specific dyes. Thus, the genome size can be determined by flow cytometry of suspensions of nuclei stained by the base independent dye, PI (Greilhuber et al. 2005). For estimation of the AT frequency, the AT-specific dyes 4,6-diamidino-2-phenylindole dihydrochloride (DAPI) and Hoechst 33342 have been used (Barow and Meister 2002; Dolezel et al. 2007; Ochatt 2008). Barow and Meister (2002) did not find any significant correlation between genome size and the dye fluorescence for DAPI when 54 plant species were investigated, which they explained by the nonrandom distribution of base pairs. Thus, similar AT/GC ratios within a plant family would result from the general similarity of the DNA sequences within a family. In the recent study with *Lathyrus* species mentioned above (Ochatt et al. 2010b), they could also be distinguished from one another according to this trait.

4.8 Conclusions

Despite the economic importance of *L. sativus*, molecular approaches for the improvement of this crop have largely been lacking. Progress has been made in other areas, particularly in germplasm evaluation and the reduction of seed ODAP content. As this main hurdle is removed, new challenges and opportunities may be found in breeding new varieties of *L. sativus* for cultivation in their traditional regions, and also in Europe and in countries such as Australia and Canada (Skiba et al. 2007).

In view of the complex relationships and the number of traits involved in stress tolerance/resistance as well as in quality or yield, pyramiding using marker-assisted breeding, induced mutation and other biotechnologies, combined with the participation of farmers, is likely to provide the best route towards the development of region-specific tolerant crop germplasm on the one hand, and to warrant the production of high yields of higher quality seed to alleviate hunger on the other (Skiba et al. 2007). The development of technologies in *L. sativus* such as cDNA library screening, microarrays, real time PCR, associated with the powerful new sequencing technique of SuperSAGE, allow that the expression of a large number of genes can be studied. These technologies can be used to more efficiently identify genes from interesting germplasm (resistant, high quality and high yielding) providing the means for accelerating variety development.

Despite the stigma of lathyrism connected with grass pea, the species can, with all the developed breeding tools, be considered a Neolithic crop with a bright future.

References

Abd El-Moneim AM, Cocks PS (1993) Adaptation and yield stability of selected lines of *Lathyrus* spp. under rainfed conditions in West Asia. Euphytica 66: 89–97.

Abd El-Moneim AM, van Dorrestein B, Baum M, Mulugeta W (2000) Improving the nutritional quality and yield potential of grasspea (*Lathyrus sativus* L.). Food Nut Bull 21: 493–496.

Addis G, Narayan RKJ (2000) Interspecific hybridisation of *Lathyrus sativus* (guaya) with wild *Lathyrus* species and embryo rescue. Afr Crop Sci J 8: 129–136.

Alba E, Polignano GB, De Carlo D, Mincione A (2001) Electrophoretic phenotypes of different enzymes in some entries of *Lathyrus sativus* L. Lathyrus Lathyrism Newsl 2: 15–20.

Ali HBM, Meister A, Schubert I (2000) DNA content, rDNA loci, and DAPI bands reflect the phylogenetic distance between *Lathyrus* species. Genome 43: 1027–1032.

Almeida NF, Caminero C, Vaz Patto MC (2008) Cross-amplification of pea (*Pisum sativum* L.) and lentil (*Lens culinaris* Medik.) microsatellites in grass pea (*Lathyrus sativus* L.). In: Proc 4th Int Conf on Legume Genomics and Genetics. 7–12 December 2008, Puerto Vallarta, Mexico, p 66.

Almeida NF, Sardinha J, Satovic Z, Rubiales D, Vaz Patto MC (2009) Molecular aspects of Iberian grass pea germplasm diversity. In: Proc 19th EUCARPIA Conf on Genetic Resources Section, 26–29 May 2009, Ljubljana, Slovenia, p 26.

Asmussen CB, Liston A (1998) Chloroplast DNA characters, phylogeny, and classification of *Lathyrus* (Fabaceae). Am J Bot 85: 387–401.

Asthana AN, Dixit GP (1998) Utilization of genetic resources in *Lathyrus*. In: PN Mathur, VR Rao, RK Arora (eds) *Lathyrus* Genetic Resources Network. Pro IPGRI-ICARDAICAR Regional Working Group Meeting, 8–10 December 1997, New Delhi, India, pp 64–70.

Badr A, El Shazly H, El Rabey H, Watson LE (2002) Systematic relationships in *Lathyrus* sect. *Lathyrus* (Fabaceae) based on amplified fragment length polymorphism (AFLP) data. Can J Bot 80: 962–969.

Barik DP, Mohapatra U, Chand PK (2005) Transgenic grasspea (*Lathyrus sativus* L.) Factors influencing *Agrobacterium*-mediated transformation and regeneration. Plant Cell Rep 24: 523.

Barik DP, Acharya L, Mukherjee AK, Chand PK (2007) Analysis of genetic diversity among selected grasspea (*Lathyrus sativus* L.) genotypes using RAPD markers. Z Naturforsch 62c: 869–874.

Barilli E, Satovic Z, Sillero JC, Rubiales D, Torres AM (2011) Phylogenetic analysis of *Uromyces* species infecting grain and forage legumes by Sequence analysis of Nuclear Ribosomal Internal Transcribed Spacer Region. J Phytopathol 159: 137–145.

Barna KS, Mehta SL (1995) Genetic transformation and somatic embryogenesis in *Lathyrus sativus*. J Plant Biochem Biotechnol 4: 67–71.

Barow M, Miester A (2002) Lack of correlation between AT frequency and genome size in higher plants and the effect of nonrandomness of base sequences on dye binding. Cytometry 47: 1–7.

Belaid Y, Chtourou-Ghorbel N, Marrakchi M, Trifi-Farah N (2006) Genetic diversity within and between populations of *Lathyrus* genus (Fabaceae) revealed by ISSR markers. Genet Resour Crop Evol 53: 1413–1418.

Ben Brahim N, Salhi A, Chtourou N, Combes D, Marrakchi M (2002) Isozymic polymorphism and phylogeny of 10 *Lathyrus* species. Genet Resour Crop Evol 49: 427–436.

Benková M, Záková M (2001) Evaluation of selected traits in grasspea (*Lathyrus sativus* L.) genetic resources. Lathyrus Lathyrism Newsl 2: 27–30.

Bisignano V, Chiaretti D, Crinò P, Polignano GB (2003) Characterization of Italian and exotic *Lathyrus* germplasm for quality traits. Genet Resour Crop Evol 50: 273–280.

Biswas AK (2007) Induced mutation in grasspea (*Lathyrus sativus* L.). In: SJ Ochatt, S Mohan Jain (eds) Breeding of Negelected and Under-utilised Crops, Spices and Herbs. Science Publishers, Enfield, NH, USA, pp 29–39.

Brunet J, Repellin A, Varrault G, Terryn N, Zuily-Fodil Y (2008) Lead accumulation in the roots of grass pea (*Lathyrus sativus* L.): a novel plant for phytoremediation systems? Com Rend Biol 331: 859–864.

Campbell CG (1997) Grass pea. *Lathyrus sativus* L. Promoting the conservation and use of under-utilized and neglected crops. 18. Institute of Plant Genetics and Crop Plant Research. Gatersleben/International Plant Genetic Resources Institute, Rome, Italy, p 91.

Campbell CG, Briggs CJ (1987) Registration of low neurotoxin content *Lathyrus* germplasm LS 8246. Crop Sci 27: 821.

Campbell CG, Mehra RB, Agrawal SK, Chen YZ, El Ali AMA, Khawaja HIT, Yadav CR, Toy J, Araya WA (1994) Current status and future strategy in breeding grass pea (*Lathyrus sativus*). Euphytica 73: 167–175.

Chaudhury D, Das A (1988) X-irradiation induced biochemical changes in the seeds of *L. sativus*. Indian Biol 20: 30–35.

Chowdhury MA, Slinkard AE (1997) Natural outcrossing in grasspea. J Hered 88: 154–156

Chowdhury MA, Slinkard AE (1999) Linkage of random amplified polymorphic DNA, isozyme and morphological markers in grasspea (*Lathyrus sativus*). J Agri Sci 133: 389–395.

Chowdhury MA, Slinkard AE (2000) Genetics of isozymes in grasspea. J Hered 91: 142–145.

Chtourou-Ghorbel N, Lauga B, Combes D, Marrakchi M (2001) Comparative genetic diversity studies in the genus *Lathyrus* using RFLP and RAPD markers. Lathyrus Lathyrism Newsl 2: 62–68.

Chtourou-Ghorbel N, Lauga B, Ben Brahim N, Combes D, Marrakchi M (2002) Genetic variation analysis in the genus *Lathyrus* using RAPD markers. Genet Resour Crop Evol 49: 363–370.

Coemans B, Matsumura H, Terauchi R, Remy S, Swennen R, Sagi L (2005) SuperSAGE combined with PCR walking allows global gene expression profiling of banana (*Musa acuminata*), a non-model organism. Theor Appl Genet 111: 1118–1126.

Cook RTA, Fox RTV (1992) *Erysiphe pisi* var. *pisi* on faba beans and other legumes in Britain. Plant Pathol 41: 506–512.

Coram TE, Pang ECK (2006) Expression profiling of chickpea genes differentially regulated during a resistance response to *Ascochyta rabiei*. Plant Biotechnol J 4: 647–666.

Croft AM, Pang ECK, Taylor PWJ (1999) Molecular analysis of *Lathyrus sativus* L. (grasspea) and related *Lathyrus* species. Euphytica 107: 167–176.

Crop Trust (2007) Global Crop Diversity Trust Strategy for the *Ex Situ* Conservation of *Lathyrus* (grass pea), with special reference to *Lathyrus sativus, L. cicera, L. ochrus:* http://www.croptrust.org/documents/cropstrategies/grasspea.pdf (Cited 9 Sept 2009).

Croser JS, Lülsdorf MM, Davies PA, Clarke HJ, Bayliss KL, Mallokarjuna N, Siddique KHM (2006) Toward doubled haploid production in the *Fabaceae*: Progress, constraints, and opportunities. Crit Rev Plant Sci 25: 139–157.

Debergh P, Aitken-Christie J, Cohen D, Grout B, Von Arnold S, Zimmerman R, Ziv M (1992) Reconsideration of the term "vitrification" as used in micropropagation. Plant Cell Tiss Org Cult 30: 135–140.

De la Rosa L, Martín I (2001) Morphological characterisation of Spanish genetic resources of *Lathyrus sativus* L. Lathyrus Lathyrism Newsl 2: 31–34.

Delgado-Montero VM, Moreno R (1985) Callus induction and culture from different explants of *Lathyrus sativus*. In: AK Kaul, D Combes (eds) Proc Int Symp on *Lathyrus* and Lathyrism. Third World Medical Research Foundation, New York, USA, pp 169–170.

Dixon GR (1978) Powdery mildews of vegetable and allied crops. In: MD Spencer (ed) Powdery Mildews, Academic Press. London, UK, pp 495–524.

Dolezel J, Greilhuber J, Suda J, (2007) Flow cytometry with plant cells. Analysis of genes, chromosomes and genomes. Wiley-VCH Verlag GmbH & Co. KgaA, Weinheim, Germany, p 455.

Durieu P, Ochatt SJ (2000) Efficient intergeneric protoplast fusion between pea (*Pisum sativum* L.) and grass pea (*Lathyrus sativus* L.). J Exp Bot 51: 1237–1242.

Elmaghrabi A, Ochatt S (2006) Isoenzymes and flow cytometry for the assessment of true-to-typeness of calluses and cell suspensions of barrel medic prior to regeneration. Plant Cell Tiss Org Cult 85: 31–43.

El-Shanshoury A (1997) The use of seed proteins revealed by SDSPAGE in taxonomy and phylogeny of some *Lathyrus* species. Biol Plant 39: 553–559.

Emeran AA (2003) Estudios de la resistancia genética y métodos alternativos de control de la roya de las habas. PhD Dissertation. Univ de Córdoba. Córdoba, Spain, p 184.

Emeran AA, Sillero JC, Niks RE, Rubiales D (2005) Morphology of infection structures help to distinguish among rust fungi infecting leguminous crops. Plant Dis 89: 17–22.

Emre I (2009) Electrophoretic analysis of some *Lathyrus* L. species based on seed storage proteins. Genet Resour Crop Evol 56: 31–38.

Farr DF, Rossman AY, Palm ME, McCray EB (2008) Fungal Databases, Systematic Botany & Mycology Laboratory, Agricultural Research Service, United States Department of Agriculture: http://nt.ars-grin.gov/fungaldatabases/ (Cited 13 May 2008).

Fernández-Aparicio (2005) Resistencia a jopo (*Orobanche crenata*) en *Lathyrus* spp. MSc Dissertation, Univ de Córdoba. Córdoba, Spain, p 69.

Fernández-Aparicio M, Rubiales D (2010) Characterisation of resistance to crenate broomrape (*Orobanche crenata* Forsk.) in *Lathyrus cicera* L. Euphytica 173: 77–84.

Fernández-Aparicio M, Sillero JC, Flores F, Rubiales D (2007). Screening *Lathyrus cicera* and *L. sativus* resistance against crenate broomrape. In: 6th Eur Conf on Grain Legumes:

Integrating legume biology for sustainable agriculture, 12–16 November 2007, Lisbon, Portugal, p 160.

Fernández-Aparicio M, Flores F, Rubiales D (2009). Field response of *Lathyrus cicera* germplasm to crenate broomrape (*Orobanche crenata*). Field Crops Res 113: 321–327.

Fondevilla S, Satovic Z, Rubiales D, Moreno MT, Torres AM (2008) Mapping of quantitative trait loci for resistance to *Mycosphaerella pinodes* in *Pisum sativum* subsp *syriacum*. Mol Breed 21: 439–454.

Fratini R, Ruiz ML (2003) A rooting procedure for lentil (*Lens culinaris* Medik.) and other hypogeous legumes (pea, chickpea and *Lathyrus*) based on explant polarity. Plant Cell Rep 21: 726–732.

Gaspar T, Kevers C, Debergh P, Maene L, Pâques M, Boxus P (1987) Vitrification: morphological, physiological and ecological aspects. In: JM Bonga, DJ Durzan (eds) Cell and Tissue Culture in Forestry, vol 1. Martinus Nijhoff Publishers. Dordrecht, The Netherlands, pp 152–166.

Gautam PL, Singh IP, Karihaloo JL (1998) Need for a crop network on *Lathyrus* genetic resources for conservation and use. In: PN Mathur, VR Rao, RK Arora (eds) *Lathyrus* Genetic Resources Network. Proc IPGRI-ICARDA-ICAR Regional Working Group Meeting, New Delhi, India, pp 15–21.

Getahun H, Mekonnen A, Teklehaimanot R, Lambein F (1999) Epidemic of neurolathyrism in Ethiopia. Lancet 354: 306–307.

Getahun H, Lambein F, Van der Stuyft P (2002a) ABO blood groups, grass pea preparation, and neurolathyrism in Ethiopia. Trans Roy Soc Trop Med Hyg 96: 700–703.

Getahun H, Lambein F, Vanhoorne M (2002b) Neurolathyrism in Ethiopia: assessment and comparison of knowledge and attitude of health workers and rural inhabitants. Soc Sci Med 54: 1513–1524.

Getahun H, Lambein F, Vanhoorne M, Van der Stuyft P (2003) Food-aid cereals to reduce neurolathyrism related to grass-pea preparations during famine. Lancet 362: 1808–1810.

Getahun H, Lambein F, Vanhoorne M, Van der Stuyft P (2005) Neurolathyrism risk depends on type of grass pea preparation and on mixing with cereals and antioxidants. Trop Med Int Health 10: 169–178.

Gharyal PK, Maheshwari SC (1980) Plantlet formation from callus cultures of a legume, *Lathyrus sativus* cv. L.S.D.-2. Z Pflanzenphysiol 100: 358–362.

Granati E, Bisignano V, Chiaretti D, Crinò P, Polignano GB (2003) Characterization of Italian and exotic *Lathyrus* germplasm for quality traits. Genet Resour Crop Evol 50: 273–280.

Greilhuber J, Doležel J, Lysàk MA, Bennett MD (2005) The origin, evolution and proposed stabilization of the terms 'genome size' and 'C-value' to describe nuclear DNA contents. Ann Bot 95: 255–260.

Grewal R, Lulsdorf M, Croser J, Ochatt S, Vandenberg A, Warkentin TD (2009) Doubled haploid production in chickpea (*Cicer arietinum* L) – Role of stress treatments. Plant Cell Rep 28: 1289–1299.

Guha S, Maheshwari SC (1964) *In vitro* production of embryos from anthers of *Datura*. Nature 204: 497.

Gurung AM, Pang ECK, Taylor PWJ (2002) Examination of *Pisum* and *Lathyrus* species as sources of ascochyta blight resistance for field pea (*Pisum sativum*). Aust Plant Path 31: 41–45.

Gutierrez-Marcos JF, Vaquero F, Sáenz de Miera LE, Vences FJ (2006) High genetic diversity in a world-wide collection of *Lathyrus sativus* L. revealed by isozymatic analysis. Plant Genet Resour 4: 159–171.

Haimanot TR, Abegaz BM, Wuhib E, Kassina A, Kidane Y, Kebede N, Alemu T, Spencer P(1993) Pattern of *Lathyrus sativus* (grass pea) consumption and b-N-oxalyl-a,b-diaminopropionic acid (b-ODAP) content of food samples in the lathyrism endemic region of northwest Ethiopia. Nutr Res 13: 1113–26.

Haimanot RT, Kidane Y, Wuhib E, Kalissa A, Alemu T, Zein ZA, Spencer PS (1990) Lathyrism in rural northwestern Ethiopia: a highly prevalent neurotoxic disorder. Int J Epidemiol 19: 664–672.

Haimanot RT, Lambein F, Feleke A (2005) Is lathyrism still endemic in northern Ethiopia? - The case of Legambo Woreda (district) in the South Wollo Zone, Amhara National Regional State. Ethiop J Health Dev 19: 230–236.

Hamwieh A, Udupa SM, Choumane W, Sarker A, Dreyer F, Jung C, Baum M (2005) A genetic linkage map of *Lens* sp. based on microsatellite and AFLP markers and the localization of fusarium vascular wilt resistance. Theor Appl Genet 110: 669–677.

Hanada H, Hirai M (2003) Development of genetic marker linked to the tendril trait of sweet pea (*Lathyrus odoratus* L.). Breed Sci 53: 7–13.

Hanbury CD (2000) *Lathyrus* grain as quality animal feed. Grain Legumes 30: 10–11.

Hanbury CD, Sarker A, Siddique KHM, Perry MW (1995) Evaluation of *Lathyrus* germplasm in a Mediterranean type environment in south-western Australia. CLIMA Occasional Publication No. 8, Perth, Australia, p 23.

Hanbury CD, Siddique KHM, Galwey NW, Cocks PS (1999) Genotype-environment interaction for seed yield and ODAP concentration of *Lathyrus sativus* L. and *L. cicera* L. in Mediterranean-type environments. Euphytica 110: 445–460.

Hanbury CD, White CL, Mullan BP, Siddique KHM (2000) A review of the potential of *Lathyrus sativus* L. and *L. cicera* L. grain for use as animal feed. Anim Feed Sci Technol 87: 1–27.

Haque A, Hossain M, Khan JK, Kuo YH, Lambein F, De Reuck J (1994) New findings and symptomatic treatment for neurolathyrism, a motor neuron disease occurring in north west Bangladesh. Paraplegia 32: 193–195.

Haque A, Hossain M, Wouters G, Lambein F (1996) Epidemiological study of lathyrism in northwestern districts of Bangladesh. Neuroepidemiology 15: 83–91.

Haque A, Hossain M, Lambein F, Bell EA (1997) Evidence of osteolathyrism among patients suffering from neurolathyrism in Bangladesh. Nat Toxins 5: 43–46.

Harlan JR, de Wet JMJ (1971) Toward a rational classification of cultivated plants. Taxon 20: 509–517.

Heller FO (1973) DNA Measurement of *Vicia faba* L. with the pulse cytophotometry. Ber Deutsch Bot Gesellsch Bd 86: 437–441.

Heywood V, Casas A, Ford-Lloyd B, Kell S, Maxted N (2007) Conservation and sustainable use of crop wild relatives. Agri Ecosyst Environ 121: 245–255.

Hoque R, Hussain M, Kuo YH, Lambein F (1996) Salinity tolerance and accumulation of neurotoxin and excitatory amino acids in *Lathyrus sativus*. Bangladesh J Biochem 2: 15–27.

Hugon J, Ludolph A, Roy DN, Schaumburg HH, Spencer PS (1988) Studies on the etiology and pathogenesis of motor neuron diseases. II. Clinical and electrophysiologic features of pyramidal dysfunction in macaques fed *Lathyrus sativus* and IDPN. Neurology 38: 435–442.

Hugon J, Ludolph A, Spencer PS, Gimenez Roldan S, Dumas JL (1993) Studies of the etiology and pathogenesis of motor neuron diseases. III. Magnetic cortical stimulation in patients with lathyrism. Acta Neurol Scand 88: 412–416.

Hull R, Selman IW (1965) The incidence and spread of viruses in sweet peas (*Lathyrus odoratus* L.) in relation to variety and the use of systemic insecticides. Ann Appl Biol 55: 39–50.

International Board for Plant Genetic Resources (IBPGR) (1985) Forages for Mediterranean and Adjacent Arid/Semi-Arid Areas: Report of a Working Group Meeting Held at Limassol, Cyprus. IBPGR, Rome, Italy, p 29.

Islam MS, Ali MS, Shaikh MAQ, Saha CS (1989) Variability and path-coefficient analysis in grasspea (*Lathyrus sativus*). Indian J Agri Sci 59: 123–124.

Jackson MT, Yunus AG (1984) Variation in the grass pea (*Lathyrus sativus* L.) and wild species. Euphytica 33: 549–559.

Jahan K, Ahmad K (1993) Studies on neurolathyrism. Environ Res 60: 259–266.

Jansen RC, Nap JP (2001) Genetical genomics: the added value from segregation. Trends Genet 17: 338–391.

Kahl G, Winter P, Horres R, Rotter B, Jungling R, the LEGRESIST Consortium (2009) Pathogenesis-related genes and genetic variation in potential resistance genes of major European legumes. The LEGRESIST Project. In: Proc Second Int Ascochyta Workshop, 28 June-2 July 2009. Washington State University, Pullman, Washington, USA, p 47.

Kapuscinski J (1995) DAPI: a DNA-specific fluorescent probe. Biotech Histochem 70: 220–233.

Kaul AK, Islam MQ, Begum K (1982) Variability for various agronomic characters and neurotoxin content in some cultivars of khesari (*Lathyrus sativus* L.) in Bangladesh. Bang J Bot 11: 158–167.

Kaul AK, Islam MQ, Hamid A (1986) Screening of *Lathyrus* germplasm of Bangladesh for BOAA content and some agronomic characters. In: AK Kaul, D Combes (eds) *Lathyrus* and Lathyrism. Third World Medical Research Foundation, New York, USA, pp 130–141.

Kenicer GJ, Kajita T, Pennington RT, Murata J (2005) Systematics and biogeography of *Lathyrus* (Leguminosae) based on internal transcribed spacer and cpDNA sequence data. Am J Bot 92: 1199–1209.

Kessler A (1947) Lathyrismus. Monatsschr Psychiatr Neurol 13: 76–92.

Khan JK, Kebede N, Kuo YH, Lambein F, De Bruyn A (1993) Analysis of the neurotoxin beta-ODAP and its alpha-isomer by precolumn derivatization with phenylisothiocyanate. Anal Biochem 208: 237–240.

Khan JK, Kuo YH, Kebede N, Lambein F (1994) Determination of nonprotein amino-acids and toxins in *Lathyrus* by High-Performance Liquid-Chromatography with precolumn phenyl isothiocyanate derivatization. J Chromatogr A 687: 113–119.

Khan JK, Kuo YH, Haque A, Lambein F (1995) Inhibitory and excitatory amino acids in cerebrospinal fluid of neurolathyrism patients, a highly prevalent motorneurone disease. Acta Neurol Scand 91: 506–510.

Kislev ME (1989) Origins of the Cultivation of *Lathyrus sativus* and *Lathyrus cicera* (Fabaceae). Econ Bot 43: 262–270.

Kobe B, Deisenhofer J (1995) A structural basis of the interactions between leucine-rich repeats and protein ligands. Nature 374: 183–186.

Koné M, Patat-Ochatt EM, Conreux C, Sangwan RS, Ochatt SJ (2007) *In vitro* morphogenesis from cotyledon and epicotyl explants and flow cytometry distinction between landraces of Bambara groundnut (*Vigna subterranea* (L.) Verdc) an underutilised grain legume. Plant Cell Tiss Org Cult 88: 61–75.

Krause D, Krause I (2003) New green manuring *Lathyrus sativus* variety AC Greenfix available in USA. Lathyrus Lathyrism Newsl 3: 13.

Kumar G, Tripathi R (2007) Anomalous nucleolar and chromosomal organization in induced phenodeviants of grasspea. Cytologia 72: 345–350.

Kumari V (2001) Field evaluation of grasspea (*Lathyrus sativus* L.) germplasm for its toxicity in the northwestern hills of India. Lathyrus Lathyrism Newsl 2: 82–84.

Kumari V, Mehra RB (1991) Nature of genetic variation and character association for yield components in *Lathyrus sativus* L. In Proc Symp on Grain Legumes, Indian Society of Genetics and Plant Breeding, IARI, New Delhi, India, pp 193–197.

Kunig G, Hartmann J, Niedermeyer B, Deckert J, Ransmayr G, Heinsen H, Beckmann H, Riederer P (1994) Excitotoxins L-beta-oxalyl-amino-alanine (L-BOAA) and 3,4,6-trihydroxyphenylalanine (6-OH-DOPA) inhibit [3H] alpha-amino-3- hydroxy-5-methyl-4-isoxazole-propionic acid (AMPA) binding in human hippocampus. Neurosci Lett 169: 219–22.

Kuo YH, Ikegami F, Lambein F (1998) Metabolic routes of beta-(isoxazolin-5-on-2-yl)-L-alanine (BIA), the precursor of the neurotoxin ODAP (beta-N-oxalyl-L-alpha,beta,-diaminopropionic acid), in different legume seedlings. Phytochemistry 49: 43–48.

Kuo YH, Bau HM, Rozan P, Chowdhury B, Lambein F (2000) Reduction efficiency of the neurotoxin beta-ODAP in low-toxin varieties of *Lathyrus sativus* seeds by solid state

fermentation with *Aspergillus oryzae* and *Rhizopus microsporus* var *chinensis*. J Sci Food Agri 80: 2209–2215.
Kupicha FK (1983) The infrageneric structure of *Lathyrus*. Notes Royal Botanic Garden Edinburgh 41: 209–244.
Kusama-Eguchi K, Ikegami F, Kusama T, Lambein F, Watanabe K (1996) Effects of beta-ODAP and its biosynthetic precursor on the electrophysiological activity of cloned glutamate receptors. Environ Toxicol Pharmacol 2: 339–342.
Kusama-Eguchi K, Ikegami F, Kusama T, Suda A, Ogawa Y, Igarashi K, Watanabe K (2005) A rat model of neurolathyrism: repeated injection of L: -beta-ODAP induces the paraparesis of the hind legs. Amino Acids 28: 139–43.
Lal MS, Agrawal I, Chitale MW (1986) Genetic improvement of chickling vetch (*Lathyrus sativus* L.) in Madhya Pradesh, India. In: AK Kaul, D Combes (eds) *Lathyrus* and Lathyrism. Third World Medical Research Foundation, New York, USA, pp 146–160.
Lambein F, Kuo YH (2004) Somaclonale variatie in *Lathyrus sativus*. *Abstract*, Autumn meeting Netherlands Society for Plant Biotechnology and Tissue culture NVPW, Leiden, The Netherlands: http://www.nvpw.nl/pdf/NVPW04-2.pdf (Cited Sept 2009).
Lambein F, Haque R, Khan JK, Kebede N, Kuo YH (1994) From soil to brain: zinc deficiency increases the neurotoxicity of *Lathyrus sativus* and may affect the susceptibility for the motorneurone disease neurolathyrism. Toxicon 32: 461–466.
Lambein F, Diasolua Ngudi D, Kuo YH (2001) Vapniarca revisited: Lessons from an inhuman human experience. Lathyrus Lathyrism Newsl 2: 5–7.
Lambein F, Kuo YH, Kusama-Eguchi K, Ikegami F (2007) 3-N-oxalyl-L-2,3-diaminopropanoic acid, a multifunctional plant metabolite of toxic reputation. Arkivoc IX: 45–52.
Leitch IJ, Bennett MD (2007) Genome size and its uses: the impact of flow cytometry. In: J Doležel, J Greilhuber, J Suda (eds) Flow cytometry with plant cells. Analysis of genes, chromosomes and genomes. Wiley-VCH Verlag GmbH & Co. KgaA, Weinheim, Germany, pp 153–176.
Linke KH, Abd El-Monein AM, Saxena MC (1993). Variation in resistance of some forage legumes species to *Orobanche crenata* Forsk. Field Crops Res 32: 277–285.
Loridon K, McPhee K, Morin J, Dubreuil P, Pilet-Nayel ML, Aubert G, Rameau C, Baranger A, Coyne C, Lejeune-Hènaut I, Burstin J (2005) Microsatellite marker polymorphism and mapping in pea (*Pisum sativum* L.). Theor Appl Genet 111: 1022–1031.
Ludolph AC, Spencer PS (1996) Toxic models of upper motor neuron disease. J Neurol Sci 139 Suppl: 53–59.
Ludolph AC, Hugon J, Dwivedi MP, Schaumburg HH, Spencer PS (1987) Studies on the aetiology and pathogenesis of motor neuron diseases. 1. Lathyrism: clinical findings in established cases. Brain 110: 149–165.
Mackenzie GG, Zago MP, Aimo L, Oteiza PI (2007) Zinc deficiency in neuronal biology. IUBMB Life 59: 299–307.
Malek MA (1998) Genetic resources of grass pea (*Lathyrus sativus* L.) in Bangladesh. In: PN Mathur, VR Rao, RK Arora (eds) *Lathyrus* Genetic Resources Network: Proc IPGRI-ICARDA-ICAR Regional Working Group Meeting. National Bureau of Plant Genetic Resources, IPGRI Office for South Asia, New Delhi, India, pp 1–6.
Malik KA, Ali-Khan ST, Saxena PK (1993) High frequency organogenesis from direct seed culture in *Lathyrus*. Ann Bot 72: 629–637.
Maluszynski M, Kasha KJ, Szarejko I (2003) Published double haploid protocols in plant species. In: M Maluszynski, KJ Kasha, BP Forster, I Szarejko (eds) Doubled haploid production in crop plants. A manual. Kluwer Acad Publ, Dordrecht, The Netherlands, Boston, USA, London, UK, pp 309–33.
Mantri NL, Ford R, Coram TE, Pang ECK (2007) Transcriptional profiling of chickpea genes differentially regulated in response to high-salinity, cold and drought. BMC Genomics 8: 30.

Matsumura H, Reich S, Ito A, Saitoh H, Kamoun S, Winter P, Kahl G, Reuter M, Kruger DH, Terauchi R (2003) Gene expression analysis of plant host–pathogen interactions by SuperSAGE. Proc Natl Acad Sci USA 100: 15718–15723.

McCutchan JS, Larkin PJ, Stoutjesdijk PA, Morgan ER, Taylor PWJ (1999) Establishment of shoot and suspension cultures for protoplast isolation in *Lathyrus sativus* L. J Breed Genet 31: 43–50.

Mehra RB, Raju DB, Himabindu K (1995) Status paper on *Lathyrus* Research. Division of Genetics, IARI, New Delhi, India, pp 1–7.

Mehta SL (1997) Plant biotechnology for removal of ODAP from *Lathyrus*. In: RT Haimanot, F Lambein (eds) *Lathyrus* and Lathyrism, A Decade of Progress. University of Ghent, Ghent, Belgium, pp 103–104.

Melka A, Haimanot RT, Lambein F (1997) Symptomatic treatment of neurolathyrism with Tolperizone HCL (Mydocalm): A randomized double blind and placebo controlled drug trial. Ethiop Med J 35: 77–91.

Mera M, Montenegro A, Espinoza N, Gaete N, Barrientos L (2003a) Heritability of seed weight in an inbred population of large-seeded *Lathyrus sativus*. Lathyrus Lathyrism Newsl 3: 24.

Mera M, Tay J, France A, Montenegro A, Espinoza N, Gaete N (2003b) Luanco-INIA, a large-seeded cultivar of *Lathyrus sativus* released in Chile. Lathyrus Lathyrism Newsl 3: 26.

Milczak M, Pedzinski M, Mnichowska H, Szwed-Urbas K, Rybinski W (2001) Creative breeding of grasspea (*Lathyrus sativus* L.) in Poland. Lathyrus Lathyrism Newsl 2: 85–88.

Mishra G, Shukla R, Hasan M, Khanna SK, Das M (2009) Potentiation of neurotoxicity of *Lathyrus sativus* by manganese: alterations in blood-brain barrier. Toxicol Mech Method 19: 318–326.

Mishra RP, Kotasthane SR, Khare MN, Gupta OM, Tiwari SP (1986) Screening of germplasm collections of *Lathyrus* spp. against *Cercospora pisi-sativae* f. sp. *lathyrisativae* Mishra. Indian J Mycol Plant Pathol 16: 302.

Misra M, Addis G, Narayan RKJ (1994) Methods for callus induction and differentiation of *Lathyrus sativus* and embryo rescue in interspecific crosses by tissue culture. J Univ Wales Agri Soc 74: 129–144.

Mitra J, Mehra RB (2001) Comparison of single seed descent, pedigree and random bulk methods in grasspea (*Lathyrus sativus* L.). Legume Res 24: 44–47.

Molina C, Rotter B, Horres R, Udupa SM, Besser B, Bellarmino L, Baum M, Matsumura H, Terauchi R, Kahl G, Winter P (2008) SuperSAGE: the drought stress-responsive transcriptome of chickpea roots. BMC Genomics 9: 553.

Mukhopadhyay A, Bhojwani SS (1978) Shoot-bud differentiation in tissue cultures of leguminous plants. Z Pflanzenphysiol 88: 263–268.

Murti VVS, Seshadri TR, Venkitasubramanian TA (1964) Neurotoxic compounds of the seeds of *Lathyrus sativus*. Phytochemistry 3: 73–78.

Nandini AV, Murray BG, O'Brien IEW, Hammett KRW (1997) Intra- and interspecific variation in genome size in *Lathyrus* (*Leguminosae*). Bot J Linn Soc 125: 359–366.

Narayan RKJ (1982) Discontinuous DNA variation in the evolution of plant species – the genus Lathyrus. Evolution 36: 877–891.

Nerkar YS (1972) Induced variation and responses to selection for low neurotoxin content in *Lathyrus sativus*. Indian J Genet 32: 175–180.

Nerkar YS (1976) Mutation studies in *Lathyrus sativus*. Indian J Genet 36: 223–229.

Nunn PB, Perera CKPW, Bell EA, Lambein F (1994) Appearance of the neurotoxin beta-N-oxalyl-L-alpha,beta-diaminopropionic acid (beta-ODAP) and of homoarginine in the systemic blood and urine following the consumption of cooked *Lathyrus sativus* by human volunteers. Arch Int Physiol Biochem 102: B89.

Ochatt SJ (2006) Flow cytometry: ploidy determination, cell cycle analysis, DNA content per nucleus. In: *Medicago truncatula* Handbook, p 13: http://www.noble.org/medicagohandbook/index.html (Cited Nov 2006).

Ochatt SJ (2008) Flow cytometry in plant breeding. Cytom Part A 73: 581–598.

Ochatt S, Jain SM (eds) (2007) Breeding of Neglected and Under-utilized Crops, Spices and Herbs. Science Publishers, Enfield, NH, USA, p 468.

Ochatt SJ, Pontécaille C, Rancillac M (2000) The growth regulators used for bud regeneration and shoot rooting affect the competence for flowering and seed set in regenerated plants of protein pea. *In Vitro* Cell Dev-Plant 36: 188–193.

Ochatt S, Durieu P, Jacas L, Pontécaille C (2001) Protoplast, cell and tissue cultures for the biotechnological breeding of grass peas. Lathyrus Lathyrism Newsl 2: 35–38.

Ochatt SJ, Muneaux E, Machado C, Jacas L, Pontécaille C (2002) The hyperhydricity of *in vitro* regenerants is linked with an abnormal DNA content in grass pea (*Lathyrus sativus* L.). J Plant Physiol 159: 1021–1028.

Ochatt SJ, Sangwan RS, Marget P, Assoumou Ndong Y, Rancillac M, Perney P (2004) New approaches towards the shortening of generation cycles for faster breeding of protein legumes. Plant Breed 121: 436–440.

Ochatt SJ, Abirached-Darmency M, Marget P, Aubert G (2007) The *Lathyrus* Paradox: "poor men's diet" or a remarkable genetic resource for protein legume breeding? In: S Ochatt, SM Jain (eds) Breeding of Neglected and Under-utilised Crops, Spices and Herbs. Science Publishers, Enfield, NH, USA, pp 41–60.

Ochatt S, Pech C, Grewal R, Conreux C, Lülsdorf M, Jacas L (2009) Abiotic stress enhances androgenesis from isolated microspores of some legume species (Fabaceae). J Plant Physiol 166: 1314–1328.

Ochatt SJ, Conreux C, Jacas L (2010a) *In Vitro* Production of sweet peas (*Lathyrus odoratus* L.) via axillary shoots. In: SM Jain, SJ Ochatt (eds) Methods in Molecular Biology, vol 589: Protocols for *In Vitro* Propagation of Ornamental Plants. The Humana Press, Springer, New York, USA, pp 293–301.

Ochatt SJ, Conreux C, Jacas L (2010b) Flow cytometry assessment of true-to-typeness of *Lathyrus in vitro* regenerants, distinction between species and between landraces within *Lathyrus* species. 5th Int Food Legumes Research Conf & 7th Eur Conf on Grain Legumes, April 26–30, Antalya, Turkey, p 198.

Pai KS, Ravindranath V (1993) L-BOAA induces selective inhibition of brain mitochondrial enzyme, NADH-dehydrogenase. Brain Res 621: 215–21.

Paleacu D, Cohn DF, Rabey JM (1999) Cognitive evaluation of patients with chronic neurolathyrism. Parkinsonism Relat Disord 5: 55–8.

Pandey RL, Chitale MW, Sharma RN, Rastogi N (1996) Status of *Lathyrus* research in India. In: RK Arora, PN Mathur, KW Riley, Y Adham (eds) *Lathyrus* Genetic Resources in Asia: Proc Regional Workshop. Indira Gandhi Agricultural University, Raipur, India. IPGRI Office for South Asia, New Delhi, India, pp 45–52.

Pandey RL, Sharma RN and Chitale MW (1998) Status of *Lathyrus* genetic resources in India. In: PN Mathur, VR Rao, RK Arora (eds) *Lathyrus* Genetic Resources Network. Proc IPGRI-ICARDA-ICAR Regional Working Group Meeting. ICARDA, New Delhi, India, pp 7–14.

Parlevliet JE (1979) Components of resistance that reduce the rate of epidemic development. Annu Rev Phytopathol 17: 203–222.

Parniske M, Hammond-Kosack KE, Golstein C, Thomas CM, Jones DA, Harrison K, Wulff BBH, Jones JDG (1997) Novel disease resistance specificities result from sequence exchange between tandemly repeated genes at the *CfK4/9* locus of tomato. Cell 91: 821–32.

Patil VD, Kulkarni UG, Harkal SR, Nerkar YS (1997) Studies on *in vitro* plant regeneration in grass pea (*Lathyrus sativus* L.). In: RT Haimanot, F Lambein (eds) *Lathyrus* and Lathyrism, A Decade of Progress. University of Ghent, Ghent, Belgium, pp 125–127.

Pecket RC, Selim ARAA (1965) Embryo culture in *Lathyrus sativus*. J Exp Bot 16: 325–328.

Peoples MB, Herrdge DF, Ladha JK (1995) Biological nitrogen fixation: An efficient source of nitrogen for sustainable agricultural production?. Am J Altern Agric 10: 88–96.

Pérez-de-Luque A, Fondevilla S, Pérez-Vich B, Aly R, Thoiron S, Simier P, Castillejo MA, Fernández-Martínez JM, Jorrín J, Rubiales D, Delevault P (2009) Understanding *Orobanche* and *Phelipanche*- host plant interaction and developing resistance. Weed Res 48: 8–22.

Pérez-de-Luque A, Eizenberg H, Grenz JH, Sillero JC, Ávila CM, Sauerborn J, Rubiales D (2010) Broomrape management in faba bean. Field Crops Res 115: 319–328.

Poulter R, Harvey L, Burritt DJ (2003) Qualitative resistance to powdery mildew in hybrid sweet peas. Euphytica 133: 349–358.

Przybylska J, Zimniak-Przybylska Z, Krajewski P (1998) Diversity of seed albumins in the grasspea (*Lathyrus sativus* L.): an electrophoretic study. Genet Resour Crop Evol 45: 423–431.

Przybylska J, Zimniak-Przybylska Z, Krajewski P (1999) Diversity of seed albumins in some *Lathyrus* species related to *L. sativus* L.: an electrophoretic study. Genet Resour Crop Evol 46: 261–266.

Przybylska J, Zimniak-Przybylska Z, Krajewski P (2000) Diversity of seed globulins in *Lathyrus sativus* L. and some related species. Genet Resour Crop Evol 47: 239–246.

Quader MS (1987) Male sterility in *Lathyrus sativus* L. Bang J Bot 16: 9–13.

Quader MS, Ramanujam S, Barat GK (1986) Genetics of flower colour, BOAA content and their relationship in *Lathyrus sativus* L. In AK Kaul, D Combes (eds) *Lathyrus* and lathyrism: Proc Colloque *Lathyrus*, 1985. Third World Medical Research Foundation, New York, USA, pp 93–97.

Quader MS, Singh SP, Mehra RB (1988) Correlation between protein and neurotoxin content and some morphological characters of *Lathyrus sativus* L. (grasspea). Bang J Agri 13: 231–239.

Rahman MA, Rahman MM, Sarkar MA (2001) Progress in isolation and purification of *Lathyrus sativus* breeding lines. Lathyrus Lathyrism Newsl 2: 39–40.

Rahman MM, Kumar J, Rahman MA, Afzal MA (1995) Natural outcrossing in *Lathyrus sativus L.* Indian J Genet 55: 204–207.

Rao SLN, Adiga PR, Sarma PS (1964) The isolation and characterization of β-N-oxalyl-L-α,β,-diaminopropionic acid: a neurotoxin from the seeds of *Lathyrus sativus*. Biochemistry 3: 432–436.

Razdan MK, Cocking EC, Power JB (1980) Callus regeneration from mesophyll protoplasts of sweet pea (*Lathyrus odoratus* L.). Z Pflanzenphysiol 96: 181–183.

Ressler C (1962) Isolation and identification from common vetch of the neurotoxin beta-cyano-L-alanine, a possible factor in neurolathyrism. J Biol Chem 237: 733–735.

Robertson JR, Abd El-Moneim AM (1996) *Lathyrus* Germplasm Collection, Conservation and Utilization for Crop Improvement at ICARDA. In: RK Arora, PN Mathur, RW Riley, Y Adham (eds) *Lathyrus* Genetic Resources in Asia: Proc Regional Workshop. Indira Gandhi Agricultural University, Raipur, India. IPGRI, Office for South Asia, New Delhi, India, pp 97–111.

Ross SM, Roy DN, Spencer PS (1989) Beta-N-oxalylamino-L-alanine action on glutamate receptors. J Neurochem 53: 710–5.

Roth JA, Garrick MD (2003) Iron interactions and other biological reactions mediating the physiological and toxic actions of manganese. Biochem Pharmacol 66: 1–13.

Roy PK, Singh B, Mehta SL, Barat GK, Gupta N, Kirti PB, Chopra VL (1991) Plant regeneration from leaf discs of *Lathyrus sativus*. Indian J Exp Biol 29: 327–330.

Roy PK, Barat GK, Mehta SL (1992) *In vitro* plant regeneration from callus derived from root explants of *Lathyrus sativus*. Plant Cell Tiss Org Cult 29: 135–138.

Roy PK, Ali K, Gupta A, Barat GK, Mehta SL (1993) *β*-Noxalyl-L-*α*,*β*-diaminopropionic acid in somaclones derived from internode explants of *Lathyrus sativus*. J Plant Biochem Biotechnol 2: 9–13.

Rubiales D (2003) Parasitic plants, wild relatives and the nature of resistance. New Phytol 160: 459–461.

Rubiales D, Pérez-de-Luque A, Fernández-Aparicio M, Sillero JC, Román B, Kharrat M, Khalil S, Joel DM, Riches C (2006) Screening techniques and sources of resistance against parasitic weeds in grain legumes. Euphytica 147: 187–199.

Rubiales D, Fernández-Aparicio M, Pérez-de-Luque A, Prats E, Castillejo MA, Sillero JC, Rispail N, Fondevilla S (2009a) Breeding approaches for crenate broomrape (*Orobanche crenata* Forsk.) management in pea (*Pisum sativum* L.). Pest Manag Sci 65: 553–559.

Rubiales D, Fernández-Aparicio M, Wegmann K, Joel D (2009b) Revisiting strategies for reducing seedbank of *Orobanche* and *Phelipanche* spp. Weed Res 49: 23–33.

Rubiales D, Fernández-Aparicio M (2011) Innovations in crop protection: parasitic weeds infecting legume crops as a case study. Agron Sust Dev, in press.

Rudra MN, Chaudhury LM (1952) Preliminary note on methionine excretion in lathyrism. Curr Sci 4: 99–100.

Rudra MN, Chowdhury LM, Sinha SP (1952) A preliminary report on the treatment of lathyrism with parenteral methionine. Ind Med Gaz 87: 89–91.

Rybinski W (2003) Mutagenesis as a tool for improvement of traits in grasspea (*Lathyrus sativus* L.). Lathyrus Lathyrism Newsl 3: 27–31.

Rybinski W, Szot B, Rusinek R (2008) Estimation of morphological traits and mechanical properties of grasspea seeds (*Lathyrus sativus* L.) originating from EU countries. Int Agrophys 22: 261–275.

Sachdev A, Sharma M, Johari RP, Mehta SL (1995) Characterization and cloning of ODAP degrading gene from a soil microbe. J Plant Biochem Biotechnol 4: 33–36.

Sáenz de Miera LE, Ramos J, Pérez de la Vega M (2008) A comparative study of convicilin storage protein gene sequences in species of the tribe Vicieae. Genome 51: 511–523.

Sardinha J, Almeida N, Vaz Patto MC (2007) Molecular characterization of *Lathyrus sativus* using AFLP markers. In: Proc 6th Eur Conf on Grain Legumes. Lisbon, Portugal, p 131.

Selye H (1957) Lathyrism. Rev Can Biol 16: 1–82.

Sen NK, Ghosh AK (1962) Genetics of grass pea (*Lathyrus sativus* L.). J Genet 58: 51–64.

Siddique KHM, Loss SP, Herwig SP, Wilson JM (1996) Growth, yield and neurotoxin (ODAP) concentration of three *Lathyrus* species in Mediterranean-type environments of Western Australia. Aust J Exp Agri 36: 209–218.

Siddique KHM, Hanbury C, Sarker A (2006) Registration of 'Ceora' Grass Pea. Crop Sci 46: 986.

Sillero JC, Cubero JI, Fernández-Aparicio M, Rubiales D (2005) Search for resistance to crenate broomrape (*Orobanche crenata*) in *Lathyrus*. Lathyrus Lathyrism Newsl 4: 7–9.

Sillero JC, Fondevilla S, Davidson J, Vaz Patto MC, Warkentin T, Thomas J, Rubiales D (2006) Screening techniques and sources of resistance to rusts and mildews in grain legumes. Euphytica 147: 255–272.

Sinha RR, Das K, Sen SK (1982) Nutritional requirements of tissue cultures of some tropical legume crops. Indian J Exp Biol 21: 113–119.

Sinha RR, Das K, Sen SK (1983) Plant regeneration from stem-derived callus of the seed legume *Lathyrus sativus* L. Plant Cell Tiss Org Cult 2: 67–76.

Skiba B, Ford R, Pang ECK (2003) Amplification and detection of polymorphic sequence-tagged sites in *Lathyrus sativus*. Plant Mol Biol Rep 21: 391–404.

Skiba B, Ford R, Pang ECK (2004) Construction of a linkage map based on a *Lathyrus sativus* backcross population and preliminary investigation of QTLs associated with resistance to ascochyta blight. Theor Appl Genet 109: 1726–1735.

Skiba B, Ford R, Pang E (2005) Construction of a cDNA library of *Lathyrus sativus* inoculated with *Mycosphaerella pinodes* and the expression of potential defense-related expressed sequence tags (ESTs). Physiol Mol Plant Pathol 66: 55–67.

Skiba B, Gurung AM, Pang ECK (2007) Genome Mapping and Molecular Breeding in *Lathyrus*. In: C Kole (ed) Genome Mapping and Molecular Breeding in Plants, vol 3: Pulses, Sugar and Tuber Crops. Springer, Berlin, Heidelberg, Germany, pp 123–132.

Skrzypek E, Czyczyło-Mysza I, Marcinska I, Wedzony M (2008) Prospects of androgenetic induction in *Lupinus* spp. Plant Cell Tiss Org Cult 94: 131–137.

Smartt J (1984) Evolution of Grain Legumes. I. Mediterranean Pulses. Exp Agric 20: 275–296.

Spencer PS (1995) Lathyrism. Handbook of Clinical Neurology 21: 1–20.

Spencer PS (1999) Food toxins, ampa receptors, and motor neuron diseases. Drug Metab Rev 31: 561–87.
Spencer PS, Roy DN, Ludolph A, Hugon J, Dwivedi MP, Schaumburg HH (1986) Lathyrism: evidence for role of the neuroexcitatory aminoacid BOAA. Lancet 8515: 1066–1067.
Sriram K, Shankar SK, Boyd MR, Ravindranath V (1998) Thiol oxidation and loss of mitochondrial complex I precede excitatory amino acid-mediated neurodegeneration. J Neurosci 18: 10287–10296.
Tadesse W (1997) Grasspea (*Lathyrus sativus* L.): The hardy pulse crop. FABIS Newsl 40: 1.
Tadesse W, Bekele E (2001) Isozymes variability of grasspea (*Lathyrus sativus* L.) in Ethiopia. Lathyrus Lathyrism Newsl 2: 43–46.
Tadesse W, Bekele E (2003a) Variation and association of morphological and biochemical characters in grass pea (*Lathyrus sativus* L.). Euphytica 130: 315–324.
Tadesse W, Bekele E (2003b) Phenotypic diversity of Ethiopian grass pea (*Lathyrus sativus* L.) in relation to geographical regions and altitudinal range. Genet Resour Crop Evol 50: 497–505.
Tanksley SD, Young ND, Paterson AH, Bonierbale MW (1989) RFLP mapping in plant breeding: New tools for old science. Biotechnology 7: 257–264.
Tavoletti S, Iommarini L (2007) Molecular marker analysis of genetic variation characterizing a grasspea (*Lathyrus sativus*) collection from central Italy. Plant Breed 126: 607–611.
Tavoletti S, Iommarini L, Crinó P, Granati E (2005) Collection and evaluation of grasspea (*Lathyrus sativus* L.) germplasm of central Italy. Plant Breed 124: 388–391.
Tay J, Valenzuela A, Venegas F (2000) Collecting and evaluating Chilean germplasm of grasspea (*Lathyrus sativus* L.). Lathyrus Lathyrism Newsl 1: 21.
Tewari S, Brown SM, Fristensky B (2003) Plant defense multigene families: I. Divergence of *Fusarium solani*—Induced expression in *Pisum* and *Lathyrus*: http://arXiv.org/q-bio. PE/0310003 (Cited Sept 2009).
Tiwari KR, Campbell CG (1996a) Inheritance of neurotoxin (ODAP) content, flower and seed coat colour in grass pea (*Lathyrus sativus* L.). Euphytica 91: 195–203.
Tiwari KR, Campbell CG (1996b) Inheritance of seed weight in grasspea (*Lathyrus sativus* L.). FABIS Newsl 38/39: 30–33.
Tiwari KR, Campbell CG (1997) Male sterility in grasspea. FABIS Newsl 40: 10–12.
Turk M, Albayrak S, Çelik N (2007) Estimates of broad-sense heritability for seed yield and yield components of grass pea (*Lathyrus sativus* L.). Turk J Agri For 31: 155–158.
Tyagi A, Santha IM, Mehta SL (1995) Molecular response to water stress in *Lathyrus sativus*. J Plant Biochem Biotechnol 4: 47–49.
Tyagi A, Santha IM, Mehta SL (1999) Effect of water stress on proline content and transcript levels in *Lathyrus sativus*. Indian J Biochem Biophys 36: 207–210.
Van Dorrestein B, Baum M, Abd El Moneim AM (1998) Use of somaclonal variation in *Lathyrus sativus* (Grass pea) to select variants with low β-ODAP concentration. 3rd Eur Conf on Grain Legumes. Valladolid, Spain, p 364.
Vaz Patto MC, Rubiales D (2009) Identification and characterization of partial resistance to rust in a germplasm collection of *Lathyrus sativus*. Plant Breed 128: 495–500.
Vaz Patto MC, Fernández-Aparicio M, Moral A, Rubiales D (2006a) Characterization of resistance to powdery mildew (*Erysiphe pisi*) in a germplasm collection of *Lathyrus sativus*. Plant Breed 125: 308–310.
Vaz Patto MC, Skiba B, Pang ECK, Ochatt SJ, Lambein F, Rubiales D (2006b) *Lathyrus* improvement for resistance against biotic and abiotic stresses: from classical breeding to marker assisted selection. Euphytica 147: 133–147.
Vaz Patto MC, Fernández-Aparicio M, Moral A, Rubiales D (2007) Resistance reaction to powdery mildew (*Erysiphe pisi*) in a germplasm collection of *Lathyrus cicera* from Iberian origin. Genet Resour Crop Evol 54: 1517–1521.
Vaz Patto MC, Fernández-Aparicio M, Moral A, Rubiales D (2009) Pre and posthaustorial resistance to rusts in *Lathyrus cicera* L. Euphytica 165: 27–34.

Vinogradov AE (1994) Measurement by flow cytometry of genomic AT/GC ratio and genome size. Cytom Part A 16: 34–40.
Wang X, Warkentin TD, Briggs CJ, Oomah BD, Campbell CG, Woods S (1998) Trypsin inhibitor activity in field pea (*Pisum sativum* L.) and grass pea (*Lathyrus sativus* L.). J Agri Food Chem 46: 2620–2623.
Warren BA, Patel SA, Nunn PB, Bridges RJ (2004) The *Lathyrus* excitotoxin beta-N-oxalyl-L-alpha,beta-diaminopropionic acid is a substrate of the L-cystine/L-glutamate exchanger system xc. Toxicol Appl Pharmacol 200: 83–92.
Xiong YC, Xing GM, Li FM, Wang Sm, Fan XW, Li ZX, Wang YF (2006) Abscisic acid promotes accumulation of toxin ODAP in relation to free spermine level in grass pea seedlings (*Lathyrus sativus* L.). Plant Physiol Biochem 44: 161–169.
Yadav CR (1996) Genetic evaluation and varietal improvement of grasspea in Nepal. In: RK Arora, PN Mathur, KW Riley, Y Adham (eds) *Lathyrus* Genetic Resources in Asia: Proc Regional Workshop. Indira Gandhi Agricultural University, Raipur, India. IPGRI Office for South Asia, New Delhi, India, pp 21–28.
Yadav VK, Mehta SL (1995) *Lathyrus sativus*: a future pulse crop free of neurotoxin. Curr Sci 68: 288–292.
Yamamoto K, Fujiwara T, Blumenreich ID (1984) Karyotypes and morphological characteristics of some species in the genus *Lathyrus* L. Jpn J Breed 34: 273–284.
Yang HM, Zhang XY (2005) Considerations on the reintroduction of grass pea in China. Lathyrus Lathyrism Newsl 4: 22–26.
Yunus AG, Jackson MT (1991) The gene pools of the grasspea (*Lathyrus sativus* L.). Plant Breed 106: 319–328.
Yunus AG, Jackson MT, Janet PC (1991) Phenotypic polymorphism of six enzymes in the grasspea (*Lathyrus sativus* L.). Euphytica 55: 33–42.
Zambre M, Chowdhury B, Kuo YH, van Montagu M, Angenon G, Lambein F (2002) Prolific regeneration of fertile plants from green nodular callus induced from meristematic tissues in *Lathyrus sativus* L. (grass pea). Plant Sci 163: 1107–1112.
Zohary D, Hopf M (2000) Domestication of Plants in the Old World. 3rd edn. Oxford University Press, New York, USA, p 316.

5

Chickpea

Josefa Rubio,[1] *Juan Gil,*[2] *Maria José Cobos*[1] and *Teresa Millán*[2,a,*]

ABSTRACT

Chickpea (*Cicer arietinum* L.) is cultivated in more than 40 countries and it is mainly used for human consumption. In the last few years, increase in productivity has been achieved, but it is still necessary to obtain new productive varieties, which include superior levels of resistance or tolerance to biotic and abiotic stresses. The search for resistance to Ascochyta blight continues to be a major task in this crop because it should allow expanding the cycle of the crop in many regions and, consequently, it could increase final chickpea yield. Moreover, breeding for quality is becoming an important challenge in a world of consumers and industry looking for new food preferences. Molecular markers, and particularly sequence tagged microsatellite site (STMS), have significantly increased in chickpea and they are very useful to evaluate genetic resources as well as to develop genetic maps using both intra- and interspecific crosses. Moreover, cross-genome markers from model species *Medicago truncatula* were mapped in chickpea, which permitted to establish synteny relationships between both species. This information together with a correct characterization of agronomic traits will help to target new genes and to simplify the study of quantitative traits.

Keywords: breeding, stresses, marker-assisted selection (MAS), diversity, linkage maps

[1]Área de Mejora y Biotecnología. IFAPA. Apdo 3092. 14080 Córdoba, Spain.
[2]Dpto. de Genética. Univ. de Córdoba. Campus Rabanales Edif. C-5, 14071 Córdoba, Spain.
[a]e-mail: *ge1mivat@uco.es*
*Corresponding author

5.1 Basic Information on the Plant

5.1.1 General Aspects

Chickpea (*Cicer arietinum* L.) is, together with cowpea, the second most important cultivated grain legume in the world following common bean (*Phaseolus vulgaris* L.) in terms of harvested area (FAO 2010). Like other annual legumes, it is a significant contributor to agricultural sustainability by improving soil fertility through N_2–fixation and, as a rotation crop, by allowing the diversification of agricultural production systems (Fig. 5-1a). Among temperate pulses, chickpea is the most tolerant crop to heat and drought and is considered to be an important low-input crop in cropping systems of arid and semi-arid regions around the world (Saxena 1987).

Figure 5-1 a) Chickpea advanced lines in the field, b) chickpea kabuli type and c) desi type.
Color image of this figure appears in the color plate section at the end of the book.

This species is cultivated mostly in the Mediterranean basin, the Near East, Central and South Asia, East Africa, South America, North America and, more recently, in Australia. The main producers are India, Turkey and Pakistan, contributing with 65, 5.9 and 5.4%, respectively to the world harvest. Average world chickpea yield has improved during the past decades increasing from 504 kg/ha in 1980 to 771 kg/ha in 2008, but it is still quite low despite its agronomical importance. Total world production was approximately 8.7 Mt from 11.5 Mha in 2008 (FAO 2010).

Chickpea is mainly used for human consumption. It is used as a whole dry seed, decorticated dry split cotyledons (dhal) and flour, and it is processed by different methods including mainly boiling, roasting and frying. The nutritional value of chickpea seeds is well known: the protein content ranges from 20.3 to 28.2%, starch content from 35 to 43%, and oil content only from 3–6% (Gil et al. 1996). Seed oil contains a high proportion of unsaturated fat (oleic and linoleic acids) rather than saturated fats that have been linked to heart and circulatory diseases. Besides, chickpea seeds are a good source of calcium, magnesium, potassium, phosphorus, iron, zinc and manganese, and in comparison with other grain legumes, antinutritive components are nearly absent (Williams and Singh 1987; Rincon et al. 1998). Consequently, chickpea could be considered as functional or nutraceutical food. While it is a cheap source of protein and energy in the developing world, it has the added value of being a healthy food in the First World.

5.1.2 Brief History of the Crop

Chickpea is an Old World pulse, probably domesticated in the Near East. The accepted wild progenitor of chickpea (*Cicer reticulatum* Ladz.) was discovered and collected in an area of southeastern Turkey (Ladizinsky and Adler 1976). Early botanists had postulated several different origins. Initially, its origin was located in an area in south of the Caucasus and north of Persia; later, two primary centers of diversity in Southwest Asia and the Mediterranean, and one secondary center of origin in Ethiopia were postulated (de Candolle 1883 and Vavilov 1926, in van der Maesen 1987).

Carbonized chickpea seeds, which have been unearthed in various archeological sites throughout the Middle East, indicate the use of this legume in the early days of the Neolithic Revolution together with cereals and other pulses (Ladizinsky and Adler 1976). Therefore, the cultivated chickpea, *C. arietinum,* was one of the first grain legumes domesticated in the Old World along with other crops such as wheat, barley, rye, pea, lentil, flax and vetch (Abbo et al. 2003a). Probably mutations like pod indehiscence, erect habit and non-dormant seeds have played an important role in its domestication process. After domestication, chickpea spread from its center of origin to the Indian subcontinent in the east, Ethiopia in the south through Nile valley, the Caucasus in the north and through both the Mediterranean and the Danuvio basins in the west (van der Maesen 1972). In its evolutionary pathway as a cultivated crop, new mutations like white flower, photoperiod insensitivity, thinner seed coat, an increased erect habit, etc., were selected in the different environments, generating a high phenotypic variation. Since early times, high phenotype diversity for morphological and agronomic characters has been reported in this cultivated species (Moreno and Cubero 1978; Upadhyaya 2003). Despite this high morphological diversity, a narrow genetic base has been reported regarding isozyme and seed protein. This fact together with the limited area where its ancestor is distributed suggests a monophyletic origin of this cultivated species (Zohary 1999). A series of genetic bottlenecks in the domestication of chickpea has been described, including the restricted distribution of the wild progenitor, the founder effect associated with domestication and the shift from winter to summer cropping (Abbo et al. 2003b).

For breeding purposes this species is divided into "desi" and "kabuli" types (Fig. 5-1b and 5-1c). This distinction is made mainly on the basis of seed shape and color. White flower, thin seed coat, large and cream colored seeds are present in kabuli chickpea, while pink flower, thick seed coat, small angular and dark seeds are characteristics of desi ones. This classification overlaps, to a certain extent, with the *macrosperma* and *microsperma* races proposed by Moreno and Cubero (1978) studying quantitative as well as qualitative traits. These types differ also on a series of traits from agronomic values, some of which are more frequently found in kabuli types like cold tolerance, resistance to Ascochyta blight and erect growth habit; whereas heat and drought tolerance, resistance to Fusarium wilt and early flowering are prevalent in desi types (Singh 1987). Kabuli chickpeas are grown mainly in the Mediterranean basin, the Near East, central Asia and America, while desi types are cultivated mostly in India, Pakistan and East Africa. Kabuli probably evolved from the desi type in the Mediterranean basin and oligogenic traits like flower color, seed coat thickness and seed size seem to have played an important role in its evolution (Moreno and Cubero 1978; Gil and Cubero 1993).

5.1.3 Taxonomic Position

Chickpea is a self-pollinated diploid ($2n = 2x = 16$) annual grain legume with a genome size of approximately 738 Mbp (DNA content of 1.53 pg/2C) (Arumuganathan and Earle 1991). The genus *Cicer* is a member of the monogeneric tribe Cicereae Alef, subfamily Papilionoideae, family Leguminosae. It includes nine annual and 35 perennial species. Only one of these is cultivated on a large scale: the chickpea (*C. arietinum*). This genus has traditionally been classified into four sections (*Monocicer, Chamaecicer, Polycicer* and *Acanthocicer*) based on morphological traits and geographical distribution (van der Maesen et al. 1987, 2007). Annual species, including the cultivated chickpea, belong to the section *Monocicer*. While the cultivated species is distributed worldwide, the eight wild annual species are endemic to an area reaching from Turkey to central Asia and to zones of Egypt, Ethiopia and the Sudan (Robertson et al. 1997). The perennial *Cicer* species generally show a distant relationship with the cultivated species (Tayyar and Wainess 1996; Iruela et al. 2002).

Chickpea wild relatives have attracted much attention as they are potential sources of a wide range of agronomically favorable traits for the improvement of the genepool of the cultigen (Muehlbauer et al. 1994; Singh et al. 1998; Croser et al. 2003a). Therefore, based on their crossability with chickpea, the wild *Cicer* annual species have been classified into genepools that reflect their distance from the cultivated species. Van der Maesen (1972) and Ladizinsky and Adler (1976) adopted the classical definition of the primary, secondary and tertiary genepools as proposed by Harlan and de Wet (1971). According to this definition, the primary genepool includes all species that hybridize freely with chromosome pairing and gene exchange, and produces viable hybrids. In this model, the primary annual genepool of *Cicer* consists of *C. arietinum*, the wild annual progenitor *C. reticulatum*, and the closely related *C. echinospermum* P.H. Davis. Moreno and Cubero (1978) proposed *C. arietinum* as a species with two subspecies: ssp. *reticulatum* (Lad.) including the wild forms, and ssp. *arietinum* L. including the cultivated forms. *Cicer bijugum* K.H. Rech, *C. pinnatifidum* Jaup and Spp. and *C. judaicum* Boiss. compose the secondary genepool where sterility is associated with the first-generation hybrids due to genetic barriers or chromosome alterations (Verma et al. 1990; Singh et al. 1999a, b). Members of the tertiary genepool are difficult to utilize and sterility is always associated with hybrids. In chickpea, the tertiary genepool consists of: *C. chorassanicum* (Bge.) M. Pop., which has never been successfully crossed with chickpea; *C. cuneatum* Hoechst ex. Rich, which has been crossed only twice with chickpea and both times the hybrid has either not flowered or failed to germinate; and *C. yamashitae* Kitamura, which has been crossed only once with hybrids completely sterile (Singh et al. 1999a, b).

5.1.4 Cytogenetic Description

The cultivated chickpea karyotype consists of a long chromosome pair (>3.5 μm) that is submetacentric and distinctly satellited, six pairs of medium-sized (1.6–2.75 μm) metacentric to submetacentric chromosomes and a pair of very short (<1.5 μm) metacentric chromosomes (Ocampo et al. 1992; Ahmad 2000).

Meiosis in *Cicer* species is characterized by a rather diffuse prophase. The shorter chromosomes frequently form ring bivalents, whereas the longer chromosomes form rod bivalents (Ahmad and Chen 2000).

Five out of the eight pairs of chickpea chromosomes, named A, B, C, G and H, could be identified by flow cytometric analysis. They were identified by their size, relative DNA content and the presence of telomeric and 5S rDNA sequences. Chromosomes D, E and F could not be discriminated individually due to their similarities in shape and size. PCR reaction on flow-sorted chromosomes with specific sequence tagged microsatellite site

(STMS) markers permitted to assign chromosome H to the linkage group (LG) 8 of the chickpea genetic map (Vláčilová et al. 2002).

Artificial induction of tetraploidy has been reported in chickpea (Sohoo et al. 1970; Pundir et al. 1983). The induced tetraploidy was accompanied by larger leaves, flowers, stomata, pollen grains and seeds. Pundir et al. (1983) reported tetraploid plants with relatively normal meiosis and normal production, indicating their potential for chickpea improvement. In addition, it would be interesting to use polyploidization in related wild species in order to break crossability barriers with the cultigen.

5.2 Breeding Achievements

Increasing and stabilizing seed yield while minimizing inputs is the major aim of chickpea breeding. This goal can be achieved by developing cultivars better adapted to stresses in local environments. Especially Ascochyta blight, Fusarium wilt, pod borer, drought and cold are major constraints to yield improvement and adoption of this crop by farmers. Therefore, improving resistance to biotic and tolerance to abiotic stresses as well as a general increase in dry matter are major aims of chickpea breeders around the world (Gaur et al. 2007).

As it was mentioned before (see Section 5.1.2), desi and kabuli types represent two different genetic backgrounds. Crosses between kabuli and desi are used in many chickpea breeding programs in order to obtain genotypes combining desirable traits from both types. High variation and transgressive segregation for agronomic traits related to yield in these crosses have been reported (Singh 1987; Maynez et al. 1993). On the other hand, the knowledge of the genetic system controlling agronomic characters determines the choice of the most suitable breeding method. In general, important agronomic characters like yield components (pods and seeds per plant, seeds per pod and seed weight) show quantitative distribution, some of them positively correlated with yield (Moreno and Cubero 1978; Kharrat et al. 1991; Maynez et al. 1993). Thus, indirect selection via pod per plant and seed weight has been reported to be more efficient than direct selection for yield (Kumar and Bahl 1992). In addition, some characters with single Mendelian inheritance have played an important role in chickpea breeding as flowering time, double pod per node and growth habit (Rubio et al. 2004; Upadhyaya et al. 2007).

Breeding efforts have contributed substantially to improve chickpea yield potential. However, the lack of stable production continues to be a major concern in this crop. Therefore, major constraints to increase chickpea yield are agronomic traits related to the adaptation to a cropping system. For instance, in the Mediterranean region, chickpea is traditionally a spring crop and, consequently, plants develop poor biomass during their growing period. However, winter sowing is limited by the susceptibility of chickpea to blight, a disease caused by *Ascochyta rabiei* (Pass.) Lab. Today, the development of blight resistant lines has made possible the introduction of winter sowing in this region with the prospect of increasing chickpea production that could be doubled (Singh and Reddy 1996).

Wild *Cicer* species closely related to the cultigen are being used in breeding programs to expand the genetic diversity. Some of these wild species have been identified as sources for improved tolerance/resistance to both biotic and abiotic stresses (Singh et al. 1998; Croser et al. 2003a). However, the wide crosses will inevitably also introduce undesirable traits, such as a reticulated and thick seed coat, small seed size, prostrate growth habit and pod dehiscence, although these can, in principle, be eliminated through backcrossing and selection.

5.2.1 Morphological and Phenological Traits

The improvement of the adaptability of chickpea to different environments is influenced by modifications in plant architecture and phenological traits. Thus, bushy growth habit seems to have a positive effect on yield and yield stability in chickpea crop under Mediterranean conditions (Rubio et al. 1998, 2004). Nevertheless, a high yielding erect cultivar has been released in Mexico (Soltero-Díaz et al. 2008). The double podding natural mutation has been reported to have influence on yield stability in the Mediterranean region (Rubio et al. 1998, 2004) and to contribute to higher yield in semi-arid tropics (Kumar et al. 2000). Both traits, double pod and growth habit, are each controlled by a single gene: simple/double pod (*S/s* or *Sfl/sfl*) and erect/bushy habit (*Gh/gh*) (Muehlbauer and Singh 1987).

An appropriate time for flowering is also a major task to improve crop adaptation influenced by photoperiod and temperature. Early flowering associated with early maturity is preferred to escape from terminal drought, high temperature or frost at the end of the season (Gaur et al. 2007). Major genes (*Ppd/ppd* equal to *Efl2/elf2* and *Efl1/elf1*), which control photoperiodic response have been described using the lines ICC5810 and ICCV2 (Or et al. 1999; Kumar and van Rheenen 2000; Hegde 2010), even though other minor genes could affect the control of this trait. Early flowering has a positive effect on yield in winter and spring sowing under Mediterranean conditions (Rubio et al. 2004). This trait has also been reported to be highly associated with high growth vigor in recombinant inbred lines (RILs) and backcross populations, playing an important role in the establishment of the crop (Sabaghpour et al. 2003). In addition, Hovav et al. (2003) reported a moderate positive genetic correlation between days to flowering and seed weight, and suggested that it might be difficult to breed early-flowering cultivars without compromising seed weight. However, two extra-large kabuli chickpea from Mexican origin were evaluated as very early maturating at ICRISAT, which suggests that it is possible to breed early varieties with extra-large seeds (Gaur et al. 2006). Seed size, determined by seed weight, is a key quality determinant in the market of the cool-season food legume chickpea because it fetches three times higher prices than other chickpea cultivars.

Leaf shape may also affect agronomic performance. Several leaf shape mutations have been described but the most promising is the simple leaf trait controlled by a single gene (*Slv/slv*) (Muehlbauer and Singh 1987). It was introduced with success into modern Ascochyta-resistant kabuli cultivars released by Muehlbauer et al. (1998a, b) and derived from unifoliate cultivar "Surutato-77" developed in Mexico.

5.2.2 Biotic Stresses

Breeding chickpea for resistance to Ascochyta blight has become one of the most important objectives of chickpea improvement programs. This fungal disease is caused by *Ascochyta rabiei* (Pass) Lab., which affects all aerial parts of the plant. It reduces chickpea seed yield significantly when favorable conditions such as cool (5–15°C) and wet weather favor the disease development. Susceptible cultivars may reach 100% of losses in yield (Pande et al. 2005).

Pathogenic variation has been described and it seems to be the cause of the resistance breakdown. The presence of two *Ascochyta* mating types in most chickpea production regions has been described and it can contribute to increase the levels of genotype variation of the pathogen (Kaiser 1997). Different pathogenic groups, races, virulence forms and pathotypes have been reported in *A. rabiei* reaching up to 14 classes (Chongo et al. 2004; Basandrai et al. 2005; Jayakumar et al. 2005; Pande et al. 2005). This profusion of classifications for variation in *A. rabiei* could be due to both different assay procedures and different sets of differential cultivars used. Furthermore, pathogenicity or aggressiveness within the same isolate/line combination could vary according to the environmental conditions. Udupa et al. (1998) used, in an experiment under controlled conditions, a three pathotypes (I, II and III) system based on pathogenicity on three differential lines: ILC1929, which was susceptible to all pathotypes; ILC482, with resistance to pathotype I but susceptibility to II and III; and ILC3279, resistant to pathotypes I and II, but susceptible to III. Chen et al. (2004) found, in a study using both a high number of known chickpea differentials and previously identified pathotypes and races of *A. rabiei*, a similar reaction of differentials to inoculation with each isolate and similar levels of disease of races on the differentials. Their results indicate that the number of differentials can be reduced without sacrificing accuracy in describing pathogenic variation of *A. rabiei* on chickpea.

Different results about the genetic control of resistance have been reported considering this trait as qualitative or quantitative (Table 5-1). These differences in results depend on the screening techniques and the genotype used as a source of resistance. Sources of resistance to Ascochyta blight have been identified at the International Center for Agricultural Research in the Dry Areas (ICARDA) (Singh et al. 1991; Singh and Reddy 1993). The use of different resistance sources in breeding programs is needed to diversify the genetic basis of resistance in elite chickpea germplasm and/or to increase the level of resistance through gene pyramiding.

Table 5-1 Different results on the genetic control of chickpea resistance to Ascochyta blight.

Resistance genes	References
Single gene	Vir et al. 1975; Singh and Reddy 1983; Singh and Reddy 1989; Tewari and Pandey 1986; Udupa and Baum 2003; Bhardwaj et al. 2010.
2–3 genes	Tewari and Pandey 1986; Dey and Singh 1993; Udupa and Baum 2003; Bhardwaj et al. 2010.
2–3 major genes plus minor genes	Tekeoglu et al. 2000; Danehloueipour et al. 2007.
Quantitative inheritance	Muehlbauer and Kaiser 1994; Santra et al. 2000; Flandez-Galvez et al. 2003b; Collard et al. 2003; Udupa and Baum 2003; Cho et al. 2004; Cobos et al. 2006; Lichtenzveig et al. 2006.

Development of blight resistant lines makes possible the introduction of winter sowing in the Mediterranean basin with the prospect of doubling chickpea production (Singh and Reddy 1996). However, an integrated pest management approach, including the use of cultural practices, monitoring field populations of the pathogen, seed treatments, foliar sprays of fungicide and the use of resistant or tolerant cultivars, should be practiced and promoted in chickpea in order to control Ascochyta blight (Gan et al. 2006).

Fusarium wilt, caused by the vascular pathogen *Fusarium oxysporum* Schlechtend: Fr. f. sp. *ciceris*, is along with Ascochyta blight the most important fungal disease in chickpea. Persistence of the pathogen in soil and its capacity to survive for years even in the absence of the host render its control difficult (Haware et al. 1996). To date, eight races have been reported from India, Spain and the United States (Haware and Nene 1982; Jimenez-Diaz et al. 1993; Jimenez-Gasco et al. 2004) and have showed differences in disease symptoms: yellowing syndrome (races 0 and 1B/C) and wilting syndrome (races 1A, 2, 3, 4, 5 and 6). About yellowing and wilting syndromes, emerging evidence indicates that both types of symptoms can be caused by a single race depending on the host genotype used (Sharma and Muehlbauer 2007). A late wilting phenomenon has been described in some genotypes susceptible to race 1 (Upadhyaya et al.

1983a). This phenomenon refers to those susceptible genotypes showing a prolonged latent period. A different phenomenon described as slow wilting has also been observed for races 2, 3, 5 and 0 (Sharma et al. 2005; Cobos et al. 2009; Halila et al. 2009a). Slow wilting differs from late wilting on the latent period, the disease progress rate and the final disease severity (Sharma and Muehlbauer 2007).

Resistance to chickpea wilt has been described to be race-specific and governed by major resistance genes, the majority of which are recessive in nature. Early studies showed that disease reaction to race 1 was governed by three independent genes whose combination can confer resistance, late wilting or early wilting reaction (Upadhyaya et al. 1983a, b). Another genetic system based also on three independent genes was found to confer resistance to race 2 (Kumar 1998), whereas two genes were reported as controlling resistance to race 4 and 0 (Tullu et al. 1999; Rubio et al. 2003). A single resistance gene to each race was reported for race 3 and race 5, respectively (Tekeoglu et al. 2000; Sharma et al. 2004). Apart from these major genes conferring vertical resistance, other genes could be related to slow wilting resistance, although the genetics of this phenomenon in chickpea have not been determined (Sharma and Muehlbauer 2007). Sources of resistance to all races are available in cultivated chickpea. The use of resistant elite cultivars pyramiding resistance genes is expected to provide durable resistance against this disease.

Botrytis grey mold (BGM), caused by *Botrytis cinerea* Pers. ex. Fr., is another crucial biotic constraint to chickpea production in several South Asian countries causing complete crop loss (Pande et al. 2002). There is not an adequate level of genetic resistance to BGM in the cultivated genotypes. Nevertheless, higher levels of resistance have been found in wild *Cicer* species, including *C. judaicum*, *C. bijugum*, *C. echinospermum*, and *C. pinnnatifidum* (Singh et al. 1991; Pande et al. 2002). Thus, several wide and intraspecific hybridizations have been carried out to transfer the identified disease resistance in wild types and landraces to commonly adopted and widely grown chickpea cultivars. Hence, an integrated disease management using the available options is essential to successfully control the disease and mitigate yield losses (Pande et al. 2006a; Johansen et al. 2008).

Another fungal disease considered of local importance is chickpea rust, caused by the fungus *Uromyces ciceris arietini*, which can be a problem in central Mexico and Italy (Ragazzi 1982; Díaz-Franco and Pérez-García 1995). The resistance is controlled by a single gene (*Uca1/uca1*) (Madrid et al. 2008) and moderate levels of incomplete and partial resistance are available (Rubiales et al. 2001). Broomrape (*Orobanche crenata*) could also be a problem in winter sowing under Mediterranean conditions (Rubiales et al. 1999). An evaluation of germplasm collections permitted to detect sources of resistance to this parasitic plant (Rubiales et al. 2003).

Also insect pests are affecting the production of chickpea. Pod borer (*Helicoverpa armigera* Hübner) in India and Australia, and leafminer (*Liriomyza cicerina* Rondani) in West Asia and North Africa (WANA) and southern Europe seem to be the most serious problems. Efforts have been made to identify sources of resistance to both the pests in the cultivated and wild species at ICRISAT and ICARDA (Gaur et al. 2007).

5.2.3 *Abiotic Stresses*

Terminal drought is globally the most serious abiotic stress to chickpea productivity. Development of early-maturing varieties has been the most effective conventional approach to escape from terminal drought (Toker et al. 2007). Morphological traits such as a large root system (density and length), that is more efficient for extraction of available soil moisture, and smaller leaf area, in order to reduce transpiration water losses, should be taken into account to control drought. A promising drought tolerant accession (ICC4958) with a good root volume has been evaluated at ICRISAT (Saxena et al. 1993). More recently, two desi chickpea genotypes, ACC 316 and ACC 317, were identified by Cancy and Toker (2009) as drought and heat resistant. Genetic engineering is currently being explored for enhancing the levels of drought tolerance in chickpea (Millán et al. 2006). Thus, transgenic plants have been developed at ICRISAT having either a dehydration responsive element or a gene that increases proline accumulation in the plant (Gaur et al. 2007) (see also Chapter 9).

Both freezing (< –1.5°C) and chilling (between –1.5 and 15°C) are known to affect chickpea at various stages of development from germination to maturation (Croser et al. 2003b) and should be considered in chickpea breeding for winter sowing in Mediterranean environments. Chilling affecting pod setting has been described in the field and under controlled conditions (Clarke and Siddique 2004; Nayyar et al. 2005). Clarke and Siddique (2004) demonstrated that pollen tubes derived from chilling tolerant phenotypes grew faster down the style to the ovary than pollen derived from chilling-sensitive plants. Pollen selection as a breeding tool has been applied in chickpea, and two chilling tolerant cultivars ("Sonali" and "Rupali") have been released in Australia (Clarke et al. 2005) and they should be included in winter sowing breeding programs. On the other hand, ICARDA and ICRISAT breeding programs developed cultivars tolerant to cold adapted to winter sowing (Gaur et al 2007).

Chickpea, together with faba bean and field pea, are more sensitive to salinity than other legumes. Salt stress imposes a significant limitation of productivity related to the adverse effects on the dry weights of both shoots and roots and also on nodulation and nitrogen fixation (Manchanda and Garg 2008). Several tolerant sources have been identified in India, Pakistan and Australia (Millán et al. 2006).

5.3 Diversity Analysis

An efficient and effective conservation of crop and wild relative germplasm is required to succeed in plant breeding programs. In chickpea, an extensive collection with more than 20,000 accessions, including desi, kabuli and wild *Cicer* sp. from 59 countries, is maintained at ICRISAT (http://www.icrisat.or). ICARDA (http://www.icarda.or) also preserve chickpea genetic resources with a collection of 12,500 entries, mainly kabuli types. Another important chickpea collection is conserved by the USDA (United States Department of Agriculture, http://www.ars-grin.gov) with approximately 6,000 accessions. Other collections in Australia (CLIMA, Center for Legumes in Mediterranean Agriculture and ATFCC Australian Temperate Field Crops Collection), India (NBPGR, National Bureau of Plant Genetic Resources) and Russia (VIR, Vavilov Research Institute of Plant Industry) are also significant. More information about chickpea collection is available at: http://www.croptrust.org/documents/we/CicerStrategy_FINAL_2Dec08.pdf.

In spite of this large number of germplasm accessions available at the genebanks, there has been a very limited use of these accessions in chickpea breeding (Upadhyaya et al. 2008).

A good characterization of collections is time-consuming and expensive. Thus, core and mini core collections (about 10 and 1% of the total accessions, respectively), representative of the entire chickpea collection, have been obtained. ICRISAT developed a reference set consisting of 300 accessions representing diversity from the entire spectrum of a composite collection made between ICRISAT and ICARDA (Upadhyaya et al. 2008). Minicore collections have been useful to identify accessions with resistance to different diseases and good agronomic traits (Pande et al. 2006b).

Diversity studies should also include wild *Cicer* sp. The limitations of allelic variability in the cultigens are well known for important traits due to "evolutionary bottlenecks" that affected this cultivated species (Abbo et al. 2003b). Abbo et al. (2007) pointed out the potential use of wild *Cicer* taxa to find contrasting genotypes for both research and practical breeding purposes. These authors mentioned the few examples with success in utilization of *C. reticulatum* reported until today. Furthermore, they indicated the limitations of wide crosses using *C. reticulatum* and *C. echinospermum*, both with a limited adaptive allelic variation. Abbo et al. (2007) proposed using wild species with more distantly related genepools, for comparative genetic and physiological studies. However, to be able to exploit wild diversity, a bigger effort should be done to maintain wild *Cicer* species in collections (Berger et al. 2003) and to solve difficulties in the success of interspecific crosses, especially with the secondary genepool.

5.3.1 Molecular Markers Applied to Diversity Analysis

Molecular markers analysis increases the accuracy of classification present in germplasm collections. Thus, DNA analyses permit detecting redundancy of identical samples, to identify new alleles and consequently to deepen in the genetic structure of existing plant material.

Initially, diversity studies were achieved using seed storage protein electrophoresis (Ahmad and Slinkard 1992) and isozymes (Labdi et al. 1996; Tayyar and Wainess 1996) but the level of detected polymorphism was low. Afterwards DNA markers such as restriction fragment length polymorphism (RFLP), random amplified polymorphic DNA (RAPD), amplified fragment length polymorphism (AFLP), internal transcribed spacer (ITS), and their derivatives were used (Serret et al. 1997; Iruela et al. 2002; Nguyen et al. 2004; Singh et al. 2008) but still most of them revealed a limited variability within *C. arietinum*.

At present, microsatellite markers (STMS or SSR, simple sequence repeats) are preferred for diversity assessments in most species including chickpea (Sethy et al. 2006a, b; Upadhyaya et al. 2008). These markers are highly polymorphic, repeatable and reproducible among laboratories and, thanks to the use of fluorescence dye-labeled primers, the automation of results using capillary electrophoresis is possible. An additional advantage of those markers in chickpea is that many STMS have been included in genetic maps and consequently it is possible to know their distribution among the genome (Nayak et al. 2009; Millán et al. 2010).

Upadhyaya et al. (2008) published a description of the genetic structure, diversity and allelic richness in a composite collection of nearly 3,000 accessions using 48 SSR markers. This relevant study presents information about the polymorphic information content (PIC) values for different markers and the distribution of rare, common and unique alleles in desi, kabuli and pea-shaped chickpeas as well as for the variability in wild related species. Imtiaz et al. (2008) also use SSR markers distributed across the chickpea genome to assess the variability in different elite genotypes and breeding lines. Twenty-one SSR markers amplified a total of 370 alleles confirming their power to detect polymorphism in very closely related accessions. Microsatellites have also been described as an effective molecular tool for the study of new germplasm acquisition activities (Jomová et al. 2009). All previous studies demonstrated that when using SSR it is possible to detect a considerable molecular variation in cultivated chickpea formerly considered as a monotone species from the molecular point of view.

Molecular markers have also been used to detect favorable alleles related to particular agronomic traits of interest in chickpea germplasm. Thus, SSR and sequence characterized amplified regions (SCAR) markers linked to quantitative trait loci (QTLs) for resistance to Ascochyta blight have been employed to characterize sources of resistance (Tar`an et al. 2007a; Imtiaz et al. 2008). These results propose that makers linked to Ascochyta blight resistance may help in choosing parents to pyramid different sources of resistance.

Microsatellites are also useful for varietal identification. STMS distributed across all linkage groups of the chickpea genetic map were used to characterize Spanish commercial chickpea varieties (Castro et al. 2011). These molecular profiles can be used as a DNA fingerprint for each registered variety, avoiding redundancy of identical cultivars as well as protecting breeders' rights.

5.4 Molecular Linkage Maps

DNA marker technology has made it possible to generate genetic maps that ensure the use of marker-assisted selection (MAS) and positional cloning of genes of interest. In chickpea, efforts to develop genetic maps and to identify markers close to genes or genomic regions with agronomical importance are underway. Even though maps are still incomplete, present achievements are the framework necessary to be able to develop a map of the chickpea genome useful for breeding purposes.

5.4.1 Brief History of Mapping Efforts

First genetic maps published in chickpea were developed analyzing isozymes in F_2 populations derived from interspecific crosses involving *C. arietinum* and the wild relatives *C. reticulatum* and *C. echinospermum* (Gaur and Slinkard 1990; Kazan et al. 1993). At that time, some morphological traits could be located in linkage groups but the number of polymorphic isozymes was very limited. Consequently, DNA markers were required to be able to expand those genetic maps. RFLP were the first DNA molecular markers widely used to develop maps in many plant species but they did not succeed in *C. arietinum* due to the low level of polymorphism detected in intraspecific crosses (Udupa et al. 1993). Later, DNA markers based on PCR techniques like RAPDs, inter simple sequence repeats (ISSRs), DNA amplification fingerprinting (DAF), SCAR or gene-specific markers showed a moderate level of polymorphism within cultivated species. However, these markers permitted to improve previous genetic maps and it was possible to detect genes or QTLs of interest principally related to Ascochyta blight and Fusarium wilt resistance (Simon and Muehlbauer 1997; Ratnaparkhe et al. 1998; Santra et al. 2000; Millán et al. 2003; Pfaff and Khal 2003; Rakshit et al. 2003).

Essentially, chickpea genetic maps had the greatest benefit with the incorporation of STMS or SSR markers. The first chickpea STMS were developed by Huettel et al. (1999) and Winter et al. (1999) and provided the skeleton of a genetic map analyzed in an RIL population derived from a cross between *C. arietinum* and *C. reticulatum* (Winter et al. 2000). In this map loci conferring resistance to Fusarium wilt race 4 and 5 were located in LG 2. New STMS markers were later obtained in *Cicer* species thus increasing the possibilities to saturate chickpea maps (Sethy et al. 2003, 2006a, b; Lichtenzveig et al. 2005; Choudhary et al. 2006). Collard et al. (2003) developed the only map published using a population derived from a cross between *C. arietinum* x *C. echinospermum*.

Many of those maps focused in targeting areas controlling the two major chickpea diseases, Ascochyta blight and Fusarium wilt, and also rust resistance, quality components and agronomic traits (Tables 5-2 and 5-3). Due to the possibility to exchange information among maps based on STMS markers, a consensus map of chickpea genome has been constructed (Millán et al. 2010). On the other hand, Nayak et al. (2009) completed the reference map published by Winter et al. (2000) adding 175 new markers. Both these maps provided anchor points for comparing chickpea linkage groups with *Medicago truncatula* chromosomes.

Even with the previous efforts to achieve a saturated chickpea map, much work is still needed. New markers like expressed sequence tag-SSR (EST-SSR) increase the probability to target transcribed sequences (Chouddary et al. 2009). Besides, thousands of ESTs derived from expression studies related to biotic or abiotic stresses have been recently obtained and could be included in genetic maps (Mantri et al. 2007; Molina et al. 2008). The existence of several bacterial artificial chromosome (BAC) libraries, will facilitate map-based gene/QTL cloning, genome sequencing and physical map construction (Rajesh et al. 2004; Lichtenzveig et al. 2005).

5.4.2 Mapping and Tagging of Simply Inherited Traits

Not many single locus controlling phenotypic traits have been located in chickpea genetic maps. A summary of simply inherited traits mapped in chickpea is shown in Table 5-2. The first traits mapped were related to the pigmentation of the flower (pink/white=*P/p=B/b*), brown/yellow seed testa (*T3/t3*) and purple/green epicotyl (*Gst/gst*) and those were linked (Kazan et al. 1993). We know now that they are located in LG 4 of the chickpea map due to the linkage between *P/p* gene and marker GAA47 (Tekeoglu et al. 2002; Cobos et al. 2005). Flower and seed color differentiate the two cultivated chickpea types: kabuli and desi, as previously mentioned (see Section 5.1.2).

Table 5-2 Qualitative traits and indicative markers localized in different linkage groups of the chickpea genetic map.

LG[1]	Qualitative Trait	Gene	Indicative Markers	Distance[2]	References
LG 2	Fusarium wilt race 0	*Foc-0_2/ foc-0_2*	TA59, TS47	1.3	Halila et al. 2009b[a]
	Fusarium wilt race 1	*foc-1*	H3A12, TA110	3.9; 2.1	Gowda et al. 2009[a]
	Fusarium wilt race 2	*foc-2*	TA96, H3A12	0.2; 2.7	Gowda et al. 2009[a]
	Fusarium wilt race 3	*foc-3*	H1B06y, TA194	0.2; 0.7	Gowda et al. 2009[a]
	Fusarium wilt race 4	*foc-4*	TA96, CS27	2.5; 3.7	Winter et al. 2000[b]
			R-2609-1	2	Benko-Iseppon et al. 2003[b]
	Fusarium wilt race 5	*foc-5*	TA59, TA96	2	Cobos et al. 2009[b]
			TA27	3.5	Winter et al. 2000[b]
	Ascochyta blight	*ar1*	GA16	23	Udupa and Baum 2003[a]
LG 3	Growth habit	*Hg*	Pgd-c	18	Kazan et al. 1993[b]
		Hg/hg	$OPB17_{789}$, $OPAI09_{1651}$	11.9; 11.7	Cobos et al. 2009[b]
LG 4	Seed testa color	*T3*	*P*	10	Kazan et al. 1993[b]
	Epicotyl color	*Gst*	*P*	10	Kazan et al. 1993[b]
	Seed surface	*Rs*	*T3/t3*	10	Kazan et al. 1993[b]
	Anthocyanin pigmentation	*P*	ubc181a	18.4	Santra et al. 2000[b]
	Flower color	*P*	*Gst, Rs*	10	Kazan et al. 1993[b]
		B/b	TA61	10.3	Cobos et al. 2009[b]
	Seed coat thickness	*Tt/tt*	*B/b*	23	Cobos et al. 2005[a]
LG 5	Fusarium wilt race 0	*Foc-0_1/ foc-0_1*	$OPJ20_{600}$, TR59	3; 2	Cobos et al. 2005[a]
LG 6	Single/Double pod	*s*	TA80	4.84	Rajesh et al. 2002[a]
		s	Tr44	7.8	Cho et al. 2002[a]
LG 7	Rust	*Uca1/ uca1*	TA18, TA180	2.3; 1.6	Madrid et al. 2008[b]

[1]Linkage groups.
[2]Distance in cM between locus and marker.
[a]Population derived from a narrow cross.
[b]Population derived from a wide cross.

Seed coat thickness (*Tt/tt*), considered as a qualitative trait in narrow crosses (Gil and Cubero 1993; Cobos et al. 2005) is located in LG 4 and linked to flower color. This trait could be related to seed hardness, and if so, it is undesirable for whole seed consumption. This trait should be considered as quantitative in wide crosses because the variation fits to a normal distribution. In fact, a QTL located in LG 4 within a region coincident with previous results explained 20% of the phenotypic variance in this trait (Table 5-3) (Cobos et al. 2009). Seed surface reticulation (*Rs/rs*) has been reported to be controlled by one or two genes (Muehlbauer and Singh 1987). This trait seems to be related to seed coat thickness in wide crosses (Cobos et al. 2009) but until now it could not be included in the chickpea genetic map.

Erect/prostrate plant growth habit (*Hg/hg*) is another trait with simple inheritance, erect being dominant (Muehlbauer and Singh 1987). This trait was linked on early chickpea maps to the loci detected by the isozyme PGD (phosphogluconate dehydrogenase) (Kazan et al. 1993), and this isozyme was later located on LG 3 by Winter et al. (2000). Cobos et al. (2009) confirmed the linkage between *Hg/hg* and STMS markers in LG 3 of the chickpea map. However, this trait has always been mapped in populations derived from interspecific crosses. The location of this gene in cultivated chickpea could help select appropriate genotypes for mechanical harvesting.

Double pod is an interesting mutation controlled by a single recessive gene designated as "*s*" or "*sfl*" (Muehlbauer and Singh 1987). It was detected to be linked to the STMS marker, TA80, in a study using near-isogenic lines (NILs) (Rajesh et al. 2002), and this marker is located in LG 6 of the chickpea reference map (Cho et al. 2002; Cobos et al. 2005). As previously mentioned (see Section 5.2.1), this trait has a positive effect on yield stability (Rubio et al. 1998, 2004).

Some resistances to fungal diseases are oligogenically controlled as it happens with Fusarium wilt in chickpea. In the chickpea genetic map, genes controlling resistance reaction to races 0, 1, 2, 3, 4 and 5 (*foc-0, foc-1, foc-2, foc-3, foc-4* and *foc-5*) have been mapped in LG 2 arranged as a gene cluster (Mayer et al. 1997; Ratnaparkhe et al. 1998; Tullu et al. 1998; Winter et al. 2000; Sharma et al. 2004; Gowda et al. 2009; Halila et al. 2009b). However, not all Fusarium wilt resistance genes are located in LG 2. For example, a resistance gene for race 0 ($foc\text{-}0_1$) was found in LG 5 (Cobos et al. 2005) confirming that the resistance is controlled by two independent genes ($foc\text{-}0_1$ and $foc\text{-}0_2$), as Rubio et al. (2003) reported. A similar situation could occur with still-unmapped genes controlling resistance to other pathogenic races. Resistance gene analog (RGA) markers have been previously included in LG 2 although no highly significant association with any resistance gene has been found (Huettel et al. 2002; Flandez-Galvez et al. 2003a, Palomino et al. 2009). An analysis of new progenies together with targeting at the more variable leucine-rich repeat (LRR) coding region has been proposed for future prospects involving RGA (Palomino et al. 2009). In addition, a gene that controls resistance to chickpea rusts (*Uca1/uca1*) has been located in LG 7 tightly flanked by two STMS markers (Madrid et al. 2008).

5.4.3 Molecular Mapping of Complex Traits

The complex traits currently located on the chickpea genetic map are related to yield improvement and resistance to Ascochyta blight. The QTLs located in different linkage groups are summarized in Table 5-3.

A QTL related to yield has been positioned in LG 4 but with a low LOD value, probably due to the high environmental influence on this trait. However, a QTL related to seed size, a yield component with high heritability, has also been detected in this linkage group by several authors (Cho et al. 2002; Abbo et al. 2005; Cobos et al. 2007, 2009). Even though the positions of QTLs are slightly different, it might be considered that they are located in the same genomic region because all of them have the same indicative markers. Other minor QTLs have been reported in LG 1, LG 2 and LG 8, which should be validated in further studies.

Early flowering time seems to have a positive effect on yield under the Mediterranean environment (Siddique et al. 2003; Rubio et al. 2004). QTLs controlling this trait have been detected in LGs 1, 2, 3, 4 and 6. The QTLs with high LOD values in LG 3 have special interest for breeding applications as they were validated in different environments and in both intra- and interspecific populations (Cho et al. 2002; Cobos et al. 2009).

Until now complete resistance to Ascochyta blight has not been reported in any germplasm. However, several QTLs involved in blight resistance have been located in LGs 2, 3, 4, 6 and 8 in the chickpea genetic map (Table 5-3). The most significant ones are those sited in LGs 2 and 4 with high LOD values and corroborated by different authors (Santra et al. 2000; Tekeoglu et al. 2002; Iruela et al. 2006, 2007; Tar'an et al. 2007b; Kottapalli et al. 2009). Particularly, two genomic regions in LG 4 enclose two clearly differentiated QTLs (QTL1 and QTL2) more than 30 cM apart. Recently, an attempt to saturate QTL1 has been performed developing markers from bacterial artificial chromosome (BAC) clones (Rajesh and Muehlbauer 2008). In the same genomic region, microsynteny was established between chickpea and *M. truncatula*, finding candidate genes for QTL1 (Rajesh et al. 2008). In the same way, efforts to find candidate genes for QTL2 have been done by a genome walking approach using the SCAR marker, $SCK13_{603}$, as an anchored sequence and it was located in a region of a putative retrotransposon (Iruela et al. 2009).

A few QTLs related to quality components have been published. Two QTLs were associated with β-carotene content in LG 3 (LOD > 3) and an additional QTL associated with lutein concentration in LG 1 with a low LOD of 2.4 (Abbo et al. 2005). It would be necessary to expedite research to detect genomic regions with QTLs controlling quality-related traits in this crop.

Table 5-3 Quantitative traits and indicative markers localized in different linkage groups of the chickpea genetic map.

LG[1]	Quantitative Trait	QTL	LOD	Indicative Marker	References
LG 1	β-carotene	QTL3	2.1	GA11, TA122	Abbo et al. 2005[c]
	Seed weight	QTL2	3.2	GA11	Abbo et al. 2005[c]
	Days to first flower	QTL1	10	H1F022, GAA40	Lichtenzveig et al. 2006[b]
LG 2	Seed weight	QTL_{SW3}	4.4	TA110-TAA60	Cobos et al. 2009[c]
	Days to first flower	QTL2	3.7/4.4[a]	H4B09, H1B06	Lichtenzveig et al. 2006[b]
	Ascochyta blight	QTLar2a	7	GA16	Udupa and Baum 2003[b]
		QTL	5	$UBC881_{621}$, $OPAI09_{746}$	Cobos et al. 2006[c]
		QTL_{AR3}	2.5/5.92	TA194	Iruela et al. 2007[b]
		QTL1	14.32	TR19	Anbessa et al. 2009[b]
LG 3	β-carotene	QTL 2	3.9/3/2.1	TA64, STMS28	Abbo et al. 2005[c]
	Days to flower	QTL	3.03	TS57, TA127	Cho et al. 2002[b]
		QTL_{DF3}	16/5.9[a]	TA142	Cobos et al. 2009[c]
	Ascochyta blight	QTL1	2.89	STMS28, TS12b	Flandez Galvez et al. 2003b[b]
		QTL	4.27	TA64	Ta`ran et al. 2007b[b]
		QTL2	4.27/3.16	TA64	Anbessa et al. 2009[b]
		QTL1	2.03	TR58	Kottapalli et al. 2009[b]
LG 4	Yield	QTL_{YD}	2.5/2.41[a]	$OPS01_{870}$-$OPAF08_{1125}$	Cobos et al. 2007[b]
	Seed number	QTL	5.7	TA130	Cho et al. 2002[b]
	Seed weight	QTL1	3.8	GA24, STMS11, GA2	Abbo et al. 2005[c]
		QTL_{SW1}	7/9.6	$OPS01_{870}$-$OPAF08_{1125}$/STMS11	Cobos et al. 2009[c]
		QTL	11.7	TA130	Cho et al. 2002[b]
	Seed coat thickness	QTL_{Tt}	6	*B/b*	Cobos et al. 2009[c]
	Days to flower	QTL_{DF1}	2.4/4.4/4[a]	GAA47	Cobos et al. 2007[b]
	Ascochyta blight	QTL-1	17.8/16.9[a]	UBC733b, UBC181a	Santra et al. 2000[c]

Table 5-3 contd....

Table 5-3 contd....

LG[1]	Quantitative Trait	QTL	LOD	Indicative Marker	References
				GAA47	Tekeoglu et al. 2002[c]
			>5	OPS06-1	Rakshit et al. 2003[c]
			19.98	20(T)112, SCAR733b	Rajesh and Muehlbauer 2008[c]
		QTL-2	7.27/6.52[a]	UBC836b, Dia	Santra et al. 2000[c]
				TA146	Tekeoglu et al. 2002[c]
		QTLar2b	3.5	TA72	Udupa and Baum, 2003[b]
		QTL 4-6	~11.25	TA130	Flandez-Galvez et al. 2003b[b]
		QTL_{AR2}	2.1/5.6[a]	TA72,TA146	Iruela et al. 2006[b]
		QTL_{AR1}	5.13	GAA47	Iruela et al. 2006[b]
		QTL4-1	11	TA2	Lichtenzveig et al. 2006[b]
		QTL4-2	11	H1G20	Lichtenzveig et al. 2006[b]
		QTL	9.58	TS54	Ta´ran et al. 2007[b]
		QTL3	4.09/5.53	TA132, TS54	Anbessa et al. 2009[b]
		QTL2	2.68	TS54	Kottapalli et al. 2009[b]
		QTL3	2.15	TA2	Kottapalli et al. 2009[b]
LG 6	Seed weight	QTL3	2.4	TA120,TR40	Abbo et al. 2005[c]
	Ascochyta blight	QTL	3.88	TA176	Ta´ran et al. 2007b[b]
		QTL4	3.88	TA176	Anbessa et al. 2009[b]
LG 8	Lutein concentration	QTL1	2.4	TA25	Abbo et al. 2005[c]
	Seed weight	QTL_{SW2}	3.4	$OPE09_{1594}$-$MER05_{1645}$	Cobos et al. 2007[b]
	Ascochyta blight	QTL8	3.5	TA3, H3C11a	Lichtenzveig et al. 2006[b]
		QTL5	2.92	TS45	Anbessa et al. 2009[b]
		QTL2-3	3.38	TA3 TS45	Flandez-Galvez et al. 2003b[b]

[1]Linkage groups.
[a]LOD values for the same QTL in different environments.
[b]Population derived from a narrow cross.
[c]Population derived from a wide cross.

5.5 Molecular Breeding

DNA markers are widely accepted as potentially valuable tools in breeding of many crop species, and a greater adoption of MAS in the future is inevitable (Collard and Mackill 2008). Nonetheless, the efficacy of MAS relies on the saturation of genomic areas of interest with robust and easy to interpret markers (Collard et al. 2005). At present, microsatellite-based markers seem to be the most useful ones for MAS in chickpea. Reference genetic maps that were already obtained may help to choose adequate markers (Nayak et al. 2009; Millán et al. 2010).

Nevertheless, markers linked to blight resistance (QTL1, QTL2 in LG 4, and QTL_{AR3} in LG 2) and Fusarium wilt race 5 are being used to select genotypes in a current breeding program (Castro et al. 2009, 2010a). Besides, the use of the SCAR markers, $SCY17_{590}$ and $SCAE19_{336}$, has been reported to tag a source of Ascochyta blight resistance in a collection of chickpea genotypes (Imtiaz et al. 2008). The application of these markers is the first step to incorporate favorable alleles in breeding lines.

Markers could also help to select traits that are negatively correlated, for example, yield and seed size with a negative relation that could be explained for the presence of coincident QTLs (QTL_{YD} and QTL_{SW1} respectively) located in LG 4 of the chickpea map (Cobos et al. 2007). However, MAS could help in selecting genotypes with high production and high seed size using markers linked to no coincident QTLs controlling these traits located in different genomic regions like LG 2 and LG 8 (Cobos et al. 2007, 2009) (Table 5-3).

To date, RIL populations have been the most useful plant material to target interesting genomic regions in chickpea. It would be interesting to use NIL populations in order to increase efficiently the density of markers in a particular region and obtain high resolution maps. NILs have the advantage that segregation is limited to a little target genome region. Consequently, the genetic background noise can be eliminated. It has been extensively used for genetic studies, including the investigation of gene effects, gene expression and gene isolation (Zhou et al. 2005). These lines have been traditionally generated by consecutive backcrossing, followed by a self-pollination but this method is laborious and time-consuming. In chickpea, NILs for double pod and Fusarium wilt resistance have been developed using segregant RILs as an alternative method (Rajesh et al. 2002; Castro et al. 2010b).

A possibility to obtain a new genetic combination with advantages in production is gene introgression from wild species (Croser et al. 2003a). But wide crosses imply the inclusion of undesirable traits in cultivated material. Molecular markers linked to genes controlling pod dehiscence, prostrate growth habit, small seed size etc., could be very useful to carry out a selection against these undesirable traits. Thus, these adverse traits should be excluded from the introgression process of useful genes.

5.6 Limitations and Prospects of MAS

Successful results in MAS require highly saturated maps with easy to handle and cost-effective markers. We have previously mentioned that in chickpea, numerous useful markers have been obtained and many of them have been located in genetic maps yielding highly saturated areas. Furthermore, automation of those markers is becoming easy and, consequently, it is decreasing the price of MAS process, which is being routinely applied nowadays. For instance, a breeding program using MAS to select Ascochyta blight resistant materials has been developed with positive results (Castro et al. 2009). Likewise, markers were used to assist a backcross program to introduce Fusarium wilt resistance in "Spanish white" cultivars using STMS markers (J. Gil, pers. comm.).

Nevertheless, it is still necessary to target many interesting genes controlling simple inherited traits and determine candidate genes for different QTLs. Examples are genes determining adaptation to a particular environment such as flowering time or chilling stress tolerance that still remain unmapped (see Section 5.2.1 and 5.2.3). Similarly, complex abiotic constraints as drought and cold tolerance, essential to improve chickpea productivity, should be studied in detail. Regarding biotic stresses, pyramiding resistance genes still require the location of race-specific genes (e.g., for Fusarium wilt resistance) or, in the case of Ascochyta blight resistance, the consideration of the importance of pathotype differentiation for differences in QTL location. Furthermore, minor diseases should receive more attention and further studies are also necessary to determine the genetics and location of beneficial phenotypes that have been identified in wild species closely related to chickpea.

Despite the previously mentioned limitations, there are optimistic future prospects with the advent of functional markers that increase the probability to target interesting transcribed sequences. Currently, more than 34,000 EST-derived from expression studies in *Cicer* species have been deposited in public plant sequence databases (http://www.ncbi.nlm.nih.gov/Taxonomy).

The possibility to develop diagnostic markers, or the determination of the importance of different transcripts related to particular phenotypes, will now be easier with the detection of single nucleotide polymorphisms (SNPs) using massive sequencing technologies together with new computational tools. Moreover, the development of new RIL populations segregating for unmapped traits together with the development of NILs should also contribute to further advances in MAS. Finally, the progress achieved in the genetics and genomics of model legumes such as *M. truncatula* and *Lotus japonicus* should be taken into account in future chickpea breeding programs.

5.7 Synteny Studies

Two major clades have been described in crop legumes: Galegoid (cool season legumes) and Phaseoloid (tropical season legumes). Chickpea is the single genus within the tribe Cicereae included within the Galegoid clade together with the tribes Vicieae and Trifolieae, the latter including the model species *M. truncatula* (Choi et al. 2004). The first macrosynteny studies in chickpea were made by Simon and Muehlbauer (1997) comparing chickpea with pea and lentil. These authors, using an isozyme analysis, found five regions of the chickpea map in the same order as those found in the pea genome. By contrast, lack of hybridization among *Lens* and chickpea revealed a divergence between chickpea and lentil. Subsequent macrosynteny studies based on EST sequences established a high level of correspondence between chickpea and the model species *M. truncatula* (Zhu et al. 2005).

Mapping studies using cross-genome markers confirmed that *M. truncatula* linkage groups 1, 2, 3, 4, 7, and 8 corresponded to chickpea LGs 4, 1, 5, 6, 3, and 7, respectively (Nayak et al. 2009; Millán et al. 2010). In contrast to the conserved synteny noted for LG pairs 1–4, 2–1, 3–5, 4–6, 7–3, and 8–7, one-to-one relationships do not occur in *M. truncatula* LGs 5 and 6 and chickpea LGs 2 and 8. LG 8 in *Cicer* species appears to be derived entirely from one arm of *M. truncatula* LG 5, consistent with its short genetic distance and with its assignment to chromosome H, which is the smallest chickpea chromosome (Vláčilová et al. 2002; Nayak et al. 2009). Interestingly, intrachromosomal rearrangements in *M. truncatula* chromosome 6 have been detected in relation to pea and chickpea (Zhu et al. 2005). Recent efforts are dedicated to define the conserved and syntenic gene content of ancestral legume genomes. The approach is based on sequencing of both low copy genes and NBS-LRR disease resistance genes (Penmetsa et al. 2010).

Microsynteny studies conducted by Rajesh et al. (2008), using 500 kb from 11 BAC clones of chickpea revealed that similarities with *M. truncatula* genomic sequences varied depending on the region analyzed. Significant similarity of genes representing *QTL*1 for Ascochyta blight resistance (protein phosphatase, 40S ribosomal protein, flavonoid galactosyl transferase and eukaryotic transcription factor) was observed. Nevertheless, 21 out of 51 predicted genes in chickpea did not show homology to genome and EST sequences of the model species.

Acknowledgements

Chickpea research has been mostly financed by the projects AGL2005-07497-CO2-C5/VEG, GEN2006-277798-CO6-C5/VEG and RTA2007-00030, co-financed by FEDER funds and by the EU project GLIP (FP6-FOOD-CT-

2004-506223). The authors thank the members of the chickpea breeding group at IFAPA and University of Córdoba (E Madrid, M Iruela, MD Fernandez Romero, and P Castro) for their enthusiastic work.

References

Abbo S, Shtienberg D, Lichtenzveig J, Lev-Yadun S, Gopher A (2003a) The chickpea, summer cropping, and a new model for pulse domestication in the ancient near east. Quart Rev Biol 78: 435–448.

Abbo S, Berger J, Turner NC (2003b) Evolution of cultivated chickpea: four bottlenecks limit diversity and constrain adaptation. Funct Plant Biol 30: 1081–1087.

Abbo S, Molina C, Jungmann R, Grusak MA, Berkovitch Z, Reifen R, Kahl G, Winter P, Reifen R (2005) Quantitative trait loci governing carotenoid concentration and weight in seeds of chickpea (*Cicer arietinum* L.). Theor Appl Genet 111: 185–195.

Abbo A, Redden RJ, Yadav SS (2007) Utilization of wild relatives. In: SS Yadav, R Redden, WB Chen, B Sharma (eds) Chickpea Breeding and Management. CABI Publ, Wallingford, UK, pp 338–354.

Ahmad F (2000) A comparative study of chromosome morphology among the nine annual species of *Cicer* L. Cytobios 101: 37–53.

Ahmad F, Chen Q (2000) Meiosis in *Cicer* L. species: the relationship between chiasma frequency and genome length. Cytologia 65: 161–166.

Ahmad F, Slinkard AE (1992) Genetic relationships in the genus *Cicer* L. as revealed by polyacrylamide gel electrophoresis of seed storage proteins. Theor Appl Genet 84: 688–692.

Anbessa Y, Taran B, Warkentin TD, Tullu A, Vandenberg A (2009) Genetic analyses and conservation of QTL for ascochyta blight resistance in chickpea (*Cicer arietinum* L.). Theor Appl Genet 119: 757–765.

Arumuganathan K, Earle ED (1991) Nuclear DNA content of some important plant species. Plant Mol Biol Rep 9: 208–218.

Basandrai AK, Pande S, Kishore GK, Crouch JH, Basandrai D (2005) Cultural, morphological and pathological variation in Indian isolates of *Ascochyta rabiei*, the chickpea blight pathogen. Plant Pathol J 21: 207–213.

Benko-Iseppon AM, Winter P, Huettel B, Staginnus C, Muehlbauer FJ, Kahl G (2003) Molecular markers closely linked to fusarium resistance genes in chickpea show significant alignments to pathogenesis-related genes located on *Arabidopsis* chromosomes 1 and 5. Theor Appl Genet 107: 379–386.

Berger J, Abbo S, Turner NC (2003) Ecogeography of annual wild *Cicer* species: The poor state of the world collection. Crop Sci 43: 1076–1090.

Bhardwaj R, Sandhu JS, Kaur L, Gupta SK, Gaur PM, Varshney R (2010) Genetics of ascochyta bligtht resistance in chickpea. Euphytica 171: 337–343.

Cancy H, Toker C (2009) Evaluation of yield criteria for drought and heat resistance in chickpea (*Cicer arietinum* L.). J Agron Crop Sci 195: 47–54.

Castro P, Fernández MD, Millán T, Gil J, Rubio J (2009) A comparison of phenotypic and marker-assisted selection for ascochyta blight resistance in chickpea. In: Ascochyta 2009, Proc 2nd Int Ascochyta Workshop, Pullman, Washington, USA, p 30.

Castro P, Gil J, Cubero JI, Madrid E, Millán T, Rubio J (2010a) Marker-Assisted Selection for wilt resistance in a chickpea breeding program. In: 5th Int Food Legumes Research Conf (IFLRC V) & 7th Eur Conf on Grain Legumes (AEP VII). Antalya p 176: http://www.iflrc-ecgl.org/BookofAbstracts.pdf (Cited 21 July 2010).

Castro P, Pistón F, Madrid E, Millán T, Gil J, Rubio J (2010b) Development of chickpea near-isogenic lines for fusarium wilt. Theor Appl Genet 121: 1519–1526.

Castro P, Millán T, Gil J, Mérida J, García ML, Rubio J, Fernández-Romero MD (2011) Identification of chickpea cultivars by microsatellite markers. J Agri Sci (in prees): doi: 10.1017/S0021859610001061

Chen W, Coyne CJ, Peever TL, Muehlbauer FJ (2004) Characterization of chickpea differentials for pathogenicity assay of ascochyta blight and identification of chickpea accessions resistant to *Didymella rabiei*. Plant Pathol 53: 759–769.

Cho S, Kumar J, Shultz JL, Anupama K, Tefera F, Muehlbauer FJ (2002) Mapping genes for double podding and other morphological traits in chickpea. Euphytica 128: 285–292.

Cho S, Chen W, Muehlbauer FJ (2004) Pathotype-specific genetic factors in chickpea (*Cicer arietinum* L.) for quantitative resistance to ascochyta blight. Theor Appl Genet 109: 733–739.

Choi HK, Mun JH, Kim DJ, Zhu H, Baek JM, Mudge J, Roe B, Ellis N, Doyle J, Kiss GB, Young ND, Cook DR. (2004) Estimating genome conservation between crop and model legume species. Proc Natl Acad Sci USA 101: 15289–15294.

Chongo G, Gossen BD, Buchwaldt L, Adhikari T, Rimmer SR (2004) Genetic diversity of *Ascochyta rabiei* in Canada. Plant Dis 88: 4–10.

Choudhary S, Sethy NK, Shokeen BA, Bhatia S (2006) Development of sequence-tagged microsatellite site markers for chickpea (*Cicer arietinum* L). Mol Ecol Notes 6: 93–95.

Choudhary S, Sethy NK, Shokeen B, Bhatia S (2009) Development of chickpea EST-SSR markers and analysis of allelic variation across related species. Theor Appl Genet 118: 591–608.

Clarke HJ, Siddique KHM (2004) Response of chickpea genotypes to low temperature stress during reproductive development. Field Crops Res 90: 323–334.

Clarke HJ, Siddique KHM, Khan TN (2005) Chickpea improvement in Southern Australia: breeding for tolerance to chilling at flowering. Indian J Pulses Res 18: 1–8.

Cobos MJ, Fernández MJ, Rubio J, Kharrat M, Moreno MT, Gil J, Millán T (2005) A linkage map of chickpea (*Cicer arietinum* L.) based on populations from Kabuli x Desi crosses: location of genes for resistance to fusarium wilt race 0. Theor Appl Genet 110: 1347–1353.

Cobos MJ, Rubio J, Strange RN, Moreno MT, Gil J, Millán T (2006) A new QTL for Ascochyta blight resistance in a RIL population derived from an intraspecific cross in chickpea. Euphytica 149: 105–111.

Cobos MJ, Rubio J, Fernández-Romero MD, Garza R, Moreno MT, Millán T, Gil J (2007) Genetic analysis of seed size, yield and days to flowering in a chickpea recombinant inbred line population derived from a Kabuli x Desi cross. Ann Appl Biol 151: 33–42.

Cobos MJ, Winter P, Kharrat M, Cubero JI, Gil J, Millan T, Rubio J (2009) Genetic analysis of agronomic traits in a wide cross of chickpea. Field Crops Res 111: 130–136.

Collard BCY, Mackill DJ (2008) Marker-assisted selection: an approach for precision plant breeding in the twenty-first century. Phil Trans Roy Soc B 363: 557–572.

Collard BCY, Pang ECK, Ades PK, Taylor PWJ (2003) Preliminary investigation of QTLs associated with seedling resistance to ascochyta blight from *Cicer echinospermum*, a wild relative of chickpea. Theor Appl Genet 107: 719–729.

Collard BCY, Jahufer MZZ, Brouwer JB, Pang ECK (2005) An introduction to markers, quantitative trait loci (QTL) mapping and marker-assisted selection for crop improvement: the basic concepts. Euphytica 142: 169–96.

Croser JS, Admad F, Clarke HJ, Siddike KHM (2003a) Utilisation of wild in chickpea improvement-progress, constraints, and prospect. Aust J Agri Res 54: 429–444.

Croser JS, Clarke HJ, Siddique KHM, Khan TN (2003b) Low temperature stress: implications for chickpea (*Cicer arietinum* L.) improvement. Crit Rev Plant Sci 22: 185–219.

Danehloueipour N, Yan G, Clarke HJ, Siddique KHM (2007) Diallel analyses reveal the genetic control of resistance to ascochyta blight in diverse chickpea and wild *Cicer* species. Euphytica 154: 195–205.

Dey SK, Singh G (1993) Resistance to ascochyta blight in chickpea-Genetic basis. Euphytica 68: 147–153.

Díaz-Franco A, Pérez-García P (1995) Control químico de la roya y la rabia del garbanzo y su influencia en el rendimiento de grano. Rev Mex Fitopatol 13: 123–125.

FAO (2010): http://faostat.fao.org/ (Cited 23 June 2009).

Flandez-Galvez H, Ford R, Pang ECK, Taylor PWJ (2003a) An intraspecic linkage map of the chickpea (*Cicer arietinum* L.) genome based on sequence tagged microsatellite site and resistance gene analog markers. Theor Appl Genet 106: 1447–1456.

Flandez-Galvez H, Ades PK, Ford R, Pang ECK, Taylor PWJ (2003b) QTL analysis for ascochyta blight resistance in an intraspecic population of chickpea (*Cicer arietinum* L.). Theor Appl Genet 107: 1257–1265.

Gan YT, Siddique KHM, MacLeod WJ, Jayakumar P (2006) Management options for minimizing the damage by ascochyta blight (*Ascochyta rabiei*) in chickpea (*Cicer arietinum* L.). Field Crops Res 97: 121–134.

Gaur PM, Slinkard AE (1990) Genetic control and linkage relations of additional isozyme markers in chickpea. Theor Appl Genet 80: 648–656.

Gaur PM, Pande S, Upadhyaya HD, Rao BV (2006) Extra-large kabuli chickpea with high resistance to fusarium wilt. Chickpea Pigeonpea Newsl 13: 5–7.

Gaur PM, Gowda CLL, Knights EJ, Warketin T, Acikgöz N, Yadav SS, Kumar J (2007) Breeding achievements. In: SS Yadav, R Redden, W Chen, B Sharma (eds) Chickpea Breeding and Management. CABI Publ, Wallingford, UK, pp 391–416.

Gil J, Cubero JI (1993) Inheritance of seed coat thickness in chickpea (*Cicer arietinum* L.) and its evolutionary implications. Plant Breeding 111: 257–260.

Gil J, Nadal S, Luna D, Moreno MT, De Haro A (1996) Variability of some physico-chemical characters in Desi and Kabuli chickpea types. J Sci Food Agri 71: 179–184.

Gowda SJM, Radhika P, Kadoo NY, Mhase LB, Gupta VS (2009) Molecular mapping of wilt resistance genes in chickpea. Mol Breed 24: 177–183.

Halila I, Rubio J, Millán T, Gil J, Kharrat M, Marrakchi M (2009a) Resistance in chickpea (*Cicer arietinum*) to fusarium wilt race 0. Plant Breed 129: 563–566.

Halila I, Cobos MJ, Rubio J, Millán T, Kharrat M, Marrakchi M, Gil J (2009b) Tagging and mapping a second resistance gene for fusarium wilt race 0 in chickpea. Eur J Plant Pathol 124: 87–92.

Harlan JR, de Wet JMJ (1971) Towards a rational classification of cultivated plants. Taxon 20: 509–517.

Haware MP, Nene YL (1982) Races of *Fusarium oxysporum* f. sp. *ciceris* . Plant Dis 66: 809–810.

Haware MP, Nene YL, Natarajan M (1996) Survival of *Fusarium oxysporum* f. sp. *ciceris* in the soil in the absence of chickpea. Pytopathol Medit 35: 9–12.

Hegde VS (2010) Genetics of flowering time in chickpea in a semi-arid environment. Plant Breed 129: 683–687.

Hovav R, Upadhyaya KC, Beharav A, Abbo S (2003) Major flowering time gene and polygene effects on chickpea seed weight. Plant Breed 122: 539–541.

Huettel B, Winter P, Weising K, Choumane W, Weigand F, Kahl G (1999) Sequence-tagged microsatellite-site markers for chickpea (*Cicer arietinum* L.). Genome 42: 210–217.

Huettel B, Santra D, Muehlbauer FJ, Kahl G (2002) Resistance gene analogues of chickpea (*Cicer arietinum* L.): isolation, genetic mapping and association with a *Fusarium* resistance gene cluster. Theor Appl Genet 105: 479–490.

Imtiaz M, Martene M, Hobson K, van Ginkel M, Malhotra RS (2008) Molecular genetic diversity and linked resistance to ascochyta blight in Australian chickpea breeding materials and their wild relatives. Aust J Agri Res 59: 554–560.

Iruela M, Rubio J, Cubero JI, Gil J, Millán T (2002) Phylogenetic analysis in the genus *Cicer* and cultivated chickpea using RAPD and ISSR markers. Theor Appl Genet 104: 643–651.

Iruela M, Rubio J, Barro F, Cubero JI, Millán T, Gil J (2006) Detection of two quantitative trait loci for resistance to ascochyta blight in an intra-specific cross of chickpea (*Cicer arietinum* L.): development of SCAR markers associated with resistance. Theor Appl Genet 112: 278–287.

Iruela M, Castro P, Rubio J, Cubero JI, Jacinto C, Millán T, Gil J (2007) Validation of a QTL for resistance to ascochyta blight linked to resistance to fusarium wilt race 5 in chickpea (*Cicer arietinum* L.). Eur J Plant Pathol 119: 20–37.

Iruela M, Pistón F, Cubero JI, Millán T, Barro F, Gil J (2009) The marker $SCK13_{603}$ associated with resistance to ascochyta blight in chickpea is located in a region of a putative retrotransposon. Plant Cell Rep 28: 53–60.

Jayakumar P, Gossen BD, Gan YT, Warkentin TD, Banniza S (2005) Ascochyta blight of chickpea: infection and host resistance mechanisms. Can J Plant Pathol 27: 499–509.

Jimenez-Diaz RM, Alcala-Jimenez AR, Hervas A, Trapero-Casas JL (1993) Pathogenic variability and hosts resístanse in the *Fusarium oxysporum* f. sp. *ciceris*/*Cicer arietinum* pathosystem. Proc. Eur. Sem. *Fusarium* mycotoxins, taxonomy, pathogenicity and host resistance, 3rd Hodowsla Roslin Aklimatyazacja i Nasiennictwo. Plant Breeding and Acclimatization Institute, Radzikov, pp 87–94.

Jimenez-Gasco MM, Navas-Cortes JA, Jimnez-Diaz RM (2004) The *Fusarium oxysporum* f. sp. *ciceri/Cicer arietinum* pathosystem: a case study of the evolution of plant-pathogenic fungi into races and pathotypes. Int Microbiol 7: 95–104.

Johansen C, Bakr MA, Sirajul Islam M, Mondal NA, Afzal A, MacLeod WJ, Pande S, Siddique KHM (2008) Integrated crop management of chickpea in environments of Bangladesh prone to botrytis grey mould. Field Crops 108: 238–249.

Jomová K, Benková M, Czech KJ (2009) Enrichment of chickpea genetic resources collection monitored by microsatellites. J Genet Plant Breed 45: 1–7.

Kaiser WJ (1997) Inter- and intranational spread of *Ascochyta* pathogens of chickpea, faba bean, and lentil. Can J Plant Pathol 19: 215–224.

Kazan K, Muehlbauer FJ, Weeden NF, Ladizinsky G (1993) Inheritance and linkage relationships of morphological and isozyme loci in chickpea (*Cicer arietinum* L.). Theor Appl Genet 86: 417–426.

Kharrat M, Gil J, Cubero JI (1991) Genetics of grain yield components in chickpea (*Cicer arietinum* L.). J Genet Breed 45: 87–92.

Kottapalli P, Gaur PM, Katiyar SK, Crouch JH, Buhariwalla HK, Pande S, Gali KK (2009) Mapping and validation of QTLs for resistance to an Indian isolate of Ascochyta blight pathogen in chickpea. Euphytica 165: 79–88.

Kumar S (1998) Inheritance of resistance to *Fusarium* wilt (race 2) in chickpea. Plant Breed 117: 139–142.

Kumar J, Bahl PN (1992) Direct and indirect selection for yield in chickpea. Euphytica 60: 197–199.

Kumar J, van Rheenen HA (2000) A major gene for time of flowering in chickpea. J Hered 91: 67–68.

Kumar J, Srivastava RK, Ganesh M (2000) Penetrance and expresivity of the gen for double podding in chickpea. J Hered 91: 234–236.

Labdi M, Robertson LD, Singh KB, Charrier A (1996) Genetic diversity and phylogenetic relationships among the annual *Cicer* species as revealed by isozyme polymorphism. Euphytica 88: 181–188.

Ladizinsky G, Adler A (1976) The origin of chickpea *Cicer arietinum* L. Euphytica 25: 211–217.

Lichtenzveig J, Scheuring C, Dodge J, Abbo S, Zhang HB (2005) Construction of BAC and BIBAC libraries and their applications for generation of SSR markers for genome analysis of chickpea, *Cicer arietinum* L. Theor Appl Genet 110: 492–510.

Lichtenzveig J, Bonfil DJ, Zhang HB, Shtienberg D, Abbo S (2006) Mapping quantitative trait loci in chickpea associated with time to flowering and resistance to *Didymella rabiei* the causal agent of Ascochyta blight. Theor Appl Genet 113: 1357–1369.

Madrid E, Rubiales D, Moral A, Moreno MT, Millán T, Gil J, Rubio J (2008) Mechanism and molecular markers associated with rust resistance in a chickpea interspecific cross (*Cicer arietinum* × *Cicer reticulatum*). Eur J Plant Pathol 121: 43–53.

Manchanda G, Garg N (2008) Salinity and its effects on the functional biology of legumes. Acta Physiol Plant 30: 595–618.
Mantri NL, Ford R, Coram TE, Pang ECK (2007) Transcriptional profiling of chickpea genes differentially regulated in response to high-salinity, cold and drought. BMC Genom 8: 303–317.
Mayer MS, Tullu A, Simon CJ, Kumar J, Kaiser WJ, Kraft JM, Muehlbauer FJ (1997) Development of a DNA marker for fusarium wilt resistance in chickpea. Crop Sci 37: 1625–1629.
Maynez M, Moreno MT, Gil J (1993) Desi/kabuli introgression for yield improvement in chickpea (*Cicer arietinum* L.). J Genet Breed 47: 61–64.
Millán T, Rubio J, Iruela M, Daly K, Cubero JI, Gil J (2003) Markers associated with ascochyta blight resistance in chickpea and their potential in marker-assisted selection. Field Crops Res 84: 373–384.
Millán T, Clarke HJ, Siddique KHM, Buhariwalla HK, Gaur PM, Kumar J, Gil J, Kahl G, Winter P (2006) Chickpea molecular breeding: new tools and concepts. Euphytica 147: 81–103.
Millán T, Winter P, Jüngling R, Gil J, Rubio J, Cho S, Cobos MJ, Iruela M, Rajesh PN, Tekeoglu M, Khal G, Muehlbauer FJ (2010) A consensus genetic map of chickpea (*Cicer arietinum* L.) based on 10 mapping populations. Euphytica 175: 175–189.
Molina C, Rotter B, Horres R, Udupa SM, Besser B, Bellarmino L, Baum M, Matsumura H, Terauchi R, Kahl G, Winter P (2008) SuperSAGE: the drought stress-responsive transcriptome of chickpea roots. BMC Genom 9: 553–581.
Moreno MT, Cubero JI (1978) Variation in *Cicer arietinum* L. Euphytica 27: 465–485.
Muehlbauer FJ, Kaiser WJ (1994) Using host plant resistance to manage biotic stresses in cool season grain legumes. Euphytica 73: 1–10.
Muehlbauer FJ, Singh KB (1987) Genetic of chickpea. In: MC Saxena, KB Singh (eds) The Chickpea. CABI Publ, Wallingford, UK, pp 99–125.
Muehlbauer FJ, Kaiser WJ, Simon CJ (1994) Potential for wild species in cool season food legume breeding. Euphytica 73: 109–114.
Muehlbauer FJ, Kaiser WJ, Kusmenoglu I (1998a) Registration of 'Sanford' chickpea. Crop Sci 38: 282.
Muehlbauer FJ, Kaiser WJ, Kusmenoglu I (1998b) Registration of 'Dwelley' chickpea. Crop Sci 38: 282–283.
Nayak SN, Zhu H, Varghese N, Datta S, Choi HK, Horres R, Jüngling R, Singh J, Kavi Kishor PB, Sivaramakrishnan S, Hoisington DA, Kahl G, Winter P, Cook DR, Varshney RK (2009) Integration of novel SSR and gene-based SNP marker loci in the chickpea genetic map and establishment of new anchor points with *Medicago truncatula* genome. Theor Appl Genet 120: 1415–1441.
Nayyar H, Bains T, Kumar S (2005) Low temperature induced floral abortion in chickpea: relationship to abscisic acid and cryoprotectants in reproductive organs. Env Exp Bot 53: 39–47.
Nguyen TT, Taylor PW, Redden RJ, Ford R (2004) Genetic diversity estimates in *Cicer* using AFLP analysis. Plant Breed 123: 173–179.
Ocampo B, Venora G, Errico A, Singh KB, Saccardo F (1992) Karyotype análisis in the genus *Cicer*. J Genet Breed 46: 229–240.
Or E, Hovav R, Abbo S (1999) A major gene for flowering time in chickpea. Crop Sci 39: 315–322.
Palomino C, Fernández-Romero MD, Rubio J, Torres A, Moreno MT, Millán T (2009) Integration of new CAPS and dCAPS-RGA markers into a composite chickpea genetic map and their association with disease resistance. Theor Appl Genet 118: 671–682.
Pande S, Singh G, Rao JN, Bakr MA, Chaurasia PCP, Joshi S, Johansen C, Singh SD, Kumar J, Rahman MM, Gowda CLL (2002) Integrated management of botrytis gray mold of chickpea. Information Bull No 61, ICRISAT, Andhra Pradesh, India.
Pande S, Siddique KHM, Kishore GK, Bayaa B, Gaur PM, Gowda CLL, Bretag T, Crouch JH (2005) Ascochyta blight of chickpea biology, pathogenicity and disease management. Aust J Agri Res 56: 317–332.

Pande S, Galloway J, Gaur PM, Siddique KHM, Tripathi HS, Taylor P, MacLeod MWJ, Basandrai AK, Bakr A, Joshi S, Kishore GK, Isenegger DA, Rao JN, Sharma M (2006a) Botrytis grey mould of chickpea: a review of biology, epidemiology, and disease management. Aust J Agri Res 57: 1137–1150.

Pande S, Kishore GK, Upadhyaya HD, Rao JN (2006b) Identification of sources of multiple disease resistance in mini-core collection of chickpea. Plant Dis 90: 1214–1218.

Penmetsa RV, Carrasquilla-Garcia N, Rosen B, Gao J, Sarma BK, Datta S, Vaill S, Garzon L, Bett K, Vandenberg B, Woodward J, Blair M, Bertioli D, May GD, He G, Bruening GE, Rajeev K, Cook DG (2010) Studies of genome evolution in Papilionoid legumes. In: 5th Int Food Legumes Res Conf (IFLRC V) & 7th Eur Conf on Grain Legumes (AEP VII). Antalya, p 40: http://www.iflrc-ecgl.org/BookofAbstracts.pdf. Cited 21 July 2010.

Pfaff T, Kahl G (2003) Mapping of gene-specific markers on the genetic map of chickpea (*Cicer arietinum* L.). Mol Genet Genom 269: 243–251.

Pundir RPS, Rao NK, van der Maesen LJG (1983) Induced autotetraploididy in chickpea (*Cicer arietinum* L.). Theor Appl Genet 65: 119–122.

Ragazzi A (1982) Un grave attacco di ruggine su foglie discese. (A serious attack of rust on *Cicer arietinum* leaves). Informatore Fitopatológico 2: 41–43.

Rajesh PN, Muehlbauer FJ (2008) Discovery and detection of single nucleotide polymorphism (SNP) in coding and genomic sequences in chickpea (*Cicer arietinum* L.). Euphytica 162: 291–300.

Rajesh PN, Tullu A, Gil J, Gupta VS, Ranjekar PK, Muehlbauer FJ (2002) Identification of an STMS marker for double podding gene in chickpea. Theor Appl Genet 105: 604–607.

Rajesh PN, Coyne C, Meksem K, Sharma KD, Gupta V, Muehlbauer FJ (2004) Construction of a *Hind*III bacterial artificial chromosome library and its use in identification of clones associated with disease resistance in chickpea. Theor Appl Genet 108: 663–669.

Rajesh PN, Bleness M, Roe BA, Muehlbauer FJ (2008) Analysis of genome organization, composition and microsynteny using 500 kb BAC sequences in chickpea. Theor Appl Genet 117: 449–458.

Rakshit S, Winter P, Tekeoglu M, Juarez Muñoz J, Pfaff T, Benko-Iseppon AM, Muehlbauer FJ, Kahl G (2003) DAF marker tightly linked to a major locus for ascochyta blight resistance in chickpea (*Cicer arietinum* L.). Euphytica 132: 23–30.

Ratnaparkhe MB, Santra DK, Tullu A, Muehlbauer FJ (1998) Inheritance of inter-simple-sequence-repeat polymorphisms and linkage with a fusarium wilt resistance gene in chickpea. Theor Appl Genet 96: 348–353.

Rincón F, Martínez B, Ibáñez MV (1998) Proximate composition and antinutritive substances in chickpea (*Cicer arietinum* L.) as affected by the biotype factor. J Sci Food Agri 78: 382–388.

Robertson LD, Ocampo B, Singh KB (1997) Morphological variation in wild annual *Cicer* species in comparison to the cultigen. Euphytica 95: 309–319.

Rubiales D, Sillero JC, Moreno MT (1999) Resistance to *Orobanche crenat*a in chickpea. In: JI Cubero, MT Moreno, D Rubiales, JC Sillero (eds) Resistance to *Orobanche*: The state of the art. Junta de Andalucia, Sevilla, Spain, pp 55–62.

Rubiales D, Moreno I, Moreno MT, Sillero JC (2001) Identification of partial resistance to chickpea rust (*Uromyces ciceris-arietini*). Proc 4th Eur Conf on Grain Legumes. Cracow, Poland, pp 194–195.

Rubiales D, Pérez de Luque A, Joel DM, Alcántara C, Sillero JC (2003) Characterization of resistance in chickpea to broomrape (*Orobanche crenata*). Weed Sci 51: 702–707.

Rubio J, Moreno MT, Cubero JI, Gil J (1998) Effect of the gene for double pod in chickpea on yield, yield components and stability of yield. Plant Breed 117: 585–587.

Rubio J, Hajj-Moussa E, Kharrat M, Moreno MT, Millan T, Gil J (2003) Two genes and linked RAPD markers involved in resistance to *Fusarium oxysporum* f. sp. *ciceris* race 0 in chickpea. Plant Breed 122: 188–191.

Rubio J, Flores F, Moreno MT, Cubero JI, Gil J (2004) Effects of the erect/bushy habit, single/double pod and late/early flowering genes on yield and seed size and their stability in chickpea. Field Crops Res 90: 255–262.

Sabaghpour SH, Kumar J, Rao TN (2003) Inheritance of growth vigour and its association with other characters in chickpea. Plant Breed 122: 542–544.

Santra DK, Tekeoglu M, Ratnaparkhe M, Kaiser WJ, Muehlbauer FJ (2000) Identification and mapping of QTLs conferring resistance to ascochyta blight in chickpea. Crop Sci 40: 1606–1612.

Saxena MC (1987) Effects of climatic stress and soil chemical toxicities on productivity of chickpea in West Asia and North Africa. In: NP Saxena, C Johansen (eds) Adaptation of Chickpea and Pigeonpea to Abiotic Stresses. ICRISAT, Patancheru, AP, India, pp 135–141.

Saxena NP, Krishnamurthy L, Johansen C (1993) Registration of drought-resistant chickpea germplasm. Crop Sci 33: 1424.

Serret MD, Udupa SM, Weigand F (1997) Assessment of genetic diversity of cultivated chickpea using microsatellite-derived RFLP markers: implications for origin. Plant Breed 116: 573–578.

Sethy NK, Shokeen B, Bhatia S (2003) Isolation and characterization of sequence-tagged microsatellite sites markers in chickpea (*Cicer arietinum* L.). Mol Ecol Notes 3: 428–430.

Sethy NK, Choudhary S, Shokeen B, Bhatia S (2006a) Identification of microsatellite markers from *Cicer reticulatum*: molecular variation and phylogenetic analysis. Theor Appl Genet 112: 347–357.

Sethy NK, Shokeen B, Edwards KJ, Bhatia S (2006b) Development of microsatellite markers and analysis of intraspecific genetic variability in chickpea (*Cicer arietinum* L.). Theor Appl Genet 112: 1416–1428.

Sharma KD, Muehlbauer FJ (2007) Fusarium wilt of chickpea: physiological specialization, genetics of resistance and resistance gene tagging. Euphytica 157: 1–14.

Sharma KD, Winter P, Kahl G, Muehlbauer FJ (2004) Molecular mapping of *Fusarium oxysporum* f. sp. *ciceris* race 3 resistance gene in chickpea. Theor Appl Genet 108: 1243–1248.

Sharma KD, Chen W, Muehlbauer FJ (2005) Genetics of chickpea resistance to five races of fusarium wilt and a concise set of race differentials for *Fusarium oxysporum* f. sp. *ciceris*. Plant Dis 89: 385–390.

Siddique KHM, Loss SP, Thomson BD (2003) Cool season grain legumes in dryland Mediterranean environments of Western Australia: Significance of early flowering. In: NP Saxena (ed) Management of Agricultural Drought. Science Publishers, Enfield, NH, USA, pp 151–161.

Simon CJ, Muehlbauer FJ (1997) Construction of a chickpea linkage map and its comparison with maps of pea and lentil. J Hered 88: 115–119.

Singh A, Singh NP, Asthana AN (1999a) Genetic potential of wide crosses in chickpea. Legume Res 22: 19–25.

Singh A, Singh N, Asthana AN (1999b) Studies on inter-specific crossability barriers in chickpea. Indian J Pulses Res 12: 13–19.

Singh A, Devarumath RM, RamaRao S, Singh VP, Raina SN (2008) Assessment of genetic diversity, and phylogenetic relationships based on ribosomal DNA repeat unit length variation and Internal Transcribed Spacer (ITS) sequences in chickpea (*Cicer arietinum*) cultivars and its wild species Genet Resour Crop Evol 55: 65–79.

Singh G, Kaur L, Sharma YR (1991) Ascochyta blight and gray mold resistance in wild species of *Cicer.* Crop Improv 18: 150–151.

Singh KB (1987) Chickpea breeding. In: MC Saxena, KB Singh (eds) The Chickpea. CABI Publ, Wallingford, UK, pp 127–162.

Singh KB, Reddy MV (1983) Inheritance of resistance to ascochyta blight in chickpea. Crop Sci 23: 9–10.

Singh KB, Reddy MV (1989) Genetics of Resistance to ascochyta Blight in four chickpea lines. Crop Sci 29: 657–659.

Singh KB, Reddy MV (1993) Resistance to six races of *Ascochyta rabiei* in the world germplasm collection of chickpea. Crop Sci 33: 186–189.

Singh KB, Reddy MV (1996) Improving chickpea yield by incorporating resistance to ascochyta blight. Theor Appl Genet 92: 509–515.

Singh KB, Holly L, Vejiga G (1991) A catalog of kabuli chickpea germplasm. Internacional Center for Agricultural Research in the Dry Areas (ICARDA), Aleppo, Syria.

Singh KB, Ocampo B, Robertson LD (1998) Diversity for abiotic and biotic stress resistance in the wild annual *Cicer* species. Genet Resour Crop Evol 45: 9–17.

Sohoo MS, Athwal DS, Chandra S (1970) Colchicine induced polyploidy in chickpeas (*Cicer arietinum* L.). Theor Appl Genet 40: 163–168.

Soltero-Diaz L, Andrade E, Grageda O, Pérez J (2008) San Antonio 05, forage chickpea cultivar for the Cienega de Chapala region, México. Agri Técn Méx 34: 263–265.

Tar'an B, Warketin T, Tullu A, Vandenberg A (2007a) Genetic relationships among chickpea (*Cicer arietinum* L.) genotypes based on the SSRs at the quantitative trait loci for resistance to ascochyta blight. Eur J Plant Pathol 119: 39–51.

Tar'an B, Warkentin TD, Tullu A, Vandenberg A (2007b) Genetic mapping of ascochyta blight resistance in chickpea (*Cicer arietinum* L.) using a simple sequence repeat linkage map. Genome 50: 26–34.

Tayyar R, Wainess JG (1996) Genetic relationships among annual species of *Cicer* (Fabaceae) using isozyme variation. Theor Appl Genet 92: 245–254.

Tekeoglu M, Tullu A, Kaiser WJ, Muehlbauer FJ (2000) Inheritance and linkage of two genes that confer resistance to fusarium wilt in chickpea. Crop Sci 40: 1247–1251.

Tekeoglu M, Rajesh PN, Muehlbauer FJ (2002) Integration of sequenced tagged microsatellite sites to the chickpea genetic map. Theor Appl Genet 105: 847–854.

Tewari SK, Pandey MP (1986) Genetics of resistance to ascochyta blight in chickpea (*C. arietinum* L.). Euphytica 35: 211–215.

Toker C, Lluch C, Tejera NA, Serraj R, Siddique KHM (2007) Abiotic stresses. In: SS Yadav, R Redden, W Chen, B Sharma (eds) Chickpea Breeding and Management. CABI Publ, Wallingford, UK, pp 474–496.

Tullu A, Kaiser WJ, Kraft JM, Muehlbauer FJ (1999) A second gene for resistance to race 4 of fusarium wilt in chickpea and linkage with a RAPD marker. Euphytica 109: 43–50.

Tullu A, Muehlbauer FJ, Simon CJ, Mayer MS, Kumar J, Kaiser WJ, Kraft JM (1998) Inheritance and linkage of a gene for resistance to race 4 of fusarium wilt and RAPD markers in chickpea. Euphytica 102: 227–232.

Udupa SM, Baum M (2003) Genetic dissection of pathotype-specific resistance to ascochyta blight disease in chickpea (*Cicer arietinum* L.) using microsatellite markers. Theor Appl Genet 106: 1196–1202.

Udupa SM, Sharma A, Sharma AP, Pai RA (1993) Narrow genetic variability in *Cicer arietinum* L. as revealed by RFLP analysis. J Plant Biochem Biotechnol 2: 83–86.

Udupa SM, Weigand F, Saxena MC, Kahl G (1998) Genotyping with RAPD and microsatellite markers resolves pathotype diversity in the ascochyta blight pathogen of chickpea. Theor Appl Genet 97: 299–307.

Upadhyaya HD (2003) Geographical patterns of variation for morphological and agronomic characteristic in the chickpea germplasm collection. Euphytica 132: 343–352.

Upadhyaya HD, Haware M, Kumar J, Smithson JB (1983a) Resistance to wilt in chickpea. I. Inheritance of late wilting in response to race 1. Euphytica 32: 447–452.

Upadhyaya HD, Smithson JB, Haware MP, Kumar J (1983b) Resistance to wilt in chickpea. II. Further evidence for two genes for resistance to race 1. Euphytica 32: 749–755.

Upadhyaya HD, Salimath PM, Gowda CLL, Singh S (2007) New early-maturing germplasm lines for utilization in chickpea improvement. Euphytica 157: 195–208.

Upadhyaya HD, Dwivedi SL, Baum M, Varshney RK, Udupa SM, Gowda CLL, Hoisington D, Singh S (2008) Genetic structure, diversity and allelic richness in composite collection and reference set in chickpea (*Cicer arietinum* L.). BMC Plant Biol 8: 106–118.

van der Maesen LJG (1972) *Cicer* L., a monograph of the genus, with special reference to the chickpea (*Cicer arietinum* L.) its ecology and distribution. Mendeligen Landbouwhogeschool Wageningen, Wageningen, The Netherlands, pp 1–341.

van der Maesen LJG (1987) Origin, history and taxonomy of chickpea. In: MC Saxena, KB Singh (eds) The Chickpea. CABI Publ, Wallingford, UK, pp 11–34.

van der Maesen LJG, Maxted N, Javadi F, Coles S, Davies AMR (2007) Taxonomy of the Genus *Cicer* Revisited . In: SS Yadav, R Reeden, W Chen, B Sharma (eds) Chickpea Breeding and Management. CABI Publ, Wallingford, Uk, pp 14–46.

Verma MM, Sandhu JS, Rrar IIS, Brar JS (1990) Crossability studies in different species of *Cicer* (L). Crop Improv 17: 179–181.

Vir S, Grewal JS, Gupta VP (1975) Inheritance of resistance to ascochyta blight in chickpea. Euphytica 24: 209–211.

Vláčilová K, Ohri D, Vrána J, Číhalíková J, Kubaláková M, Kahl G, Dolezel J (2002) Development of flow cytogenetics and physical genome mapping in chickpea (*Cicer arietinum* L.). Chrom Res 10: 695–706.

Williams PC, Singh U (1987) Nutritional quality and the evaluation of quality in breeding programmes. In: MC Saxena, KB Singh (eds) The Chickpea. CABI Publ, Wallingford, UK, pp 143–166.

Winter P, Pfaff T, Udupa SM, Hüttel B, Sharma PC, Sahim S, Arreguin-Espinoza R, Weigand F, Muehlbauer FJ, Kahl G (1999) Characterization and mapping of sequence-tagged microsatellite sites in the chickpea (*Cicer arietinum* L.) genome. Mol Gen Genet 262: 90–101.

Winter P, Benko-Iseppon AM, Hüttel B, Ratnaparkhe M, Tullu A, Sonnante G, Pfaff T, Tekeoglu M, Santra D, Sant VJ, Rajesh PN, Kahl G, Muehlbauer FJ (2000) A linkage map of the chickpea (*Cicer arietinum L.*) genome based on recombinant inbred lines from a *C. arietinum* × *C. reticulatum* cross: Localization of resistance genes for *fusarium* wilt races 4 and 5. Theor Appl Genet 101: 1155–1163.

Zhou R, Zhu Z, Kong X, Huo N, Tian Q, Li P, Jin C, Dong Y, Jia J (2005) Development of wheat near-isogenic lines for powdery mildew resistance. Theor Appl Genet 110: 640–648.

Zohary D (1999) Monophyletic vs. polyphyletic origin of the crops on which agriculture was founded in the near east. Genet Resour Crop Evol 46: 133–142.

Zhu H, Choi HK, Cook DR, Shoemaker RC (2005) Bridging model and crop legumes through comparative genomics. Plant Physiol 137: 1189–1196.

6

Functional Genomics–Transcriptomics for Legumes: Background, Tools and Insights

Günter Kahl,[1,*] *Carlos Molina Medina*[2] and *Peter Winter*[3]

ABSTRACT

Functional genomics describes the whole repertoire of large-scale and high-throughput techniques and subsequent computational analyses for deciphering the roles of DNA and RNA in the progression from information to function. The present chapter reviews the functional genomics of cool season food legumes, though many insights into the biology of a specific legume derive from so-called model plants such as *Medicago truncatula* and *Lotus japonicus*. As compared to model legumes and to highly researched crops such as soybean, rice and maize, the volume of transcriptomic information for the cool season food legumes in the databases still is poor, in some cases practically non-existing. The continuously improving high-throughput procedures are contributing to a leap in the quantity and quality of the transcriptomic data, even at the cellular level, and with a welcome reduction in costs. Future prospects in cool season food legumes are: 1) the development and use of expression markers; 2) the analysis of the expression of quantitative trait loci; and 3) the understanding of the epigenetic code. All these developments, genome and transcriptome sequencing included, will certainly enrich the technological and informational space of plant research, and especially research on cool season food legumes .

Keywords: Legumes, transcriptome, transcriptomics, expression profiling, deepSuperSAGE, chickpea, salt stress

[1]Molecular BioSciences, Biocenter, University of Frankfurt am Main, D-60438 Frankfurt am Main, Germany; e-mail: *kahl@em.uni-frankfurt.de*

[2]Unité de Recherche en Légumineuses, INRA-URLEG, 17 Rue Sully, 21000 Dijon, France.

[3]GenXPro GmbH, Frankfurt Innovation Center Biotechnology (FIZ), Altenhöferallee 3, D-60438 Frankfurt am Main, Germany.

*Corresponding author

6.1 Introduction

The main topic of the present chapter builds around functional genomics of legumes, more specifically cool season food legumes, though many insights into the biology of a specific legume derives from the so-called model plants such as *Medicago truncatula* and *Lotus japonicus*. Functional genomics describes the whole repertoire of large-scale and high-throughput techniques and subsequent computational analyses for deciphering the roles of DNA and RNA in the progression from information (DNA) to function (both RNA as well as protein). For example, specific gene disruption (gene knock-down, gene knock-in, gene knock-out) allows to reveal the function of a gene, the determination of gene expression patterns (i.e., expressed genes in a given cell, tissue, organ, or organism at a time by e.g., microarrays) relates gene function(s) to cellular processes, and the transfer and integration of foreign genes allow in studying their influence(s) on the activity of other resident genes. An ever-increasing area of functional genomics focuses on post-transcriptional events such as messenger RNA association with carrier and chaperon proteins, stability, export and compartmentalization, turnover rate and frequency of translation, but also the stability of the protein product and the many protein-protein interactions of a target protein with other proteins (as, e.g., detected by yeast two-hybrid analysis), to name a few. However, the real complexity cannot be caught by these parameters at all. On top of the genetic code (i.e., the sequence of bases in DNA), other codes emerged. The nucleosome positioning code decides what regions in a genome can easily be accessed by proteins to turn on transcription of a gene. Additionally, nucleosomal DNA is more highly methylated than flanking DNA. So, nucleosome positioning influences DNA methylation patterning throughout, in this case, the *A. thaliana* genome, and DNA methyltransferases preferentially target nucleosome-bound DNA (see Chodavarapu et al. 2010).

Also superimposed on the genetic code is a structural DNA code, which means the various two- and three-dimensional configurations of the DNA like hairpins, bulges, foldbacks, kinks, bends, pulsations, Z-DNA conformations, and others, that influence the transcriptional competence of genes. Another DNA code is encrypted in the methylation of cytosine residues (**epigenetic code**), which is instrumental for the change from euchromatin to heterochromatin and vice versa (Zhang et al. 2006; Zilberman et al. 2007; Lister et al. 2008). The chromatin condensation states also define active versus inactive regions of a genome. Importantly, the **second genetic code** represents the genome-wide pattern of transcription factor binding sites. Transcription cannot start without the interaction(s) of various transcription factors with their cognate sequence motifs, mostly in promoters or enhancer regions, a prerequisite for the recruitment of the transcription machinery with its more than 100 different proteins (Kim and Ren 2006). The second genetic code also encircles the genome-wide binding patterns of proteins, especially proteins with a function in transcription, as, e.g., DNA-dependent RNA polymerases (Johnson et al. 2007). A parallel world to all these codes is the histone side chain modification code (**histone code**), one of the most complex parameters in gene transcription. Histones are low molecular weight basic proteins that fall into several classes (e.g., H1, H2A, H2B, H3 and H4, to name the most prevalent ones). Pairs of histones H2A and B, H3 and H4 forms a core by protein-protein interactions, around which about two turns of DNA are wrapped to form a nucleosome. During the last decade it became clear that the nucleosome does not only structure and condense the huge amount of DNA in a nucleus of a eukaryotic cell, but additionally functions as a fine tuner of transcriptional activity by various spatial and temporal posttranslational modifications of side chains of histones (Fillingham and Greenblatt 2008). Though this field of gene regulation has matured into a discipline of its own, only a summary of the potential modifications and their consequences for gene function is presented here. If, for example, a yeast gene is activated, the following histones become modified around the transcription start site (TSS): H3K4me1, H3K4me2, H3K4me3, H3K9me1 (where H3 = histone 3, K4 = lysine at position 4 from the N-terminus, me 1, 2, 3 = 1, 2 or 3 methyl groups covalently attached to the lysine). The main body of the activated gene carries H2BK5me1, H3K36me3, H3K27me1, and H4K20me1. In contrast, the histone code of silent genes is quite different and characteristic. So, the mono-, di- and tri-methylated histone H3K4 occurs at low levels at promoter sites, whereas high levels of H3K27me3 and H3K79me3, and low levels of H3K36me3, H3K27me1, H3K9me1, and H4K20me1, respectively, prevail in the gene body. Other side chain modifications, as e.g., acetylation, phosphorylation, and ubiquitinylation of various residues add to the immense complexity of the chromatin configuration around eukaryotic genes (Henikoff and Ahmad 2005; Shilatifard 2006; Berger 2007). To give only a few examples, H3K56Ac (where Ac = acetyl moiety) plays a key role in chromatin formation after DNA replication and regulates replication-coupled nucleosome assembly, H3K9Ac and H3K14Ac both recruit basal transcription factor TF IID and remove a nucleosome from the TATA box, and generally mark euchromatin, and H3S10p (S = serine, p = phosphoryl residue) inhibits K9me, enhances K9 and K14 acetylation, a crosstalk coined "methyl-phos switch". Therefore, the specific histone code at a particular location within a genic sequence can be taken as signature for active or inactive genes, respectively (Barski et al. 2007). Since similar signatures have been found in other organisms, it is likely that a histone code also determines the activity status of plant genes. Additionally, a linker histone code emerges, that is based on various modification(s) of histone H1, but whose importance is still obscure (Godde and Ura 2008). The involved enzymes, histone methyltransferases and demethylases, acetyl transferases and deacetylases, phopho-transferases and phosphatases, and ubiquitin transferases and ubiquitinases, catalyzing specific histone modifications, in turn are regulated by a variety of intrinsic and environmental factors and as such are important players in gene activation and silencing. It comes as no surprise, that e.g., histone deacetylases are prominent targets for pharmaceutical research (Talbert and Henikoff 2010).

In recent years, another complexity was added: the RNA *cis*-regulatory code (the sequence motifs in 3′-untranslated regions, 3′-UTRs of messenger RNAs), the **microRNA code** (He and Hannon 2004; Jones-Rhoades et al. 2006; Chen and Rajewsky 2007; Slezak-Prochazka et al. 2010), and the mRNP code, i.e., the specific association of an mRNA and cognate proteins. These codes are not dealt with here. All these different levels, that govern the activity of a gene, have hardly or not even at all been touched in the major legume crops.

Therefore, it is quite obvious that a careful and detailed consideration of functional genomics in this broad sense is beyond the scope of this chapter, especially since most of the relevant information was collected from mammals, *Drosophila*, *Caenorhabditis elegans* and the yeast *Saccharomyces cerevisiae*, but not plants, some model plants notwithstanding (mostly *Arabidopsis thaliana*, *Oryza sativa*). Instead, we focus on only one single aspect: **transcription** itself and the **function of some selected transcripts**.

6.1.1 Transcripts, Transcriptomes, Transcriptomics

A few definitions should ease understanding and avoid misunderstanding. The term **transcription** describes the synthesis of an RNA molecule on a DNA or RNA template, catalyzed by DNA-dependent or RNA-dependent RNA polymerases, respectively. The recently discovered **pervasive transcription** of almost all DNA in human genomes produces a great variety of RNAs with different structures, lengths and functions (Kapranov et al. 2007), and there is reason to expect a similar spectrum of RNAs in plants, and legumes specifically. All products of transcription are coined **transcripts**. Transcripts in eukaryotes are single-stranded RNA molecules produced by RNA polymerase I (A) on rDNA (transcript: ribosomal RNA), by RNA polymerase II (B) on structural genes (transcript: messenger RNA), and by RNA polymerase III (C) on tDNA (transcript: transfer RNA). All the different RNA polymerases additionally transcribe a series of other RNAs. Within a genome, transcriptional desert (transcript-poor regions, TPRs) regions containing relatively few transcribed genes, and transcriptional forests (transcriptionally active regions, TARs, transcript-rich regions, TRRs) whose DNA is transcribed from both strands without gaps are intermingled.

Genes themselves, either protein-coding or non-protein-coding genes are embedded in regulatory sequences such as enhancers, silencers, or promoters, to name a few. Most eukaryotic genes are split genes (mosaic genes), that contain 5′-untranslated regions (5′-UTRs), 3′-UTRs, exons and introns with a whole series of cognate sequence motifs for the recognition and binding of nuclear proteins. The reader is probably aware of these sequences, and these are not explained here further.

A highly complex mechanism involving many environmental and intrinsic triggers and molecules leads to the activation of a previously inactive gene, and vice versa, the inactivation of a previously active gene. One of the prime events for gene activation is the removal or at least relaxation of a nucleosome on a strategic sequence motif of the promoter (formation of a lexosome) such that specific proteins, so-called **transcription factors** (TFs; trans-acting factors, trans-acting proteins, nuclear factors, transcriptional activators) can bind to cognate sequences (binding sites). Transcription factors may interact with upstream regulatory sequences (so-called upstream binding factors, UBFs), with TATA boxes or also with sequences within the coding region (e.g., in the case of class III genes), provided these regions are accessible. The so-called general transcription factors are highly conserved and interchangeable between mammals, *Drosophila*, yeast and plants. In the case of TATA-box-containing promoters, transcription activation factors (TAFs; TATA-box binding protein associated factors, TBP-associated factors), a series of nuclear transcription co-factors, are recruited to the TATA box-binding protein (transcription factor II D) to form a large heterogeneous multi-subunit protein complex (pre-initiation complex), that is a prerequisite for the formation of a transcription initiation complex, and mediates activation (or repression) signals from regulatory proteins bound to enhancer (or silencer) sequences. The transcription factor binding site (TFBS, transcription factor address site, transcription factor recognition element, transcription factor recognition site) represents a 4–25 bp conserved sequence element in promoters that functions as address module for the specific recognition and binding by the corresponding transcription factors. Such sites are also present in different promoters. For example, the octamer element 5′-ATGCAAAT-3′ is part of immunoglobulin gene enhancers, but also present in promoters of house-keeping genes such as e.g., histone genes and snRNA genes. Moreover, one promoter may contain up to 10 (or more) such address sites.

Once the interaction(s) between DNA sequence and TF take place, the transcription initiation complex (Transcription Complex, TC) is formed. This two MDa complex between more than 100 different proteins and DNA-dependent RNA polymerase II (B) assembles at the cap site of eukaryotic genes, and directs the polymerase to the start codon for correct initiation of gene transcription. A first step in the assembly process in class II gene promoters is the binding of transcription factor IID (TF IID), consisting of the TATA-box binding protein (TBP) and a series of so-called TBP-associated factors (TAFs), to the TATA box, which leads to protection of a region spanning base pairs -42 to -17 from DNase I attack. The bound TF IID nucleates the core complex, to which transcription factors TF IIA and TF IIB (and possibly other proteins, e.g., so-called co-activators) bind serially. This multi-protein complex extends from -80 to +10 and is recognized by RNA polymerase II (B) and TF IIE, that binds downstream of the complex and extends the protected region to +30. Then TF IIF, TF IIH and TF IIJ join in. The whole complex interacts with other proteins bound to the GC box (e.g., SP 1), the CAAT box (e.g., CTF), and other sequence motifs in distal promoter regions (e.g., enhancers). These proteins interfere positively (enhancers) or negatively (silencers) with the activity of the transcription initiation complex. This amply demonstrates that transcription factors play key roles in the transcription of genes. In specific combinations they also exert cooperative synergistic effects on the transcription of the adjacent gene. This synergism, called **transcriptional synergy** leads to a significantly higher level of transcription than the sum of the effects of all individual factors.

Transcription ensues at the transcription initiation sequence (transcription initiator sequence; initiator, Inr; initiator element; initiator box, transcription initiation site, mRNA initiation site, transcription start site, cap site), a 17 bp sequence element (consensus sequence: 5′-CTCA-3′, 5′-PyPyCA$(Py)_5$-3′ (animals), and 5′-PyPyCA$(Py)_n$-3′ (plants), more generally 5′-PyPyC**A**PyPyPyPyPy-3′ (where A marks + 1) of promoters of RNA polymerase II (B) genes, that is located at + 1 to + 11 and contains the transcription start site. The Inr acts as a major transcription promoter in TATA box-less genes. The term "cap site" refers to the addition of a 7-methyl-guanosine cap to the first nucleotide (mostly an A) of the primary transcript during its processing. It is also necessary to nucleate the assembly of the various proteins of the transcription initiation complex, and drives accurate basal transcription initiation of various genes. The stop of the transcription of a gene into the corresponding messenger RNA or RNA generally (transcription termination) in eukaryotes is largely unknown, but involves different signals for the different DNA-dependent RNA polymerases. For example, RNA polymerase I in mice recognizes an 18 bp termination sequence (5′-ACCGTCGACCAG[T/A][A/T]-NTCCG-3′) preceded by T stretches, to which an auxiliary protein binds, which terminates ribosomal RNA gene transcription. RNA polymerase II leaves the template strand after contact with a specific secondary structure at the termination site which has a specific sequence (5'-TTTTTATA-3'), and RNA polymerase III terminates transcription at a U_4-sequence embedded in a GC-rich region. In yeast, transcription termination of polymerase II genes occurs at AT-rich sequences, e.g., a 38 bp region including the sequences 5′-TTTTTATA-3′,

5′-TATATA-3′ (very effective), 5′-TACATA-3′, and 5′-TAGTAGTA-3′. None of these sequences function as transcriptional terminators in the reverse orientation. Transcription termination mechanisms in different organisms therefore vary considerably.

Transcription of protein-encoding genes produces transcripts, that are immediately processed, e.g., trimmed, or 3′-polyadenylated, 5′-capped, and spliced. Splicing normally leads to the removal of introns and the correct ligation of exons, but may considerably deviate from this normality: exon skipping, intron retention, exon repetition, exon and intron shuffling and other post-transcriptional processes lead to a pronounced **transcript heterogeneity**, which may also reflect protein variability. Transcript variability may also be encoded at the DNA level, for example, through the use of multiple transcriptional start sites. This is true only for protein-coding genes, but about 50% of the coding sequences in the human genome encode non-protein-coding transcripts (Cheng et al. 2005). While these figures have never been demonstrated for plants, the transcriptional landscape in eukaryotes, and therefore also plants, is extremely complex. Today, about 40 functionally and structurally different **RNA superfamilies** are known, and more are to be expected. Among one category of these RNAs, the small RNAs (sRNAs), a whole series of structurally and functionally different components are identified, mostly in animals including humans. For example, enhancer RNAs (eRNAs) have been identified recently (Kim et al. 2010), but a series of sRNAs with partly unknown function are known, the most recently discovered being 1) circular interfering RNA (ciRNA), or circularized interfering RNA, whose hairpins at both extremities confer stability against cellular ribonucleases, 2) circulating microRNA, any one of a series of relatively stable microRNAs, that circulate in the bloodstream of an organism, 3) 26G-RNA, small RNAs of *Caenorhabditis elegans* (and probably other organisms), that are 26 nucleotides in length, own a strong bias for a 5′ guanine and map to several clusters in inter-genic regions of different chromosomes, and 4) 22G-RNA, a class of abundant endogenous small RNAs of *C. elegans*, that are primarily 22 nucleotides in length (therefore 22), with a 5′-triphosphate and a strong bias for 5′-guanosine (therefore G), some of which interact with the Argonaute protein CSR-1 and function to silence transposons, pseudogenes, cryptic elements, and certain protein-coding genes. However, this is only the tip of the iceberg of the RNA world, because already the variety of RNAs is enormous and bewildering: activator RNA, adaptor RNA, anti-sense RNA, anti-sense siRNA, aRNA, ambisense RNA, catalytic RNA, ccRNA, cell cycle RNA, chromosomal RNA, *cisR*, cryptic RNA, circulating RNA, *cis*RNA, cRNA, competitor RNA, copy RNA, dasRNA, decoy RNA, degradation-resistant signal RNA, diverse autonomously expressed small RNA, double-stranded RNA, efferent RNA, fatal RNA, guide RNA, hairpin RNA, immunostimulatory RNA, informational RNA, intron-containing hairpin RNA, intronic snoRNA, iRNA, isoacceptor tRNA, isRNA, large RNA, L RNA, lipophylic silencing RNA, masked messenger RNA, medium RNA, micro RNA, mirtron, M RNA, non-coding RNA, non-stop transcript, nuclear regulatory RNA, nuclear RNA, PIN RNA, piRNA, Piwi-interacting RNA, primer RNA, pRNA, promoter-associated long RNA, promoter-associated small RNA, promoter RNA, pyranosyl-RNA, repeat-associated small interfering RNA, redox RNA, repression RNA, RNA-OUT, scan RNA, scrambled small interfering RNA, secondary small interfering RNA, sense RNA, sense siRNA, short hairpin RNA, short interfering RNA, short RNA, short stop RNA, small auxin up RNA, small cytoplasmic RNA, small endogenous RNA, small interfering RNA, small interfering stable RNA, small non-coding RNA, small non-mRNA, small nuclear RNA, small nucleolar RNA, small regulatory RNA, small temporal RNA, spatial development RNA, spliced leader RNA, stress-response RNA, subgenomic RNA, TAR RNA, tectoRNA, therapeutic RNA, TIN RNA, tmRNA, tncRNA, telomerase RNA, telomeric repeat-containing RNA, TERRA, TIN RNA, tiny RNA, tiny expressed RNA, tiny non-coding RNA, tmRNA, transacting RNA, trans-acting small interfering RNA, trans-activation response region RNA, transcription start site RNA, transitive interference RNA, translational control RNA, translationally induced messenger RNA, translationally repressed messenger RNA, trans-splicing RNA, Tsix-RNA, 21U-RNA, tx RNA, U-RNA, Xist RNA. For a guide through this jungle we recommend RNomenclature (Brosius and Tiedge 2004; Kapranov et al. 2007). It is expected that more RNAs with differing structure and function will be discovered, making the **transcriptome** one of the most complex cellular spaces. The irony is that only few of such RNAs are known from plants, fewer from legumes as a whole, and almost none from cool season food legumes.

An even higher level of complexity can be inferred by the fact that the concerted activation (or silencing) of whole batteries of genes (frequently not clustered on the same chromosome, but distributed all over the genome) creates a **transcriptional network** of nucleus-wide dimensions. While we are just beginning to understand more of this type of networking in humans, to our knowledge no sound experimental approach has been reported for plants, not to speak of legumes. However, systems biology, which is more than the transcriptional network, will nevertheless require a comprehensive understanding of the transcriptome and the transcriptome networks, also between different cell organelles (see for e.g., Kyung et al. 2010).

The transcriptome, the entirety of all expressed genes of a genome (also called "expressed genome"), or the full complement of messenger RNAs or transcripts of a specific cell, tissue, organ or organism at a given time, can be profiled at different levels. The process of transcription profiling produces an expression profile. A whole series of techniques have been developed over time to detect single genes, or at least a low number of genes, a more historic, though still used approach. A well-known example is Northern blotting. However, the last decades have seen unprecedented progress in **transcriptomics,** the whole repertoire of techniques to analyze and characterize the transcriptome of an organelle or a cell, towards a whole-cell transcript profiling at appreciable depth, using next- and next-next-generation sequencing platforms.

6.2 The Tools of Transcriptomics

One of the kick-off technologies for higher throughput transcriptomics was the development of **expressed sequence tags** (ESTs), short oligonucleotides of 300–500 bp, complementary to the 5′ or 3′ end of a specific messenger RNA and usually derived from a cDNA library by random sequencing. ESTs represent tags for the state of expression of genes at a given time and cell or tissue type. Hundreds of thousands of ESTs have been sequenced and deposited in databases for gene discovery. Any collection of ESTs cloned into a cloning vector (e.g., a plasmid) is coined EST library. Also for plants, a vast number of EST sequences from various organs or tissues of different plants crowd the corresponding databases. The universal dbEST (database of ESTs) is a division of GenBank that contains sequence data from single-pass cDNA sequences, or ESTs from a number of organisms, available from NCBI by anonymous ftp and through Entrez (NCBI dbEST). PlantGDB downloads all *Viridiplantae* sequence data deposited in GenBank and UniProt every 4 months, parses them by species, and makes data available for search, download or BLASTing. The PlantGDB-assembled Unique Transcripts (PUTs) (Table 6-1) are annotated with UniProt BLAST hits and corresponding GO annotations. The EST assemblies are accessible via http://www.plantgdb.org/

Table 6-1 PlantGDB unique transcript (PUT) assembly.

Species	Total	Assembled	Total number of PUTs
Zea mays	2,099,908	1,293,797	181,717
Glycine max	1,440,234	1,379,625	259,849
Oryza sativa japonica	1,025,015	1,015,010	146,642
Phaseolus coccineus	391,153	352,161	221,903
Medicago truncatula	273,411	264,186	65,628
Lotus japonicus	246,295	246,249	56,525
Vigna unguiculata	187,637	183,799	34,736
Phaseolus vulgaris	84,238	77,093	21,361
Cicer arietinum	34,542	34,349	14,386
Glycine soja	18,548	18,523	11,332

As compared to the EST repositories of the model legumes *Medicago truncatula* and *Lotus japonicus*, more so the highly researched soybean, rice and maize, the volume of EST sequences for the cool season food legumes in the databases still is poor, in some cases such as lentil (see Chapter 3), practically non-existing, mainly because EST sequencing is expensive and the latter legumes are systemically under-researched. Also, ESTs are far from perfection: they represent a mere sample of the complete transcriptome, contain a high sequencing error rate (up to 1–3%), are relatively short (average: 400 bp), contain artifacts such as vector and bacterial sequence contaminants, and represent only 5′- and/or 3′-ends of transcripts. Moreover, splice variants involving exons located in the center of long transcripts are underrepresented in dbEST depositories.

Among the plethora of other databases the TIGR Plant Transcript Assemblies bank is frequented (http://plantta.jcvi.org/). The plant transcript assemblies (TAs) in this data assembly are expressed transcripts collected from dbEST and NCBI GenBank nucleotide database (full length and partial cDNAs). Only plant species for which more than 1,000 ESTs or cDNA sequences are available, are included. Other data bases are restricted to species, still others represent a collection of e.g., species-oriented full-length cDNAs (search for e.g., www.rsoy.psc.riken.jp/ or www.pgb.kazusa.or.jp/kaftom/ or www.amber.gsc.riken.jp/cassava).

Field pea (*Pisum sativum* L.), lentil (*Lens culinaris* Medik.), faba bean (*Vicia faba* L.), chickpea (*Cicer arietinum* L.), and grass pea (*Lathyrus sativus* L.) are collectively known as cool season food legumes, the sequence information in the databases becomes scarce, as already outlined above. However, some genic markers were developed and mapped onto a genetic map of chickpea earlier on (Hüttel et al. 2002; Benko-Iseppon et al. 2003; Pfaff and Kahl 2003). More importantly, a collection of more than 2,800 ESTs to the already existing minimal number of chickpea ESTs secured a moderate annotation of transcripts from this crop, the most prevalent transcripts putatively encoding proteases, T6P synthase, non-specific lipid transfer proteins, MRP-like ABC transporters, HSP70 chaperone, TCP-1-alpha, bZIP transcription factor, calcium ATPases, protein kinases, MRP4 glutathione-conjugate transporter, glutathione S-transferase, phosphoenol pyruvate carboxylase, and S-adenosyl methionine synthetase (Buhariwalla et al. 2005; Jayashree et al. 2005). Also, suppression subtractive hybridization (SSH) libraries (Diatchenko et al. 1996), which per se do not allow the quantification of expression of differentially expressed genes, were used to generate macroarrays for an analysis of gene expression (Boominathan et al. 2004). The same approach held for e.g., *Lathyrus sativus*, to name an example (Skiba et al. 2005). However, the relatively laborious technique does not yield the number of expressed genes that is needed (101 dehydration-responsive transcripts in chickpea roots). Later on, the same authors enriched the EST resource by the generation and analysis of drought- and salinity-responsive ESTs. A total of 20,162 (18,435 high quality) such ESTs from 10 different root tissue cDNA libraries of chickpea were sequence-edited, clustered and assembled into 6,404 unigenes (1,590 contigs and 4,814 singletons). Unigene annotation via BLASTX revealed that 46.3% (2,965) had significant similarity to sequences in the non-redundant UniProt database. BLASTN analysis of unique sequences with ESTs from the four other legumes *Medicago*, *Lotus*, soybean and groundnut, and the model plant species rice, *Arabidopsis* and poplar discovered conserved genes across legumes and novel chickpea transcripts. Of 2,965 unigenes, only 2,071 could be functionally categorized via gene ontology (GO) terms (see Varshney et al. 2009). Also other groups contributed to the EST sequence collection for chickpea (Gao et al. 2008; Ashraf et al. 2009; Jain and Chattopadhyay 2010), and still others exploited the EST data bases for *Cicer* [e.g., for the development of EST-SSR markers and the detection of allelic variation in related species (Choudhary et al. 2009)]. A web page was opened some time ago by scientists from ICRISAT, the ICRISAT chickpea EST database (http://www.icrisat.org/gt1/cpest/home.asp). Additionally, EST arrays (for the initial idea see Richmond and Somerville 2000), the ordered alignment of different ESTs on supports of small dimensions (e.g., nylon membranes, glass or quartz slides, silicon chips), allow the simultaneous detection of thousands of expressed genes in a particular cell, tissue, organ, or organism at a given time by hybridization of fluorochrome-labeled cDNA preparations to the array. Any hybridization event between an EST and cDNA is then detected by fluorescence. Most use of ESTs is made of their polymorphisms (ESTPs). ESTPs describe any difference in DNA sequence between two (or more) ESTs that can be detected by either restriction digestion of the ESTs, or by separation of polymorphic sequences using denaturing gradient gel electrophoresis. ESTPs can be used to screen for DNA polymorphisms in populations, or serve as markers in mapping and comparative mapping procedures, generating EST maps, graphical depictions of the positions of (preferentially all) ESTs (representing expressed genes) along a chromosome or, generally DNA, that is generated by EST mapping.

ESTs are usually clustered to derive minimal sets of tentative consensus sequences (TCs) representing virtual genes. Based on the EST distribution, in silico analyses ("electronic Northerns") can detect collections of genes with a potentially differential expression. In particular for species where a whole-genome sequence is not available, TCs are not only good for gene identification, but also constitute a valuable resource for the construction of expression microarrays. So, the use of ESTs is quite rewarding for these applications. However, in general, EST sequencing is still very costly, and even the most comprehensive EST sequence bank will not cover all possible transcripts and their isoforms. Therefore, tag-based approaches to transcriptome profiling, emerging over the last two decades, are favored.

One of the most successful transcriptome technologies are **microarrays**, microscale solid supports (e.g., nylon membranes, nitrocellulose, glass, quartz, silicon, or other synthetic material), onto which either DNA fragments, cDNAs, oligonucleotides, genes, open reading frames, peptides or proteins (e.g., antibodies) are spotted in an ordered pattern ("array") at extremely high density. Such microarrays (laboratory jargon: "chips") are increasingly used for high-throughput expression profiling. There are hundreds of different microarray formats existing, so that any description of all the variants becomes inhibitory. For expression genomics, the most commonly employed microarray platforms are all-exon arrays, antisense genome arrays, cDNA expression microarrays, chromosomal region expression arrays, double-stranded DNA microarrays, EST arrays, exon arrays, exon junction microarrays, exon tiling arrays, expression arrays, gene expression microarrays, gene interaction arrays, high-density oligonucleotide arrays, microRNA microarrays, nucleosomal arrays, proximal promoter arrays, RNA expression microarrays, splice oligonucleotide arrays, and splice variant monitoring arrays, to name a few. Microarrays can further be specified by their physico-chemical features. For example, bead-based arrays, bioelectronic arrays, cantilever arrays, diffusion sorting arrays, electrochemical microarrays, electronic microarrays, electrophoresis chips, enamel chips, entropic trap arrays, evanescent resonator chips, fiber bead arrays, fiber-optic DNA arrays, filter arrays, flow-through chips, microcantilever arrays, microchips, microelectrode arrays, microelectronic arrays, microelectrophoresis chips, microfluidics chips, microsphere arrays, microtube microarrays, nanoarrays, nanoparticle chips, nanowire sensor arrays, planar waveguide chips, PNA arrays, reverse capture microarrays, suspension arrays, and ultra-high density microarrays can be discriminated from each other by their specific design and detection physics or chemistry.

6.2.1 *Microarray-Based Expression Profiling*

Microarray-based expression profiling (transcript profiling) for the simultaneous detection of the expression of thousands of genes (the so-called expressome), whose complementary DNA sequences are immobilized on the array, requires the hybridization of labeled cDNAs from target tissue(s). In short, the poly(A)$^+$- mRNAs of two (or more) tissues are separately reverse-transcribed into cDNAs using α-^{32}PdATP or α -^{32}PdCTP (or e.g., fluorophore-labeled nucleotides) and reverse transcriptase. These labeled cDNA probes are then hybridized to the microarray (following standard procedures described by the microarray provider). After hybridization and high-stringency washing, the hybridization patterns can be visualized by autoradiography and quantified by phosphorimaging (or fluorescence detection and intensity measurement of hybridization events).

Very early on, planar low-, medium- and high-density microarrays have been integrated into transcription profiling schemes in legumes. In the course of the European Grain Legumes Integrated Project (GLIP), entitled New Strategies to Improve Grain Legumes for Food and Feed (www.eugrainlegumes.org) aiming at developing new strategies to enhance the use of grain legume crops for animal feed in Europe, macroarrays of very low density were designed (e.g., a 2.5 k cDNA macroarray representing transcript sequences from arbuscular mycorrhizal roots, and a 6k macroarray from root interactions, designed as Mt6k-RIT). On the basis of these macroarrays, microarrays of initially low density (e.g., 8 k, called Mt8k) evolved, later on superseded by medium density (16 k, means 16,000 genes; Mt16kOLI1Plus microarray) and quasi-high density microarrays (22 k), that altogether were predominantly used to study expression profiles at different stages of root nodule development. The synthesis of a 16 k 70mer oligonucleotide collection representing all TCs from the TIGR *M. truncatula* Gene Index version 5 was later extended to encompass 384 probes targeted against transcription factors and other regulators. These tools are referred to as Mt16kOLI1 and Mt16kOLI1Plus, respectively. Based on the average gene density in the *M. truncatula* genome, one estimates that about 35–40% of all *M. truncatula* genes are represented on Mt16kOLI1Plus microarrays. Also, companies joined in and a 70-mer oligonucleotide microarray has been marketed by Affymetrix (51k *Medicago* GeneChip).

In combination with subtractive hybridization techniques, a series of 826 regulatory genes in root apexes under salt stress were identified. Among them, 95 sequences (11.5%) corresponded to transcription factors (TF). In summary, 107 TF-encoding genes were identified, including members of the HD-ZIP, NAC/NAM, WRKY, Dof Zn-finger, MYB, EREBP and PR/ERF families, as well as several new elements like receptor kinases and hormone receptors, that all were supposed to be linked to the management of salt stress (Küster et al. 2004; Manthey et al. 2004; Parniske 2004; Frenzel et al. 2005; Harrison 2005; Hohnjec et al. 2005, 2006; Küster et al. 2007b). These microarrays were additionally employed in many different aspects of plant development such as differentiation, flower and pod formation (Firnhaber et al. 2005), drought tolerance in seeds of *M. truncatula* (Buitink et al. 2006), nodulation (El Yahyaoui et al. 2004; Geurts et al. 2005; Lohar et al. 2005; Stacey et al. 2006; Küster et al. 2007a), and expression profiling in *Medicago sativa* (Tesfaye et al. 2006).

Large-scale cDNA macro- and microarrays were also used to better understand the molecular responses of *Lotus japonicus* plants to arbuscular mycorrhizal (AM) fungi. For example, Deguchi et al. (2007) analyzed the differential gene expression patterns of *Lotus* and the colonizing fungus in the presence of contaminating microorganisms in the soil inoculants. After the colonization by AM fungi, i.e., *Glomus mosseae* and *Gigaspora margarita*, was established, four cysteine protease genes were induced. *In situ* hybridization revealed that these cysteine protease genes were specifically expressed in arbuscule-containing inner cortical cells of AM roots.

Microarrays of various designs also played a substantial role in transcriptional profiling of responses of cool season food legumes to biotic stresses. For example, the transcriptional responses of chickpea genotypes with different levels of resistance to *Ascochyta rabiei* were probed with salicylic acid (SA), methyl jasmonate (MeJA) and aminocyclopropane carboxylic acid (ACC) and microarrays. These 727-feature arrays consisted of 559 chickpea cDNAs (Coram and Pang 2005a, b, 2006), 156 grass pea cDNAs, and 12 controls. Overall, 425 transcripts were differentially expressed. Generally, the *A. rabiei*-resistant chickpea genotypes responded to an *A. rabiei* infection with a more substantial and rapid induction of defense-related genes (as for e.g., genes encoding PR proteins, proteins managing oxidative burst, proteins involved in cell wall support and antimicrobial activity, and many proteins with unknown functions) under all situations, which was interpreted as a stronger potential in defense against the fungal pathogen. These experiments were validated with qPCR (Coram and Pang 2007; see also Brian et al. 2007). For pea (*Pisum sativum*), a 70-mer oligonucleotide microarray (Ps6kOLI1) was developed, that (in contrast to Mt16kOLI1Plus microarray) is based on a collection of cDNAs from various tissues, and that was primarily designed to identify genes relevant for seed formation.

Ascochyta blight resistance in lentil (*Lens culinaris*, see Ford et al. 2007) has also been probed in the highly resistant (ILL7537) and highly susceptible (ILL6002) lentil varieties inoculated with *Ascochyta lentis*. Ninety genes were differentially expressed in ILL7537, and 95 genes were differentially expressed in ILL6002. The expression profiles of the two varieties were substantially different, both as the type of genes and the temporal expression and level of expression in response to the pathogen is concerned. The resistant variety boasted early up-regulation of for e.g., PR4 and 10 proteins and other defense-related genes. The susceptible genotype showed early down-regulation of defense-related genes. Real-time RT-PCR verified the microarray expression ratios (Mustafa et al. 2009).

Without elaborating further on the employment of microarrays for gene expression profiling under various situations of legumes, cool season food legumes in particular, it is commended that microarrays played a pioneering role in transcriptomics of these plants, and beyond (Rensink and Buell 2005). However, their role in transcriptomics is fading (see Section 6.2.2 and Shendure 2008).

In addition to microarrays that rely on nucleic acid hybridization, **high-throughput real-time RT-PCR** (qPCR) is nowadays a popular alternative that, if controlled properly, allows an ultra-sensitive measurement of transcription by using gene-specific primers and probes in a PCR-based assay of up to 384 samples at a time, or even more, if multiplexing is possible (Bustin et al. 2009). Specific requirements in qPCR have led to the publication of a wealth of papers discussing various parameters for a secure protocol and recommending improvement of existing procedures. For example, normalization and the choice of reference genes are technical problems (Hugget et al. 2005; Hu et al. 2009; Garg et al. 2010), and a persisting difficulty is the RNA template integrity (Fleige and Pfaffl 2006; Fleige et al. 2006). Golden rules for qPCR were developed (Udvardi et al. 2008) and so-called MIQE (pronounced "mykee") guidelines designed and published (Bustin 2002, 2010; Bustin and Nolan 2004; Bustin et al. 2005, 2009; Nolan et al. 2006; Derveaux et al. 2010). The minimum information for publication of MIQE guidelines portray a set of rules for the precise and reliable performance and standardization of quantitative RT-PCR experiments, the appropriate data analysis, and the minimum technical information required for publication of the results. In short, the experimental design, collection and handling of the sample, nucleic acid extraction from the sample, reverse transcription, qPCR target information, qPCR oligonucleotides, the qPCR protocol, qPCR validation and data analysis are described in detail, and subordinate to analytical specificity, accuracy, repeatability (intra-assay variance), and reproducibility (inter-assay variance). With all these guidelines available, qPCR experiments are now matured into daily experiences, and are, for example, necessary for the validation of SuperSAGE experimental results in chickpea (e.g., Molina et al. 2008). Also, specific constitutively expressed genes as reference genes for qPCR have been identified for this crop (Garg et al. 2010).

Among other techniques, qPCR platforms have been used to quantify gene expression in legumes. Here only one approach is described briefly. However, no cool season food legume, except for *Medicago truncatula* as a model legume was employed. The approach aimed at characterizing some **transcription factors** (TFs).

Transcriptomic studies, using arrays of cDNAs or oligonucleotides, have almost automatically found TFs in *M. truncatula*. Actually, the first discovered and functionally characterized legume TF gene was pea *FLO*, isolated via sequence homology to *FLO* and *LFY* of snapdragon (*Antirrhinum majus*) and *Arabidopsis*, respectively. Both TFs control floral development in snapdragon and *Arabidopsis*, and the aberrant floral and leaf development in the pea *unifoliata* (*uni*) mutant were consequently attributed to a defect in pea *FLO* (Hofer et al. 1997). Other TFs were characterized the same way (e.g., Tattersall et al. 2005). In many cases, however, bioinformatics has had a major impact on TF discovery. Putative TF genes can be identified by DNA sequence motifs within the gene that encode known DNA-binding domains, using BLAST searches (Guo et al. 2005; Iida et al. 2005). Specific databases, as for e.g., the International Medicago Gene Annotation Group (IMGAG) dataset, additionally assist the search for sequences encoding DNA-binding and other TF domains and therefore these are candidate sequences. Once the sequences are known, gene-specific primers can be designed that allow high-throughput quantitative reverse transcription (qRT)-PCR analysis of TFs.

Besides the more bioinformatically driven approaches, also degenerate primers complementary to specific TF domains have been designed and tested in qPCRs on legume DNA, leading to the identification of sequences corresponding to TF-encoding genes (Chern et al. 1996a, b; Heard et al. 1997; Zucchero et al. 2001; Tucker et al. 2002).

In the context of this review it is noteworthy, that several legume TFs are involved in abiotic stress tolerance. For example, alfalfa *Mszpt2-1* was induced by salt in roots. If *Mszpt2-1* is down-regulated by antisense RNA, increased sensitivity of transgenic plants to salinity ensues (Merchan et al. 2003; Zhang et al. 2005). These few examples do not reflect the abundant literature on *M. truncatula* TFs and their genes, but are taken as proof that research on a specific class of genes and the encoded proteins, here the important TFs, provides insights into the regulation of cognate genes. This may be taken as an example, of how the massive data from the high-throughput transcriptome technologies can be processed to further functional genomics of legumes (see the excellent review on TFs of legumes by Udvardi et al. 2007), and also lead to the establishment of a gene expression atlas (Benedito et al. 2008).

It can be expected that qPCR will remain an important technology for legume transcriptomics and in future be converted into a high-throughput format so as to meet the increasing needs of experimenters.

6.2.2 The Dawn of Microarrays and Advancement of Tag-based Expression Platforms

By all their virtue, microarrays in whatever format suffer from a series of devaluating insufficiences. In fact, the poor correlation between different microarray platforms stands out (relatively large differences in data obtained in different laboratories using the same platform), but - equally important - its closed architecture format allows to detect only transcription of genes that are spotted on the array. Therefore, microarrays cannot detect novel genes. They require large amounts of input RNA for robust answers, are at the most semi-quantitative and at their best with more abundant mRNAs. Microarrays are also prone to cross-hybridization of a single probe to different target RNAs, and the experimenter has no reliable predictor for on-chip hybridization efficiency. Ambiquity exists in data analysis and interpretation, and in some cases defective oligonucleotides prior to printing were reported. The widely different fluorescence intensity signals generated by different probes targeting the same gene confuses the experimenter. More information on this important aspect of microarrays can be consulted in numerous references (Yauk et al. 2004; Barnes et al. 2005; Larkin et al. 2005; Canales et al. 2006; De Reynies et al. 2006; Koltai and Weingarten-Baror 2008; only few of a plethora of publications). While many of these disadvantages can be at least partly overcome, the general value of microarrays surely persists in only two areas. First, the isolation of interesting sequences by hybridization (**sequence capture**), especially exome sequence capture. This is the enrichment of (preferentially) all exons of a genome prior to their sequencing by one of the next-generation sequencing technologies. In short, genomic DNA is first isolated and purified, then fragmented and the ends polished. Linkers are subsequently ligated onto the polished ends, and the fragments hybridized to a microarray containing all the exons (as spotted oligonucleotides) of the target genome. After washing away the unbound fragments, the bound (target) fragments are eluted, and the exome-enriched capture library directly sequenced (as an example see Nautiyal et al. 2010). Second, the design of **focussed microarrays**, that contain only previously identified target sequences and serve to screen whole populations or cohorts for expression of these genes. Also for cool season food legumes, these attributes were exploited recently (P. Winter, pers. comm.).

All these inadequacies of microarray technologies led to the development and perfection of **alternative gene expression technologies**, that are open (do not need any previous sequence knowledge), high-throughput (can be performed in massively parallel format) and are truly quantitative (digital readout of transcript numbers). For an illustrative review on this topic consult Forrest and Carninci (2009). One of the pioneering techniques of this kind was coined serial analysis of gene expression (SAGE), a term that persisted over almost two decades (Velculescu et al. 1995, 1997, and myriads of subsequent publications). SAGE is a high-throughput technique for the simultaneous detection, identification and quantization of virtually all genes expressed in a given cell at a given time, that additionally allows to identify unknown genes, novel genes, up- or down-regulated genes, to monitor patterns of gene expression at various developmental stages and define disease marker transcripts. SAGE outperformed the traditional EST (Sun et al. 2004) and microarray analysis as well (for example, Lu et al. 2004; Dinel et al. 2005; Griffith et al. 2005; Ibrahim et al. 2005), and is based on the isolation of a short, 9–14 bp so-called SAGE tag from a defined location within a transcript, that contains unique and sufficient information to identify specifically this transcript ("diagnostic tag"). Such tags from various transcripts were originally concatenated serially into a single long DNA molecule for efficient sequencing and for identification of the multiple tags simultaneously. The expression pattern of any transcript population can be quantified by determining the abundance of individual tags and identifying the gene corresponding to each tag. The sequence data is analyzed by special software to identify each gene expressed in the cell, and to determine its expression level. Today, the technique of concatenation and cloning is superseded by simply ligating sequence adaptors ("linkers") to both ends of the tags, and direct sequencing of millions of tags by one of the next-generation sequencing technologies. Without meandering into technical details, SAGE initiated the design of a series of improvement and led to the development of complementary technologies. For example, long serial analysis of gene expression (LongSAGE, LS) has been designed, that generates 19–21 bp tags derived from the 3'-ends of messenger RNAs (mRNAs) rather than the 14 bp in the original SAGE protocol. Simply linkers containing an *Mme*I recognition site are ligated to the 3'ends of the cDNAs. Linker-tag molecules are then released from the cDNA using the type IIS restriction enzyme *Mme*I. The resulting tags are then directly ligated with DNA ligase. Tag concatemers are sequenced, and the longer tags analyzed and matched to genomic sequence data. Matching of tags to genomic sequences in many cases allows localization of genes, from which the tags ultimately are derived (Saha et al. 2002; Wahl et al. 2005). More recently, we developed SuperSAGE (Matsumura et al. 2005a, b; also called HT-SuperSAGE by Matsumura et al. 2010). SuperSAGE also allows the genome-wide and quantitative gene expression profiling of cells, tissues, organs and organisms. SuperSAGE basically follows the original SAGE protocol, but involves the type III restriction endonuclease *Eco*P15I, that cleaves the cDNA template most distantly from its recognition site. Therefore, the resulting tags are 26 bp long, and much longer than the tags from traditional SAGE (13–14 bp) or LongSAGE (19–21 bp). The advantages of SuperSAGE are twofold. First, the information content of a SuperSAGE tag ("SuperTag") of 26 bp is higher than the conventional tags and allows to identify a gene directly from the Genbank databases. Second, the ends of linker-tag fragments generated by SuperSAGE are blunt-ended to ensure random association of the tags to form ditags. SuperSAGE has the additional benefit of discovering host and pathogen messages simultaneously from the same infected material without physically separating them, as was mandatory in the past (Matsumura et al. 2005a, b; 2006, 2007, 2008; Terauchi et al. 2008).

A jungle of very similar, partly competing and partly advanced techniques derived from the original SAGE bewilder the innocent beginner. From the whole spectrum of different SAGE methods only a reference to few modifications should suffice: microSAGE (Datson et al. 1999), miniSAGE

(Ye et al. 2000), SAGE-LITE (Peters et al. 1999), small amplified RNA-SAGE (SAR-SAGE, Vilain et al. 2003), robust LongSAGE (Gowda et al. 2004), 5′-SAGE (Hashimoto et al. 2004), 3′-SAGE, deep SAGE (Ruzanov and Riddle 2010), but also techniques with spin-off titles such as cap analysis of gene expression (CAGE; Kodzius et al. 2006) and its variants nanoCAGE and CAGEscan (Plessy et al. 2010), sequence tag analysis of genomic enrichment (STAGE; Kim et al. 2005), serial analysis of chromatin occupancy (SACO; Impey et al. 2004), and serial analysis of binding elements (SABE; Chen 2009), among many others crowd the relevant literature. For a review of the older techniques the reader is referred to Lorkowski and Cullen (2003), tag-based techniques of transcription profiling is dealt with in detail by Harbers and Carninci (2005), more comprehensive information on SAGE and its variants together with protocols and applications can be found from http://www.sagenet.org/.

Other competing high-throughput transcript profiling technologies were designed, but did not survive. A good example is massively parallel signature sequencing (MPSS), capitalizing on the sequencing of millions of cDNAs conjugated to oligonucleotide tags on the surface of 5 mm diameter microbeads, that avoids separate cDNA isolation, template processing and robotic procedures. In short, 32-mer capture oligonucleotides are attached to the surface of separate microbeads by combinatorial synthesis, such that each microbead has a unique tag for its complementary cDNA. Then messenger RNA is reverse transcribed into cDNA using oligo(dT) primers, restricted at both ends with for e.g., *Dpn*I, complements of the capture oligonucleotides are attached to the poly(A) tail of each cDNA molecule and the construct cloned into an appropriate vector containing PCR handles, which serve as primer-binding sites for PCR-based amplification of the tagged cDNA. The cDNA is now amplified with a fluorochrome-labeled primer, denatured, and the single-stranded address tag-containing fragments annealed ("cloned") to the surface of microbeads containing address tag sequences as hybridization anchors, and then ligated ("*in vitro* cloning"). Each microbead displays about 100,000 identical copies of a particular cDNA ("microbead library"). The fluorescent microbeads (all containing a cDNA) are then separated from non-fluorescent ones (not containing a cDNA) by a fluorescence-activated cell sorter (FACS). Each single microbead in the library harbors multiple copies of a cDNA derived from different mRNA molecules. If a particular mRNA is highly abundant in the original sample, its sequence is represented on a large number of microbeads, and vice versa. In the original version of MPSS, 16–20 bases at the free ends of the cloned templates on each microbead are sequenced ("signature sequences"). First, millions of template-containing microbeads are assembled in a densely packed planar array at the bottom of a flow cell such that they remain fixed as sequencing reagents are pumped through the cell, and their fluorescence can be monitored by imaging. Then the fluorophore at the end of the cDNA is removed, and the sequence at the end of the cDNA determined in repetitive cycles of ligation of a short adaptor carrying a restriction recognition site for a class IIS restriction endonuclease (binding within the adaptor and cutting the cDNA remotely, producing a four nucleotide overhang; e.g., *Bbv*I). Next, a collection of 1,024 specially encoded adaptors are ligated to the overhangs, and the coded tails interrogated by the successive hybridization of 16 different fluorescent decoder oligonucleotides. This process is repeated several times to determine the signature of the cDNA on the surface of each bead in the flow cell. The abundance of each mRNA in the original sample is estimated by counting the number of clones with identical signatures (Brenner et al. 2000; Reinartz et al. 2002). We detail this procedure to show that the complex protocol (which could only be processed by a company) prohibited its adoption by the scientific community (exception: Meyers et al. 2004 a, b). Consequently it was no longer used. However, interested researchers can use other transcription profiling techniques (Kim et al. 2007; Morrissy et al. 2009 to name a few).

Here, we will capitalize on SuperSAGE or its variant deepSuperSAGE only, because it has proven potential for ultra-deep transcriptome analysis of cool season food legumes. The latter, deepSuperSAGE, is in fact the combination of the powerful SuperSAGE technique with now- and next-generation sequencing, which produces massive tag data (up to 5 millions of tags per situation) and warrants a deep transcriptome analysis. Therefore, we first introduce some of these techniques in some detail to illustrate their potential.

To begin with, our deepSuperSAGE technology in a comparison to predecessor tag-based transcriptome profiling methods like SAGE and LongSAGE is preferred. As can be recalled, SAGE produces tags only 14 bp in length, and the improved LongSAGE protocol works with tags of 18 bp, both much too short to allow unequivocal annotation of tags to sequences in the databases. Figure 6-1 compares an autoBLAST of libraries with tags of different lengths. As it is obvious (see black, blue, and also red columns), the 26 bp SuperTags are superior to the shorter SAGE or RL-SAGE fragments, because longer tags increase the probability to discriminate highly similar transcripts within the same dataset. In libraries consisting of 26bp tags (as in deepSuperSAGE), more than 70% of the UniTags have no other highly similar BLAST hit within the own dataset, whereas in normal SAGE libraries (9–14 bp long tags) this proportion is reduced to well below 50%. Thus, very important information about e.g., transcript variants is lost when short tags are used.

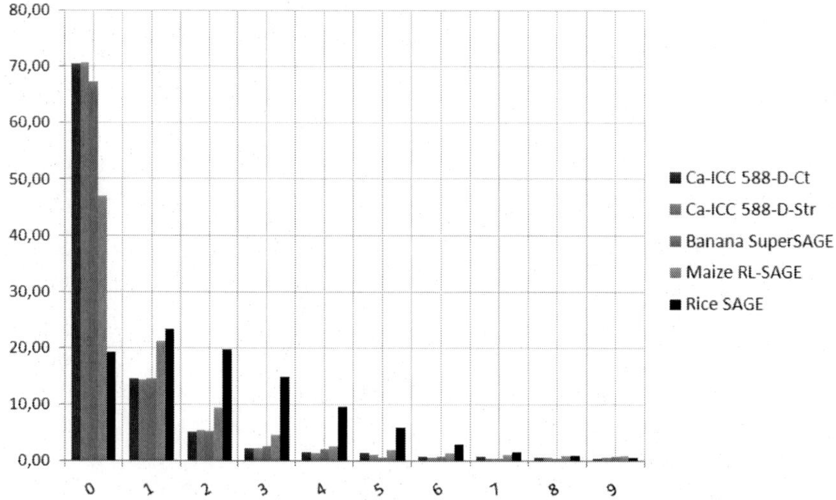

Figure 6-1 Resolution of SuperSAGE and other tagging techniques. Proportion of total UniTags (ordinate) versus number of similar hits for a given tag (abscissa) after self-BLASTing of the chickpea ICC588 datasets (blue columns). Results from ICC588 were compared to other self-BLASTed SuperSAGE as well as Long-SAGE and SAGE libraries retrieved from the NCBI gene expression omnibus. Close to 70% of the SuperSAGE tags did not find high similarity hits, whereas 15 and 4% found high similarity hits (> 20 bp) with one and two other UniTags, respectively, within the own datasets. 0: Percentage of UniTags with low similarity hits within the own dataset; 1–9: Number of similar hits for a given UniTag (in percentage of total tags).

Color image of this figure appears in the color plate section at the end of the book.

6.2.3 The Impact of Next and Next-Generation Sequencing Technologies on Legume Transcriptomics

The advance in the knowledge of biosciences and elsewhere is intimately linked to an advance in technology. Our knowledge of the genome and transcriptome of many organisms literally exploded during the past decade, largely catalyzed by an unprecedented boost of new and novel, in essence advanced technologies. For example, SuperSAGE was recently merged with one of the next-generation sequencing technologies, and the result, coined deepSuperSAGE, illustrates well the advancement of genome science (see below).

A paradigm for this advance is the revolutionary development of several DNA sequencing technologies, that altogether have superseded the traditional, but not at all ineffective Sanger sequencing. In fact, the dideoxy Sanger method is still being used worldwide, but competing techniques evolved quickly, and are still evolving. Amongst them, basically two classes of sequencing procedures contributed to the market value: the longer read Roche 454 FLX sequencer with its Titanium chemistry on one, and the shorter reads sequencers from Illumina (GAII) as well as ABI's SOLiD system. These platforms were initially called next-generation sequencers, but then titled now-generation sequencing machines, to cope for the speed of change. The once **next-generation sequencing** (NGS; next generation sequencing technology, nextGen DNA sequencing) with the potential to sequence a human genome for 100.000, or even only US $ 1.000, that are not based on the conventional Sanger dideoxy sequencing procedure (Sanger et al. 1977), relies on extremely high-throughput procedures, mostly based on massively parallel reactions. For example, in sequencing by oligonucleotide ligation and detection (SOLiD™), each run produces at least 700 million reads, covering 40 billion bases. Next-generation sequencing technologies fall into two broad categories: clonal cluster sequencing, and single molecule sequencing. A novel category is being developed right now, coined third-generation sequencing platforms.

In this limited context, only two widely used next-, or better now-generation sequencing technologies are briefly portrayed. First, the **fiber-optic reactor sequencing** marketed by ROCHE 454 Life Sciences for the rapid and highly parallel sequencing of any DNA in multiple samples by combining the capture of genomic DNA fragments on beads with a variant of the emulsion-polymerase chain reaction (emPCR) for the isolation and *in vitro* amplification of these DNA fragments and their simultaneous sequencing by a variant of the pyrosequencing procedure (Margulies et al. 2005). In short, entire genomes are first isolated, randomly fragmented by nebulization ("shotgun fragmentation") into pieces of 500–1,000 bp, which are polished (blunt-ended). These blunted fragments are then ligated to short, specially designed common adaptors A and B, that contain sequences complementary to primers for subsequent amplification and sequencing steps. Adaptor B harbors a 5′-biotin tag that allows the immobilization of the library onto streptavidin-coated beads. After nick repair, the non-biotinylated strand is released to form a single-stranded template DNA library (sstDNA library). Following a purification and quantization step, individual single-stranded DNA fragments are bound to beads by limiting dilution (favoring the binding of only one single fragment per bead), and the individual DNA fragments clonally amplified within droplets of an oil emulsion by emPCR. These droplets contain all amplification reagents (including DNA polymerase) and act as closed microreactors. This step results in millions of beads each carrying 10–20 millions of copies of a unique DNA template. Subsequently, the emulsion is broken, and the released beads with the single-stranded DNA "clones" deposited into the wells ("picolitre reactors") of a 70 x75 mm fiber-optic slide by centrifugation. The diameter of the 3.6 million wells of a single slide each permits accommodation of only one single bead of 28 μm diameter per well. About 90% of the beads do not carry DNA. Therefore, an enrichment step for DNA-bound beads follows. Then smaller beads with immobilized enzymes for the pyrosequencing protocol (i.e., ATP sulfurylase and luciferase) are added to the wells, and the sequencing reaction started by pumping deoxynucleotides across the reactor array. Extension reactions can then simultaneously occur on all the bead-bound single strands in the open wells by convective and diffusive transport of the nucleotides. After the flow of one nucleotide, the panel is washed with solution containing apyrase (degradation of residual nucleotide triphosphates) prior to the addition of the next nucleotide, and the waste collected in a receptacle. The nucleotides are added in the series T→A→C→G. Nucleotide incorporation occurs, when the template strand carries the complementary base, and results in the release of inorganic pyrophosphate and the generation of photons. These photons emitted from the bottom of each well are captured by fiber-optic imaging bundles bonded to a large format charge-coupled device (CCD) camera. The images are processed into sequence information simultaneously for all wells with beads carrying template DNA. Normally, sequence reads comprise 400 bases with at least 99% accuracy in a single run, generating over 400 million bases with a Phred quality score of 20 or more. The resulting sequences are then assembled by powerful software of an onboard computer. For example, the assembler consists of various modules, of which the Overlapper identifies overlaps between different reads, the Unitigger constructs larger contigs of overlapping sequences, the Multialigner generate consensus calls and quality scores for each base in each contig, and the FlowMapper maps individual reads to a reference genome. Fiber-optic reactor sequencing avoids robotics for colony picking and for handling of microtiter plates, cloning into bacterial vectors and sub cloning, and processing of individual clones. The technique allows to sequence bacterial and lower eukaryotic genomes in days, to identify open reading frames, to compare the sequenced genomes with other sequenced genomes of the same species, to identify conserved sequence elements, mutational hotspots and rare mutations (see the seminal publication by Margulies et al. 2005, that was followed by a myriad of reports on different applications of the technology). On the other extreme, SOLiD, based on sequencing by ligation for the massively parallel sequencing of clonally amplified DNA fragments, generates DNA by measuring the serial ligation of oligonucleotides. In short, SOLiD starts with genomic DNA that is first fragmented by random shearing. The resulting fragments are size-selected, synthetic oligonucleotide adaptors ligated onto their ends, and the ends brought into close proximity by re-circularization on the adaptor. A series of steps then generate a so-called mate-paired library, which is clonally amplified by emPCR on paramagnetic beads supported with primers and all reaction components in 34 femtoliter reactors. After emPCR, the clonally amplified templates are denatured, 3′-modified and covalently bound to a glass slide. These so-called bead arrays are then placed into an automated flow cell, where primers hybridize to the adaptor sequences. Then a set of four-color fluorochrome-labeled 8-mer oligonucleotide probes, comprising all 16 potential combinations of two neighboring bases, are present simultaneously and compete for incorporation. The complementary (fitting) probes are ligated, the fluorescence signal is measured, the probe surplus removed by washing, the fluorophore removed from the incorporated probe, and the next set of probes added for another round of ligations. These probes are offset by one base in the 5′-direction. This procedure of serial stripping and sequencing the template with 5 offset sequencing primers is repeated several fold to generate 25 base pairs of sequence. Mate-pair sequencing requires additional determination of the bases of the second tag. Every fourth and fifth base of the oligo probes is interrogated. This procedure determines all bases of a template with accuracy (based on a two-base encoding) and generates data of at least 700 million sequence tags (read length: 50 bp) in a single run. Mate-pairs can then be mapped onto known genome sequences, and pairs mapping in the same orientation ("normal mate-pairs") used to produce a scaffold of the genome. Mate-pairs, that map in reverse polarity and/or otherwise differ from expectation, can be used to identify and localize structural rearrangements of the genome. SOLiD is used for large-scale resequencing, expression analysis, genome methylation detection, and Chromatin ImmunoPrecipitation (ChIP) experimentation (see Nutter 2008).

Variations in the techniques add to their versatility. An example: **mate-pair sequencing**. This term describes the determination of the sequence of about 100 bp flanking each side of a loop of genomic DNA up to 30 kb in length by next-generation sequencing. In short, genomic DNA is first isolated and purified, then randomly fragmented by shearing (specifically hydroshearing) or sonication and the resulting fragments (laboratory slang: "frags") size-selected (into 3, 8, 12, or 20 kb) by gel electrophoresis. The sized fragments are then end-polished, and a specific biotinylated hairpin adaptor ligated onto the polished ends. Then the adaptor-positive fragments are enriched by streptavidin capture, and circularized, which is facilitated by the adaptor sequences. The circularized DNA is again fragmented, adaptors ("library adaptors") ligated to the ends of each fragment after end-polishing, and adaptored frags isolated and sequenced. This procedure generates paired-end reads with two end-tags flanking genomic DNA of either 3, 8, 12, or 20 kb in length. Paired end reads improve the assembly of shotgun reads, allowing to order contigs into larger scaffolds and to span regions with extended repeats.

There is no doubt, that whatever NGS procedures contributed to a leap in sequencing depth and a welcome reduction in costs, also compared to updated microarray technologies (see for e.g., Bradford et al. 2010). This is amply witnessed by a whole series of reviews and critical overviews of the field (e.g., Mitchelson 2007; Mardis 2008, to name a few). NGS is also possible with low input material and with single cells (e.g., mouse oocytes; Tang et al. 2010). As another example for the progress made by the advent of these technologies, we list the number of deepSuperSAGE tags sequenced per situation over a time of only six years (Table 6-2). It becomes apparent that the time of the good old gene-by-gene analyses as a starting point in transcriptomics is history.

Table 6-2 The impact of next-generation sequencing technologies.

Species	Target	Number of Tags[b]	Partner
Rice/*Magnaporthe*	Resistance	12,000	IBRC, Japan
Human stem cells	Protein production	500,000	Customer
Pig (*Sus scrofa*)	Meat quality	650,000	Customer
Chickpea[a]	Salt/Drought	850,000	ICARDA/GLIP
6 Legumes/Pathogens[a]	Biotic stress	1.4 Million	LEGRESIST
Boechera holboellii	Apomixis	1.6 Million	IPK, Gatersleben
Olives	Oil Quality	2.0 Million	Customer
Medicago truncatula[a]	Salt stress/Antisense RNA	2.0 Million	CNRS Gif
Physcomitrella patens	Homologous recombination	3.5 Million	UF/MPIZ Cologne
Mouse spleen	IL18 –receptor knock-out	4.5 Million	Univ. Colorado
Sugar beet	Vernalization response	9.0 Million	Univ. Kiel/GABI
Pancreatic cancers	Oncogenesis	15.0 Million	Customer

[a]Legume-related projects.
[b]Increase in tag numbers over time.

6.2.4 Next-Next Transcriptomics: RNA-seq and dRNA-seq

A new development should not go unnoticed, because it adds another level of sequence depth to the repertoire of transcriptomics: The fashionably named **RNA-seq** technique, also direct cDNA sequencing, mRNA-Seq or short read cDNA analysis (Marioni et al. 2008; Mortazavi et al. 2008; Nagalakshmi et al. 2008; Wang et al. 2009; Trapnell et al. 2010).

It capitalizes on the identification and quantification of transcripts (also allele-specific transcripts) and their splice isoforms, RNA editing products, and mapping of transcribed regions of a genome. In short, poly(A)$^+$- RNA is isolated from target cells (by preferably several rounds of oligo[dT] capture on paramagnetic beads), and fragmented (either by shearing or hydrolysis in 500 mM potassium acetate/150 mM magnesium acetate at 94°C). Fragmentation of the mRNA reduces the extent of its secondary structure, which otherwise interferes with the randomness of subsequent priming, and also diminishes the priming preference for 5′ ends of transcripts. After removal of the fragmentation ions by exclusion chromatography, the mRNA is reverse-transcribed into double stranded cDNA with either random hexamers or oligo (dT) primers. Since dT priming produces a bias in favor of the 3′-end of the transcript, random priming with hexamers is preferred. The double stranded cDNA fragments are then directly sequenced by next generation sequencing platforms. One of these (Illumina Genome Sequencer platform) determines a 35 bp sequence from one end of the fragments. These sequence reads are analyzed with an algorithm that maps unique (i.e., single-copy) sequences to the corresponding genome. For example, RNA-Seq analyses reveal that about 75% of the yeast genome is expressed as RNA-Seq tags, and more reads from the 3′- rather than the 5′-ends of annotated genes are detected, presumably a consequence of enrichment of 3′ sequences during poly(A)-purification as well as enhanced priming at 3′ends. For a 5′ end mapping of genes, the 5′-ends of the fragments first are determined by 5′-RACE. Since RNA-Seq detects low-abundance messenger RNAs, it is better than standard microarrays or tiling DNA microarrays. Moreover, the dynamic range of RNA-Seq is at least 8,000-fold; compared to ~60-fold for DNA microarrays. RNASeq allows accurate determination of exon boundaries, and can be used to generate a high-resolution transcriptome map of a genome. The term RNA-Seq is therefore a misnomer, since cDNA, not RNA is sequenced. If massively parallel high-throughput extensive sequencing ("deep sequencing") is employed to characterize the transcriptome, the term deep RNA sequencing is also used. **Differential RNA-seq** (dRNA-seq): A variant of the RNA-seq technology that is selective for the 5′ end of primary transcripts and provides a genome-wide map of transcriptional start sites and operons in bacteria. Moreover, it permits to differentiate between primary transcripts with native 5′-triphosphate (5′PPP) ends (as is the case with most messenger RNAs and small RNAs), and processed transcripts with 5′-monophosphate ends (5′P; such as mature ribosomal and transfer RNAs) by sequencing differential cDNA library pairs (one from control, and another from experimental bacterial culture). In short, total RNA is first extracted from bacterial (e.g., *Helicobacter pylori*) cultures under two (or more) regimes, one remains untreated, and another one is treated with terminator™-5′-phosphate-dependent exonuclease (Epicentre Biotechnologies), that degrades 5′P but not 5′PPP RNA. This enzyme is a processive 5′→3′ exonuclease, that digests RNA carrying a 5′-monophosphate (for example, bacterial ribosomal RNAs) to deplete processed RNAs and to enrich for primary transcripts. As a consequence, bacterial RNA is substantially free of 16S and 23S rRNA. Then as many as possible (minimum 250,000–500,000) cDNAs per library are sequenced with one of the next-generation sequencing technologies (e.g., on a Roche 454 Life/APG GS FLX Titanium sequencer), and mapped to the bacterial genome (Sharma et al. 2010).

Neither RNA-seq nor its derivatives have been applied to leguminous plants or their associated rhizobia, not to speak of cool season food legumes. We therefore focus on few aspects of **deepSuperSAGE transcriptome** (or better sub-transcriptome) profiling in response to abiotic stresses (here: salt and drought stress). However, it is emphasized that deepSuperSAGE recommends itself to profile the transcriptomes of a pathogen, a parasite, or a commensal on one, and the respective host on the other hand (biotic stresses). The charm of this technology lies in the fact, that the transcriptomes of both interacting organisms can be characterized synchronously without the tedious isolation and separate culture of the interaction partners and their independent analysis, which, of course, does not at all reflect the situation *in situ* (Matsumura et al. 2005a).

6.3 The Harvest of Transcriptome Sequencing: Stress Transcriptomes of the Cool Season Food Legume Crop Chickpea

As demonstration of the power of deepSuperSAGE, the salt and the drought stress transcriptomes of roots and nodules of the third most important legume crop chickpea (*Cicer arietinum* L.) will be briefly portrayed here. Both high salinity and extensive drought are responsible for crop losses of millions of tons of various legume (and other) crops, chickpea included. Though in its infancy, transcriptomics has already added to our understanding of the molecular responses of chickpeas to salt and drought stress.

Seed-grown three weeks old chickpea plants were transferred to 40 L hydroaeroponics buckets with freshly prepared medium. Then 25mM NaCl was added to the experimental sample (salt-stress induction), while the control remained salt-free. The salt stress responses of 86,919 transcripts representing 17,918 unique 26 bp deepSuperSAGE tags (UniTags) from roots of the salt-tolerant chickpea variety INRAT-93, and of 57,281 transcripts representing 13,115 UniTags in nodules of the same plants were characterized two hours after treatment. From a total of 144,200 26 bp tags, 21,401 unique transcripts were identified. Of these, only 363 and 106 specific transcripts, respectively, were commonly up- or down-regulated (> 3.0-fold) under salt stress in both organs. Therefore, a **differential organ-specific response to salt stress** can be inferred from the data. Moreover, close to, or more than 99% of all transcripts are present at very low abundances in both organs. For example, in roots less than 1% percent of deepSuperSAGE tags were present in very high copy numbers (> 5,000 copies million^{-1}), 9% of the transcripts were present between 100 and 1,000, and about 90% were represented by less than 100 copies million^{-1}. In nodules of the same chickpea plants, again less than 1% of the transcripts were present in very high copy numbers (> 5,000 copies million^{-1}), and this distribution does not change during salt or drought stress (Fig. 6-2).

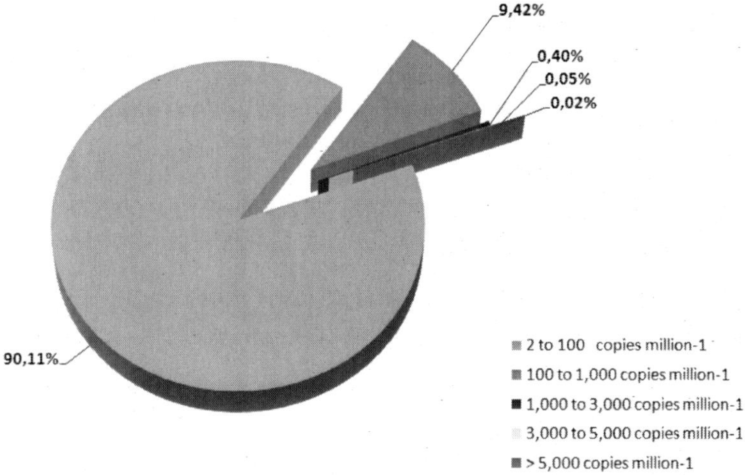

Figure 6-2 Abundancy classes of deepSuperSAGE tags in chickpea roots (copies million^{-1}).
Color image of this figure appears in the color plate section at the end of the book.

The vast amount of information generated by deepSuperSAGE forces to exemplarily consider only transcripts associated with one or few pathways. Here we embark on the generation and scavenging of **reactive oxygen species (ROS)** as well as with ROS signaling (for a review see Zhu 2002; Apel and Hirt 2004; Laloi et al. 2004; Boudsocq and Lauriere 2005). Reactive oxygen species (superoxide [O_2^-], hydrogen peroxide [H_2O_2], and hydroxyl radicals [OH^-]) are continuously generated in considerable amounts by mitochondria, less so in plastids (Moller 2001), and/or peroxisomes, and interact with almost all molecules of a cell, mostly detrimentally. However, ROS also function as signal molecules and activate response genes via sensors (e.g., some receptor-like kinases, RLKs, mitogen-activated protein kinases, MAPKs, and Ca^{2+}-sensing proteins; from an extensive literature see Dat et al. 2000; Laloi et al. 2004; Samuel et al. 2005). Since mitochondria produce the largest amount of ROS (mostly $.O_2^-$), it is probably essential to scavenge them, especially after salt stress. Therefore, a 26-fold up-regulation of a UniTag encoding $NADP^+$-dependent isocitrate dehydrogenase (ICDH) is characteristic for salt-stressed chickpea roots. ICDH generates NADPH to keep antioxidants like glutathione in a reduced state (Moller 2001). In contrast, four different tag isoforms encoding mitochondrial alternative oxidases (AOXs) were at least 6-fold down-regulated under salt stress. Similar results are reported from a *Vigna unguiculata* salt-tolerant variety (Costa et al. 2007). Transcripts encoding superoxide dismutases (SODs) were among the most up-regulated ones (e.g., 40-fold). In INRAT-93 roots, six SOD-annotated UniTags were present, of which four were constitutive, and two others were more than 2.7-fold up-regulated. ROS-scavenging starts with the dismutation of (O_2^-) radicals to hydrogen peroxide (H_2O_2) catalyzed by SODs. Hydrogen peroxide is decomposed via the ascorbate- and glutathione-cycles, or by catalases. The expression levels of APX, DHAR, some transcript variants of GPX and CAT demonstrate that the complete battery of defense against oxidative stress is deployed in chickpea roots only 2 hours after onset of salt stress. Additionally, glutathione S-transferases involved in the detoxification of lipid hydroxyperoxides are represented by seven transcript variants constitutively expressed, four and two were at least 2.5- and 8-fold up-regulated, respectively, whereas two other transcript variants were 4.5-fold down-regulated. Metallothionein-like proteins MT1 and MT2, metal ion-sequestering enzymes, that prevent the Fenton reaction ($Fe^{2+}+H_2O_2 \rightarrow Fe^{3+}OH+OH$), and generate hydroxyl radicals ($OH^{\cdot}$), are encoded by 29 UniTags with levels between 20-fold down- and 10-fold up-regulation.

Symbiotic N_2-fixation (SNF) induces intensive mitochondrial respiration and therefore high ROS production rates. In chickpea, UniTags annotated to ROS-scavengers (i.e., SOD, catalase [CAT], glutathione-S-transferase [GST], glutathione-peroxidase [GPX], and dehydroascorbate-reductase [DHAR]) are more expressed in nodules than in roots, even under control conditions.

DeepSuperSAGE revealed an unanticipated **salt stress-induced differential gene expression** pattern in chickpea roots and nodules as well. In roots **(root transcriptome)** of the salt-tolerant variety INRAT-93, 35% of the tags were at least 2.7-fold up- or down-regulated, respectively, and among these, > 2,000 tags (= 11%) were at least 8-fold down-regulated. These striking changes occur only 2 hours after the onset of salt treatment. A quantitative picture of salt stress up- and down-regulated tags at different thresholds can be viewed in Fig. 6-3.

To illustrate this drastic change even further, Table 6-3 contains the top 20 salt stress up-regulated transcripts from chickpea roots. At the extreme, a deepSuperSAGE tag annotated to a specific nodulin class 40 (Enod40) was most dramatically induced (> 250-fold induced) and differentially expressed. Other transcripts were co-regulated with Enod40. For example, tags encoding a superoxide dismutase (SOD), a trypsin protein inhibitor-3, an extensin, a dormancy-associated protein, an NADP-dependent isocitrate dehydrogenase I, an acetyl-CoA synthetase, a cysteine synthase, a putative extracellular dermal glycoprotein, a 14-3-3 protein, a disease resistance response protein DRRG49-C, the *Arabidopsis* gene *AAD20160.1*-encoded protein, a fiber protein Fb19 and an isoflavone 3′-hydroxylase were among the most up-regulated ones. The tag with the second highest up-regulation corresponds to a transcript encoding superoxide dismutase, pointing to an involvement of ROS scavenging enzymes in the first defense reaction of the plant.

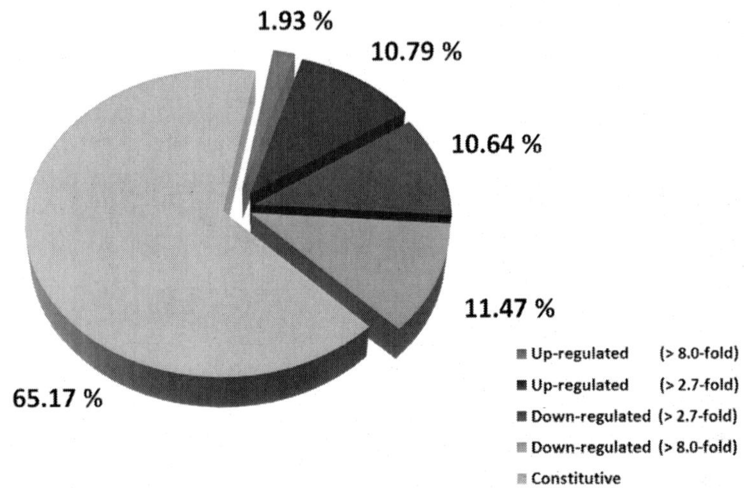

Figure 6-3 Up- and down-regulated transcripts (tags) in chickpea roots after 2 hours of salt stress at two different thresholds.

Color image of this figure appears in the color plate section at the end of the book.

Table 6-3 The 20 most up-regulated annotatable deepSuperSAGE tags from chickpea INRAT-93 roots 2 hours after salt stress.

Tag ID	Encoded protein	$R_{(ln)}$	Uniprot ID
STCa-18884	Early nodulin 40*	5.69	NO40_SESRO
STCa-7896	Superoxide dismutase	3.70	Q9ZNQ4_CICAR
STCa-318	Trypsin protein inhibitor 3	3.59	Q5WM51_CICAR
STCa-19021	Extensin	3.40	O65760_CICAR
STCa-17087	Dormancy-associated protein	3.38	O22611_PEA
STCa-7166	NADP-dependent isocitrate dehydrogenase I	3.25	Q6R6M7_PEA
STCa-1381	Acetyl-CoA synthetase	3.19	Q9ZR69_SOLTU
STCa-2982	Cysteine synthase	3.15	O65747_CICAR
STCa-15648	Mitochondrial 24S mt-RNL ribosomal gene	3.10	No Uniprot link
STCa-20215	Putative extracellular dermal glycoprotein	3.08	Q9FSZ9_CICAR
STCa-20066	14-3-3-like protein A	3.03	1433A_VICFA
STCa-15159	Disease resistance protein DRRG49-C	2.98	DRR4_PEA
STCa-17434	Gb I AAD20160.1	2.92	Q9FYR1_ARATH
STCa-22427	Fiber protein Fb19	2.88	Q6T7D1_GOSBA
STCa-4531	Isoflavone 3′-hydroxylase	2.88	Q2ENF7_ASTME
STCa-14437	60S acidic ribosomal protein P1	2.83	RLA1_MAIZE
STCa-1385	1-aminocylopropane-1-carboxylate oxidase	2.83	Q41681_9FABA
STCa-12309	Ankyrin-like protein	2.83	Q9FMJ2_ARATH
STCa-23197	Hypothetical protein	2.78	Q9LEN3_CICAR
STCa-8459	UDP-glucose pyrophosphorylase	2.78	Q8W557_9FABA

If non-stressed sub-transcriptomes of roots and nodules are compared, an **organ-specific and stress-induced differential gene expression** emerges. A total of 51,545 tags from both libraries represented 11,525 different UniTags, and from these 7,941 showed < 3.0-fold differential expression between both organs. Being more prevalent in nodules, 2,098 UniTags revealed > 3.0-fold differential expression. With a higher threshold, 140 transcripts were still more than 8.0-fold prevalent in the nodules. The most differentially expressed tags are listed in Table 6-4. Among these, a tag encoding a class I chitinase emerged. Also, transcripts involved in ROS detoxification, as for e.g., encoding a SOD, a GST, and a metal metallothionein-like MT2 protein are differentially induced after salt stress perception.

Table 6-4 The 20 most differentially expressed, annotatable deepSuperSAGE tags between normal nodules and roots of chickpea variety INRAT-93.

Tag code	Encoded protein	$R_{(ln)}$	Uniprot ID
STCa-23843	Class I chitinase	3.21	Q76KW5_PEA
STCa-7896	Superoxide dismutase	2.94	Q9ZNQ4_CICAR
STCa-22470	Glutathione S-transferase	2.94	Q948X4_MEDSA
STCa-21852	Protein phosphatase 2A	2.94	Q8L5L1_MUSAC
STCa-3424	Metallothionein-like protein 2	2.94	MT2_CICAR
STCa-21953	Protein-transport protein SEC61	2.84	SC61G_ORYSJ
STCa-21691	ADP-ribosylation factor	2.72	ARFG3_HUMAN
STCa-19859	Translationally controlled tumor-like protein	2.72	Q8LPE3_CICAR
STCa-16288	F6A14.17 protein	2.72	Q9M9U3_ARATH
STCa-22330	Beta-amylase	2.59	Q9LIR6_ARATH
STCa-21330	TGF-beta receptor-interacting protein 1	2.59	Q94KS2_PHAVU
STCa-24180	Signal recognition particle receptor-L protein	2.51	Q9M0A0_ARATH
STCa-8706	Histone H2B	2.43	H2B_GOSHI
STCa-15890	Formin binding protein 3-like	2.43	Q5JM35_ORYSJ
STCa-3014	F17O7.4 protein	2.43	O64594_ARATH
STCa-11675	F22C12.5 protein	2.43	Q9SH69_ARATH
STCa-10302	Poly(A)-binding protein	2.43	Q9M6E6_TOBAC
STCa-23601	Cytokinin-regulated kinase 1	2.43	Q9FUK3_TOBAC
STCa-3289	WRKY DNA-binding protein 11	2.35	WRK11_ARATH
STCa-23197	Hypothetical protein	2.25	Q9LEN3_CICAR

Nodules already have elevated levels of transcripts encoding ROS scavengers prior to any salt treatment (i.e., seem to be in a state of increased stress by ROS). Both nodules and roots rapidly respond to salt stress by inducing common and organ-specific genes encoding ROS-scavengers. Generally, from the 2,098 tags prevalent in non-stressed nodules, 515 (= 24.5%) were 3-fold up-regulated in roots under salt stress. These 515 tags represented 23.3% of the root transcripts > 3-fold up-regulated by salt. However, only 10 out of the 2,098 tags were more than 3-fold up-regulated in salt-stressed nodules. In both salt-stressed roots and nodules, 363 common tags were more than 3-fold up-regulated (16.7% from nodules, and 16.4% from roots). Down-regulation of genes, mostly disregarded in transcript profiling studies, is also a striking feature in both organs. So, 1,729 out of 1,936 tags prevalent in non-stressed roots were more than 3-fold down-regulated in salt-treated roots, and 275 tags were more than 3-fold down-regulated in both roots and nodules under salt-stress. Two tags were > 20-fold induced by salt stress in both organs, of which an Enod40-encoding mRNA was most up-regulated in both organs upon salt stress (> 250-fold in roots; 60-fold in nodules). Tags shared by salt-stressed nodules and roots and at least 8-fold up-regulated in response to salt stress are listed in Table 6-5.

Table 6-5 Overlapping salt stress responses of highly up-regulated (> 8.0-fold up- regulation) deepSuperSAGE tags in roots and nodules.

Tag ID	Encoded Protein	$R_{(ln)}$ [a]	$R_{(ln)}$ [b]	Uniprot ID
STCa-18884	Early nodulin 40	5.69	4.12	NO40_SESRO
STCa-24417	Lipoxygenase	3.19	2.50	Q9M3Z5_CICAR
STCa-15648	Mitochondrial 26S ribosomal gene	3.11	3.17	no ID
STCa-17434	AAD20160.1	2.93	2.61	Q9FYR1_ARATH
STCa-1958	Gibberellin-stimulated protein	2.68	2.61	Q53AN3_ORYSA
STCa-8434	Fiber protein Fb2	2.36	2.14	Q8GT87_GOSBA
STCa-7166	NADP-dependent isocitrate dehydrogenase I	3.26	1.92	Q6R6M7_PEA
STCa-199	Nodulin-like protein	2.36	1.92	Q6NMB7_ARATH
STCa-10656	Putative UDP-glycose	2.32	1.92	Q9M3H8_CICAR
STCa-542	Prolyl 4-hydroxylase	2.19	1.92	Q9FKX6_ARATH
STCa-15530	NADH ubiquinone oxidoreductase	2.09	1.92	Q9SP38_LUPLU
STCa-13688	O-methyltransferase	2.09	1.92	Q96424_GLYEC
STCa-10316	NtEIG-E80 protein	1.99	2.33	Q9FXS6_TOBAC

[a]$R_{(ln)}$ of UniTags expression in salt-stressed roots.
[b]$R_{(ln)}$ of UniTags expression in salt-stressed nodules.

In contrast to salt-stressed chickpea roots, the **response of nodules to salt** is relatively modest. Only 95 and 72 UniTags were at least 8.0-fold up- or down-regulated, respectively. The top 20 most up-regulated nodular transcripts after 2 hours of salt stress are accumulated in Table 6-6. Of the most differentially expressed transcripts, a tag encoding Enod40 was the most up-regulated one. Down the line are tags for the 40S ribosomal proteins RSSA and RS92, respectively, and a cystein proteinase inhibitor.

Table 6-6 The 20 most up-regulated, annotatable UniTags from INRAT-93 nodules 2 hours after salt stress induction.

Tag ID	Encoded Protein	$R_{(ln)}$	Associated process	Uniprot ID
STCa-18884	Eearly nodulin 40	4.11	Nodulation	NO40_SESRO
STCa-15648	24S Mitochondrial ribosomal mt-RNL gene	3.17	Translation	X55832
STCa-11090	40S ribosomal protein SA	2.73	Protein biosynthesis	RSSA_CICAR
STCa-17434	AAD20160.1 protein	2.61	No associated term	Q9FYR1_ARATH
STCa-1958	Gibberellin-stimulated protein	2.61	No associated term	Q53AN3_ORYSA
STCa-3760	Cysteine proteinase inhibitor	2.48	Inhibition of proteolysis	O04720_SOYBN
STCa-89	Drought-induced protein	2.48	Response to stress	Q941N0_9FABA
STCa-16482	40S ribosomal protein S9-2	2.48	Protein biosynthesis	RS92_ARATH
STCa-10316	NtEIG-E80 protein	2.33	No associated term	Q9FXS6_TOBAC
STCa-3321	Leghemoglobin	2.33	Oxygen transport	Q42928_MEDSA
STCa-1263	Benzoyltransferase-like protein	2.33	No associated term	Q9FLM5_ ARATH
STCa-13055	Nonspecific lipid-transfer protein precursor	2.33	Transport (lipids)	NLTP_CICAR
STCa-22149	Acyl carrier protein	2.33	Lipid biosynthesis	ACPM_ARATH
STCa-10862	F6N18.8 protein	2.33	No associated term	Q9LPJ4_ARATH
STCa-21007	Two-component response regulator PRR37	2.33	Regulation of transcription	PRR37_ORYSA
STCa-4833	T13M11_21 protien	2.14	Regulation of transcription	Q94JL3_ARATH
STCa-8434	Fiber protein Fb2	2.14	No associated term	Q8GT87_GOSBA
STCa-23572	F7K24_140 protein	2.14	Signal transduction	Q8RWQ4_ ARATH
STCa-7572	Protein phosphatase 2A	2.14	Signal transduction	Q9FVD5_ MEDVA
STCa-1895	GDP-mannose 3,5-epimerase	2.14	Ascorbic acid biosynthesis	GME_ARATH

Comparative qRT-PCR on RNAs from chickpea varieties with differing tolerance to salt (INRAT-93: salt-tolerant; Amdoun1: salt-sensitive, ICC-4958: sensitive, and ICC-6098: weakly tolerant) help to identify genes involved in salt stress management. As an example among many others, the lipoxygenase-annotated deepSuperSAGE tag STCa-7252 reacts upon salt with a 2.5-fold up-regulation in INRAT-93, but a much stronger activation in salt-sensitive Amdoun1, suggesting a more vigorous response in sensitive varieties.

If one looks at more **global transcriptome differences** between both organs and visualizes them by Gene Ontology (GO) categories, it is seen that transcripts are channeled into various pathways, and that this process is augmented after salt stress. For example, in the likely order of importance for the stressed cell, biological processes like response to stimulus (GO:0050896), intracellular protein transport (GO:0006886), translation (GO:0006412), generation of precursor metabolites and energy (GO:0006091), catabolic process (GO:0009056), cellular carbohydrate catabolic process (GO:0044275), and electron transport (GO:0006118), macromolecule biosynthetic process (GO:0009059), cellular component organization and biogenesis (GO:0016043) were over-represented ($P < 1.0^{e-12}$). Moreover, at least 11 categories indicated either metabolism or catabolism, five biological processes reflected responses to stimuli (oxidative stress, chemicals, stress generally), and four other processes were related to transport.

As far as cellular components are concerned, intrinsic to membrane (GO:0031224), nucleus (GO:0031224), membrane part (GO:0044425), ribonucleoprotein complex (GO:0030529), integral to membrane (GO:0016021), intracellular non-membrane-bound organelle (GO:0043232), protein complex (GO:0043234), outer membrane (GO:0019867), and ribosome (GO:0005840) were most highly over-represented ($P < 1.\ 0^{e-12}$).

To sum up the few observations on the general flow of transcripts into the various pathways (as identified by GO categories), the whole complex spectrum of transcripts representative of the various GO categories is revealed (exemplified for normal chickpea roots in Fig. 6-4), and an unprecedented change of the different categories becomes visible after onset of salt stress (Fig. 6-5). Whereas many pathways in the intact root are represented at almost equal proportions, i.e., are balanced, only 2 hours of 25 mM salt changes the picture dramatically. Those pathways, that are obviously not very important in the stress situation, are turned off or reduced in importance (for e.g., cell cycle, energy, protein biosynthesis, ribosome, seed storage, transcription, translocation), and the pathways of importance for coping with the salt stress (as for e.g., ion homoeostasis, metabolism, biosynthesis, cell wall, oxidative stress, also pathogenesis, though only abiotic stress was involved), are strikingly over-represented. Therefore, the pathway image of the stressed root simplifies considerably, potential proof for a trial of salt management to keep the membranes intact and functioning, to protect DNA, RNA and proteins from oxygen radicals, and to restore salt-damaged structures and functions by enhanced biosynthetic and energetic capacities (Fig. 6-5).

In summary, one of the major insights into the primary steps of salt stress response in chickpea plants is that 1) normal nodules exhibit elevated levels of transcripts encoding ROS scavengers prior to any salt treatment,

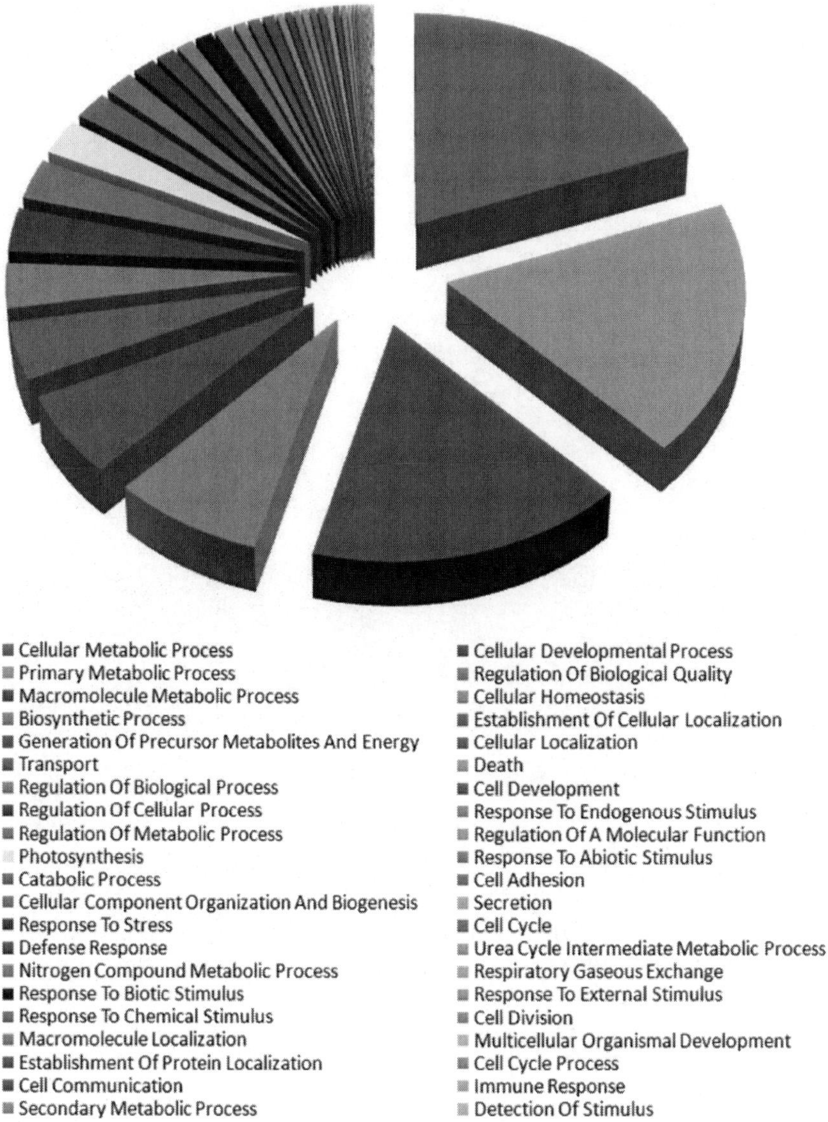

Figure 6-4 Representation of tags from unstressed chickpea roots assigned to different GO categories (Biological Processes).

Color image of this figure appears in the color plate section at the end of the book.

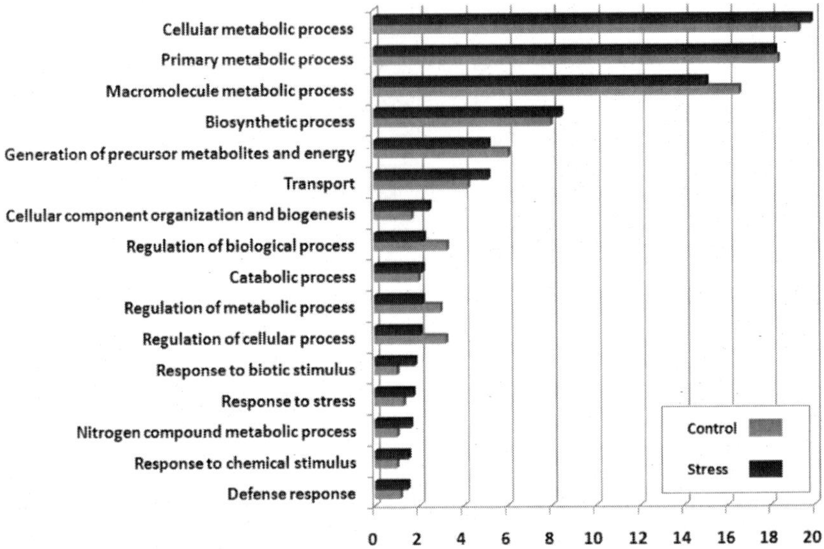

Figure 6-5 Representiation of chickpea roots tags assigned to different GO categories (control versus salt stress). Abscissa: Percentage of assigned deepSuperSAGE tags. Ordinate: GO Biological Processes.

and 2) both nodules and roots rapidly (already 2 hours after addition of 25 mM NaCl) respond to salt stress by transcription of genes encoding ROS-scavengers. None of these results clash with previous findings (Coram et al. 2007; Mantri et al. 2007; Merchan et al. 2007; Jain and Chattopadhyay 2010).

The present chapter demonstrates the potential high-throughput deepSuperSAGE coupled to the next-generation sequencing platform Roche 454 Life/APG GS FLX Titanium for a genome-wide quantitative gene expression profiling of plants or plant organs under stress. The same technology also allowed characterizing the drought-responsive transcriptome of chickpea leaves (Molina et al. 2008). The main obstacle for exploiting all the data from deepSuperSAGE is tag annotation because no genome sequence background is yet available (the sequencing of the chickpea genome at 30x coverage already began in the National Institute of Plant Genome Research in New Delhi, India). In other plants, as e.g., *Arabidopsis thaliana*, whose genome sequence is almost complete, transcriptome analysis is more comprehensive and secure (Fizames et al. 2004). However, re-annotation improves the quantity and quality of structural and functional annotations, allows a more comprehensive Gene Ontology-based categorization, and improves pathway coverage (van den Berg et al. 2010).

6.4 eQTLs, methQTLs, pQTLs: A Future Prospect

As a concluding remark on future prospects in cool season food legumes, we envisage several potential directions. First, the development and use of expression markers, based on sequence polymorphisms in expressed transcripts or genes, will complement the traditional anonymous mutation markers as they were generated by many different, partly outdated marker technologies such as isozyme patterning, protein profiling, amplified fragment length polymorphism (AFLP), random amplified polymorphic DNA (RAPD), DNA amplification fingerprinting (DAF), sequence-tagged microsatellite sites (STMSs), also short simple repeats (SSRs), amplified consensus gene markers (ACGMs), amplified restriction fragment length polymorphism technique (cDNA-AFLP), single nucleotide polymorphism (SNP) technique, and many others. Such DNA markers reflect mutations in a specific region of one, but not another related genome, which may or may not contain one (or more) gene(s), and therefore only rarely imply a functional consequence. In contrast, transcriptome-derived markers (transcriptome markers) are transcript sequences detected by cDNA-AFLP, restriction fragment differential display polymerase chain reaction (RFDD-PCR), restriction fragment length polymorphism (RFLP), the above mentioned SAGE technology and its multiple variants or also microarrays in their different versions, that allow in identifying polymorphisms between parents (and the progeny from a cross between these parents) and can therefore be mapped. Transcriptome markers are not anonymous, but derived from active genes. Therefore, transcript marker maps or transcriptome atlases (comprehensive collections of data on the patterns of expressed genes, their expression levels, and their tissue-specific regulation at a particular time point in the life of an organism) depict the active genes in a genome. Several such markers may identify a haplotype, and a set of expression markers may allow selecting a preferable genotype or also phenotype from a population of organisms (e.g., a gene bank or germplasm collection).

Second, by analyzing the expression levels of genes or gene clusters within a segregating population, the inheritance of expression patterns can be mapped (Varshney et al. 2005). Any quantitative trait locus (QTL), identified by quantitative expression data and supported by phenotyping data are called expression quantitative trait locus (**eQTL**). An eQTL (also transcriptional QTL, tQTL) is characterized by both a clinically or agronomically proven association phenotype (e.g., disease or yield), linked DNA markers (as for e.g., microsatellite markers, see Sharma et al. 2007) and the expression level of underlying genes. Therefore, eQTL analysis combines genetic mapping of QTLs and genome-wide expression profiling by for e.g., expression microarrays or tag-based techniques such as serial analysis of gene expression (SAGE), or better deepSuperSAGE, to name a few. Any statistically significant correlation between both parameters suggests, that the gene (or genes) within the chromosomal region where the sequence variation resides, could account for some (in extreme cases, most) of the variation in phenotype (Schadt et al. 2003). In this case, the eQTL is coined cis-eQTL. If the eQTL maps far away from its parent gene or the target QTL, then their effect is composed of two loci (trans-eQTL). Usually, trans-eQTLs contain clusters of genes that simultaneously regulate a large fraction of the transcriptome. For example, in *Arabidopsis thaliana*, most trans-eQTLs clustered into regions defined as trans-eQTL hotspots that control the levels of a larger number of transcripts but only a small fraction of the variation for each transcript. The e-traits affected by a trans-eQTL hotspot also showed directionality: most transcripts were either up- or down-regulated by the same parental allele at the trans-eQTL (De Koning et al. 2007; Haley and deKoning 2007).

Any co-localization of a candidate gene (or gene) with QTLs controlling a particular phenotype allows in developing a gene marker bona fide ("perfect marker") to select the phenotype in marker-assisted breeding ("genomics-assisted breeding", see Varshney et al. 2005).

Third, while it is apparent that the expression of a gene alone is an excellent parameter, it has to be supported with other genomic features to be complete. One of these parameters is the so-called **epigenetic code**, either the **DNA epigenetic code** (the methylation of C5 atoms of cytosyl residues in DNA, mostly, but not at all exclusively in the CpG context) or the **histone code** (the various post-translational modifications of histone proteins at a given time, that are recognized by other proteins involved in chromatin modeling, chromatin re-modeling and transcriptional regulation). The acetylation of 13 different lysine residues in all core histones, methylation of lysine and arginine residues on histone H3 and H4, phosphorylation and ubiquitinylation of all histones are such histone "codes". For example, lysine residue 4 and 9 (K9) in histone H3 and lysine 20 in histone H4 are methylated by histone methyltransferase $SU(VAR)_{39}$ (in mammals) or Clr4 (in yeast). This methylated lysine is the only binding site for heterochromatin protein HP_1, which is associated with silent heterochromatic regions of a genome. Phosphorylation of the adjacent S10 residue by Aurora kinase B loosens this association. Acetylation at H3K14, catalyzed by histone acetyltransferases (HATs) prevents HP1 binding (for a review see Jenuwein and Allis 2001, also Koch et al. 2007 for a genome-wide study). While genome-wide C-methylation analyses at a single base resolution will soon become commonplace in genomics, also plant gemomics, histone modifications genome-wide and comprehensively will still remain a very demanding task, but will certainly be a tool for plant genetics (see for e.g., the successful characterization of the *Arabidopsis* and the human methylome by Lister et al. 2008, 2009).

The C-methylation screening can lead to so-called CpG methylation quantitative trait loci (**methQTLs**). Joint methQTL-eQTLs may become a tool to detect the classical inverse and causative relationship between CpG methylation and expression (see Gibbs et al. 2010). A major lesson to be learnt for plant geneticists is that CpG methylation quantitative trait loci occur at higher frequency at CpG sites outside of CpG islands.

An upcoming technology identifies protein quantitative trait loci (**pQTLs or PQLs**), preferably genome-wide, and generates accessory information relating to gene expression and transcript processing to translation into protein (an important aspect, because pQTLs reflect the complexity and composition of a cell proteome). By now, transcriptomics is almost totally disconnected from proteomics, and therefore one does not know whether a protein-encoding transcript is in fact translated into a protein (Melzer et al. 2008).

All these developments, genome and transcriptome sequencing included, will certainly enrich the technological and informational space of plant research, and research on cool season food legumes especially, as we hope and expect.

Acknowledgements

Research was supported by grants from the EU (Aquarhiz project, INCO-CT-2004-509115) to GK, and Grain Legumes project (Food-CT-2004-506223) to GK and GenXPro GmbH, Volkswagenstiftung (grant I/78 296) to GK, and Deutsche Forschungsgemeinschaft (Grant Ka 332/22-1) to GK. We dedicate this article to the teams of the Johann Wolfgang Goethe-University (Molecular BioSciences, Biocenter, Frankfurt University) and GenXPro GmbH (Frankfurt Innovation Center Biotechnology, Frankfurt am Main).

References

Apel K, Hirt H (2004) Reactive oxygen species: metabolism, oxidative stress, and signal transduction. Annu Rev Plant Biol 55: 373–399.

Ashraf N, Ghai D, Barman P, Basu S, Gangisetty N, Mandal MK, Chakraborty N, Datta A, Chakraborty S (2009) Comparative analyses of genotype dependent expressed sequence tags and stress-responsive transcriptome of chickpea wilt illustrate predicted and unexpected genes and novel regulators of plant immunity. BMC Genom 10: 415.

Barnes M, Freudenberg J, Thompson S, Aronow B, Pavlidis P (2005) Experimental comparison and cross-validation of the Affymetrix and Illumina gene expression analysis platforms. Nucl Acids Res 33: 5914–5923.

Barski A, Cuddapah S, Cui K, Roh T-Y, Schones DE, Wang Z, Wei G, Chepelev I, Zhao K (2007) High-resolution profiling of histone methylations in the human genome. Cell 129: 823–837.

Benedito VA , Torres-Jerez I, Murray JD, Andriankaja A, Allen S, Kakar K, Wandrey M, Verdier J, Zuber H, Ott T, Moreau S, Niebel A, Frickey T, Weiller G, He J, Dai X, Zhao PX, Tang

Y, Udvardi MK (2008) A gene expression atlas of the model legume *Medicago truncatula*. Plant J 55: 504–513.

Benko-Iseppon AM, Winter P, Huettel B, Staginnus CM, Muehlbauer FJ, Kahl G (2003) Molecular markers closely linked to fusarium resistance genes in chickpea show significant alignment to pathogenesis-related genes located on *Arabidopsis* chromosomes 1 and 5. Theor Appl Genet 107: 379–386.

Berger SL (2007) The complex language of chromatin regulation during transcription Nature 447: 407–412.

Boominathan P, Shukla R, Kumar A, Manna D, Negi D, Verma PK, Chattopadhyay D (2004) Long term transcript accumulation during the development of dehydration adaptation in *Cicer arietinum*. Plant Physiol 135: 1608–1620.

Boudsocq M, Lauriere C (2005) Osmotic signaling in plants: multiple pathways mediated by emerging kinase families. Plant Physiol 138: 1185–1194.

Bradford JR, Hey Y, Yates T, Li Y, Pepper SD, Miller CJ (2010) A comparison of massively parallel nucleotide sequencing with oligonucleotide microarrays for global transcription profiling. BMC Genom 11: 282.

Brenner S, Johnson M, Bridgham J, Golda G, Lloyd DH, Johnson D, Luo S, McCurdy S, Foy M, Ewan M, Roth R, George D, Eletr S, Albrecht G, Vermaas E, Williams SR, Moon K, Burcham T, Pallas M, DuBridge RB, Kirchner J, Fearon K, Mao J, Corcoran K (2000) Gene expression analysis by massively parallel signature sequencing (MPSS) on microbead arrays. Nat Biotechnol 18: 630–634.

Brian XY, Ford R, Han TR, Coram TE, Pang ECK, Taylor PWJ (2007) Approaching chickpea quantitative trait loci conditioning resistance to *Ascochyta rabiei* via comparative genomics. Australas Plant Pathol 36: 419–423.

Brosius J, Tiedge H (2004) RNomenclature. RNA Biol 1: 81–88.

Buhariwalla HK, Jayashree B, Eshwar K, Crouch JH (2005) Development of ESTs from chickpea roots and their use in diversity analysis of the *Cicer* genus. BMC Plant Biol 5: 16.

Buitink J, Leger JJ, Guisle I, Ly Vu B, Wuillème S, Lamirault G, Le Bars A, Le Meur N, Becker A, Küster H, Leprince O (2006) Transcriptome profiling uncovers metabolic and regulatory processes occurring during the transition from desiccation-sensitive to desiccation-tolerant stages in *Medicago truncatula* seeds. Plant J 47: 735–750.

Bustin SA (2002) Quantification of mRNA using real-time reverse transcription PCR (RT-PCR): trends and problems. J Mol Endocrinol 29: 23–39.

Bustin SA (2010) Why the need for qPCR publication guidelines?—The case for MIQUE. Methods 50: 217–226.

Bustin SA, Nolan T (2004) Pitfalls of quantitative real-time reverse transcription polymerase chain reaction. J Biomol Tech 15: 155–166.

Bustin SA, Benes V, Nolan T, Pfaffl MW (2005) Quantitative real-time RT-PCR—a perspective. J Mol Endocrinol 34: 597–601.

Bustin SA, Benes V, Garson JA, Hellemans J, Huggett J, Kubista M, Mueller R, Nolan T, Pfaffl MW, Shipley GL, Vandesompele J, Wittwer CT (2009) The MIQE Guidelines: Minimum information for publication of quantitative real-time PCR experiments. Clin Chem 55: 611–622.

Canales RD, Luo Y, Willey JC, Austermiller B, Barbacioru CC, Boysen C, Hunkapiller K, Jensen RV, Knight CR, Lee KY, Ma Y, Maqsodi B, Papallo A, Peters EH, Poulter K, Ruppel PL, Samaha RR, Shi L, Yang W, Zhang L, Goodsaid FM (2006) Evaluation of DNA microarray results with quantitative gene expression platforms. Nat Biotechnol 24: 1115–1122.

Chen J (2009) Serial analysis of binding elements for transcription factors. Meth Mol Biol 567: 113–132.

Chen K, Rajewsky N (2007) The evolution of gene regulation by transcription factors and microRNAs. Nat Rev Genet 8: 93–103.

Cheng J, Kapranov P, Drenkow J, Dike S, Brubaker S, Patel S, Long J, Stern D, Tammana H, Helt G, Sementchenko V, Piccolboni A, Bekiranov S, Bailey DK, Ganesh M, Ghosh S, Bell

I, Gerhard DS, Gingeras TR (2005) Transcriptional maps of 10 human chromosomes at 5-nucleotide resolution. Science 308: 1149–1154.

Chern MS, Bobb AJ, Bustos MM (1996a) The regulator of MAT2 (ROM2) protein binds to early maturation promoters and represses PvALF-activated transcription. Plant Cell 8: 305–321.

Chern MS, Eiben HG, Bustos MM (1996b) The developmentally regulated bZIP factor ROM1 modulates transcription from lectin and storage protein genes in bean embryos. Plant J 10: 135–148.

Chodavarapu RK, Feng S, Bernatavichute YV, Chen P-Y, Stroud H, Yu Y, Hetzel JA, Kuo F, Cokus SJ, Casero D, Bernal M, Huijser P, Clark AT, Krämer U, Merchant SS, Zhang X, Jacobsen SE, Pellegrin M (2010) Relationship between nucleosome positioning and DNA methylation. Nature 466: 388–392.

Choudhary S, Sethy NK, Shokeen B, Bathia S (2009) Development of chickpea EST-SSR markers and analysis of allelic variation across related species. Theor Appl Genet 118: 591–608.

Coram TE, Pang ECK (2005a) Isolation and analysis of candidate ascochyta blight defence genes in chickpea. Part I. Generation and analysis of an expressed sequence tag (EST) library. Physiol Mol Plant Pathol 66: 192–200.

Coram TE, Pang ECK (2005b) Isolation and analysis of candidate ascochyta blight defence genes in chickpea, part II. Microarray expression analysis of putative defence-related ESTs. Physiol Mol Plant Pathol 66: 201–210.

Coram TE, Pang ECK (2006) Expression profiling of chickpea genes differentially regulated during a resistance response to *Ascochyta rabiei*. Plant Biotechnol J 4: 647–666.

Coram TE, Pang ECK (2007) Transcriptional profiling of chickpea genes differentially regulated by salicylic acid, methyl jasmonate and aminocyclopropane carboxylic acid to reveal pathways of defence-related gene regulation. Funct Plant Biol 34: 52–64.

Coram TE, Mantri N, Ford R, Pang ECK (2007) Functional genomics in chickpea: An emerging frontier for molecular-assisted breeding. Funct Plant Biol 34: 861–873.

Costa JH, Jolivet Y, Hasenfratz-Sauder MP, Orellano EG, da Guia Silva Lima M, Dizengremel P, Fernandes de Melo D (2007) Alternative oxidase regulation in roots of *Vigna unguiculata* cultivars differing in drought/salt tolerance. J Plant Physiol 164: 718–727.

Dat J, Vandenabeele S, Vranova E, Van Montagu M, Inze D, Van Breusegem F (2000) Dual action of the active oxygen species during plant stress responses. Cell Mol Life Sci 57: 779–795.

Datson NA, van der Perk-de Jong J, van den Berg MP, de Kloet ER, Vreugdenhil E (1999) MicroSAGE: a modified procedure for serial analysis of gene expression in limited amounts of tissue. Nucl Acids Res 27: 1300–1307.

Deguchi Y, Banba M, Shimoda Y, Chechetka SA, Suzuri R, Okusako Y, Ooki Y, Toyokura K, Suzuki A, Uchiumi T, Higashi S, Abe M, Kouchi H, Izui K, Hata S (2007) Transcriptome profiling of *Lotus japonicus* roots during arbuscular mycorrhiza development and comparison with that of nodulation. DNA Res 14: 117–133.

De Koning D J, Archibald A, Haley CS (2007) Livestock genomics: bridging the gap between mice and men. Trends Biotechnol 25: 483–489.

De Reynies A, Geromin D, Cayuela J-M, Petel F, Dessen P, Sigaux F, Rickman DS (2006) Comparison of the latest commercial short and long oligonucleotide microarray technologies. BMC Genom 7: 51.

Derveaux S, Vandesompele J, Hellemans J (2010) How to do successful gene expression analysis using real-time PCR. Methods 50: 227–230.

Diatchenko L, Lau YF, Campbell AP, Chenchik A, Moqadam F, Huang B, Lukyanov S, Lukyanov K, Gurskaya N, Sverdlov ED, Siebert PD (1996) Suppression subtractive hybridization: a method for generating differentially regulated or tissue-specific cDNA probes and libraries. Proc Natl Acad Sci USA 93: 6025–6030.

Dinel S, Bolduc C, Belleau P, Boivin A, Yoshioka M, Calvo E, Piedboeuf B, Snyder EE, Labrie F, St-Amand J (2005) Reproducibility, bioinformatic analysis and power of the SAGE method to evaluate changes in transcriptome. Nucl Acids Res. 33: e26.

El Yahyaoui F, Küster H, Ben Amor B, Hohnjec N, Pühler A, Becker A, Gouzy J, Vernié T, Gough C, Niebel A, Godiard L, Gamas P (2004) Expression profiling in *Medicago truncatula* identifies more than 750 genes differentially expressed during nodulation, including many potential regulators of the symbiotic program. Plant Physiol 136: 3159–3176.

Fillingham J, Greenblatt JF (2008) A histone code for chromatin assembly. Cell 134: 206–208.

Firnhaber C, Pühler A, Küster H (2005) EST sequencing and time course microarray hybridizations identify more than 700 *Medicago truncatula* genes with developmental expression regulation in flowers and pods. Planta 222: 269–283.

Fizames C, Muños S, Cazettes C, Nacry P, Boucherez J, Gaymard F, Piquemal D, Delorme V, Commes T, Doumas P, Cooke R, Marti J, Sentenac H, Gojon A (2004) The *Arabidopsis* root transcriptome by serial analysis of gene expression. Gene identification using the genome sequence. Plant Physiol 134: 67–80.

Fleige S, Pfaffl MW (2006) RNA integrity and the effect on the real-time qRT-PCR performance. Mol Aspects Med 27: 126–139.

Fleige S, Walf V, Huch S, Prgomet C, Sehm J, Pfaffl MW (2006) Comparison of relative mRNA quantification models and the impact of RNA integrity in quantitative real-time RT- PCR. Biotechnol Lett 28: 1601–1613.

Ford R, Mustafa B, Inder P, Rubeena S, Materne M, Taylor P (2007) Lensomics: advances in genomics and molecular techniques for lentil breeding and management. In: SS Yadav, D McNeil, PC Stevenson (eds) Lentil. An Ancient Crop for Modern Times. Spinger, Dordrecht, The Netherlands, pp 275–289.

Forrest ARR, Carninci P (2009) Whole genome transcriptome analysis. RNA Biol 6: 107–112.

Frenzel A, Manthey K, Perlick AM, Meyer F, Pühler A, Krajinski F, Küster H (2005) Combined transcriptome profiling reveals a novel family of arbuscular mycorrhizal-specific *Medicago truncatula* lectin genes. Mol Plant-Microbe Interact 18: 771–782.

Gao WR, Wang XS, Liu QY, Peng H, Chen C, Li JG, Zhang JS, Hu SN, Ma H (2008) Comparative analysis of ESTs in response to drought stress in chickpea (*C. arietinum* L.). Biochem Biophys Res Comm 376: 578–583.

Garg R, Sahoo A, Tyagi AK, Jain M (2010) Validation of internal control genes for quantitative gene expression studies in chickpea (*Cicer arietinum* L.). Biochem Biophys Res Comm 396: 283–288: doi: 10.1016/j.bbrc.2010.04.079.

Geurts R, Fedorova E, Bisseling T (2005) Nod factor signaling genes and their function in the early stages of *Rhizobium* infection. Curr Opin Plant Biol 8: 346–352.

Gibbs JR, van der Brug MP, Hernandez DG, Traynor BJ, Nalls MA, Lai SL, Arepalli S, Dillman A, Rafferty IP, Troncoso J, Johnson R, Zielke HR, Ferrucci L, Longo DL, Cookson MR, Singleton AB (2010) Abundant quantitative trait loci exist for DNA methylation and gene expression in human brain. PLoS Genet 6: e1000952.

Godde JS, Ura K (2008) Cracking the enigmatic linker histone code. J Biochem 143: 287–293.

Gowda M, Jantasuriyarat C, Dean RA, Wang G-L (2004) Robust-LongSAGE, RL-SAGE: a substantially improved LongSAGE method for gene discovery and transcriptome analysis. Plant Physiol 134: 890–897.

Griffith OL, Pleasance ED, Fulton DL, Oveisi M, Ester M, Siddiqui AS, Jones SJM (2005) Assessment and integration of publicly available SAGE, cDNA microarray, and oligonucleotide microarray expression data for global coexpression analyses. Genomics 86: 476–488.

Guo A, He K, Liu D, Bai S, Gu X, Wei L, Luo J (2005) DATF: a database of *Arabidopsis* transcription factors. Bioinformatics 21: 2568–2569.

Haley CS, de Koning DJ (2007) Towards *in vitro* genetics. Trends Genet 23: 382–386.

Harbers M, Carninci P (2005) Tag-based approaches for transcriptome research and genome annotation. Nat Meth 2: 495–502.

Harrison MJ (2005) Signaling in the arbuscular mycorrhizal symbiosis. Annu Rev Microbiol 59: 19–42.

Hashimoto S, Suzuki Y, Kasai Y, Morohoshi K, Yamada T, Sese J, Morishita S, Sugano S, Matsushima K (2004) 5′-end SAGE for the analysis of transcriptional start sites. Nat Biotechnol 22: 1146–1149.

He L, Hannon GJ (2004) MicroRNAs: small RNAs with a big role in gene regulation. Nat Rev Genet 5: 522–531.

Heard J, Caspi M, Dunn K (1997) Evolutionary diversity of symbiotically induced nodule MADS box genes: characterization of nmhC5, a member of a novel subfamily. Mol Plant-Microbe Interact 10: 665–676.

Henikoff S, Ahmad K (2005) Assembly of variant histones into chromatin. Annu Rev Cell Dev Biol 21: 133–153.

Hofer J, Turner L, Hellens R, Ambrose M, Matthews P, Michael A, Ellis N (1997) Unifoliata regulates leaf and flower morphogenesis in pea. Curr Biol 7: 581–587.

Hohnjec N, Vieweg MF, Pühler A, Becker A, Küster H (2005) Overlaps in the transcriptional profiles of *Medicago truncatula* roots inoculated with two different *Glomus* fungi provide insights into the genetic program activated during arbuscular mycorrhiza. Plant Physiol 137: 1283–1301.

Hohnjec N, Henckel K, Bekel T, Gouzy J, Dondrup M, Goesmann A, Küster H (2006) Transcriptional snapshots provide insights into the molecular basis of arbuscular mycorrhiza in the model legume *Medicago truncatula*. Funct Plant Biol 33: 737–748.

Hu R, Fan C, Li H, Zhang Q, Fu Y-F (2009) Evaluation of putative reference genes for gene expression normalization in soybean by quantitative real-time RT-PCR. BMC Mol Biol 10: 93.

Hugget J, Dheda K, Bustin S, Zumla A (2005) Real-time RT-PCR normalisation; strategies and considerations. Genes Immunol 6: 279–284.

Hüttel B, Santra D, Muehlbauer F, Kahl G (2002) Resistance gene analogues of chickpea (*Cicer arietinum* L.): isolation, genetic mapping and association with a *Fusarium* resistance gene cluster. Theor Appl Genet 105: 479–490.

Ibrahim AF, Hedley PE, Cardle L, Kruger W, Marshall DF, Muehlbauer GJ, Waugh R (2005) A comparative analysis of transcript abundance using SAGE and Affymetrix arrays. Funct Integr Genom 5: 163–174.

Iida K, Seki M, Sakurai T, Satou M, Akiyama K, Toyoda T, Konagaya A, Shinozaki K (2005) RARTF: database and tools for complete sets of *Arabidopsis* transcription factors. DNA Res 12: 247–256.

Impey S, McCorkle SR, Cha-Molstad H, Dwyer JM, Yochum GS, Boss JM, McWeeney S, Dunn JJ, Mandel G, Goodman RH (2004) Defining the CREB regulon: a genome-wide analysis of transcription factor regulatory regions. Cell 119: 1041–1054.

Jain D, Chattopadhyay D (2010) Analysis of gene expression in response to water deficit of chickpea (*Cicer arietinum* L.) varieties differing in drought tolerance. BMC Plant Biol 10: 24: doi: 10.1186/1471-2229-10-24.

Jayashree B, Hutokshi K, Buhariwalla HK, Shinde S, Crouch JH (2005) A legume genomics resource: The Chickpea Root Expressed Sequence Tag Database. Electronic Journal

Jenuwein T, Allis CD (2001) Translating the histone code. Science 293: 1074–1080.

Johnson DS, Mortazavi A, Myers RM, Wold B (2007) Genome-wide mapping of *in vivo* protein-DNA interactions. Science 316: 1497–1502.

Jones-Rhoades MW, Bartel DP, Bartel B (2006) MicroRNAs and their regulatory roles in plants. Annu Rev Plant Biol 57: 19–53.

Kapranov P, Cheng J, Dike S, Nix DA, Duttagupta R (2007) RNA maps reveal new RNA classes and a possible function for pervasive transcription. Science 316: 1484–1488.

Kim J, Bhinge AA, Morgan XC, Iyer VR (2005) Mapping DNA-protein interactions in large genomes by sequence tag analysis of genomic enrichment. Nat Meth 2: 47–53.

Kim JB, Porreca GJ, Song L, Greenway SC, Gorham JM, Church GM, Seidman CE, Seidman JG (2007) Polony multiplex analysis of gene expression (PMAGE) in mouse hypertrophic cardiomyopathy. Science 316: 1481–1484.

Kim TH, Ren B (2006) Genome-wide analysis of protein-DNA interactions. Annu Rev Genom Hum Genet 7: 81–102.

Kim TK, Hemberg M, Gray JM, Costa AM, Bear DM, Wu J, Harmin DA, Laptewicz M, Barbara-Hailey K, Kuersten S, Markenscoff-Papadimitriou E, Kuhl D, Bito H, Worley PF, Kreiman G, Greenberg ME (2010) Widespread transcription at neuronal activity-regulated enhancers. Nature 465: 182–187.

Koch CM, Andrews RM, Flicek P, Dillon SC, Karaöz U, Clelland GK, Wilcox S, Beare DM, Fowler JC, Couttet P, James KD, Lefebvre GC, Bruce AW, Dovey OM, Ellis PD, Dhami P, Langford CF, Weng Z, Birney E, Carter NP, Vetrie D, Dunham I (2007) The landscape of histone modifications across 1% of the human genome in five human cell lines. Genome Res 17: 691–707.

Kodzius R, Kojima M, Nishiyori H, Nakamura M, Fukuda S, Tagami M, Sasaki D, Imamura K, Kai C, Harbers M, Hayashizaki Y, Carninci P (2006) CAGE: cap analysis of gene expression. Nat Meth 3: 211–222.

Koltai H, Weingarten-Baror C (2008) Specificity of DNA microarray hybridization: characterization, effectors and approaches for data correction. Nucl Acids Res 36: 2395–2405.

Küster H, Hohnjec N, Krajinski F, El Yahyaoui F, Manthey K, Gouzy J, Dondrup M, Meyer F, Kalinowski J, Brechenmacher L, van Tuinen D, Gianinazzi-Pearson V, Pühler A, Gamas P, Becker A (2004) Construction and validation of cDNA-based Mt6k-RIT macro- and microarrays to explore root endosymbioses in the model legume *Medicago truncatula*. J Biotechnol 108: 95–113.

Küster H, Becker A, Firnhaber C, Hohnjec N, Manthey K, Perlick AM, Bekel T, Dondrup M, Henckel K, Goesmann A, Meyer F, Wipf D, Requena N, Hildebrandt U, Hampp R, Nehls U, Krajinski F, Franken P, Pühler A (2007a) Development of bioinformatic tools to support EST-sequencing, *in silico*- and microarray-based transcriptome profiling in mycorrhizal symbioses. Phytochemistry 68: 19–32.

Küster H, Vieweg MF, Manthey K, Baier MC, Hohnjec N, Perlick AM (2007b) Identification and expression regulation of symbiotically activated legume genes. Phytochemistry 68: 1–18.

Kyung T, Hemberg M, Gray JM, Costa AM, Bear DM, Wu J, Harmin DA, Laptewicz M, Barbara-Haley K, Kuersten S, Markenscoff-Papadimitriou E, Kuhl D, Bito H, Worley PF, Kreiman G, Greenberg EM (2010) Widespread transcription at neuronal activity-regulated enhancers. Nature 465: 182–187.

Laloi C, Apel K, Danon A (2004) Reactive oxygen signalling: the latest news. Curr Opin Plant Biol 7: 323–328.

Larkin JE, Frank BC, Gavras H, Sultana R, Quackenbush J (2005) Independence and reproducibility across microarray platforms. Nat Meth 2: 337–344.

Lister R, O'Malley RC, Tonti-Filippini J, Gregory BD, Berry CC, Millar AH, Ecker JR (2008) Highly integrated single-base resolution maps of the epigenome in *Arabidopsis*. Cell 133: 523–536.

Lister R, Pelizzola M, Dowen RH, Hawkins RD, Hon G, Tonti-Filippini J, Nery JR, Lee L, Ye Z, Ngo Q-M, Edsall L, Antosiewicz-Bourget J, Stewart R, Ruotti V, Millar AH, Thomson JA, Ren B, Ecker JR (2009) Human DNA methylomes at base resolution show widespread epigenomic differences. Nature 462: 315–322.

Lohar DP, Sharopova N, Endre S, Peñuela S, Samac D, Town C, Silverstein KAT, VandenBosch KA (2005) Transcript analysis of early nodulation events in *Medicago truncatula*. Plant Physiol 140: 221–234.

Lorkowski S, Cullen P (2003) Analyzing Gene Expression. A Handbook of Methods (Possibilities and Pitfalls), vol 1& 2. Wiley-VCH, Weinheim, Germany.

Lu J, Lal A, Merriman B, Nelson S, Riggins G (2004) A comparison of gene expression profiles produced by SAGE, longSAGE, and oligonucleotide chips. Genomics 84: 631–636.

Manthey K, Krajinski F, Hohnjec N, Firnhaber C, Pühler A, Perlick AM, Küster H (2004) Transcriptome profiling in root nodules and arbuscular mycorrhiza identifies a collection of novel genes induced during *Medicago truncatula* root endosymbioses. Mol Plant-Microbe Interact 17: 1063–1077.

Mantri NL, Ford R, Coram TE, Pang ECK (2007) Transcriptional profiling of chickpea genes differentially regulated in response to high-salinity, cold and drought. BMC Genom 8: 303.

Mardis ER (2008) The impact of next-generation sequencing technology on genetics. Trends Genet 24: 142–149.

Margulies M, Egholm M, Altman WE, Attiya S, Bader JS, Bemben LA, Berka J, Braverman MS, Chen YJ, Chen Z, Dewell SB, Du L, Fierro JM, Gomes XV, Godwin BC, He W, Helgesen S, Ho CH, Irzyk GP, Jando SC, Alenquer ML, Jarvie TP, Jirage KB, Kim JB, Knight JR, Lanza JR, Leamon JH, Lefkowitz SM, Lei M, Li J, Lohman KL, Lu H, Makhijani VB, McDade KE, McKenna MP, Myers EW, Nickerson E, Nobile JR, Plant R, Puc BP, Ronan MT, Roth GT, Sarkis GJ, Simons JF, Simpson JW, Srinivasan M, Tartaro KR, Tomasz A, Vogt KA, Volkmer GA, Wang SH, Wang Y, Weiner MP, Yu P, Begley RF, Rothberg JM (2005) Genome sequencing in microfabricated high-density picolitre reactors. Nature 437: 376–380.

Marioni JC, Mason CE, Mane SM, Stephens M, Gilad Y (2008) RNA-seq: An assessment of technical reproducibility and comparison with gene expression arrays. Genome Res 18: 1509–1517.

Matsumura H, Reich S, Ito A, Saitoh H, Kamoun S, Winter P, Kahl G, Reuter M, Krüger DH, Terauchi R (2005a) Gene expression analysis of plant host-pathogen interactions by SuperSAGE. Proc Natl Acad Sci USA 100: 15718–15723.

Matsumura H, Ito A, Saitoh H, Winter P, Kahl G, Reuter M, Krüger DH, Terauchi R (2005b) SuperSAGE. Cell Microbiol 7: 11–18.

Matsumura H, Nasir KHB, Yoshida K, Ito A, Kahl G, Krüger DH, Terauchi R (2006) SuperSAGE-array: The direct use of 26-base-pair transcript tags in oligonucleotide arrays. Nat Meth 3: 469–474.

Matsumura H, Reuter M, Kruger DH, Winter P, Kahl G, Terauchi R (2007) SuperSAGE. In: KL Nielsen (ed) Methods in Molecular Biology 387. Humana Press, Totowa, NJ, USA, pp 55–70: doi: 10.1007/978-1-59745-454-4_4.

Matsumura H, Krüger DH, Kahl G, Terauchi R (2008) SuperSAGE: A modern platform for genome-wide quantitative transcript profiling. Curr Pharm Biotechnol 9: 368–374.

Matsumura H, Yoshida K, Luo S, Kimura E, Fujibe T, Albertyn Z, Barrero RA, Krüger DH, Kahl G, Schroth GP, Terauchi R (2010) High-throughput SuperSAGE for digital gene expression analysis of multiple samples using Next Generation Sequencing. PloS ONE 5: e12010.

Melzer D, Perry JR, Hernandez D, Corsi AM, Stevens K, Rafferty I, Lauretani F, Murray A, Gibbs JR, Paolisso G, Rafiq S, Simon-Sanchez J, Lango H, Scholz S, Weedon MN, Arepalli S, Rice N, Washecka N, Hurst A, Britton A, Henley W, van de Leemput J, Li R, Newman AB, Tranah G, Harris T, Panicker V, Dayan C, Bennett A, McCarthy MI, Ruokonen A, Jarvelin MR, Guralnik J, Bandinelli S, Frayling TM, Singleton A, Ferrucci L (2008) A genome-wide association study identifies protein quantitative trait loci (pQTLs). PLoS Genet 4: e1000072.

Merchan F, Breda C, Hormaeche JP, Sousa C, Kondorosi A, Aguilar OM, Megias M, Crespi M (2003) A kruppel-like transcription factor gene is involved in salt stress responses in *Medicago* spp. Plant Soil 257: 1–9.

Merchan F, de Lorenzo L, Gonzalez Rizzo S, Niebel A, Manyani H, Frugier F, Sousa C, Crespi M (2007) Identification of regulatory pathways involved in the reacquisition of root growth after salt stress in *Medicago truncatula*. Plant J 51: 1–17.

Meyers BC, Vu TH, Tej SS, Ghazal H, Matvienko M, Agrawal V, Ning J, Haudenschild CD (2004a) Analysis of the transcriptional complexity of *Arabidopsis thaliana* by massively parallel signature sequencing. Nat Biotechnol 22: 1006–1011.

Meyers BC, Tej SS, VuTH, Haudenschild CD, Agrawal V, Edberg SB, Ghazal H, Decola, S (2004b) The use of MPSS for whole-genome transcriptional analysis in *Arabidopsis*. Genome Res 14: 1641–1653.

Mitchelson KR (ed) (2007) New High-throughput Technologies for DNA Sequencing and Genomics. Perspectives in Bioanalysis, vol. 2. Elsevier, Amsterdam, The Netherlands, pp 3–44.

Molina C, Rotter B, Horres R, Udupa S, Besser B, Bellarmino L, Baum M, Matsumura H, Terauchi R, Kahl G, Winter P (2008) SuperSAGE: The drought stress-responsive transcriptome of chickpea roots. BMC Genom 9: 553–581.

Moller IM (2001) Plant mitochondria and oxidative stress: Electron transport, NADPH turnover, and metabolism of Reactive Oxygen Species. Annu Rev Plant Physiol Plant Mol Biol 52: 561–591.

Morrissy A, Morin R, Delaney A, Zeng T, McDonald H, ZhaoY, Jones S, Hirst M, Marra M (2009) Next-generation tag sequencing for cancer gene expression profiling. Genome Res 19: 1825–1835.

Mortazavi A, Williams BA, McCue K, Schaeffer L, Wold B (2008) Mapping and quantifying mammalian transcriptomes by RNA-Seq. Nat Meth 5: 621–628.

Mustafa BM, Coram TE, Pang ECK, Taylor PWJ, Ford R (2009) A cDNA microarray approach to decipher lentil (*Lens culinaris*) responses to *Ascochyta lentis*. Australas Plant Pathol 38: 617–631.

Nagalakshmi U, Wang Z, Waern K, Shou C, Raha D, Gerstein M, Snyder M (2008) The transcriptional landscape of the yeast genome defined by RNA sequencing. Science 320: 1344–1349.

Nautiyal S, Carlton VEH, Lu Y, Ireland JS, Flaucher D, Moorhead M, Gray JW, Spellman P, Mindrinos M, Berg P, Faham M (2010) High-throughput method for analyzing methylation of CpGs in targeted genomic regions. Proc Natl Acad Sci USA 107: 12587–12592.

Nolan T, Hands RE, Bustin SA (2006) Quantification of mRNA using real-time RT PCR. Nat Protoc 1: 1559–1582.

Nutter RC (2008) New frontiers in plant functional genomics using next generation sequencing technologies. In: G Kahl, K Meksem (eds) The Handbook of Plant Functional Genomics. Wiley-VCH, Weinheim, The Netherlands, pp 431–444.

Parniske M (2004) Molecular genetics of the arbuscular mycorrhizal symbiosis. Curr Opin Plant Biol 7: 414–421.

Peters DG, Kassam AB, Yonas H, O'Hare EH, Ferrell RE, Brufsky AM (1999) Comprehensive transcript analysis in small quantities of mRNA by SAGE-Lite. Nucl Acids Res 27: e39.

Pfaff T, Kahl G (2003) Mapping of gene-specific markers on the genetic map of chickpea (*Cicer arietinum* L.). Mol Genet Genom 269: 243–251.

Plessy C, Bertin N, Takahashi H, Simone R, Salimullah M, Lassmann T, Vitezic M, Severin J, Olivarius S, Lazarevic D, Hornig N, Orlando V, Bell I, Gao H, Dumais J, Kapranov P, Wang H, Davis CA, Gingeras TR, Kawai J, Daub CO, Hayashizaki Y, Gustincich S, Carninci P (2010) Linking promoters to functional transcripts in small samples with nanoCAGE and CAGEscan. Nat Meth 7: 528–534.

Reinartz J, Bruyns E, Lin J-Z, Burcham T, Brenner S, Bowen B, Kramer M, Woychik R (2002) Massively parallel signature sequencing (MPSS) as a tool for in-depth quantitative gene expression profiling in all organisms. Brief Funct Genom Proteom 1: 95–104.

Rensink WA, Buell CR (2005) Microarray expression profiling resources for plant genomics. Trends Plant Sci 10: 603–609.

Richmond T, Somerville S (2000) Chasing the dream: plant EST microarrays. Curr Opin Plant Biol 3: 108–116.

Ruzanov P, Riddle DL (2010) Deep SAGE analysis of the *Caenorhabditis elegans* transcriptome. Nucl Acids Res 38: 3252–3262.

Saha S, Sparks AB, Rago C, Akmaev V, Wang CJ, Vogelstein B, Kinzler KW, Velculescu VE (2002) Using the transcriptome to annotate the genome. Nat Biotechnol 20: 508–512.

Samuel M, Miles G, Ellis B (2005) MAPK signalling and plant cell survival in response to oxidative environmental stress. In: K Omasa, I Nouchi (eds) Plant Responses to Air Pollution and Global Change. Springer, Tokyo, Japan, pp 125–132.

Sanger F, Nicklen S, Coulson AR (1977) DNA sequencing with chain-terminating inhibitors. Proc Natl Acad Sci USA 74: 5463–5467.
Schadt EE, Monks SA, Drake TA, Lusis AJ, Che N, Colinayo V, Ruff TG, Milligan SB, Lamb JR, Cavet G, Linsley PS, Mao M, Stoughton RB, Friend SH (2003) Genetics of gene expression surveyed in maize, mouse and man. Nature 422: 297–301.
Sharma CM, Hoffmann S, Darfeuille F, Reignier J, Findeiss S, Sittka A, Chabas S, Reiche K, Hackermüller J, Reinhardt R, Stadler PF, Vogel J (2010) The primary transcriptome of the major human pathogen *Helicobacter pylori*. Nature 464: 250–255.
Sharma PC, Grover A, Kahl G (2007) Mining microsatellites in eukaryotic genomes. Trends Biotechnol 25: 490–498.
Shendure J (2008) The beginning of the end for microarrays? Nat Methods 5: 585–587.
Shilatifard A (2006) Chromatin modifications by methylation and ubiquitination: implications in the regulation of gene expression. Annu Rev Biochem 75: 243–269.
Skiba B, Ford R, Pang ECK (2005) Construction of a cDNA library of *Lathyrus sativus* infected with *Mycosphaerella pinodes* and the identification of Expressed Sequence Tags (ESTs). Physiol Mol Plant Pathol 66: 55–67.
Slezak-Prochazka I, Durmus S, Kroesen B-J, van den Berg A (2010) MicroRNAs, macrocontrol: Regulation of miRNA processing. RNA 16: 1087–1095.
Stacey G, Libault M, Brechenmacher L, Wan J, May GD (2006) Genetics and functional genomics of legume nodulation. Curr Opin Plant Biol 9: 110–121.
Sun M, Zhou G, Lee S, Chen J, Shi RZ, Wang SM (2004) SAGE is far more sensitive than EST for detecting low-abundance transcripts. BMC Genom 5: 1.
Talbert PB, Henikoff S (2010) Histone variants—ancient wrap artists of the epigenome. Nat Rev Mol Cell Biol 11: 264–275.
Tang F, Barbacioru C, Nordman E, Li B, Xu N, Bashkirov VI, Lao K, Surani MA (2010) RNA-Seq analysis to capture the transcriptome landscape of a single cell. Nat Protoc 5: 516–535.
Tattersall AD, Turner L, Knox MR, Ambrose MJ, Ellis THN, Hofer JMI (2005) The mutant crispa reveals multiple roles for PHANTASTICA in pea compound leaf development. Plant Cell 17: 1046–1060.
Terauchi R, Matsumura H, Krüger DH, Kahl G (2008) SuperSAGE: The most advanced transcriptome technology for functional genomics. In: G Kahl, K Meksem (eds) The Handbook of Plant Functional Genomics (Concepts and Protocols). Wiley-Verlag Chemie, Weinheim, Germany, pp 37–51.
Tesfaye M, Silverstein KAT, Bucciarelli B, Samac DA, Vance CP (2006) The Affymetrix *Medicago* GeneChip® array is applicable for transcript analysis of alfalfa (*Medicago sativa*). Funct Plant Biol 33: 783–788.
Trapnell C, Williams BA, Pertea G, Mortazavi A, Kwan G, van Baren MJ, Salzberg SL, Wold BJ, Pachter L (2010) Transcript assembly and quantification by RNA-Seq reveals unannotated transcripts and isoform switching during cell differentiation. Nat Biotechnol 28: 511–515.
Tucker ML, Whitelaw CA, Lyssenko NN, Nath P (2002) Functional analysis of regulatory elements in the gene promoter for an abscission-specific cellulase from bean and isolation, expression, and binding affinity of three TGA-type basic leucine zipper transcription factors. Plant Physiol 130: 1487–1496.
Udvardi MK, Kakar K, Wandrey M, Montanari O, Murray J, Andriankaja A, Zhang J-Y, Benedito V, Hofer JME, Chueng F, Town CD (2007) Legume transcription factors: Global regulators of plant development and response to the environment. Plant Physiol 144: 538–549.
Udvardi MK, Czechowski T, Scheible W-R (2008) Eleven golden rules of quantitative RT-PCR. Plant Cell 20: 1736–1737.
Van den Berg BHJ, McCarthy FM, Lamont SJ, Burgess SC (2010) Re-annotation is an essential step in systems biology modeling of functional genomics data. PLoS ONE 5: e10642: doi: 10.1371/journal.pone.0010642.
Varshney RK, Graner A, Sorrells ME (2005) Genomics-assisted breeding for crop improvement. Trends Plant Sci 10: 621–630.

Varshney RK, Hiremath PJ, Lekha P, Kashiwagi J, Balaji J, Deokar AA, Vadez V, Xiao Y, Srinivasan R, Gaur PM, Siddique KH, Town CD, Hoisington DA (2009) A comprehensive resource of drought- and salinity- responsive ESTs for gene discovery and marker development in chickpea (*Cicer arietinum* L.). BMC Genom 10: 523.

Velculescu VE, Zhang L, Vogelstein B, Kinzler KW (1995) Serial analysis of gene expression. Science 270: 484–487.

Velculescu VE, Zhang L, Zhou W, Vogelstein J, Basrai MA, Bassett DE, Hieter P, Vogelstein B, Kinzler KW (1997) Characterization of the yeast transcriptome. Cell 88: 243–251.

Vilain C, Libert F, Venet D, Costagliola S, Vassart G (2003) Small amplified RNA-SAGE: an alternative approach to study transcriptome from limiting amount of mRNA. Nucl Acids Res 31: e24.

Wahl MB, Heinzmann U, Imai K (2005) LongSAGE analysis significantly improve genome annotation: identifications of novel genes and alternative transcripts in the mouse. Bioinformatics 21: 1393–1400.

Wang Z, Gerstein M, Snyder M (2009) RNA-Seq: a revolutionary tool for transcriptomics. Nat Rev Genet 10: 57–63.

Yauk CL, Berndt ML, Williams A, Douglas GR (2004) Comprehensive comparison of six microarray technologies. Nucl Acids Res 32: e124.

Ye SQ, Zhang LQ, Zheng F, Virgil D, Kwiterovich PO (2000) MiniSAGE: gene expression profiling using serial analysis of gene expression from 1 microg total RNA. Anal Biochem 287: 144–52.

Zhang JY, Broeckling CD, Blancaflor EB, Sledge MK, Sumner LW, Wang ZY (2005) Overexpression of WXP1, a putative *Medicago truncatula* AP2 domain-containing transcription factor gene, increases cuticular wax accumulation and enhances drought tolerance in transgenic alfalfa (*Medicago sativa*). Plant J 42: 689–707.

Zhang X, Yazaki J, Sundaresan A, Cokus S, Chan SW (2006) Genome-wide high-resolution mapping and functional analysis of DNA methylation in *Arabidopsis*. Cell 126: 1189–1201.

Zhu JK (2002) Salt and drought stress signal transduction in plants. Annu Rev Plant Biol 53: 247–273.

Zilberman D, Gehring M, Tran RK, Ballinger T, Henikoff S (2007) Genome-wide analysis of *Arabidopsis thaliana* DNA methylation uncovers an interdependence between methylation and transcription. Nat Genet 39: 61–69.

Zucchero JC, Caspi M, Dunn K (2001) *ngl9*: a third MADS box gene expressed in alfalfa root nodules. Mol Plant-Microbe Interact 14: 1463–1467.

7

Synteny and Comparative Genomics between Model and Cool Season Grain Legumes

Péter Kaló,[1,a,*] *Hong-Kyu Choi,*[2] *Noel Ellis*[3] and *György B. Kiss*[1]

ABSTRACT

Legumes are second only to the Gramineae in their importance to humans. Traditionally used as a source of proteins and nitrogen in both natural and agricultural ecosystems, legumes account for nearly 30% of the world's primary crop production. These factors have stimulated the development of legume genomic resources over the past decades. Despite the progress achieved in two model species, *Medicago truncatula* and *Lotus japonicus*, and the long history in genetic studies in pea, relatively little genomic information is available for cool season grain legumes. Among the available genomic approaches, comparative genetic mapping is especially important for species with large and complex genome organization such as legume crops. Mapping of orthologous genes has identified large scale conservation between the genomes of galegoid legumes (often referred as temperate or cool season legumes), as well as chromosomal rearrangements that, in some cases, underlie the variation in chromosome number between these species. Because of the limited availability of large-scale genomic sequences in cool season grain legumes, this chapter focuses on comparative mapping between the most important cool season grain legumes at macrosynteny level. Results achieved so far suggest that the knowledge gained from comparative mapping may have considerable utility to solve basic and applied agronomic questions of importance in the crop species.

[1]Agricultural Biotechnology Center, Gödöllő, Szent-Györgyi A. u. 4., 2100 Hungary.
[a]e-mail: *kalo@abc.hu*
[2]Dong-A University, 840 Hadan-2-Dong, Saha-Gu, Busan 604-714, Republic of Korea.
[3]John Innes Centre, Department of Crop Genetics, Norwich, Colney Lane, NR4 7UH, UK.
*Corresponding author

Keywords: model legumes, grain legumes, orthologous regions, comparative mapping, intron-targeted PCR-based markers, breeding

7.1 Introduction

Comparative genomics studies the evolutionary relationship between genes and the genomes of species from different taxa using molecular tools. Vavilov's botanical study (1922) launched comparative genetic analysis as he pointed out that homology of characters is probably associated with homologous genes. The identification of homologous genes from different species encountered difficulties at that time but modern powerful technologies make homologous gene identification easier and confirmed Vavilov's insight. Depending on the degree of their evolutionary relationship, different species preserve similarities in the content, proximity (synteny) and linear order (colinearity) of genes in their genomes. The pattern is consistent with similarity by descent; that it becomes increasingly disrupted as time passes. The preserved conservation of gene order is clear in vertebrates (Thomas and Touchman 2002; Ferguson-Smith and Trifonov 2007; Kemkemer et al. 2009) following hundreds of millions years of divergence. Comparing the genomes of flowering plants is often problematic mainly due to the frequent polyploidization events in ancient plant genomes and consecutive gene loss (Bowers et al. 2003).

Comparative mapping and genome analysis investigate conservations and differences in content and order of homologous genes among different taxa (Fig. 7-1). For comparative mapping, only homologous genes originating from the same ancestral loci, i.e., the orthologous loci, should be used. Orthologous sequences can be detected either by cross-species genetic mapping or by computational methods. The identification of orthologous positions in distantly related species is a significant output from comparative mapping although the identification of orthologous gene-pairs is sometime difficult, since the determination of the orthologous term refers to evolutionary events (lineages), which can not be seen directly. Originally comparative analyses of genomes were performed based on genetic maps developed with molecular markers but the increasing availability of large-scale genome sequences could make comparisons direct, more extensive and exact. The determination of orthologous positions may often involve the development and genetic mapping of single or low-copy gene-based genetic markers. At the beginning of the comparative mapping studies, heterologous markers (markers which were derived from one species but used in other ones) were used to compare genetic maps of different species but later the comparison of gene function of the mapped genes in different species and the analysis of their relationship came to the fore.

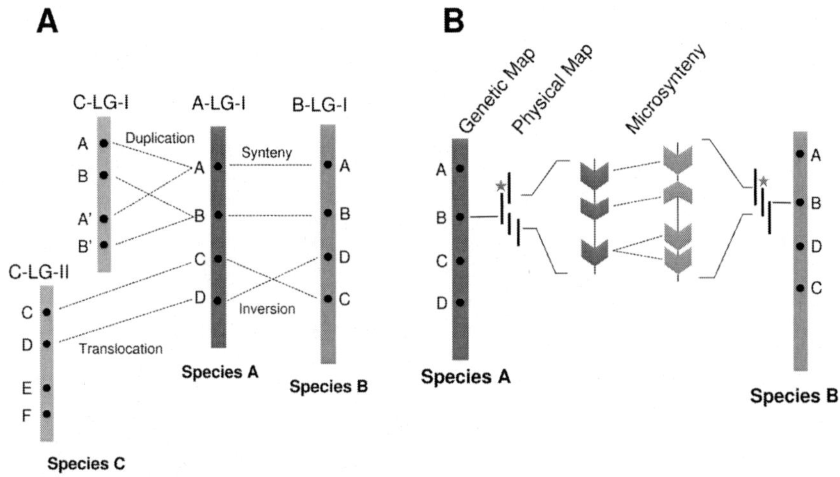

Figure 7-1 Different levels of comparative genome analysis. (**A**). Macrosynteny analysis compares linkage groups or genetic maps to identify conserved gene order and genetic rearrangements between multiple genomes. Gray vertical bars represent linkage groups of three species with homologous genetic markers. Genetic markers are designated with capital letters. Markers A, B, C and D originated from the common ancestor and therefore they are orthologs. The marker order in Species "A" resembles the order of the four markers in the common ancestor of the three species. Following some genetic rearrangements during the evolution markers A, B, C and D retained their location in species in "A and B", and markers A and B remained on the same linkage group (LG I) in all three species, that is these markers identify a chromosomal segment of conserved synteny. Markers C and D are translocated to another linkage group in species C and they are not syntenic with markers A, B, C and D in species "A and B". Markers A′ and B′ emerged by duplication within "species C" and therefore markers A-A′ and B-B′ in linkage group C-LG-I represent paralogs. (**B**) Microsynteny analysis aligns genes located in a small portion of chromosomes of distinct species. The gene order of the reference genome of "species A" facilitates the selection of a BAC clone (indicated by an asterisk) for further analysis in a less analyzed genome of an economically important crop species ("B"). Capital letters denote individual genetic markers. Solid black vertical lines indicate BAC clones and the star shows BAC clone chosen for the sequencing. LG, linkage group.

The availability of a large amount of expressed sequence tags (EST) and genomic sequence data in recent years enable computer based analysis of a vast number of genes and extended genomic regions. These analyses gave great impetus to the alignment of large-scale genome segments of different species and provided more comprehensive insights into the structure and the evolution of chromosomes. The similarity in gene content and order can be investigated within small-scale genomic regions (microsynteny) or at the level of genetic maps (macrosynteny) (Fig. 7-1) but nowadays the difference is blurred because of in silico comparative mapping methods.

The comparison of the genetic maps is especially important for species with large and complex genome organization. Multiple studies in the last three decades, in many plant taxa using linkage maps, revealed remarkable synteny, predominantly within plant families (Paterson et al. 2000). Conserved gene order between plants representing distinct plant families could be identified mostly at microsynteny level rather than among large chromosome segments (Devos et al. 1999; Oh et al. 2002; Zhu et al. 2003, 2005; Kevei et al. 2004; Stracke et al. 2004, and many others). Systematic mapping efforts in the grass family (Gramineae) detected conserved relationships between linkage groups of grass species and resulted in the development of a consensus grass comparative map (Devos and Gale 1997; Gale and Devos 1998) and the idea of grass species as a "single genetic system" (Bennetzen and Freeling 1993). The identification of large conserved segments between the genomes within the Solanaceae family (Bonierbale et al. 1988; Gebhardt et al. 1991; Tanksley et al. 1992; Livingstone et al. 1999) revealed relatively simple patterns of chromosomal rearrangements between these species and allowed a reconstruction of the ancestral genome of solanaceous plants (Doganlar et al. 2002). General conservation between the genomes of species belonging to the Brassicaceae family has also been revealed by Kowalski et al. (1994), Lagercrantz (1998), Babula et al. (2003) and Lukens et al. (2003). While large-scale macrosynteny is mainly restricted to species within the same plant family, in a few cases comparative mapping has identified colinearity between large chromosomal blocks of markers across species separated by wider taxonomic distances (Lee et al. 2001; Wu et al. 2006).

The legume family (Fabaceae) is the third largest family of the flowering plants containing many diverse and agriculturally important crops. Legume species are found throughout the world with a wide range of habits; they can grow as trees, shrubs or herbs. The estimated 20,000 species within the legume family traditionally have been divided into three subfamilies: Ceasalpinioideae, Mimosoideae and Papilionoideae (Doyle and Luckow 1993). Almost without exception, all the economically important grain and forage legumes fall into the seven clades of the Papilionoideae subfamily (Gepts et al. 2005; Cronk et al. 2006; Cannon et al. 2009; Fig. 7-2), or rather two main clades, the Galegoid and Phaseoloid contain the agricultural important crops. The exceptions are peanut (member of the clade Aeschynomenoid/Dalbergoids) and the *Lupinus* species, which belong to the Genistoids clade. Galegoid clade members are often referred as temperate or cool season legumes while the members of Phaseoloid group are termed tropical or warm season legumes. The species in Galegoid and Phaseoloid clades separated around 54 million years ago (Mya, Lavin et al. 2005; Cronk et al. 2006). The Galegoid clade is generally divided into two main groups, the Robinioids (containing the genus *Lotus*) and the so called IRLC (inverted repeat-lacking clade) group, defined by the loss of an

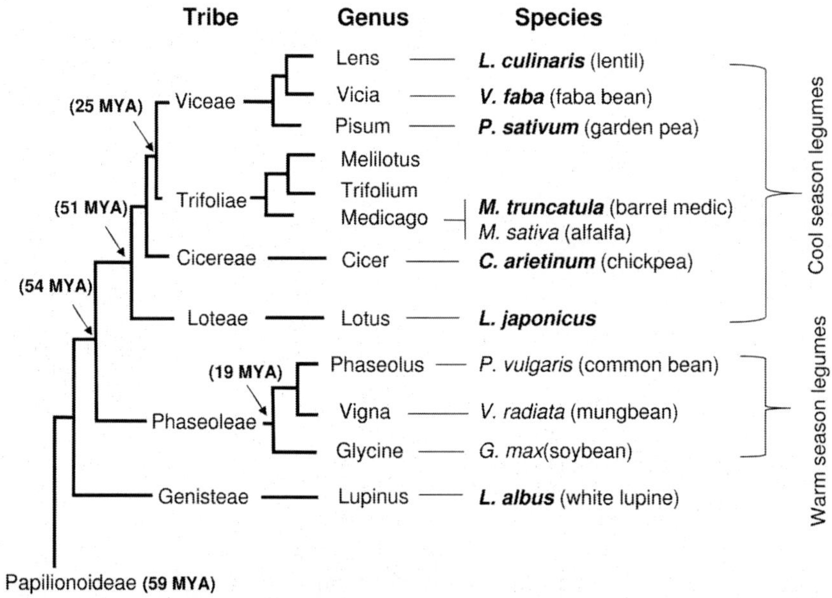

Figure 7-2 Taxonomic relationships of major cool and warm season legumes indicating the estimated time of the divergence of the lineages. Cool season legumes and the reference legume species described in this chapter are highlighted in italic and bold. MYA, million years ago. The phylogenetic tree is adapted from Choi et al. (2004a) and the estimated time of divergence of the lines is adopted from Gepts et al. (2005) and Cronk et al. (2006).

inverted repeat in their chloroplast genome, which includes the Vicieae and Trifolieae tribes. The divergence of these two lineages within the Galegoid group is estimated to have occurred about 51 Mya while the split of the Vicieae (containing the genus *Pisum*) and Trifolieae (including the genus *Medicago*) tribes is estimated at ca. 20–25 Mya (Wojciechowski 2003). This phylogenetic relationship is reflected in the detected genomic colinearity between legumes; the most conserved synteny was detected between the closely related species (same genus or tribe) and less conservation was found between species from different clades. In this chapter we summarize the results of comparative genetic and genomic studies among the cool season grain legumes and review the comparative studies between different cool season pulses and model legumes.

The cool season grain legumes are well known economically important pulse species of the Hologalegina clade such as the members of the genus *Pisum*, *Vicia*, *Lens*, *Lathyrus* (within the Vicieae) and *Cicer* (in the Cicereae, which diverges basally to the Trifolieae and Vicieae). Taxonomically speaking *Lupinus* species are not members of this clade but they are often included in cool season legumes because of their growth habit and use, therefore comparative genomics studies performed on lupins are also discussed in this chapter.

7.2 Generation of Cross-Legume Gene-based Genetic Markers

The large-scale genome comparison (macrosynteny) analysis involves generation and mapping of homologous loci or computer analysis of gene content and order in genomic sequences among various species. The comparison of the position of homologous loci identifies orthologous positions which evolved by speciation in contrast to other homologous positions (paralogs), which emerged by duplication in the same species. Therefore single or low-copy genetic loci, usually genes, can determine orthologous positions efficiently between different genomes. In order to map the orthologous genes in different species the generation of the so-called intron-targeted (IT) PCR-based genetic markers was applied in several studies (Choi et al. 2004a, b; Fredslund et al. 2006; Phan et al. 2006; Ellwood et al. 2008; Seres et al. in prep.) in legumes. IT markers are generated with gene specific PCR primer pairs complementary to exon sequences, which flank a region containing one intron or one or more exon-intron units. Targeting the sequence of highly conserved exon sequences among taxa can guarantee that the same genes or the members of a same gene family are amplified and in addition the amplification of less conserved intron sequences guarantee the highest degree of polymorphism. In order to provide the cross-species amplification usually legume consensus sequences are generated from cDNA, ESTs and tentative consensus sequences (TC) representing single and low copy number genes. In order to identify intron sequences the legume consensus sequences are aligned to the available legume, principally *Medicago truncatula* (barrel medic) genomic sequences with computer programs. If there is no legume genomic sequence available, then *Arabidopsis thaliana* genomic sequence can be used efficiently because, as found in many cases (e.g., Choi et al. 2004a), the exon-intron structure of homologous genes is maintained during evolution. Finally, PCR primers designed for the exon sequences flanking introns are used to generate genetic markers in different species. These primers are usually tested by temperature gradient or touch-down PCR for amplification and polymorphism using the parental lines of the mapping population of the legumes to be analyzed. The map positions of the polymorphic PCR fragments representing genes are determined and the markers are analyzed to reveal colinear genomic segments between species.

In the study performed by Choi et al. (2004a, b) 167 IT gene-specific PCR markers were tested for polymorphism in seven different legumes and used efficiently for comparative mapping. IT primers have been generated following a similar approach in a comparative mapping program of the Grain Legumes Integrated Project (GLIP; funded by the EU; Ellwood et al. 2008; Seres et al., in prep.) to analyze syntenic relationships between grain legumes.

Altogether 626 IT markers were tested for amplification in the study performed by Phan et al. (2006) to construct the lentil (*Lens culinaris*) and white lupin (*Lupinus albus*) genetic maps and 90% of the IT markers resulted in amplicons in *M. truncatula*, 80% in white lupin and 70% in lentil. Finally, in lentil and white lupin about 15 and 20%, respectively of the IT PCR primers tested generated genetic markers, and were used for comparative mapping analysis between these species and *M. truncatula*. IT primers from four different sets were tested on faba bean (*Vicia faba*) genomic DNA in the study carried out by Ellwood et al. (2008). Around 64% of the tested 783 primer pairs produced amplified products of which 5% showed length polymorphism. The sequences of 254 amplicons were determined and their sequence analysis revealed that 165 showed single nucleotide polymorphism. Finally a total of 151 genetic markers (19% of the 783 IT primer pairs) were positioned on the genetic map of faba bean.

7.3 Comparative Genetic Maps between Cool Season Grain Legumes

Legume comparative genomics focuses mainly on two economically important legume groups in the Papilionoideae subfamily (Doyle and Luckow 2003); 1) the Hologalegina group is represented in genomic studies mainly by the model legumes, *M. truncatula* and *Lotus japonicus*, and 2) due to their economical importance comprehensive genomic resources have been developed for soybean (*Glycine max*) (www.soybase.org) and common bean (*Phaseolus vulgaris*) (Broughton et al. 2003; Schlueter et al. 2008), as representative members of the tropical season legumes. The genome of the latter species is being sequenced by an international consortium. Despite the impressive progress achieved in the genomic studies in model legumes and the long history in genetic studies in pea, relatively less genomic information is available for cool season grain legumes. Due to their complex genome organization or in some cases (e.g., chickpea) and low level of polymorphism, mainly genetic maps have been developed and EST sequencing efforts have been launched for these species. Because of the limited availability of large-scale sequences of cool season grain legumes, which hampers the detailed microsynteny analysis, this chapter focuses on comparative mapping between cool season grain legumes at macrosynteny level.

Comparative genetic analysis among legumes species was initiated by Vavilov's study including series of similar heritable variations in related Papillionoid species. The first molecular proofs for the existence of macrosynteny between legumes were given by the comparison of genetic maps of economically important grain legumes. Isozyme and RFLP markers were first used by Weeden et al. (1992) and Simon and Muehlbauer (1997) to compare the incomplete genetic maps of lentil (*Lens culinaris*; $2n = 14$) and chickpea (*Cicer arietinum*; $2n = 16$) with the pea (*Pisum sativum*; $2n = 14$) linkage map. Consistent with their taxonomic relationships; lentil and pea belong to the Vicieae tribe while *Cicer* represents the Cicereae tribe of the IRLC clade (Cronk et al. 2006 and Fig. 7-2), these studies revealed eight and five large syntenic blocks, respectively with higher degree of genome conservation between lentil and pea compared to chickpea and pea. This study was continued by generating gene-based genetic markers to amplify orthologous sequences from pea and lentil and the detected high level of colinearity supported the previously demonstrated genome conservation between the two species (Brauner et al. 2002). The detected relationship of certain lentil and chickpea chromosomes to the updated pea linkage groups was presented by Ellis and Poyser (2002).

Since then, no direct comparative mapping studies have been done between cool season grain legumes. In the last decade the development of PCR-based genetic markers significantly advanced comparative genome analysis between legumes using model species as reference points. The most comprehensive macrosynteny analysis between legume genomes was reported by Choi et al. (2004a) and Zhu et al. (2005). Cross-species gene specific markers were used to identify homologous genome segments among eight legume species (*M. truncatula*, alfalfa, *L. japonicus*, pea, chickpea, soybean, mungbean and common bean). Using the *M. truncatula* genetic map as a reference genome the eight legume genomes were aligned and a simplified consensus map created. The degree of colinearity between the legumes reflected their phylogenetic relationship; the most conserved synteny was detected between the *M. truncatula* and *M. sativa*, and within-clade comparisons (*Medicago*–pea–chickpea) revealed a few chromosomal rearrangements while comparisons between species in different clades detected less conservation. These analyses included only two cool season grain legumes, pea and chickpea but the developed consensus legume comparative map showed several syntenic blocks between chickpea and pea chromosomes as it is shown in Figs. 7-3 and 7-4. This study was used to draw the schematic presentation of the large genetic rearrangements between the linkage groups of *Medicago*, pea and chickpea as shown in Figs. 7-3 and 7-4. All the eight linkage groups of chickpea were aligned along the linkage groups of pea and *M. truncatula* but not all the *Medicago* and pea linkage groups were covered: chromosome 6 of *M. truncatula* and linkage group VI of pea remained unpaired. The incomplete one-to-one relationship between chickpea and *Medicago* chromosomes is discussed in the next section.

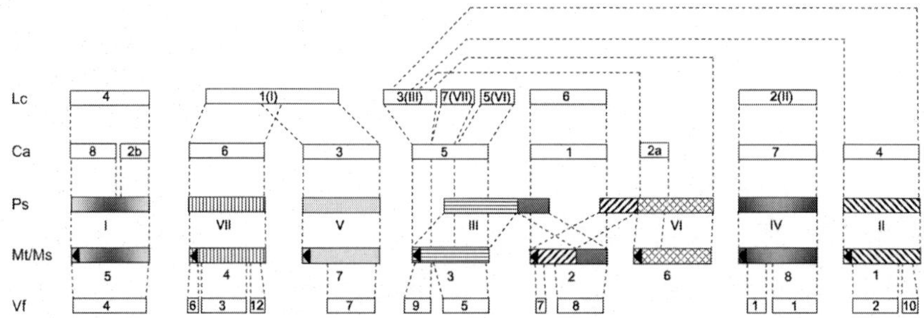

Figure 7-3 Schematic representation of large-scale synteny blocks between chromosomes and chromosome segments of *M. truncatula* (Mt) and cool season grain legumes; pea (*Pisum sativum*; Ps), faba bean (*Vicia faba*; Vf), chickpea (*Cicer arietinum*; Ca) and lens (*Lens culinaris*; Lc). Bars representing *Medicago* and pea homologous chromosomal regions are shown with the same gray intensity or pattern. Arrows in the boxes indicate the orientation of the chromosomes (short arm – long arm) in the case of *Medicago*. The corresponding synteny blocks of faba bean, chickpea and lentil are represented by blank bars. The bars do not reflect the relative sizes of chromosome or chromosome segments and the break points of chromosomes are indicated approximately. The data is integrated from the studies performed by Choi et al. (2004a), Kaló et al. (2004), Zhu et al. (2005), Nelson et al. (2006), Phan et al. (2006), Ellwood et al. (2008), and Nayak et al. (2010).

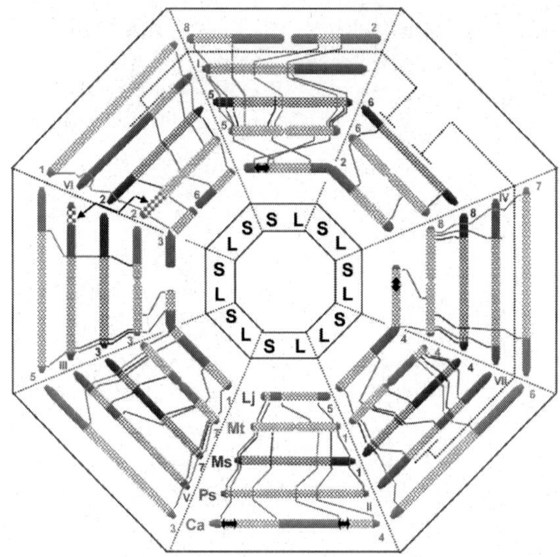

Figure 7-4 The consensus comparative map for five cool season legume species presented by octagon The figure is adapted from the data of Choi et al. (2004a) with modification. Mt, *M. truncatula*; Ms, *M. sativa*; Lj, *L. japonicus*; Ps, *Pisum sativum*; Ca; *C. arietinum*. S and L denote short and long arms of each chromosome in *M. truncatula*. Synteny blocks are drawn to scale based on genetic distance. Solid lines, postulated rearrangement; double-headed arrows, postulated inversions.

Color image of this figure appears in the color plate section at the end of the book.

Faba bean (*Vicia faba*; $2n = 2x = 12$) is one of the most important cool season legumes but its genomic manipulation is hampered by its large genome size (~ 13,000 Mbp/1C). By developing and mapping gene-specific cross-legume markers, Ellwood and his co-workers (2008) analyzed the macrosyntenic relationship between faba bean and lentil connecting their linkage groups to *M. truncatula* chromosomes. They detected shared chromosomal arrangements of faba bean and lentil compared to *Medicago*; *Vicia* Vf5 and *Lens* Lc7 (VII) and Vf9 and Lc 5(VI) were syntenic to *Medicago* Mtr 3, respectively which confirmed their closer phylogenetic relationship (Fig. 7-3).

7.4 Comparative Genomics in Relation to Legume Model Species

Historically, genetic studies performed on crop legumes were mainly focused on pea. The development of a comprehensive study of genome conservation within cool season legumes was hindered by the initial lack in genomic resources, complicated genome organization, large genome size and limited number of common genetic markers. Two species, *M. truncatula* and *L. japonicus* possessing attractive features for genetic and genomic tools (small diploid genomes, short life cycle, self-fertility, autogamy, etc.), were selected independently as legume model species (Barker et al. 1990; Handberg and Stougaard 1992; Cook et al. 1999). Genetic maps (Sandal et al. 2002; Choi et al. 2004b) and methods in functional genomics have been developed for model legumes and the determination of the sequence of the gene-rich euchromatic parts of these genomes is underway (Sato et al. 2007; Young et al., in prep.). These new resources provide the basis for macro- and microsynteny studies between model and crop legumes.

The PCR-based IT genetic markers developed for *M. truncatula* (Choi et al. 2004b) have been used to detect homologous loci in the genetic map of various legumes (Choi et al. 2004a; Zhu et al. 2005). The comparison at the chromosome level revealed large-scale synteny blocks between *M. truncatula* and two cool season grain legumes, pea (*Pisum sativum*; $2n = 14$) and chickpea (*Cicer arietinum*; $2n = 16$). The substantial conservation in the gene order between *M. truncatula* and pea identified genetic rearrangements between the two genomes, which account for the difference in their chromosome number. Consistent syntenic blocks and rearrangements were identified between alfalfa (*Medicago sativa*; $2n = 16$) and pea (Kaló et al. 2004). These structural rearrangements between *Medicago* and *Pisum* chromosomes indicated that the 10-fold difference in their genome size is not the result of large-scale genome multiplication in pea, but rather multiple transposon insertions (Vershinin et al. 2003).

Although the separation of the chickpea preceded the split of Trifolieae and Vicieae lineages, *Medicago* and chickpea have the same haploid chromosome number (*n* = 8). Because of the limited availability of cross-genome genetic markers between *Medicago* and *Cicer* no complete one-to-one relationship has been found for all linkage groups (LGs). Recent genetic studies generated a consensus linkage map of chickpea containing gene-based genetic markers suitable for genome comparison with *M. truncatula* (Millan et al. 2010; Nayak et al. 2010). Six LGs of chickpea could be completely aligned to six *M. truncatula* chromosomes but no counterpart of the *Medicago* LG 6 could be detected and the *Medicago* LG 5 was colinear with chickpea LGs 2 and 8 (Figs. 7-3 and 7-4). *Medicago* chromosome 6 has unusual features compared to other *Medicago* chromosomes. According to cytogenetic studies, chromosome 6 shows a high proportion of heterochromatin (Kulikova et al. 2004) and its euchromatic regions are interspersed with large blocks of heterochromatin. It contains a vast number of NBS-LRR type resistance genes (Zhu et al. 2002), far more than any other *Medicago* chromosome. Moreover, as discussed by Kaló et al. (2000), Choi et al. (2004b) and Bertioli et al. (2009), this *Medicago* LG has revealed less gene-specific markers than any other LGs reported so far. As a result, chromosome 6 has shown a low level of synteny with the corresponding pea LG VI, which is involved in the genetic rearrangements resulting in the different chromosome number between both species (Choi et al. 2004a; Kaló et al. 2004). Similarly, no corresponding genomic regions of *Medicago* chromosome 6 have been identified in chickpea in the study described by Millan et al. (2010) and only a single genetic marker detected on chickpea chromosome 2 has an orthologous counterpart on *Medicago* chromosome 6 in another comparative mapping study (Nayak et al. 2010).

The genomic resources developed for *M. truncatula* boosted comparative genomic analyses between *Medicago* and other legumes. Comparative mapping studies using intron-targeted gene-based markers identified simple syntenic relationship between barrel medic and lentil (Phan et al. 2006) and in another study (Ellwood et al. 2008) lentil and faba bean linkage groups were associated to the *Medicago* genetic map. The macrosynteny between faba bean and *M. truncatula* was clear; a high level of colinearity was found between *Medicago* LG 1, 5 and 8, while moderate conservation was detected in the case of *Medicago* chromosomes 2, 3, 4 and 7 (Fig. 7-3). Interestingly no macrosynteny was found between any faba bean LGs and *Medicago* chromosome 6, similar to the comparison between chickpea and *Medicago*. Based on the faba bean and *Medicago* comparison it was concluded that no segmental duplication or polyploidization is responsible for the massive (~ 25x) increase of the faba bean genome size compared to the model legume. A high level of colinearity and moderate chromosomal rearrangements was also identified between the lentil and barrel medic genomes in this study and by Phan et al. (2006). The haploid chromosome number of lentil and pea—both belonging to the same Vicieae tribe of the Galegoid legumes - is seven. These comparative genome studies revealed that the reduction of the chromosome number compared to *Medicago* is not the result of the same genomic rearrangements; for e.g., the *Medicago* LGs 4 and 7 correspond to LGs VII and V in pea but they constitute a single linkage group (LG 1) in lentil, while the *Medicago* LG 2 is involved in a genomic rearrangements with pea LG III and VI but it corresponds to lentil LG 6. The synteny blocks between certain cool season legumes and *Medicago* are presented in Figs. 7-3 and 7-4. These figures summarize the chromosomal rearrangements detected in the study of Choi et al. (2004a) Kaló et al. (2004), Zhu et al. (2005), Phan et al. (2006), Ellwood et al. (2008), Millan et al. (2010) and Nayak et al. (2010).

Using RFLP, PCR-based gene-specific and AFLP markers, the first gene-based linkage map of *Lupinus angustifolius* (narrow-leafed lupin; $2n = 40$) was constructed by Nelson et al. (2006). This study found evidence for widespread duplication in the narrow-leafed lupin genome. Additional comparative mapping studies using a similar approach—mapping intron-targeted gene-based markers in various legumes and using *M. truncatula* as a reference map–identified more complex pattern of homologous segments between *M. truncatula* and two species of the *Lupinus* genus (Nelson et al. 2006; Phan et al. 2006, 2007). Out of the 20 *L. angustifolius* LGs, 14 showed synteny with *M. truncatula* chromosome segments and only *Medicago* chromosome 6 did not show syntenic regions in *L. angustifolius*. Likewise in narrow-leafed lupin, 23 of the 50 white lupin (*Lupinus albus*) LGs revealed syntenic regions with the *Medicago* genome. Generally, two or more LGs of lupin species were syntenic to individual *M. truncatula* chromosomes as was expected on account of higher chromosome numbers in these species (*L. albus* $2n = 50$; *L. angustifolius* $2n = 40$). The complex relationships between *M. truncatula* and these two lupin species are shown in Fig. 7-5.

Generating intron-targeted gene-based anchor markers for legume species and using the *M. truncatula* as a reference genome for comparative mapping was also applied in the comparative mapping program of the Grain Legumes Integrated Project (GLIP; funded by the EU) to analyze macrosyntenic relationships between pea, chickpea, faba bean (*Vicia faba*), common bean (*Phaseolus vulgaris*), lupin (*Lupinus angustifolius*) and lentil (*Lens culinaris*). The alignment of the legume genetic maps is currently underway but preliminary data also prove that high level of macrosynteny exists between the genomes of *M. truncatula*, faba bean and chickpea (A. Torres and P. Winter, pers. comm.) as found by other studies (Choi et al. 2004a; Ellwood et al. 2008; Millan et al. 2010; Nayak et al. 2010).

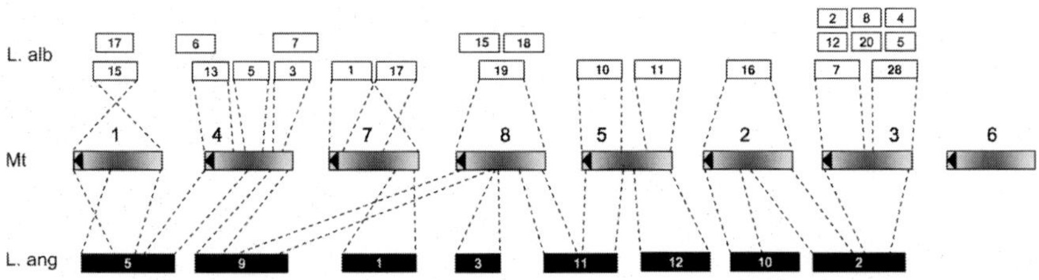

Figure 7-5 Schematic representation of large-scale synteny blocks between the genome of narrow-leafed (*Lupinus angustifolius*; L. ang.; black bars) and white lupin (*Lupinus albus*; L. alb.; white bars) and *M. truncatula*. The bars do not reflect the relative sizes of chromosome or chromosome segments and the break points of chromosomes are indicated approximately. The diagram is drawn based on the studies published by Nelson et al. (2006), Phan et al. (2006, 2007).

These recent comparative genomic studies have mainly used *M. truncatula* as a reference genome and revealed that colinearity exists between legume species to a different extent, depending on their phylogenetic distance. In order to support comparative legume biology, a few years ago the Legume Information System (LIS) was developed (Gonzales et al. 2005), which enables to integrate genetic and physical map data and to carry out in silico macrosynteny and comparative genome analyses among legumes species.

7.5 Comparative Genomics as a Breeding Tool for Cool Season Grain Legumes

The comparative mapping analyses between legumes available so far have showed that their genome organization is well conserved and the degree of synteny depends on their phylogenetic distance. In general, closely related species share a higher level of colinearity in the order of homologous genes. *M. truncatula* and the cool season pulses belong to closely related tribes (Vicieae, Trifolieae and Cicereae) while the other model legume, *L. japonicus* is not so closely related. Lupins belong to a distant group of legumes.

One of the expectations after the development of genomic tools in model species was that the high level of conservation in gene order between cool season grain legumes and reference legume species could facilitate breeding and research of pulses crops. Defining patterns of synteny and colinearity between species will help the identification of orthologous genes in genetically recalcitrant species. Given what we know now about the colinearity of legume genomes we might expect this to have been used widely in the isolation of legume genes. However, this has not yet been fully implemented and to date colinearity has simply been a clue as to which genes are likely orthologues that can be further analyzed. For example this was the case for the pea symbiotic genes *Sym2*, *Sym10* and *Sym19* (reviewed in Cullimore and Dénarié 2003), *Sym35* (Borisov et al. 2003), *Sym7* (Kaló et al. 2005) and similarly genes of flower development (Wang et al. 2008), flower color (Hellens et al. 2010), as well as those involved in the regulation of flowering and flowering time (reviewed in Hecht et al. 2005). A project aimed at the isolation of the pea *Afila* gene using comparative mapping and positional cloning has been underway for some time and, although extensive microsynteny has been found, this work has yet to be published. At the same time direct approaches to gene isolation using deletion mutants has been successful (Hofer et al. 2009), and as genomic and post genomic tools become more widely available mapping by synteny will no doubt become a tool integrated with others, facilitating the characterization of gene function in legumes. The syntenic map position of the dwarf phenotypes in diploid alfalfa (*Msdwf1*) and pea (*le*) and the genomic resources in *M. truncatula* facilitated the identification of a gene encoding a gibberellin 3-β-hydroxylase (GA3ox) required for normal growth habit in diploid alfalfa (Dalmadi et al. 2008).

These examples clearly demonstrate the two-way utility and application of molecular markers and the identified orthologous regions between the genomes of reference and crop legumes. The tools developed in model species can facilitate the identification of agronomically important genes (QTLs, genes involved in nutrient quality and quantity, biotic and abiotic stresses, etc.) and marker-assisted breeding programs in target organisms while the accumulated biological knowledge in crops species can contribute to the understanding of biological processes.

References

Babula D, Kaczmarek M, Barakat A, Delseny M, Quiros CF, Sadowski J (2003) Chromosomal mapping of *Brassica oleracea* based on ESTs from *Arabidopsis thaliana*: complexity of the comparative map. Mol Genet Genom 268: 656–65.

Barker DG, Bianchi S, Blondon F, Dattée Y, Duc G, Essad S, Flament P, Gallusci P, Génier G, Guy P, Muel X, Tourneur J, Dénarié J, Huguet T (1990) *Medicago truncatula*, a model plant for studying the molecular genetics of the *Rhizobium*-legume symbiosis. Plant Mol Biol Rep 8: 40–49.

Bennetzen JL, Freeling M (1993) Grasses as a single genetic system: genome composition, collinearity and compability. Trends Genet 9: 259–260.

Bertioli DJ, Moretzsohn MC, Madsen LH, Sandal N, Leal-Bertioli SCM, Guimarães PM, Hougaard BK, Fredslund J, Schauser L, Nielsen AM, Sato S, Tabata S, Cannon SB, Stougaard J (2009) An analysis of synteny of *Arachis* with *Lotus* and *Medicago* sheds new light on the structure, stability and evolution of legume genomes. BMC Genomics 10: 45.

Bonierbale MW, Plaisted RL, Tanksley SD (1988) RFLP maps based on a common set of clones reveal modes of chromosomal evolution in potato and tomato. Genetics 120: 1095–1103.

Borisov AY, Madsen LH, Tsyganov VE, Umehara Y, Voroshilova VV, Batagov AO, Sandal N, Mortensen A, Schauser L, Ellis N, Tikhonovich IA, Stougaard J (2003) The *Sym35* gene required for root nodule development in pea is an ortholog of *Nin* from *Lotus japonicus*. Plant Physiol 131: 1009–1017.

Bowers JE, Chapman, BA Rong J, Paterson AH (2003) Unravelling angiosperm genome evolution by phylogenetic analysis of chromosomal duplication events. Nature 422: 433–438.

Brauner S, Murphy RL, Walling JG, Przyborowski J, Weeden NF (2002) STS markers for comparative mapping in legumes. Am Soc Hort Sci 127: 616–622.

Broughton WJ, Hernandez G, Blair M, Beebe S, Gepts P, Vanderleyden J (2003) Beans (*Phaseolus* spp.) model food legumes. Plant Soil 252: 55–128.

Cannon SB, May GD, Jackson SA (2009) Three sequenced legume genomes and many crop species: rich opportunities for translational genomics. Plant Physiol 151: 970–977.

Choi HK, Mun JH, Kim DJ, Zhu H, Baek JM, Mudge J, Roe B, Ellis N, Doyle J, Kiss GB, Young ND, Cook DR (2004a) Estimating genome conservation between crop and model legume species. Proc Natl Acad Sci USA 101: 15289–15294.

Choi HK, Kim D, Uhm T, Limpens E, Lim H, Mun JH, Kaló P, Penmetsa RV, Seres A, Kulikova O, Roe BA, Bisseling T, Kiss GB, Cook DR (2004b) A sequence-based genetic map of *Medicago truncatula* and comparison of marker colinearity with *M. sativa*. Genetics 166: 1463–1502.

Cook DR (1999) *Medicago truncatula* a model in the making! Curr Opin Plant Biol 2: 301–304.

Cronk Q, Ojeda I, Pennington RT (2006) Legume comparative genomics: progress in phylogenetics and phylogenomics. Curr Opin Plant Biol 9: 99–103.

Cullimore J, Dénarié J (2003) How legumes select their sweet talking symbionts. Science 302: 575–578.

Dalmadi A, Kaló P, Jakab J, Saskői A, Petrovics T, Deák G, Kiss GB (2008) Dwarf plants of diploid *Medicago sativa* carry a mutation in the gibberellin 3-β-hydroxylase gene. Plant Cell Rep 27: 1271–1279.

Devos KM, Gale MD (1997) Comparative genetics in the grasses. Plant Mol Biol 35: 3–15.

Devos KM, Beales J, Nagamura Y, Sasaki T (1999) *Arabidopsis* - Rice: Will colinearity allow gene prediction across the eudicot-monocot divide? Genome Res 9: 825–829.

Doganlar S, Frary A, Daunay MC, Lester RN, Tanksley SD (2002) A comparative genetic linkage map of eggplant (*Solanum melongena*) and its implications for genome evolution in the Solanaceae. Genetics 161: 1697–1711.

Doyle JJ, Luckow MA (2003) The rest of the iceberg. Legume diversity and evolution in a phylogenetic context. Plant Physiol 131: 900–910.

Ellis THN, Poyser SJ (2002) An integrated and comparative view of pea genetic and cytogenetic maps. New Phytol 153: 17–25.

Ellwood SR, Phan HTT, Jordan M, Hane J, Torres AM, Avila CM, Cruz-Izquierdo S, Oliver RP (2008) Construction of a comparative genetic map in faba bean (*Vicia faba* L.); conservation of genome structure with *Lens culinaris*. BMC Genomics 9: 380.

Ferguson-Smith MA, Trifonov V (2007) Mammalian karyotype evolution. Nat Rev Genet 8: 950–62.

Fredslund J, Madsen LH, Hougaard BK, Nielsen AM, Bertioli D, Sandal N, Stougaard J, Schauser L (2006) A general pipeline for the development of anchor markers for comparative genomics in plants. BMC Genomics 7: 207.

Gale MD, Devos KM (1998) Comparative genetics in the grasses. Proc Natl Acad Sci USA 95: 1971–1974.

Gebhardt C, Ritter E, Barone A, Debener T, Walkemeier B, Schachtschabel U, Kaufmann H, Thompson RD, Bonierbale MW, Ganal MW, Tanksley SD (1991) RFLP maps of potato and their alignment with the homologous tomato genome. Theor Appl Genet 83: 49–57.

Gepts P, Beavis WD, Brummer EC, Shoemaker RC, Stalker HT, Weeden NF, Young ND (2005) Legumes as a model plant family. Genomics for food and feed report of the cross-legume advances through genomics conference. Plant Physiol 137: 1228–1235.

Gonzales MD, Archuleta E, Farmer A, Gajendran K, Grant D, Shoemaker R, Beavis WD, Waugh ME (2005) The Legume Information System (LIS): an integrated information resource for comparative legume biology. Nucl Acids Res 33: D660–D665.

Handberg K, Stougaard J (1992) *Lotus japonicus*, an autogamous diploid legume species for classical and molecular genetics. Plant J 2: 487–496.

Hecht V, Foucher F, Macknight R, Vardy ME, Ellis N, Rameau C, Weller JL (2005) Conservation of *Arabidopsis* flowering genes in model legumes Plant Physiol 127: 1420–1434.

Hellens RP, Moreau C, Lin-Wang K, Schwinn KE, Thomson SJ, Fiers MWEJ, Frew TJ, Murray SR, Hofer JMI, Jacobs JME, Davies KM, Allan AC, Bendahmane A, Coyne CJ, Timmerman-Vaughan GM, Ellis THN (2010) Identification of Mendel's white flower character. PLoS ONE 5: e13230: doi: 10:1371/journal.pone.0013230.

Hofer J, Turner L, Moreau C, Ambrose M, Isaac P, Butcher S, Weller J, Dupin A, Dalmais M, Le Signor C, Bendahmane A, Ellis N (2009) *Tendril-less* regulates tendril formation in pea leaves. Plant Cell 21: 420–428.

Kaló P, Endre G, Zimányi L, Csanádi G, Kiss GB (2000) Construction of an improved linkage map diploid alfalfa *Medicago sativa*. Theor Appl Genet 100: 641–657.

Kaló P, Seres A, Taylor SA, Jakab J, Kevei Z, Kereszt A, Endre G, Ellis THN, Kiss GB (2004) Comparative mapping between *Medicago sativa* and *Pisum sativum*. Mol Genet Genom 272: 235–246.

Kaló P, Gleason C, Edwards A, Marsh J, Mitra RM, Hirsch S, Jakab J, Sims S, Long SR, Rogers J, Kiss GB, Downie JA, Oldroyd GED (2005) Nodulation signaling in legumes requires NSP2, a member of the GRAS family of transcriptional regulators. Science 308: 1786–1789.

Kemkemer C, Kohn M, Cooper DN, Froenicke L, Högel J, Hameister H, Kehrer-Sawatzki H (2009) Gene synteny comparisons between different vertebrates provide new insights into breakage and fusion events during mammalian karyotype evolution. BMC Evol Biol 9: 84.

Kevei Z, Seres A, Kereszt A, Kaló P, Kiss P, Tóth G, Endre G, Kiss GB (2005) Significant microsynteny with new evolutionary highlights is detected between *Arabidopsis* and legume model plants despite the lack of macrosynteny. Mol Genet Genom 274: 644–657.

Kowalski SP, Lan TH, Feldmann KA, Paterson AH (1994) Comparative mapping of *Arabidopsis thaliana* and *Brassica oleracea* chromosomes reveals islands of conserved organization. Genetics 138: 499–510.

Kulikova O, Geurts R, Lamine M, Kim DJ, Cook DR, Leunissen J, de Jong H, Roe BA, Bisseling T (2004) Satellite repeats in the functional centromere and pericentromeric heterochromatin of *Medicago truncatula*. Chromosoma 113: 276–283.

Lagercrantz U (1998) Comparative mapping between *Arabidopsis thaliana* and *Brassica nigra* indicates that *Brassica* genomes have evolved through extensive genome replication accompanied by chromosome fusions and frequent rearrangements. Genetics 150: 1217–1228.

Lavin M, Herendeen PS, Wojciechowski MF (2005) Evolutionary rates analysis of Leguminosae implicates a rapid diversification of lineages during Tertiary. Syt Biol 54: 575–594.

Lee JM, Grant D, Vallejos CE, Shoemaker RC (2001) Genome organization in dicots. II. *Arabidopsis* as a "bridging species" to resolve genome evolution events among legumes. Theor Appl Genet 103: 765–773.

Livingstone KD, Lackney VK, Blauth JR, van Wijk R, Jahn MK (1999) Genome mapping in *Capsicum* and the evolution of genome structure in the Solanaceae. Genetics 152: 1183–1202.

Lukens L, Zou F, Parkin I, Lydiate D, Osborn T (2003) Comparison of the *Brassica oleracea* genetic map with the *Arabidopsis thaliana* physical map. Genetics 164: 359–372.

Millan T, Winter P, Jungling R, Gil J, Rubio J, Cho S, Cobos MJ, Iruela M, Rajesh PN, Tekeoglu M, Kahl G, Muehlbauer FJ (2010) A consensus genetic map of chickpea (*Cicer arietinum* L.) based on 10 mapping populations. Euphytica 175: 175–189.

Nayak SN, Zhu HY, Varghese N (2010) Integration of novel SSR and gene-based SNP marker loci in the chickpea genetic map and establishment of new anchor points with *Medicago truncatula* genome. Theor Appl Genet 120: 1415–1441.

Nelson MN, Phan HTT, Ellwood SR, Moolhuijzen PM, Hane J, Williams A, O'Lone CE, Fosu-Nyarko J, Scobie M,·Cakir M, Jones MGK, Bellgard M, Ksiarkiewicz M, Wolko B, Barker SJ, Oliver RP, Cowling WA (2006) The first gene-based map of *Lupinus angustifolius* L.—location of domestication genes and conserved synteny with *Medicago truncatula.* Theor Appl Genet 113: 225–238.

Oh KC, Hardeman K, Ivanchenko MG, Ellard-Ivey M, Nebenführ A, White TJ, Lomax TL (2002) Fine mapping in tomato using microsynteny with the *Arabidopsis* genome: the *Diageotropica* (*Dgt*) locus. Genome Biol 3/9:research0049.1-0049.11

Paterson AH, Bowers JE, Burow MD, Draye X, Elsik CG, Jiang CX, Katsar CS, Lan TH, Lin YR, Ming RG, Wright RJ (2000) Comparative genomics of plant chromosome. Plant Cell 12: 1523–1539.

Phan HTT, Ellwood SR, Ford R, Thomas S, Oliver R (2006) Differences in syntenic complexity between *Medicago truncatula* with *Lens culinaris* and *Lupinus albus.* Funct Plant Biol 33: 775–782.

Phan HTT, Ellwood SR, Adhikari K, Nelson MN, Oliver R (2007) The first genetic and comparative map of white lupin (*Lupinus albus* L.): identification of QTLs for anthracnose resistance and flowering time, and a locus for alkaloid content. DNA Res 14: 59–70.

Sandal N, Krusell l, Radutois S, Olbryt M, Pedrosa A, Stacke S, Sato S, Kato T, Tabata S, Parniske M, Bachmair A, Ketelsen T, Stougaard J (2002) A genetic linkage map of the model legume *Lotus japonicus* and strategies for fast mapping of new loci. Genetic 161: 1673–1683.

Sato S, Nakamura Y, Asamizu E, Isobe S, Tabata S (2007) Genome sequencing and genome resources in model legumes. Plant Physiol 144: 588–593.

Schlueter JA, Goicoechea JL, Collura K, Gill N, Lin JY, Yu Y, Kudrna D, Zuccolo A, Vallejos CE, Muñoz-Torres M, Blair MW, Tohme J, Tomkins J, McClean P, Wing RA, Jackson SA (2008) BAC-end sequence analysis and a draft physical map of the common bean (*Phaseolus vulgaris* L.) genome. Trop Plant Biol 1: 40–48.

Simon CJ, Muehlbauer FJ (1997) Construction of a chickpea linkage map and its comparison with maps of pea and lentil. J Hered 88: 115–119.

Stracke S, Sato S, Sandal N, Koyama M, Kaneko T, Tabata S, Parniske M (2004) Exploitation of colinear relationships between the genomes of *Lotus japonicus, Pisum sativum* and *Arabidopsis thaliana*, for positional cloning of a legume symbiosis gene. Theor Appl Genet 108: 442–449.

Tanksley SD, Ganal MW, Prince JP, de Vicente MC, Bonierbale MW, Broun P, Fulton TM, Giovannoni JJ, Grandillo S, Martin GB, Messeguer R, Miller JC, Miller L, Paterson AH, Pined O, Röder MS, Wing RA, Wu W, Young ND (1992) High density molecular linkage maps of the tomato and potato genomes. Genetics 132: 1141–1160.

Thomas JW, Touchman JW (2002) Vertebrate genome sequencing: building a backbone for comparative genomics. Trends Genet 18: 104–108.

Vavilov NI (1922) The law of homologous series in variation. J Genet 12: 47–88.

Vershinin AV, Allnutt TR, Knox MR, Ambrose MJ, Ellis THN (2003) Transposable elements reveal the impact of introgression, rather than transposition, in *Pisum* diversity, evolution, and domestication. Mol Biol Evol 20: 2067–2075.

Wang Z, Luo Y, Li X, Wang L, Xu S, Yang J, Weng L, Sato S, Tabata S, Ambrose M, Rameau C, Feng X, Hu X, Luo D (2008) Genetic control of floral zygomorphy in pea (*Pisum sativum* L.). Proc Natl Acad Sci USA 105: 10414–10419.

Weeden NF, Muehlbauer FJ, Ladizinsky G (1992) Extensive conservation of linkage relationships between pea and lentil genetic maps. J Hered 83: 123–129.

Wojciechowski, MF (2003) Reconstructing the phylogeny of legumes (Leguminosae): an early 21st century perspective. In: BB Klitgaard, A Bruneau (eds) Advances in Legume Systematics, part 10. Higher Level Systematics. Royal Botanic Garden, Kew, UK, pp 5–35.

Wu F, Mueller LA, Crouzillat D, Vincent Petiard V, Tanksley SD (2006) Combining bioinformatics and phylogenetics to identify large sets of single-copy orthologous genes (COSII) for comparative, evolutionary and systematic studies: a test case in the Euasterid plant clade. Genetics 174: 1407–1420.

Zhu H, Cannon SB, Young ND, Cook DR (2002) Phylogeny and genomic organization of the TIR and non-TIR NBS-LRR resistance gene family in *Medicago truncatula*. Mol Plant-Microbe Interact 15: 529–539.

Zhu H, Kim DJ, Baek JM, Choi HK, Ellis LC, Kuester H, McCombie WR, Peng HM, Cook DR (2003) Syntenic relationships between *Medicago truncatula* and *Arabidopsis* reveal extensive divergence of genome organization. Plant Physiol 131: 1018–1026.

Zhu H, Choi HK, Cook DR, Shoemaker RC (2005) Bridging model and crop legumes through comparative genomics. Plant Physiol 137: 1189–1196.

8

Proteomics and Metabolomics for Systems Biology in Legumes

Stefanie Wienkoop

ABSTRACT

Genome-based systems biology is a rapidly advancing field in life sciences. Besides the possibility to obtain high-throughput genomic sequence information of many organisms, proteomics and metabolomics emerged as fundamental techniques of this novel research area. The impact of plant proteomics on functional genomics has enhanced our understanding, not only on the plant genome(s), but also more importantly, on the functional aspect of proteins. Integrative proteomic and metabolomic studies of a variety of dicot plants, including both *Arabidopsis thaliana* and the model legume *Medicago truncatula*, have greatly helped in an efficient separation, identification and cataloging of a large number of proteins and metabolites, and thereby defining their metabolic network. This chapter focuses on proteomic and metabolomic techniques that have been successfully adopted for systems biology approaches especially of *Medicago*. First, a mass spectrometry (MS) based proteomic and metabolomic pipeline for improved genome annotation, metabolic network reconstruction and pattern recognition will be introduced. Furthermore application of these novel strategies will be described in the field of cool season legumes research.

Keywords: metabolomics, mass spectrometry, model plant species, proteomics, metabolic networks, systems biology

8.1 Introduction

Legumes provide significant sources of fatty acids and proteins for human and animal nutrition; they also have non-food uses, for example in producing industrial feedstocks and combustible fuels (Thelen and

Department of Molecular Systems Biology, University of Vienna, Althanstrasse 14, 1090 Vienna, Austria; e-mail: *stefanie.wienkoop@univie.ac.at*

Ohlrogge 2002). Conditions in almost all cultivated land are suboptimal for plant growth. Understanding the connections between initial responses and the downstream events that constitute successful adjustment to its fluctuating environment is one of the challenges of genome-based systems biology in plant research. The complete genome sequences of rice (International Rice Genome Sequencing Project 2005) and *Arabidopsis* (*Arabidopsis* Genome Initiative 2000), which are non-leguminous model species, are available and provide insight into many fundamental aspects of plant biology; however, they do not address some important aspects of legume biology. Legumes are important for maintenance of human health and as crops for sustainable agriculture due to their ability to fix atmospheric nitrogen via their symbiosis with soil rhizobia. These bacteria colonize legume roots in specialized organs called nodules. Two model species of legumes, *Lotus japonicus* (Udvardi et al. 2005) and *Medicago truncatula* (Retzel et al. 2007), have been the focus of genome sequencing and functional genomics projects. However, functional genomics studies on pea (*Pisum sativum*) are still in their infancy, because this species has genome duplications, self-incompatibility, and a long generation time. In this case, the proteomics approach may be a powerful tool for analyzing the functions of the plant genes/proteins. Gaining an understanding of the biological function of any novel gene is a more ambitious goal than merely determining its sequence. At present, the wealth of information on nucleotide sequences that is being generated through genome projects far outweighs that which is currently available on the amino acid sequences of known proteins (Lockhart and Winzeler 2000; Pandey and Mann 2000). However, proteomic and metabolomic approaches have been demonstrated to be complementary to genomic data (May et al. 2008). Bioinformatic modeling approaches revealed the identification of missing enzymatic links and allowed for the reconstruction of draft metabolic networks. Although in general, genome sequence data and inferred protein-sequence data can be used to identify proteins and to follow temporal changes in protein expression in an organism, new tools have been developed enabling also genome sequence independent analysis. These techniques are based on high accuracy of the new generation of mass spectrometers such as the so called mass accuracy precursor alignment strategy (MAPA, see also Section 8.2.2). Currently, cool season grain legumes such as pea are barely used for proteomic studies due to the above mentioned reasons. However, *Medicago* one of the model plant for legume crops has been the subject of several proteomic and metabolic studies. In this chapter an overview of some of the most common MS based strategies for systems biology will be presented. Technical requirements for targeted and non-targeted proteomic and metabolomic deliverables and current status of legumes research will be addressed.

8.2 Mass Spectrometry Based Proteomic and Metabolomic Pipeline

For systems biology, techniques are required that ideally enable an integrative overview of all regulatory relevant processes of an organism during environmental perturbation. The major goal is towards large screening and identification of multiplexed biomarkers that are interacting within a dynamic response mechanism that involves several metabolic networks within a system. Consequently, there is need to combine qualitative and quantitative techniques that can all be conduced by mass spectrometry (MS). Especially for the study on protein level, different techniques need to be combined, depending on the biological question. Due to the very high dynamic range of proteins within plant tissues and organelles no sole technical solution can be defined up to date. Initial protein pre-fractionation of profiling and 2-DE (two-dimensional gel electrophoresis) studies are more time consuming while high-throughput shotgun approaches result in less detailed information on peptide level. So far, no approach allows for the high-throughput identification of the complete proteomic, metabolomic and/or transcriptomic information of an organism. However, the more robust and reproducible information obtained the higher the resolution of complex regulatory mechanisms (Wienkoop et al. 2006, 2008b). Thus, integration of proteomics and metabolomics is a promising and powerful approach that offers a new platform for studies of complex biological functions involving large numbers of compounds and networks. Generally, MS-based techniques can be separated in descriptive (profiling) and comparative (relative and absolute quantitative) analysis used for systems biology and will be introduced in the following sections.

8.2.1 Improved Genome Annotation and Metabolic Network Reconstruction

Large-scale information of a wide array of metabolites and proteins opens the door to the study of dynamic metabolomic networks that has been previously unattainable using classical targeted methods (Weckwerth 2003). Proteomics can be used as a key tool in analyzing the gene response of non-model plants in response to several biotic and abiotic factors (Carpentier et al. 2008), particularly of legumes where their genomes have not yet been completely sequenced, such as pea. The major advantage of proteomics is that it focuses on the functional translated portion of the genome and that it can also be used independent of genomic sequence information. The analysis and identification of metabolites, on the contrary, never depends on genomic databases. However, due to various extraction and separation properties as well as high compound similarities, different MS analysis techniques are needed to enable/improve metabolite identification. Thus, at present, several techniques need to be combined for a comprehensive metabolic network prediction. For example, for the comprehensive analysis of photosynthesis and its regulation in response to a selected environmental condition, the model alga, *Chlamydomonas*, has been used as an appropriate system. The 110 Mb nuclear genome as well as the mitochondrial and chloroplast genomes of *Chlamydomonas* have been sequenced and are currently being annotated using the information generated by two large-scale expressed sequence tag (EST) projects. More than 15,000 genes have been identified and this information is accessible through the *Chlamydomonas* genome portal (Chlamydomonas Center [http://www.chlamy.org]). For improved genome annotation a combination of bioinformatics annotation methods along with metabolomics based on two-dimensional gas-chromatography-mass spectrometry (GCxGC-MS) (Kempa et al. 2009) and liquid chromatography-mass spectrometry (LC-MS) based shotgun proteomic profiling approaches under different reference growth conditions have been previously applied to characterize thousands of proteins and about 80 metabolites (May et al. 2008). This system-level methodology well demonstrated its useful recourses for metabolic engineering (May et al. 2008; Manichaikul et al. 2009), also effective for other organisms.

8.2.2 Integrative Pattern Recognition and Data Mining

In contrast to profiling strategies, quantitative approaches for systems biology will be separated in non-targeted semi-quantitative and targeted absolute quantitative analysis techniques (Fig. 8-1). This holds true for proteomics as well as metabolomics MS strategies. In systems biology, besides 2-DE based proteomics the more rapid shotgun technique has been established and will be the focus of Section 8.3.2. In general, the non-targeted approach is a crude, label-free, top-down strategy for the rapid pattern recognition and identification of regulatory relevant mechanisms and potential biomarker (Hoehenwarter et al. 2008). The non-targeted approach will again be divided into a database dependent identification and a database independent mass accuracy precursor alignment (MAPA) strategy, respectively (Fig. 8-1). For metabolomics, peak intensity is usually being used as a parameter to determine relative abundance of metabolites in a complex sample. Recently, in proteomic studies the most commonly used parameter for relative quantification turned out to be the spectral count (Liu et al. 2004). It is explained by the sum of all spectra derived from a chromatogram either belonging to the same protein in database-dependent analyses or coming from the same precursor ion (MAPA) during database independent analyses. Accordingly, MAPA is a strategy that allows for the

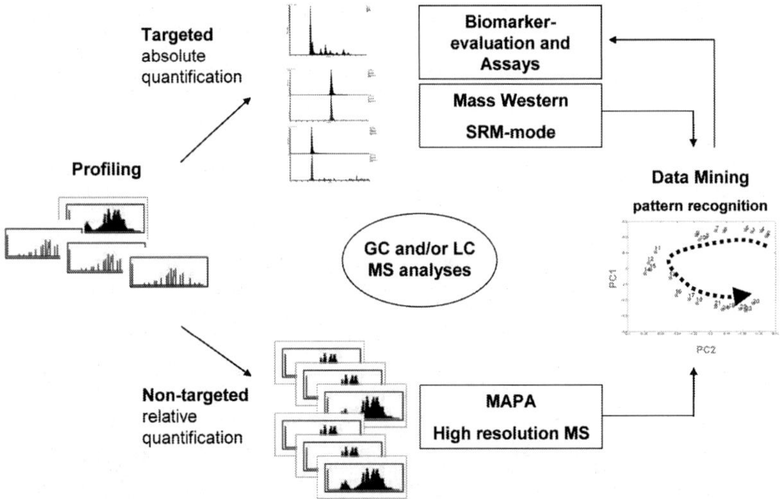

Figure 8-1 Workflow diagram of the MS based proteomic and metabolomic pipeline.
Color image of this figure appears in the color plate section at the end of the book.

unbiased and database-independent recognition of regulatory relevant precursor ions (Hoehenwarter et al. 2008). Metabolite derived ions may remain unidentified, whereas obtained peptide spectra will be picked for de novo sequencing and further protein identification. This pipeline for biomarker identification with its discovery phase, typically conducted by the non-targeted MS approach, is followed by validation, usually on an alternative targeted platform (Fig. 8-1). Thus, in contrast to the relatively crude label-free screening procedures, the accurate targeted absolute quantification based on stable isotope dilution technique is required subsequently (Wienkoop et al. 2008a). The ease of multiplexing and ability to determine protein modifications makes MS an attractive alternative to antibody-based technologies (Lehmann et al. 2008). Developments in quantitative MS, through the application of stable isotope labeling and scanning techniques, such as multiple reaction monitoring (MRM), have greatly enhanced both the specificity and sensitivity of MS-based assays to the point that they can rival immunoassays for some analytes. This MS based technique is also called Mass Western accordingly (Lehmann et al. 2008; Wienkoop et al. 2008a). By applying this stable isotope dilution technique in combination with LC-MRM, absolute contents of an isoform of sucrose synthase from *Medicago* root nodules usually not detectable by liquid chromatography-tandem quadruple mass spectrometry (LC-MS/MS) analysis on an ion trap mass spectrometer, could be detected and quantified in absolute terms. Besides validation of putative marker enzymes,

Mass Western has been used for studies on localization, stoichiometry and the targeted analysis of complete metabolic pathways including the differentiation of isoforms (Larrainzar et al. 2009). This technique has also been applied in several other plant science publications (Glinski et al. 2003; Wienkoop and Weckwerth 2006; Lehmann et al. 2008; Wienkoop et al. 2008a). Altogether, these methods are generating huge amounts of data. The critical step in working with huge and complex molecular data sets is reduction of obtained information to the most important components (Weckwerth 2008). This is possible by using multivariate data analysis tools, which can cope with specific characteristics of the molecular data: 1) usually only a limited number of samples are confronted with hundreds of variables (the molecular data). This implies problems for statistical significance. 2) Mostly the data are non-parametric, thereby influencing the selection of statistical tools. Robust strategies have been developed to address these questions and to extract as much useful biological information as possible out of the data (Weckwerth 2008). Supervised and unsupervised strategies such as independent component analysis (ICA) are employed for multivariate data mining. The application of these systems level approaches in combination with a tightly controlled experimental setup will reveal novel and completely unexpected insights. This was demonstrated within several studies in plant stress physiology, including temperature stress in *Arabidopsis* and drought stress in *Medicago* (Larrainzar et al. 2007; Morgenthal et al. 2007; Wienkoop et al. 2006, 2008b). The results demonstrate improved pattern recognition by integrating metabolite and protein data. Thus, in contrast to experiments focussing on a selected target, the application of unbiased systems approaches reveals other regulatory relevant mechanisms that may have otherwise not been taken into account. Examples are discussed in more detail later.

8.3 Target Deliverables and Current Status

8.3.1 Systems Biology Research Towards Marker Discovery

Besides the large number of investigations on the evaluation of nutritional value and industrial suitability, another very important application of MS-based metabolomics and proteomics lies in growth-limiting environmental stress response of grain legumes. Interestingly, while ambiguous projects such as the Human Proteome Organisation (HUPO) were not successful in describing new plasma marker proteins, in plant science potential marker proteins and metabolites have been characterized previously (Larrainzar et al. 2007; Wienkoop et al. 2008b). This may be attributed to reduced biological variation of plants due to highly controlled experimental growth conditions. *Medicago* has been used for many individual studies on nitrogen fixation in response to major abiotic stresses (Arrese-Igor et al. 1999; Gonzalez et al. 2001; Larrainzar et al. 2007; Ben Salah et al. 2008; Lopez et al. 2008). All these data have improved understanding of the complexity of plant response depending on the intensity and duration of homeostatic perturbation. Nevertheless, due to the diversity of the data a comparison and integration is difficult or even impossible. Given the variety of developmental and environmental factors between experiments, it is surprising that their influence on the identification of e.g., cold-responsive genes is largely unknown (Bieniawska et al. 2008). A systematic investigation of genes responding to cold treatment revealed that diurnal- and circadian-regulated genes are responsible for the majority of the substantial variation between experiments (Bieniawska et al. 2008). This is contrary to the widespread assumption that these effects are eliminated using paired diurnal controls (Bieniawska et al. 2008). For instance, the time-of-day effects induced after cold treatment as well as shifts in the developmental stage due to enhanced senescence induced after drought stress (Degenkolbe et al. 2009), reveals the dynamic complexity of the biological system that can only be visualized using systems biology approaches. In addition, standardized and robust results will be necessary for the comparison and data transfer to other important legume crops.

8.3.2 Applied Legume Specific Proteomics and Metabolomics Methodologies

Boyer (1982) estimated that about 70% of the potential yield is lost as a result of unfavorable physiochemical environments, even in developed agricultural systems. There is a crucial need to increase drought and salinity tolerance in legumes. Due to the above mentioned reasons, there are many proteomic studies on plants such as *Medicago* using various different technical strategies. The proteome of *Medicago* in cell-suspension culture has been analyzed by using 2-DE and LC-MS to yield an extensive protein reference map (Lei et al. 2005). Furthermore, some 2-DE proteomic studies of *Medicago* have been conducted to get information on for e.g. host-pathogen interaction (Mathesius et al. 2001, 2002). Their reference maps were supposed to provide information for future comparative proteomics and functional genomics studies in response to biotic and abiotic stresses. Proteomic approaches have also been applied in some other legumes such as pea (Castillejo et al. 2004; Schiltz et al. 2004; Curto et al. 2006), soybean (Hoa et al. 2004) and *Lotus* (Wienkoop and Saalbach 2003; Boukli et al. 2007) to analyze nutrition mobilization and/or to gain better understanding of the molecular basis of symbiosis in legumes. However, these initial more descriptive works are being more and more replaced by comparative studies. *Medicago* has been used for many 2-DE MS-based comparative proteomic studies on symbiosis related responses to major biotic and abiotic stresses such as Aloui et al. (2009) and Castillejo et al. (2009). One semi-quantitative shotgun LC-MS/MS (liquid chromatography coupled tandem mass spectrometry)-based database-dependent study was carried out to analyze the differential responses between plant and bacteroid to drought stress (Larrainzar et al. 2007). Several interesting gas chromatography coupled mass spectrometry (GC-MS)- based metabolite profiles were carried out to comparatively analyze for instance the response of *Medicago* cell cultures to elicitation (Broeckling et al. 2005) and the ammonium assimilation pathways in root nodules (Barsch et al. 2006). Integrative approaches on metabolomic and proteomic level have been shown to improve pattern recognition (Wienkoop et al. 2008b). In this regard, the most comprehensive systems biology approach on *Medicago* was recently carried out by Larrainzar and co-workers (Larrainzar et al. 2009). Combining GC-MS together with different LC-MS techniques for the analysis of drought stress and recovery in root nodules, complex regulatory response mechanisms have been revealed. Metabolites were grouped into four major categories differentially responding to drought and recovery (Larrainzar et al. 2009). Several metabolites and proteins involved in C/N metabolism were particularly monitored . Information gathered from this analysis showed that unlike the situation in legumes such as pea and soybean, the drought-induced inhibition of symbiotic nitrogen fixation (SNF) in *Medicago* appears to be caused by impairment of bacteroid metabolism and N_2-fixing capacity, rather than by a limitation of respiratory substrate (Larrainzar et al. 2009). One of the most interesting new regulatory relevant enzymes, a putative biomarker, identified by Larrainzar and coworkers (2007) turned out to be the methionine synthase. Its protein content was significantly decreased during drought induction. Furthermore, a sulfate transporter of the peribacteroid membrane in *Lotus* root nodules has been shown to play a crucial role for symbiotic nitrogen fixation (Krusell et al. 2005). These results suggest that sulfur metabolism may be involved in a regulatory key mechanism between plant and bacteroid interaction especially during stress response. Further investigations need to be carried out to confirm this hypothesis. However, sucrose synthase (SuSy1) has been described as a regulatory relevant key enzyme (putative biomarker) in various legumes' root nodules at least during drought stress (Arrese-Igor et al. 1999; Gordon et al. 1997). Mass Western analyses have demonstrated that a decrease in protein concentration instead of post-translational regulation is responsible for the decreased activity of this enzyme during drought stress (Larrainzar et al. 2009). Under control conditions in *Medicago* an absolute amount of about 160 fmol per µg nodule protein (in gel) and 195 fmol per µg nodule protein (in solution) has been calculated, respectively (Wienkoop et al. 2008a). These values have not yet been confirmed for other legume species, such as pea.

Further Mass Western analyses may prove transferability of data from model organisms such as *Medicago* to another organism (e.g., pea). A long-term goal will be to integrate experimental data in quantitative metabolic models. Such models, which are based on metabolic fluxes and whole pathways rather than individual reactions, have a wide array of applications such as filling gaps in metabolic networks of poorly characterized organisms, evaluating proposed functions of annotated genes, and identifying growth-limiting metabolic pathways as demonstrated previously (May et al. 2008).

8.3.3 Application in Phenomics/Genomics-Assisted Breeding

Legume crop yields have not markedly improved in comparison to those achieved with cereals. Proteins and metabolites that control the architecture of crop legumes and their tolerance or resistance to stress in a wide range of environments need to be identified and characterized to facilitate improvements in crop productivity. To meet this challenge, stronger and more adventurous breeding programs, better use of marker-assisted technologies, and emphasis on disease resistance, enhanced N fixation, and tolerance to edaphic soil constraints are required (Graham and Vance 2003). However, conventional breeding approaches have already been developed to produce cultivars with high yield (Singh et al. 2007). While most plant breeding is based on defect elimination or the selection for yield, a valuable additional approach is available through the breeding of crop ideotypes. Some cool season legume crops, such as pea, are for instance resistant to lodging and are adapted to mechanical harvesting. Nutritional value and suitability of crop legumes are determined not only by their important fatty acid composition but also by the control and reduction of antinutritional metabolites and proteins. Several examples of known key compounds can be gathered from a detailed review (Singh et al. 2007). While these targets are all based on single compound analysis, systems biology will reveal the complexity of responses resulting in a combination of targets (multiplexed biomarkers) that describe and distinguish specific steady states within a dynamic plant system.

8.4 Conclusions

Given the diversity of developmental and environmental factors between organisms and experiments, it is not surprising that proteomics and metabolomics modeling approaches are still in their infancy. This is also due to the dynamic complexity of biological systems that can only be visualized using systems biology approaches. Genome-based systems biology is still depending on model plant systems. Break-through technologies of genome-based systems biology, including the above mentioned proteomics and metabolomics strategies are needed for sustainability orientated agriculture. Data mining of metabolome, proteome as well as transcriptome analyses will in the future be connected computationally to allow for modeling of non-experimental simulations and predictions.

References

Arabidopsis Genome Initiative (2000) Analysis of the genome sequence of the flowering plant *Arabidopsis thaliana*. Nature 408: 796–815.

Aloui A, Recorbet G, Gollotte A, Robert F, Valot B, Gianinazzi-Pearson V, Aschi-Smiti S, Dumas-Gaudot E (2009) On the mechanisms of cadmium stress alleviation in *Medicago truncatula* by arbuscular mycorrhizal symbiosis: a root proteomic study. Proteomics 9: 420–433.

Arrese-Igor C, Gonzalez EM, Gordon AJ, Minchin FR, Galvez L, Royuela M, Cabrerizo PM, Aparicio-Tejo PM (1999) Sucrose synthase and nodule nitrogen fixation under drought and other environmental stresses. Symbiosis 27: 189–212.

Barsch A, Carvalho HG, Cullimore JV, Niehaus K (2006) GC-MS based metabolite profiling implies three interdependent ways of ammonium assimilation in *Medicago truncatula* root nodules. J Biotechnol 127: 79–83.

Ben Salah I, Albacete A, Martinez Andujar C, Haouala R, Labidi N, Zribi F, Martinez V, Perez-Alfocea F, Abdelly C (2008) Response of nitrogen fixation in relation to nodule carbohydrate metabolism in *Medicago ciliaris* lines subjected to salt stress. J Plant Physiol 166: 477–488.

Bieniawska Z, Espinoza C, Schlereth A, Sulpice R, Hincha DK, Hannah MA (2008) Disruption of the *Arabidopsis* circadian clock is responsible for extensive variation in the cold-responsive transcriptome. Plant Physiol 147: 263–279.

Boukli NM, Sunderasan E, Bartsev A, Hochstrasser D, Perret X, Bjourson AJ, Krause A, Broughton WJ (2007) Early legume responses to inoculation with *Rhizobium* sp. NGR234. J Plant Physiol 164: 794–806.

Boyer JS (1982) Plant Productivity and Environment. Science 218: 443–448.

Broeckling CD, Huhman DV, Farag MA, Smith JT, May GD, Mendes P, Dixon RA, Sumner LW (2005) Metabolic profiling of *Medicago truncatula* cell cultures reveals the effects of biotic and abiotic elicitors on metabolism. J Exp Bot 56: 323–336.

Carpentier SC, Panis B, Vertommen A, Swennen R, Sergeant K, Renaut J, Laukens K, Witters E, Samyn B, Devreese B (2008) Proteome analysis of non-model plants: a challenging but powerful approach. Mass Spectrom Rev 27: 354–377.

Castillejo MA, Amiour N, Dumas-Gaudot E, Rubiales D, Jorrin JV (2004) A proteomic approach to studying plant response to crenate broomrape (*Orobanche crenata*) in pea (*Pisum sativum*). Phytochemistry 65: 1817–1828.

Castillejo MA, Maldonado AM, Dumas-Gaudot E, Fernandez-Aparicio M, Susin R, Rubiales D, Jorrin JV (2009) Differential expression proteomics to investigate responses and resistance to *Orobanche crenata* in *Medicago truncatula*. BMC Genomics 10: 294.

Curto M, Camafeita E, Lopez JA, Maldonado AM, Rubiales D, Jorrin JV (2006) A proteomic approach to study pea (*Pisum sativum*) responses to powdery mildew (*Erysiphe pisi*). Proteomics 6: S163–S174.

Degenkolbe T, Do PT, Zuther E, Repsilber D, Walther D, Hincha DK, Kohl KI (2009) Expression profiling of rice cultivars differing in their tolerance to long-term drought stress. Plant Mol Biol 69: 133–153.

Glinski M, Romeis T, Witte CP, Wienkoop S, Weckwerth W (2003) Stable isotope labeling of phosphopeptides for multiparallel kinase target analysis and identification of phosphorylation sites. Rapid Comm Mass Spectrom 17: 1579–1584.

Gonzalez EM, Galvez L, Royuela M, Aparicio-Tejo PM, Arrese-Igor C (2001) Insights into the regulation of nitrogen fixation in pea nodules: lessons from drought, abscisic acid and increased photoassimilate availability. Agronomie 21: 607–613.

Gordon AJ, Minchin FR, Skot L, James CL (1997) Stress-induced declines in soybean N_2 fixation are related to nodule sucrose synthase activity. Plant Physiol 114: 937–946.

Graham PH, Vance CP (2003) Legumes: importance and constraints to greater use. Plant Physiol 131: 872–877.

Hoa LTP, Nomura M, Kajiwara H, Day DA, Tajima S (2004) Proteomic analysis on symbiotic differentiation of mitochondria in soybean nodules. Plant Cell Physiol 45: 300–308.

Hoehenwarter W, van Dongen JT, Wienkoop S, Steinfath M, Hummel J, Erban A, Sulpice R, Regierer B, Kopka J, Geigenberger P, Weckwerth W (2008) A rapid approach for phenotype-screening and database independent detection of cSNP/protein polymorphism using mass accuracy precursor alignment. Proteomics 8: 4214–4225.

International Rice Genome Sequencing Project (2005) The map-based sequence of the rice genome. Nature 436: 793–800.

Kempa S, Hummel J, Schwemmer T, Pietzke M, Strehmel N, Wienkoop S, Kopka J, Weckwerth W (2009) An automated GCxGC-TOF-MS protocol for batch-wise extraction and alignment of mass isotopomer matrixes from differential 13C-labelling experiments: a case study for photoautotrophic-mixotrophic grown *Chlamydomonas reinhardtii* cells. J Basic Microbiol 49: 82–91.

Krusell L, Krause K, Ott T, Desbrosses G, Kramer U, Sato S, Nakamura Y, Tabata S, James EK, Sandal N, Stougaard J, Kawaguchi M, Miyamoto A, Suganuma N, Udvardi MK (2005) The sulfate transporter SST1 is crucial for symbiotic nitrogen fixation in *Lotus japonicus* root nodules. Plant Cell 17: 1625–1636.

Larrainzar E, Wienkoop S, Weckwerth W, Ladrera R, Arrese-Igor C, Gonzalez EM (2007) *Medicago truncatula* root nodule proteome analysis reveals differential plant and bacteroid responses to drought stress. Plant Physiol 144: 1495–1507.

Larrainzar E, Wienkoop S, Scherling C, Kempa S, Ladrera R, Arrese-Igor C, Weckwerth W, González EM (2009) Carbon metabolism and bacteroid functioning are involved in the regulation of nitrogen fixation in *Medicago truncatula* under drought and recovery. Mol Plant-Microbe Interact 22: 1565–1576.

Lehmann U, Wienkoop S, Tschoep H, Weckwerth W (2008) If the antibody fails—a mass western approach. Plant J 55: 1039–1046.

Lei Z, Elmer AM, Watson BS, Dixon RA, Mendes PJ, Sumner LW (2005) A two-dimensional electrophoresis proteomic reference map and systematic identification of 1367 proteins from a cell suspension culture of the model legume *Medicago truncatula*. Mol Cell Proteom 4: 1812–1825.

Liu H, Sadygov RG, Yates JR III (2004) A model for random sampling and estimation of relative protein abundance in shotgun proteomics. Anal Chem 76: 4193–4201.

Lockhart DJ, Winzeler EA (2000) Genomics, gene expression and DNA arrays. Nature 405: 827–836.

Lopez M, Herrera-Cervera JA, Iribarne C, Tejera NA, Lluch C (2008) Growth and nitrogen fixation in *Lotus japonicus* and *Medicago truncatula* under NaCl stress: Nodule carbon metabolism. J Plant Physiol 165: 641–650.

Manichaikul A, Ghamsari L, Hom EF, Lin C, Murray RR, Chang RL, Balaji S, Hao T, Shen Y, Chavali AK, Thiele I, Yang X, Fan C, Mello E, Hill DE, Vidal M, Salehi-Ashtiani K, Papin JA (2009) Metabolic network analysis integrated with transcript verification for sequenced genomes. Nat Meth 6: 589–592.

Mathesius U, Keijzers G, Natera SH, Weinman JJ, Djordjevic MA, Rolfe BG (2001) Establishment of a root proteome reference map for the model legume *Medicago truncatula* using the expressed sequence tag database for peptide mass fingerprinting. Proteomics 1: 1424–1440.

Mathesius U, Imin N, Chen H, Djordjevic MA, Weinman JJ, Natera SH, Morris AC, Kerim T, Paul S, Menzel C, Weiller GF, Rolfe BG (2002) Evaluation of proteome reference maps for cross-species identification of proteins by peptide mass fingerprinting. Proteomics 2: 1288–1303.

May P, Wienkoop S, Kempa S, Usadel B, Christian N, Rupprecht J, Weiss J, Recuenco-Munoz L, Ebenhoh O, Weckwerth W, Walther D (2008) Metabolomics- and proteomics-assisted genome annotation and analysis of the draft metabolic network of *Chlamydomonas reinhardtii*. Genetics 179: 157–166.

Morgenthal K, Wienkoop S, Wolschin F, Weckwerth W (2007) Integrative profiling of metabolites and proteins: improving pattern recognition and biomarker selection for systems level approaches. Meth Mol Biol 358: 57–75.

Pandey A, Mann M (2000) Proteomics to study genes and genomes. Nature 405: 837–846.

Retzel EF, Johnson JE, Crow JA, Lamblin AF, Paule CE (2007) Legume Resources: MtDB and Medicago.org. Meth Mol Biol 406: 261–274.

Schiltz S, Gallardo K, Huart M, Negroni L, Sommerer N, Burstin J (2004) Proteome reference maps of vegetative tissues in pea. An investigation of nitrogen mobilization from leaves during seed filling. Plant Physiol 135: 2241–2260.

Singh RJ, Chung GH, Nelson RL (2007) Landmark research in legumes. Genome 50: 525–537.

Thelen JJ, Ohlrogge JB (2002) Metabolic engineering of fatty acid biosynthesis in plants. Metabol Eng 4: 12–21.

Udvardi MK, Tabata S, Parniske M, Stougaard J (2005) *Lotus japonicus*: legume research in the fast lane. Trends Plant Sci 10: 222–228.

Weckwerth W (2003) Metabolomics in systems biology. Annu Rev Plant Biol 54: 669–689.

Weckwerth W (2008) Integration of metabolomics and proteomics in molecular plant physiology—coping with the complexity by data-dimensionality reduction. Physiol Plant 132: 176–189.

Wienkoop S, Saalbach G (2003) Proteome analysis. Novel proteins identified at the peribacteroid membrane from *Lotus japonicus* root nodules. Plant Physiol 131: 1080–1090.

Wienkoop S, Weckwerth W (2006) Relative and absolute quantitative shotgun proteomics: targeting low-abundance proteins in *Arabidopsis thaliana*. J Exp Bot 57: 1529–1535.

Wienkoop S, Larrainzar E, Niemann M, Gonzalez EM, Lehmann U, Weckwerth W (2006) Stable isotope-free quantitative shotgun proteomics combined with sample pattern recognition for rapid diagnostics. J Separ Sci 29: 2793–2801.

Wienkoop S, Larrainzar E, Glinski M, Gonzalez EM, Arrese-Igor C, Weckwerth W (2008a) Absolute quantification of *Medicago truncatula* sucrose synthase isoforms and N-metabolism enzymes in symbiotic root nodules and the detection of novel nodule phosphoproteins by mass spectrometry. J Exp Bot 59: 3307–3315.

Wienkoop S, Morgenthal K, Wolschin F, Scholz M, Selbig J, Weckwerth W (2008b) Integration of metabolomic and proteomic phenotypes: analysis of data covariance dissects starch and RFO metabolism from low and high temperature compensation response in *Arabidopsis thaliana*. Mol Cell Proteom 7: 1725–1736.

9

Transgenesis

María Luisa Ruiz[1,a,*] and *Francisca Vaquero*[1]

ABSTRACT

This chapter highlights the state of the art of gene technology for genetic improvement of cool season grain legume crops including pea, chickpea, lentil faba bean and grass pea and centers around the development of transformation protocols for the production of transgenic plants. We summarize the studies on regeneration and genetic transformation in these crops and identify the hurdles being faced in the efficient recovery of transgenic plants. The review also presents a comparative account of explants used, mode of regeneration (organogenesis vs. embryogenesis), gene delivery techniques and recovery of transgenic plants in these crops.

Keywords: *Agrobacterium*, biotechnology, *in vitro* culture, regeneration, rooting, transformation

9.1 Introduction

Legumes are important crops worldwide, having a major impact on agriculture, the environment, animal/human nutrition and health. Legumes can interact symbiotically with specific soil-borne bacteria, the rhizobia, which allow the plant to fix atmospheric nitrogen and may help to protect them against some fungal pathogens (Chakraborty et al. 2003). They constitute a significant source of nitrogen and play an essential role in both the structure of ecosystems and sustainable agriculture. Among the grain crops, grain legumes (also known as pulses or food legumes) rank third in world production behind cereals and oilseeds, and constitute an important dietary component for humans and animals. Pulses are richer in

[1]Area de Genética, Departamento de Biología Molecular, Universidad de León, 240171 León, Spain.
[a]e-mail: *ml.ruiz@unileon.es*
*Corresponding author

protein than cereals and represent an important source of dietary protein, especially when intake from animal or fish sources is low or not available. Proteins of grain legumes are generally high in lysine, but low in methionine and cysteine. However, combined with cereals they offer a balanced diet of energy and protein.

Cool season grain legume crops are produced worldwide, but despite the large acreage under these crops, total productivity remains low and has been stagnating for the last few decades. The adaptability and productivity of legumes are limited by major biotic (pests and diseases) and abiotic stresses (drought, heat, frost, chilling, waterlogging, salinity and mineral toxicities). Although the type and the severity depend on the specific crop location, abiotic stresses can result in crop damages as high as those caused by biotic stresses. Furthermore, crops under abiotic stress are usually more susceptible to weeds, insects and diseases, which considerably increase the losses (Reddy et al. 2004). An additional factor relevant in legumes is the response of the symbiotic nitrogen-fixing bacteria to different stresses. Cultivars resistant to biotic and abiotic stresses with better protein quality and quantity are needed. Grain legumes have a narrow genetic base since they are essentially self-pollinated (cross-pollination occurs at low or very low frequencies, except in some *Vicia faba* accessions; see Section 9.2.1.2). Thus, there is an urgent need to widen the genetic base and incorporate desirable characters.

Model legumes are being rapidly developed as experimental systems to address a number of important biological questions unique to this family using molecular tools, including genomics. A key component of most functional genomic approaches is a high-throughput transformation system for developing various gene identification strategies. Transformation also is emerging as an important crop improvement tool. This is particularly evident in soybean (*Glycine max*), in which Roundup Ready soybean cultivars have captured a major stake in market share of soybeans planted in the US and Argentina. Transformation theoretically expands the sources of genes for plant improvement to all organisms, far beyond the genepool accessible via sexual hybridization. Transformation also offers strategies for overexpressing or suppressing endogenous genes. Thus, introducing new genes or manipulating endogenous gene expression via transformation generates new phenotypic variation useful for investigating gene function and for crop improvement (Somers et al 2003).

Although conventional plant breeding methods in the past have contributed to the improvement of pulse crops, transgenic technology has the potential to achieve this objective as an additional/supplementary technology. Crop improvement through genetic engineering has become a reality. Despite the development of transgenic technology in the 1980s, the progress achieved in legumes is not significant as compared to their counterpart crops, namely cereals, where concentrated efforts and free flow of funding support for cereal transformation have pushed transgenic cereals to the forefront of transgenic success stories (Shrawat and Lorz 2006). Meanwhile transgenic pulse crop research, although having moved forward, is still confined to the laboratory (Eapen 2008). Although it is now possible to transform many grain legumes (Chandra and Pental 2003; Somers et al. 2003), in some cases the rate of recovery of transgenic lines is still low. Besides, different factors such as recalcitrance of pulses for regeneration, low competency of regenerating cells for transformation and lack of a reproducible "in plant" transformation system have been pointed out as reasons for less progress in developing of transgenic pulse crops with high efficiency (Popelka et al. 2004; Dita et al. 2006).

9.2 Genetic Transformation

To be able to introduce foreign genes into plant genomes, it is necessary to have efficient methods of gene delivery, selection of transformed cells, and regeneration of whole plants which can be grown to maturity. The ability to regenerate whole plants is crucial to the successful application of *in vitro* methods with regard to plant breeding and to plant genetic engineering.

In crop plants, many shoot regeneration protocols have been developed in recent years specifically for exploitation in genetic manipulation experiments. However, several requirements must be fulfilled in order to produce stably transformed plants, such as, 1) a source of totipotent cells or gametes that serves as recipients of delivered DNA, 2) a means of delivering DNA into the target cells, and 3) a system for selecting or identifying transformed cells. Subsequently, the newly introduced gene(s) must be expressed in transgenic plants and, finally, the foreign DNA must be heritable and expressed reproducibly in succeeding generations. For legumes, which have been regarded as recalcitrant to transformation, regeneration *in vitro* is highly genotype-specific and only rarely cultivated varieties are amenable to regeneration. In these cases, plant regeneration remains an "art" that requires considerable training of the practitioner to develop the skills needed to generate sufficient number of transgenic plants for productive work. In addition, regeneration is often slow and the frequency of transformation (number of transformed plants generated from each explant) is often low. In species that are amenable to *in vitro* somatic embryogenesis such as alfalfa, (lucerne; *Medicago sativa*), relatively rapid and efficient transformation methods have been developed based on co-cultivation of tissue pieces (explants) with *Agrobacterium tumefaciens*. Because inducing somatic embryogenesis or organogenesis in many legume species is difficult, a variety of transformation methods have been reported that use cultures of meristematic cells as sources of totipotent cells. Regeneration of shoots from the cotyledonary node or from other meristematic explants after agrobacteria infection is emerging as a rapid and relatively efficient method of transformation. A number of legume species also have been transformed by direct DNA transfer methods including microinjection, electroporation and particle bombardment (see Eapen 2008).

Hansen and Chilton (1996) described a novel "agrolistic" system in which virulence genes from the Ti plasmid of *A. tumefaciens* were placed on one plasmid and the latter co-delivered by bombardment with a second plasmid carrying the T-DNA borders flanking the gene of interest. To date, only preliminary results have been obtained in soybean and application to other species still requires further investigation (Babaoglu et al. 2000).

The most widely used DNA delivery systems, which have potential practical applications include those based on the natural gene transfer mechanism of the Gram-negative soil bacterium *Agrobacterium*, with techniques such as particle bombardment and electroporation and/or chemical treatment of isolated protoplasts providing alternative approaches. Among these methods, *Agrobacterium*-mediated transformation of explants is the most popular method for development of transgenic plants in pulse crops (Babaoglu et al. 2000; Chandra and Pental 2003; Somers et al. 2003; Eapen 2008). While the natural virulence of agrobacteria and hence their ability to infect plants varies, the virulence of some strains can be increased by the introduction of a supervirulent plasmid carrying extra copies of some of the virulence genes, alongside with the binary vector. Superbinary vectors, in which extra copies of virulence genes are on the binary vector itself, have also proven to be useful in the transformation of cereals.

The inability of *Agrobacterium* to penetrate tissues and to adhere to the walls of cells capable of undergoing transformation is probably one of the main factors limiting the transformation of many plants. In some cases, short treatments of ultrasound in the presence of the bacteria have increased the efficiency of transformation (sonication-assisted *Agrobacterium*-mediated transformation; SAAT). Also, vacuum infiltration has been used for this purpose.

Prior to the introduction of agronomically useful genes into a crop by non-oncogenic strains of *Agrobacterium*, the plant response to bacterial inoculation is often evaluated by inoculating explants or intact plants with wild-type strains of *Agrobacterium*. These experiments provide insight into parameters such as the optimal co-cultivation period and the plant cells most competent for transformation within target tissues (De Kathen and Jacobsen 1995).

The inclusion of an antibiotic or an herbicide in the culture medium is normally used to select transformed cells and tissues from which transgenic plants are regenerated. While antibiotic marker genes were introduced to test the feasibility of transformation, other genes of economic importance such as *Cry* genes, sunflower albumin gene, α-amylase inhibitor gene and chitinase gene have also been introduced into major pulse crops. Since most legumes regenerate from young embryonic tissues, embryonic axes and cotyledonary nodes are the most preferred explants for transformation (Eapen 2008).

As a group, grain legumes are less amenable to genetic manipulation *in vitro* compared with most other dicotyledonous crop species, particularly members of the Solanaceae (De Kathen and Jacobsen 1995). While several leguminous species are susceptible to *Agrobacterium* inoculation, relatively few grain legumes have been stably transformed using disarmed vectors carried by *A. tumefaciens*. A summary of the key transformation events relating to cool season grain legumes is presented below.

9.3 Pea

Pea (*Pisum sativum* L.) is one of the main cultivated pulses worldwide. Breeders have a broad genetic diversity to use in breeding programs, with a large amount of accessions in genebank collections across the world. Pea genetics began as soon as Genetics did, through the experiments of Gregor Mendel and some of the major genes controlling morphological features of the pea plant are well known by geneticists. Knowledge and use of these genes in conventional breeding approaches have allowed breeders to change the tall, leafy and scrambling habit of the plant into one better adapted to mechanized harvesting, or to choose some desirable characters for shape and color of the seed and for pod quality.

A major challenge for breeders is to increase yield, particularly by increasing resistance and tolerance to biotic and abiotic constrains not only affecting plant development but also symbiotic relationships with nodulating bacteria. Biotic stress can be caused by fungal root and foliar diseases, viruses and pests. Genetic resistance is available for many of these pathogens and some resistance genes have been introduced into new cultivars. However, little genetic resistance to insect pests is currently available. Abiotic constrains are related to cold tolerance, soil salinity, presence of toxic minerals, soil moisture, unfavorable pH or extreme temperature.

These challenges and others such as improved protein quality, herbicide resistance, or modified nodulation can be addressed by genetic engineering. Efficient gene transfer through gene technology provides breeding programs an additional tool to overcome deficiencies in genetic variation within *Pisum* germplasm.

9.3.1 Explants and Methods

Somatic embryogenesis has been described in pea (Bencheikh and Gallais 1996; Griga 1998, 2002; Loiseau et al. 1998). Shoot organogenesis using hypocotyls explants (Ochatt et al. 2000) and immature cotyledon explants (Grant et al. 1995) have also been reported. Both types of regenerations can be either direct or indirect via a callus stage. Callus-based regeneration systems have the disadvantage that they have a much higher chance of yielding plants with somaclonal variation. However, for genetic modification this way is often preferred over direct regeneration since the callus phase is ideal for selection of genetically modified cells. Thus some protocols used callus derived from immature embryos to obtain shoots and a further multiplication for subsequent rooting or grafting (Pniewski et al. 2003). Regeneration of fertile plants from nodal tissue, which seemed to be genotype-independent and can be maintained in a cyclic fashion, was further reported (Tzitzikas et al. 2004). A review on pea regeneration and transformation has been recently published (Vignesh et al. 2010).

Pea transformation was attained by protoplast electroporation (Puonti-Kaerlas et al. 1992b), PEG-mediated delivery of DNA (Nicolaisen and Poulsen 1993), in planta electroporation (Chowrira et al. 1998), and particle bombardment (Warkentin et al. 1992). However, none of these methods have been continued. Thus, the most extended and almost exclusive method from the beginning to date is *Agrobacterium*-mediated transformation (see Table 9-1).

The earlier publications on pea genetic transformation dealt with the *Agrobacterium*-mediated protocol optimization. Susceptibility of pea to *Agrobacterium* infection and genetic transformation has been shown since 1967 (Manigault and Kurkdjian 1967; Kurkdjian et al. 1968; Bercetche et al. 1987; Puonti-Kaerlas et al. 1989). These studies described the crown gall tumor or hairy root induction observed when *A. tumefaciens* or *A. rhizogenes* were inoculated on pea, and its accompanying cellular or tissue responses. Transformation was ascertained by the appearance of tumors or by transient gene expression of reporter genes. Response variations were found; the response obtained was more dependent on the bacterial strain than on the cultivar used.

In the first *A. tumefaciens*-mediated gene transfer procedure to achieve transgenic pea plants (Puonti-Kaerlas et al. 1990), calli were derived from shoots and seedling epicotyls and selected for antibiotic resistance. Kanamycin and hygromycin were tested but only hygromycin was efficient for reproducible shoot organogenesis from calli. Later studies demonstrated that both antibiotics were useful to obtain genetic transformed plants (Grant et al. 1998).

Table 9-1 Summary of the most relevant pea genetic transformation studies.

Genotype	Method[a]/strain	Explant	Fate of explant	Transferred gene(s)	Reference
Stivo, Puget	***T***/ various strains	Epicotil segment, shoot	Plant	*hpt*	Puonti-Kaerlas et al. 1990
Madria	***T***/ GV2260, GV3850	Epicotil segment, node	Shoots	*uidA, nptii, hpt*	De Kathen and Jacobsen 1990
Gloire du Midi	***T***/ 4 strains; ***R***/ 2 strains	Root segment, root protoplast	Tumour, callus	*nptii*	Schaerer and Pilet 1991
Belamn, Filby	Electroporation	Protoplast	Callus	*uidA*	Puonti-Kaerlas et al. 1992b
Stivo, Puget	***T***/	Shoot	Plant[b]	*hpt*	Puonti-Kaerlas et al. 1992a
PI244253	Biolistic	Embryo axis, leaf, root	-	*uidA*	Warkentin et al. 1992
PI244253	***T***/	Cotyledonary node	Plant[b]	*nos, uidA, nptII*	Jordan and Hobbs 1993
Puget	***T***/	Cotyledonary node	Plant	*uidA nptii*	Davies et al. 1993
P. sativum	Polyethylene glycol	Protoplast	-	*uidA*	Nicolaisen and Poulsen 1993
Greenfeast, Rondo	***T***/ AGL1	Slices of immature embryo axis	Plant[b]	*bar, nptII*	Schroeder et al. 1993
Bolero, Trounce, Bohatyr, Huka	***T***/ AGL1	Immature cotyledonary nodes	Plant[b]	*nptII bar*	Grant et al. 1995
Greenfeast	***T***/	Slices of immature embryo axis	Plant[b]	*bar, αai*	Schroeder et al. 1995
P. sativum	***T***/ EHA101	Cotyledon, epicotil	-	*uidA*	DeKathen and Jacobsen 1995
Puget	***T***/ EHA105	Cotyledonary node from mature seed	Plant[b,c]	*bar*	Bean et al. 1997
Puget	***T***/ AGL1	Cotyledonary node from mature seed	Plant[c]	*NIb, bar*	Jones et al. 1998
7 cultivars	***T***/ C58C1, EHA105	Slices of immature embryo axis	Plant	*uidA*	Lurquin et al. 1998
7 cultivars	***T***/ AGL1	Immature cotyledonary nodes	Plant[b]	*nptII*	Grant et al. 1998

Table 9-1 contd....

Table 9-1 contd....

Genotype	Method[a]/strain	Explant	Fate of explant	Transferred gene(s)	Reference
Sparkle	In planta electroporation	Axillary bud	Plant[b]	*PEMV-CP*	Chowrira et al. 1998
Puget	***T/*** AGL1	Cotyledonary node from mature seed	Plant[c]	*psao-1*	Wismiewski and Brewin 2000
7 cultivars	***T/*** EHA105	Slices of inmature embryo axis	Plant[b]	*uidA, nptII, bar, ahat-3r*	Polowick et al. 2000
Puget, BC1:17	***T/*** EHA105	Cotyledonary node from mature seed	Plant[b,c]	*TI, uidA, bar*	Welham and Domoney 2000
Greenfeast, Laura	***T/*** AGL1	Slices of immature embryo axis	Plant[b]	*αAI-1, αAI-2*	Morton et al. 2000
94-A26, Crown, 89T46.UK13	***T/*** AGL1	Immature cotyledon	Plant[b]	*AMV-CP, nptII*	Timmerman-Vaughan et al. 2001
Bolero, Lincoln, 97-B19	***T/*** AGL1, KYTR1	Slices of immature embryo axis	Plant	*nptII uidA, β1, 3glucanase*	Grant et al. 2003
Greenfeast	***T/*** AGL1	Slices of immature embryo axis	Plant[b]	*αAI-1*	De Sousa-Majer et al. 2004
Dark Skinned Perfection	***T/*** leaf infiltration	Leaf	-	-	Constantin et al. 2004
6 cultivars	***T/*** EHA105 (SAAT)	Axilary node, in vivo and in vitro	Plant	*uidA, nptii*	Svabova et al. 2005
10 cultivars	***T/*** AGL0, AGL1, EHA105	Slices, cotyledons, cotyledonary node	Plant	*uidA, bar*	Pniewski and Kapusta 2005
30 cultivars	***T/*** AGL1, KYTR1	Immature cotyledons	Plant	*uidA, bar, nptII*	Grant and Cooper 2006
Baroness, Baccara	***T/*** EHA105	Slices of mature embryo axis	Plant[c]	*bar, pgip, Vst1*	Richter et al. 2006
Excell	***T/***	Slices of immature embryo axis	-	*αAI-1, nptII*	Collins et al. 2006
Adept, Komet, Menhir	***T/*** EHA105 (SAAT)	Axilary node, in vivo and in vitro	Plant	*uidA, nptII*	Svabova and Griga 2008
Sponsor	***T/*** EHA105	Slices of mature embryo axis	Plant[b,c]	*chit30*	Hassan et al. 2009

[a]*Agrobacterium*-mediated transformation; ***T/***, with *A. tumefaciens*; **R**, with *A. rhizogenes*.
[b]Transmission to following generations.
[c]Shoot grafted onto non-transformed rootstock.

Since then, the attainment of transgenic pea plants was mainly by means of de novo organogenesis (Puonti-Kaerlas et al. 1990, 1992a; Schroeder et al. 1993; Grant et al. 1995; Polowick et al. 2000) or by meristem cultures (De Kathen and Jacobsen 1990; Schaerer and Pilet 1991; Davies et al. 1993; Bean et al. 1997; Nadolska-Orczyk and Orczyk 2000). In the procedures described by Schroeder et al. (1993) and Polowick et al. (2000) segments of immature embryonic axis were used. Davies et al. (1993), Bean et al. (1997), and Nadolska-Orczyk and Orczyk (2000) used cotyledonary nodes as starting material. These methods differ particularly in the starting explant type, with immature tissue being the most successful. However, the use of mature seed avoided the need for growth facilities to produce immature donor pea material and therefore seeds or seedlings were chosen for some groups (Bean et al. 1997; Jones et al. 1998; Welham and Domoney 2000; Wisniewski and Brewin 2000). De Kathen and Jacobsen (1995) stated that pea cells competent for transformation were mainly restricted to the differentiating cells neighboring the vascular system of cotyledon and epicotyl explants.

High frequencies of tissue genetic transformation were often reported but, unfortunately, the rate of whole regenerated transformed pea plants was low (generally in the range of 0.2–5%). There was a high frequency of escapes and a portion of the transformed plants was chimeras (Bean et al. 1997; Grant et al. 2003). These disadvantages have been attributed to the fact that the majority of the shoots were regenerated from existing meristems. It was proposed that these problems might be reduced significantly if plants were regenerated from newly formed meristems, as produced in some protocols (Tzitzikas et al. 2004). In order to diminish the development of non-transgenic or chimeric shoots, a common practice is to remove any shoots developing during the first weeks of culture, i.e., of callus formation (Schroeder et al. 1993). However, chimeric plants can also be routinely discarded by PCR analysis (Jones et al. 1998). Another inconvenience is that sometimes the primary regenerant that produces viable seed was tetraploid (Puonti-Kaerlas et al. 1992b).

From the experience of this extended period, Grant and Cooper (2006) described a refined callus-based method and claimed the successful application over a large collection of genotypes and vectors. This method uses immature cotyledons as the explant, and one of three selectable markers, neomycin phosphotransferase, phosphinothricin or hygromycin. After transformation, which was confirmed by PCR, one-centimeter green shoots regenerated on selective medium were multiplied and rooted.

Another extended technique uses thin slices of immature embryo axis (Schroeder et al. 1993). Regeneration is achieved via organogenesis and selection is made on medium containing phosphinotricin. Developing shoots greater than 20 mm in length are rooted and transferred to soil for subsequent flowering and seed production. However, this procedure has been modified, since the use of mature seeds as donor of pea material is preferred because it can proceed without equipment facilities for maintenance of plants, and it avoids the necessity for controlling the optimum seed developmental state (Richter et al. 2006; Hassan et al. 2009).

Transformation efficiency was reported to be in the range of 0.2 to 13.5%, depending upon the pea genotype and the *Agrobacterium* strain (Grant and Cooper 2006). Although optimization of the transformation protocols is currently being attempted (Svabova and Griga 2008), unfortunately an average efficiency of approximately 2% is the most common, regardless of the method, the bacterial strain and the plant genotype. Nevertheless, a single person can process 500 seeds per lot and two repetitions in one week, thereby allowing the production of a large number of transgenic pea plants (Bean et al. 1997).

Selection of transformed plants has been achieved by introducing resistance genes to antibiotics (*nptII, hptII*) and to herbicides (*bar, ahas3r*). Selection is usually done in a two step procedure, during the shoot regeneration and during the rooting. Often primary developed shoots are discarded since they are supposed to have originated from pre-existing meristems (Schroeder et al. 1993; Bean et al. 1997). Sometimes, positively transformed shoots do not achieve a good rooting system and it can be very difficult to differentiate them from escapes. In some instances, PCR is used to retrieve positive shoots that have not properly rooted (Jones et al. 1998; Grant and Cooper 2006).

One of the difficulties with pea transformation is the long tissue culture period required for the recovery of transgenic plants. Different regeneration systems for pea tissue have been reported to require four (Bean et al. 1997), six (Polowick et al. 2000), seven (Grant et al. 1995), or nine months (Schroeder et al. 1993) or even longer (Puonti-Kaerlas et al. 1990), between co-cultivation of the explants and transplanting into soil of the rooted plants. This extended period is in part due to the difficulties of the *in vitro* rooting. The rooting is slow, erratic, with the majority of the putative transformed shoots failing to successfully produce roots. The extended culture period also resulted in the production of phenotypically abnormal plants, which showed dwarfing, early prolific flowering with reduced seed set, and premature senescence. To overcome these problems, grafting onto non-transformed rootstocks has been used (Bean et al. 1997; Jones et al. 1998; Welham and Domoney 2000; Richter et al. 2006; Hassan et al. 2009). Though grafting can save 4–6 weeks to produce T1 seeds, even for the later reports it takes at least six months from transformation until the first transgenic seed is obtained (Richter et al. 2006; Hassan et al. 2009). Nevertheless, neither *in vitro* rooting nor grafting has been generally adopted.

Apart from the expression of selectable marker genes, identification of successful transformants has been achieved by the expression of reporter genes (*uidA*) or other introduced genes of interest for breeding. PCR-based methods were used to amplify a region of the T-DNA, and DNA or RNA blotting was used for the final proof for transformation. In general, transmission to the following generation and Mendelian inheritance was demonstrated. Some studies revealed differences among transgenic lines such as complex inheritance patterns due to multiple copy insertions, chimeras, fragmented insertions, and integrated non-expressed fragments (Davies et al. 1993; Grant et al. 1995; Bean et al. 1997; Svabova et al. 2005).

9.3.2 Expressed Genes for Breeding

Transformation efficiency must be improved in pea to allow a wide application in breeding. Nevertheless, a collection of expressed genes with application in pea production and breeding can be listed among those included in transgenes, such as resistance to herbicides, fungi, viruses, and bruchids. Some of them have also been tested under field conditions (Morton et al. 2000; Timmerman-Vaughan et al. 2001).

Herbicide resistance genes have been used as selective markers and have proven to be inherited to the following generations. The most common gene is *bar* (Schroeder et al. 1993; Grant et al. 1995; Bean et al. 1997; Jones et al. 1998; Polowick et al. 2000; Pniewski and Kapusta 2005; Grant and Cooper 2006; Richter et al. 2006) being effective in "leaf painting" for selection in the laboratoray, but also when sprayed at concentrations equal to those employed in field treatments. Also *ahat-3r*, which confers chlorsulfuron resistance, has been introduced (Polowick et al. 2000).

Dry pea production is frequently affected by fungal infections. As conventional breeding does not always offer satisfying results, the heterologous expression of broad acting antifungal hydrolytic enzymes that enhance the plant defense system, might offer a solution. A β-1,3glucanase gene was included in one of the vectors (Grant et al. 2003). However, although there is abundant evidence that glucanases are involved in defense mechanism of plants against pathogen infection, Grant et al. (2003) only refer to the optimization of the *Agrobacterium* strain for transformation. Two different antifungal genes were used by Richter et al. (2006), the polygalacturonase-inhibiting protein (*PGIP*), and the stilbene synthase (*Vst1*) genes. Resveratrol produced by *Vst*1 transgenic plants, and PGIP expression were detected. Stable inheritance of both antifungal genes in the transgenic plants was demonstrated. Heterologous expression of a chimeric chitinase gene (*chit30*) in transgenic pea was obtained (Hassan et al. 2009). Using *in vitro* bioassays, an inhibition or delay of hyphal extension was reported, which might indicate enhanced antifungal activity compared with non-transformed pea plants.

Engineered resistance against viruses has been used successfully using viral coat protein (CP). Pea enation mosaic virus (PEMV) causes periodic serious losses to crops of susceptible pea and lentil cultivars, particularly in some regions. Transformation with a chimeric PEMV-coat protein was stable and inherited until the T4 generation. This transformation resulted in delayed or transient PEMV multiplication and attenuated symptoms as compared to control inoculated individuals (Chowrira et al. 1998). Coat protein of the alfalfa mosaic virus (AMV) was also engineered in pea, and five transgenic lines with improved virus resistance were identified (Timmerman-Vaughan et al. 2001). Resistance was observed only in individual plants that accumulated detectable CP product from the transgene, suggesting that resistance is CP-mediated. In the same work, to test for improved virus resistance under field conditions, plots of transgenic lines and non-transgenic controls were inoculated and it was confirmed that partial virus resistance can be produced by genetically modifying peas with AMV CP sequences. Transgenic pea lines carrying the replicase (*NIb*) gene of pea seed-borne mosaic potyvirus (PSbMV) were generated and used in experiments to determine the effectiveness of induced resistance upon heterologous isolates (Jones et al. 1998). Resistance was observed in plants, resulting in the absence of yield loss despite the initial infection. Resistance was associated with loss of both viral and transgene RNA, which is indicative of a mechanism based upon post-transcriptional gene silencing.

Bruchids are a major challenge for pea production. Transgenic pea plants were generated expressing the bean α-amylase inhibitor gene (*αAi*), which protects seeds from the attack by the pea weevil (*Bruchus pisorum*). The expression of the gene was shown to be stable at least for five generations and conferred total protection, with weevil development being completely blocked at an early larval stage (Schroeder et al. 1995). Moreover, α-amylase inhibitor provided protection from pea weevil under field conditions (Morton et al. 2000). These authors compared two genes, *αAi-1* and *αAi-2*. The primary effect of *αAi-2* appeared to be a delay in the maturation of the larvae and was partially effective in protecting field-grown transgenic peas in contrast with the effect of *αAi-1*, which resulted in larval mortality at the first or second instar. Since severe environmental factors have been reported to influence the success of genetically modified crops, the response to water deficit and high temperature was evaluated. Thus, *αAi-1* expression was not affected by water deficit, but high temperature reduced the protective capacity of the transgenic peas (De Sousa-Majer et al. 2004).

The α-amylase inhibitor 1 inhibits mammal salivary and pancreatic *α*-amylases, thus preventing starch digestion. The influence of *α*-amylase inhibitor of the transgenic pea on the digestibility of both amino acid and crude protein was tested in pigs (Collins et al. 2006). Dry matter digestibility of transgenic peas was reduced. However, the transgenic pea may still be a valuable protein source for pig diets even with the reduced energy availability, as the digestibility of crude protein and amino acid of the pea was not significantly affected by the insertion of the transgene. The transgenic pea is likely to cost less to produce owing to the reduction in insect damage and crop losses. The authors conclude that it may, however, require some additional heat processing before incorporation into pig diets to compensate for the lower energy value of the transgenic pea.

Pea genetic transformation can also be a powerful tool to understand and improve the nodulation mechanism. The role of diamine oxidase (DAO) has been analyzed in transgenic pea lines in relation to nodulation responses, indicating that putrescine derived products of DAO activity could retard nodule development. The results suggest that DAO does not have an essential role in nodule initiation. The most critical role of DAO in pea nodule development is apparently in the regulation of diamine levels in host tissues (Wisniewski and Brewin 2000).

Seed quality improvement is another goal for legume breeding, in particular the decreasing content of seed antinutritive factors. Trypsin inhibitor (TI) has been object of antisense gene silencing experiments (Welham and Domoney 2000). Seed TI activities in some lines transformed with the antisense construct were reduced significantly. A high frequency of non-transmission of transgenes from primary transformants was also described (Welham and Domoney 2000), which represents a limitation of the pea transformation methodology for antisense manipulations.

9.3.3 Other Approaches

Stably transformed plant lines provide a number of advantages, particularly the maintenance by seed production. However, there are some limitations, including low levels of expression, gene silencing and the long time required to obtain a producer line, making genetically modified plants less attractive for some purposes such as functional genomics studies and biomanufacturing. A research strategy that appears promising is the accomplishment of gene delivery by virus derived vectors, both for transient expression of proteins and for virus induced gene silencing. These vectors are delivered to plant tissues by *Agrobacterium* infiltration. The main advantage is the short time required to obtain the gene silencing or the protein production. In this way, the pea early browning virus (PEBV) has been used to construct vectors that can be applied to functional genomics study of genes involved in a wide range of biological processes in pea. Thus, silencing of phytoene desaturase, of *UNIFOLIATA*, and of a

KORRIGAN1 homolog was described (Constantin et al. 2004; Wang et al. 2008). By a similar procedure, using the AMV and the cucumber mosaic virus (CMV) vectors, delivered via *Agrobacterium*, the transient expression of foreign protein was demonstrated in three *P. sativum* varieties (Green et al. 2009). These authors were also able to express three therapeutic proteins indicating a promising future for this system in the production of biopharmaceuticals.

9.4 Chickpea

Chickpea is one of the earliest cultivated crops that originated in the Middle East. Chickpea is now grown in temperate and semi-arid regions of the world such as Asia, Europe, Australia and North America. There is a high demand for world production, exports and imports of chickpeas due to the nutritional value, but chickpea production has increased only marginally in the last three decades due to its high sensitivity to biotic and abiotic stresses such as Ascochyta blight, botrytis gray mold, dry root rot, collar rot, Fusarium wilt, pod borer, drought, salinity, and low temperature.

Chickpea is probably the cool season grain legume in which most results on genetic transformation have been achieved. Although conventional breeding methods have enhanced chickpea yield and quality, chickpea is a self-pollinated crop with an extremely narrow genetic base and variability that restricts the scope of general and heterosis breeding. Furthermore, many wild annual *Cicer* species that have desirable characters are sexually incompatible with the cultivated species *C. arietinum* (see Chapter 5) (Jaiwal and Sonia 2001). Therefore, genetic transformation represents an alternative approach to breeding for desirable characters such as resistance to insects or drought tolerance. Several authors have generated transgenic chickpeas expressing *cry1A* or *cry2A* genes from *B. thuringiensis* and have demonstrated the resistance of transformants against the pod borer *H. armigera* (Kar et al. 1997; Sanyal et al. 2005; Indurker et al. 2007; Acharjee et al. 2010). The α-amylase inhibitor gene (*αAI1*) isolated from the seeds of common bean (*Phaseolus vulgaris*) was introduced into chickpea through *Agrobacterium*-mediated transformation and the results of a bioassay study revealed a significant reduction in the survival rate of the bruchid weevil *C. maculatus* reared on transgenic chickpea seeds (Sarmah et al. 2004; Ignacimuthu and Prakash 2006).

Abiotic stresses can result in crop damages as high as those caused by biotic stresses. In chickpea, annual losses of over 3.7 million tons have been estimated to be due to water deficit conditions alone. The first report of transgenic chickpeas improved for drought tolerance used transgenic technology for the introduction of an osmoregulatory gene, *P5CSF129A*, for the overproduction of proline. The results suggest that the elevated proline reduces the level of free radicals in response to osmotic stress, thereby, significantly improving the ability of the transgenic plants to better survive under water stress (Bhatnagar-Mathur et al. 2009).

Successful gene transfer requires robust and reliable transformation and *in vitro* regeneration protocols. Chickpea was considered a recalcitrant crop for *in vitro* regeneration and genetic transformation. However, recent reports show that it is now less recalcitrant. A rapid and reproducible protocol for *in vitro* shoot induction and plantlet regeneration was described from cotyledonary nodes and axillary buds (Rekha and Thiruvengadam 2009). This protocol yielded an average of 15 plants per cotyledonary node explant in 25 days. Recently, simple protocols for successful regeneration, rooting of shoots, establishment of *in vitro* raised plantlets and transformation of chickpea has been developed (Anwar et al. 2008, 2009, 2010). Likewise, the protocol described by Bhatnagar-Mathur et al. (2009) represents a significant advance due to its high reproducibility and recovery of the transgenics in a relatively short period (90–100 days).

There are several critical factors for successful regeneration/transformation of chickpea such as choice of genotypes, suitability of explants (target plant cell), efficient interaction of *Agrobacterium* with the target plant cell and a reproducible complete plant regeneration protocol. Table 9.2 summarizes the data on gene transfer in chickpea indicating the cultivars/genotypes, explants used and some technical details about transformation procedures.

9.4.1 Explants

The observed improvement in *Agrobacterium*-mediated transformation efficiency in chickpea using excised explants may be attributed to the exposure of deep seated pluripotent cells of L2 layers (Steeves and Sussex 1989), which upon wounding get metabolically activated for synthesis of cell wall, cell division and production of *vir* genes inducing compounds (Stachel et al 1985; Sangwan et al 1992; Pitzschke and Hirt 2010). Likewise, pre-conditioning the excised chickpea explants before co-cultivation increased the transformation efficiency (Sanyal et al. 2005). In chickpea, direct organogenesis of shoots from cotyledonary nodes and embryo axes is the most common regeneration pathway (Table 9-2). The axillary meristems at the junction of the cotyledon and the embryo axes contain cells that are competent for regeneration and hence could be useful targets for gene delivery. In most instances, shoots were formed as a result of proliferation of pre-existing meristems, making these systems inefficient for transformation studies. Embryo axis explants, from which the first and second flushes of shoots were removed, were transformed using *Agrobacterium* (Patil et al 2009). Preliminary experiments wherein these authors used the shoots

Table 9-2 Summary of the *Cicer arietinum* transformation systems.

Genotype	Method[a]/strain	Explant	Fate of Explant	Transferred gene(s)	Reference
C. arietinum	***R***/ A281, R1601	Leaf, stem	Callus	*nptII*	Srinivasan et al. 1991
C. arietinum	***T***/	Decapitated embryo axis	Plant	*nptII, uidA*	Fontana et al. 1993
ICC4918	***T***/ LBA4404	Immature cotyledon	-	*nptII, uidA*	Ramana et al. 1996
ICCV-1, ICCV-6, Desi (local)	***T***/ LBA4404	Decapitated embryo axis	Plant	*nptII, uidA*	Kar et al. 1996
6153, CM72	Biolistic	Embryo axis	Embryo, epicotyl, hypocotyl	*uidA*	Husnain et al. 1997
	T/ A281, C58	Embryo axis	Embryo, callus	*nptII, uidA*	
ICCV-1, ICCV-6	Biolistic	Decapitated embryo axis	Plant	*nptII, cryIA*[b]	Kar et al 1997
Red chickpea, Canitez 87, MB-10	***T***/ LBA4404	Regenerated shoot	Plant	*nptII, uidA*	Altinkut et al. 1997
	R/ 9402	Regenerated shoot	Hairy roots	*nptII, uidA*	
6153, CM72	Biolistic	Hypocotyl	Somatic embryo, shoot	*nptII, uidA*	Husnain et al. 2000
PG1, PG12, Chafa, Turkey	***T***/ C58C1, GV2260, EHA101	Decapitated embryo axis	Plant[c]	*nptII, pat, uidA*	Krishnamurthy et al. 2000
C. arietinum	***T***/ LBA4404 and others	Embryo axis, cotyledonary node	Plant[c]	*nptII, uidA*	Sanyal et al. 2003
Gökçe, Akçin 91, Izmir92	***R***/ 15834	Shoot[d]	Shoot	*nptII*	Khawar and Ozcan 2004
Semsen	***T***/ AGL1	Cotyledon-half Decapitated embryonic axis	Plant[c]	*nptII, αAI1*	Sarmah et al. 2004
Semsen, Amethyst	***T***/ AGL1	Cotyledon-half Decapitated embryonic axis	Plant[c]	*bar, ssa, uidA*	Chiaiese et al. 2004

CDC Yuma	***T*/** EHA105	Decapitated embryo axis slice	Plant	*nptII, uidA*[b]	Polowick et al. 20004
ICCV5, H208, ICCL87322, K850, Annigeri	***T*/** C58, LBA4404, AGL1	Embryo axis slices	Plant, plant[c]	*bar*[b], *uidA*, *pgip*	Senthil et al. 2004
P-362, P-1042	Biolistic	Decapitated embryo axis	Plant	*dsAK, uidA*	Tewari-Singh et al. 2004
P-362, P-1042, P-1043	***T*/** EHA101	Decapitated embryo axis	Plant	*nptII, pat, uidA*	
C235, BG256, Pusa362, Pusa372	***T***[e,f]**/** LBA4404, EHA105, GV3850, GV2260	Cotyledonary nodes	Plant[c]	*nptII*[b], *cry1Ac, uidA*	Sanyal et al. 2005
K850	***T*/** LBA4404	Embriogenic axis[d]	Plant[c]	*nptII*[b], *αAI1, uidA*[b]	Ignacimuthu and Prakash 2006
Chafa, PG12, ICCC37, ICCC32	Biolistic	Stem, epicotyl, embryo axis	Plant	*nptII, cry1Ac, uidA*	Indurker et al. 2007
Gökçe, Er, Akçin, Uzunlu, Küsmen	***T*/** C58C1, EHA105, KYRT1	Embryo axis[d]	Seedling	*nptII, uidA*	Akbulut et al. 2008
ICC 10943, ICC 10386	***T***[f]**/** LBA4404	Decapitated embryo axis	Plant	*hptII, uidA*	Pathak and Hamzah 2008
C235, Annegiri, K850	***T*/** AgL0, LBA4404, EHA105,	Shoot meristem	Plant[c]	*bar, uidA*[b]	Singh et al. 2009
Gökce	***T***[e]**/** KYRT1	Cotyledonary node[d]	Shoot	*nptII, uidA*	Öz et al. 2009
C235	***T***[e]**/** GV3101	Cotyledon-half embryonic axis	Plant	*pmi*	Patil et al. 2009
C235	***T*/** C58	Axillary meristem	Plant	*nptII*[b], *P5CSF129A*[b], *uidA*	Bhatnagar-Mathur et al. 2009
Pusa256, KWR108, Pusa1003, one local line	***T*/** LBA4404, EHA105, AGL1	Cotyledon, cotyledonary node, embryo axis	Plant	*hpt*[b], *uidA*[b]	Bhattacharjee et al. 2010
Semsen, ICCV89314	***T*/** AGL1	Cotyledon-half Decapitated embryonic axis	Plant[c]	*nptII, cry2Aa*[b]	Acharjee et al. 2010

[a]*Agrobacterium*-mediated transformation; ***T*/**, with *A. tumefaciens*; ***R***, with *A. rhizogenes.*
[b]Mendelian inheritance pattern (3:1).
[c]Transformed shoots were grafted onto seedling.
[d]Pricked with a needle.
[e]Vacuum infiltration.
[f]Sonication-assisted *Agrobacterium* mediated transformation (SAAT).

obtained with the first and second flush produced either no or chimeric transformants. Most of the third flush of shoots that appeared after 3–4 weeks seemed to have originated de novo (Patil et al. 2009).

Two major hurdles that limit *in vitro* regeneration of chickpea are the induction and development of a strong root system and the establishment of *in vitro* raised plantlets in pots. In order to overcome these obstacles, several researchers have preferred the alternative grafting method (Krishnamurthy et al. 2000; Sarmah et al. 2004; Senthil et al. 2004; Tewari-Singh et al. 2004; Sanyal et al. 2005). Transgenic plants grafted onto rootstocks of their parental cultivar were phenotypically indistinguishable from the non-transgenic parent line. Transgenic lines that were rooted directly in tissue culture were frequently smaller than the parental line in the T0 generation, but their T1 progeny plants were of normal appearance in the glasshouse (Sarmah et al. 2004). Anwar et al. (2009) showed that *in vitro* rooting and transplantation is no longer a hurdle for chickpea transformation.

9.4.2 *Agrobacterium*

Transgenic chickpea plants have been produced using *A. tumefaciens*-mediated gene transfer and particle bombardment; although *Agrobacterium*-mediated transformation of explants is the most popular method and several protocols have been established for the production of transgenic plantlets derived from co-cultivation of different explants (Table 9-2). Early studies on transformation of chickpea reported very low frequencies of recovery of transgenic plants. Using hypervirulent *Agrobacterium* strains and/or SAAT, higher frequencies of transformation have been reported. The efficiency of *Agrobacterium*-mediated transformation and delivery of T-DNA into plant cells is influenced by numerous physicochemical and physiological conditions, and this complexity made it necessary to optimize the protocols for increased recovery of transformants.

Reports in a number of crop plants have suggested that *Agrobacterium* strains differ with respect to their capacity to transform tissues, suggesting a strong host genotype and *Agrobacterium* strain interaction. Strains LBA4404, EHA 105, AGL1, C58, EHA101, C58C1, KYRT1, A281, GV2260, GV3850, GV2260, GV3101, and AgL0 have been used for transforming chickpea (Table 9-2). The regeneration and transformation potentials of the meristematic regions of embryo axis slices have been exploited to develop a transformation method using the hypervirulent *A. tumefaciens* AGL1, and the three most susceptible chickpea genotypes, H208, ICCL87322 and ICCV5 (Senthil et al. 2004). Recently, a comparative study on the interaction of the three most used *Agrobacterium* strains (AGL1, LBA4404 and EHA105) with four genotypes of chickpea (Pusa-256, KWR-108, Pusa-1003 and a local line) and three types of explant (cotyledon, cotyledonary node and embryo axis) was performed (Bhattacharjee et al. 2010). This work showed that LBA4404 was the most effective in infecting any genotype and any explant, although the highest transformation efficiency (25.56%) was found using cotyledonary node of Pusa-256.

Several groups have reported *A. tumefaciens*-mediated transformation systems in chickpea, where transformation frequency ranged between 0.5 and 5% (Krishnamurthy et al. 2000; Polowick et al. 2004; Senthil et al. 2004; Sarmah et al. 2004; Singh et al. 2009). In chickpea, sonication and/or vacuum infiltration has been used to enhance *Agrobacterium*-mediated transformation (Sanyal et al. 2005; Akbulut et al. 2008; Pathak and Hamzah 2008; Öz et al. 2009). Thus, the transformation efficiency obtained by the SAAT method was between 24 and 26%, while without sonication it ranged between 9 and 11% (Pathak and Hamzah 2008). The enhanced transformation rates using SAAT probably resulted from sonication-induced micro-wounding that helped in the secretion of more phenolic compounds from the host target tissue to activate *vir* genes interactions (Pitzschke and Hirt 2010). Given that the infectivity of *Agrobacterium* needs a wound response, explants and media regimes that allow regeneration after wounding have been preferred for transformation.

Although efficient protocols by *Agrobacterium*-mediated transformation have been developed in which the transformation frequencies were up to 26% (Pathak and Hamzah 2008; Bhattacharjee et al. 2010), further research is necessary to understand the exact nature and cause of the reduction in plant growth and/or productivity in transgenic chickpea (Krishnamurthy et al. 2000; Sanyal et al. 2003, 2005; Polowick et al. 2004; Acharjee et al. 2010).

Transformation frequency using particle gun bombardment was very low < 1.5% (Tewari-Singh et al. 2004) or nil (Kar et al. 1997). Nevertheless, Indurker et al. (2007) showed that high frequency transformation (18%) could also be achieved in chickpea by particle bombardment, optimizing parameters such as the use of epicotyl explants, the cultivar ICCC-37, gold as microcarrier and a burst pressure of 900 psi. The co-transformation frequency for *nptII* and *cryIAc* or *cry2Aa* genes was 45 and 87%, respectively (Kar et al. 1997; Acharjee et al. 2010).

Although *A. tumefaciens* is the most commonly used vector for genetic transformation, other vectors such as *Agrobacterium rhizogenes* strains R1601 (Srinivasan et al. 1991), 9402 (Altinkut et al. 1997) or wild type 15834 (Khawar and Ozcan 2004) were used. Different groups have reported *A. rhizogenes*-mediated transformation systems in chickpea with variable degrees of hairy root formation up to 10.4% (Altinkut et al. 1997) and from 5% to 80% (Khawar and Ozcan 2004) depending on the chickpea variety and light condition. When young and wounded regenerated shoots were co-cultivated with LBA4404 (*A. tumefaciens*) and 9402 (*A. rhizogenes*) strains the best transformation response was 12.7% (cv. Canitez 87) and 10.4% (cv. MB-10), respectively (Altinkut et al. 1997).

9.4.3 Transgenes and Selectable Markers

Genetic transformation is achieved only if the transgene is stably inherited and expressed in subsequent progenies. In chickpea, Mendelian inheritance patterns (Kar et al. 1997; Polowick et al. 2004; Senthil et al. 2004; Sanyal et al. 2005; Ignacimuthu and Prakash 2006; Bhatnagar-Mathur et al. 2009; Singh et al. 2009; Acharjee et al. 2010; Bhattacharjee et al. 2010) and stable integration and expression of the transgene throughout two (Senthil et al. 2004; Singh et al. 2009; Bhattacharjee et al. 2010) or four generations (Polowick et al. 2004) have been described. Southern blot analysis of T1 individuals showed that one to three (Senthil et al. 2004; Pathak and Hamzah 2008; Bhatnagar-Mathur et al. 2009; Patil et al. 2009) or multiple copies were integrated (Krishnamurthy et al. 2000; Tewari-Singh et al. 2004; Polowick et al. 2004; Acharjee et al. 2010). Low copy inserts in transgenic plants are preferred both from the point of view of gene regulation/expression and from the perspective of plant breeders, since dealing with a trait conferred by a single, dominant gene is a convenient scenario in a plant breeding program (Sarmah et al. 2004).

Selection and recovery of transformed cells or tissues and elimination of agrobacteria from the cultures requires the use of selective agents and/or antibiotics. The effects of four selective agents (kanamycin, hygromycin, phosphinothricin and glyphosate) and four agrobacteria eliminating antibiotics (carbenicillin, cefotaxime, timentin, and augmentin) on multiple shoot and root induction in chickpea cotyledonary nodes have been investigated. While all the selective agents studied totally inhibited rooting of the regenerated shoots, the *Agrobacterium*-eliminating-antibiotics showed different effects. Timentin significantly increased and carbenicillin significantly decreased shoot induction after 4 weeks of culture (Öz et al. 2009). A progressive selection system with a stepwise increase in the concentration of the selective agent kanamycin along with selection pressure avoided at the rooting stage was found to be useful in obtaining stringently selected putative transformants in chickpea (Bhatnagar-Mathur et al. 2009). Also, a gradual increase of hygromycin in the rooting medium did not interfere with the initiation of rooting and subsequent growth of the plant (Bhattacharjee et al. 2010).

The use of antibiotic resistance markers to select transformed plants has generated widespread public concern. Therefore, alternative methods for obtaining transgenic plants without using antibiotic resistance markers and methods to selectively remove antibiotic selectable markers genes from transgenic plants have been developed. The most widely used selectable marker gene to produce transgenic chickpea plants is *nptII* (neomycin phosphotransferase), which confers resistance to antibiotics of the amino glycoside type such as kanamycin, neomycin, and geneticin (Table 9-2). Markers such as the *hpt* gene (Pathak and Hamzah 2008; Bhattacharjee et al. 2010) and herbicide resistance *pat* and *bar* genes (Chiaiese et al. 2004; Senthil et al. 2004; Tewari-Singh et al. 2004; Singh et al. 2009) have also been used but in low frequency. Another method is to use positive selectable marker genes such as *pmi/manA* (phosphomannose isomerase), which provides the ability to metabolize mannose. In this case, transgenic events can be identified early on by PCR (indicating presence of the *pmi* gene), RT-PCR (for studying expression of the *pmi* gene), or CPR (chlorophenol red assay for the PMI enzyme) (Patil et al. 2009). Tewari-Singh et al. (2004) used neomycin phosphotransferase (*nptII*), phosphinothricin-acetyltransferase (*pat*) and aspartate kinase (*AK*) genes for the production of transgenic chickpea. In this report, phosphinotricin (PPT) was found to be a better selective agent than kanamycin since transgenic plants could be identified more easily and rapidly using the former. The AK amino acid selection method via altered metabolism on media containing lysine and threonine (AK/TK) described in this study could be of considerable practical significance in transgenic crop improvement programs, as it avoids the use of antibiotic selection genes. Effectiveness of antibiotic selection regime after co-cultivation is an important factor for the elimination of escapes, which seems to be dependant on factors like type of tissue, size of explants, antibiotic concentration and time of application (Birch 1997). While 25% of escapes has been observed using kanamycin (Bhatnagar-Mathur et al. 2009), hygromycin was beneficial for recovery of the highest number of transgenic shoots, besides resulting in a 100% reduction in false transgenic shoots (Bhattacharjee et al. 2010).

Considering the different selectable markers used, the highest transformation efficiencies were 3.1% with *nptII* (Polowick et al. 2004), up to 26% using *hpt* (Pathak and Hamzah 2008; Bhattacharjee et al. 2010), 5% using *bar* as a negative selection gene (Senthil et al. 2004), 3% using the positive selectable marker gene *pmi* (Patil et al. 2009), 0.7% to 1.2% with PPT, and 1.0% with the AK/LT selection system (Tewari-Singh et al. 2004).

According to Polowick et al (2004), the frequency of recovery of rooted transgenic plants was 1.3%, with a mean tissue culture phase of 160 days, from co-cultivation to planting in soil, (range: 127–287 days). The addition of three passages on the shoot elongation medium (one with 50 mg/l kanamycin and two with 75 mg/l) increased the frequency of plant recovery to an average of 3.1%, but also increased the duration of the tissue culture phase to a mean of 217 days (range: 133–384 days) (Polowick et al.2004). Bhattacharjee et al. (2010) observed that increasing the duration of tissue culture period resulted in abnormalities such as shoot tip damage, yellowing of leaves of transgenic shoots and at a later stage, when these plants were shifted to a greenhouse, the height of the plant was lower compared with normal seed-derived plants. The number of pods/transgenic plants was also very low (only 5 or 6 per plant) compared with normal seed-derived plants.

While antibiotic marker genes were introduced to test the feasibility of transformation, other genes of economic importance such as *Cry* genes (Kar et al. 1997; Sanyal et al. 2005; Indurker et al. 2007; Acharjee et al. 2010), the *αAI1* gene (Sarmah et al. 2004; Ignacimuthu and Prakash 2006), a transgene encoding the methionine- and cysteine-rich protein sunflower seed albumin (*SSA*) (Chiaiese et al. 2004), and an osmoregulatory gene *P5CSF129A* for the overproduction of proline (Bhatnagar-Mathur et al. 2009) have also been introduced into chickpea by genetic transformation (Table 9-2). Nevertheless, few reports on genetic transformation to biotic and abiotic stress tolerance have been published. Hence, a lot still remains to be worked out in chickpea in order to get an economically superior germplasm with respect to the existing cultivated varieties.

9.5 Faba Bean

The grain legume faba bean (*Vicia faba* L.) is grown worldwide as a protein source for food and feed. The nutritional value of faba bean is high, and in some areas it is considered superior to pea or other grain legumes (Crépon et al. 2010). Even though the global average grain yield has almost doubled during the past 50 years, the total area sown with faba beans has declined by 56% over the same period. The season-to-season fluctuations in grain yield and the progressive replacement of traditional farming systems, which utilized legumes to provide nitrogen to maintain soil fertility, with industrialized, largely cereal-based systems that are heavily reliant upon fossil fuels, are some of the explanations for this decline in importance. Intercropping of faba bean with cereals may be an efficient management tool to control weeds, particularly if no appropriate herbicides are available, or in organic farming systems where herbicides cannot be used (Hauggaard-Nielsen et al. 2008). However, growth is hampered by several factors including environmental stresses and the presence of antinutritional factors. To solve these limitations, breeding programs have been initiated, although difficulties in pollination control, the limited genetic pool, the large genome size and its partial allogamy (about 50% with large variations), have slowed down the progress in varietal improvement (Selva et al. 1989) (see Chapter 2).

A major aim for any crop breeding program is the development of good quality lines with adequate resistance/tolerance to yield-reducing stresses such as frost stress and winter-specific pathogens. A major constraint in the search of genetic variability for breeding faba bean is its genetic isolation from other species in the genus *Vicia* (see Chapter 2). To improve faba bean against biotic and abiotic stress and to improve its nutritional value, traditional breeding programs could be supplemented by gene transfer technology. Integration of conventional breeding and biotechnological approaches in *Vicia faba* is difficult, since it is very recalcitrant in biotechnological studies. However, some achievements have been obtained. For example, faba bean necrotic yellows virus (FBNYV) multiplication, virion formation and virus-associated disease have been reproduced by using cloned viral DNAs for particle bombardment or agroinoculation (Timchenko et al. 2006).

9.5.1 *In Vitro* Culture

In faba bean, it is not difficult to obtain GUS-transformed tissues under *in vitro* culture but to complete the process to obtain adult plants, which allow testing of the stability and inheritance of transgene(s), is rather limited. Faba bean, like most other grain legumes, exhibits a low competence for *in vitro* culture. Regeneration in this crop has been reported from protoplasts, as well as from different explants using different modes of regeneration (Tegeder et al. 1995; Khalafalla and Hattori 1999; Bahgat et al. 2009). Unfortunately, the regeneration methods are slow, useful only on certain genotypes, and generally produce a low frequency of regenerated plants. Recently, a new reproducible system for plant regeneration via somatic embryogenesis has been described (Bahgat et al. 2009). In this work, calli were induced from shoot tips and epicotyls of two Egyptian cultivars of *V. faba*. The difference between each cultivar in the number of regenerated shoots varied due to the type of explants used and the hormonal composition of each treatment. When the interactions between the three variables of this study (cultivars, explants and treatments) were analyzed, the explant which led to the highest percentage of regeneration was the shoot tip of cv. Giza 2 using M1 treatment. Only somatic embryos derived from shoot tips of cv. Giza 2 developed to fertile plants, but somaclonal variation was detected in regenerated plants.

Additional difficulties exist during faba bean *in vitro* culture, since explant and callus cells tend to produce high levels of phenolic compounds, resulting in cell death (Selva et al. 1989). The addition of ascorbic and citric acids to the media helped to decrease or prevent the production or oxidation of the phenolic compounds, enhancing the production of somatic embryos (Bahgat et al. 2009). Alternatively (Metry et al. 2007), the cultures were incubated for one week in the dark during the shoot induction stage, and the regeneration medium was supplemented with activated charcoal to eliminate the symptoms caused by the accumulation of phenolic compounds. However, the addition of activated charcoal into the regeneration medium had a negative impact since it reduced the number of shoots per explant.

The conclusion of all these works is that *Vicia faba* still lacks a genotype-independent, efficient, and reproducible plant regeneration system that is not prone to somaclonal variation and suitable for *Agrobacterium*-mediated genetic transformation. A quick method of delivering foreign genes with high transformation efficiency for the development of transgenic faba bean, is also lacking, since it is among those legumes where the production of transgenic plants has not convincingly been reported. Table 9-3 shows a summary of the transformation events in faba bean.

9.5.2 *Transformation*

Different researchers worked on faba bean regeneration and transformation (Ramsay and Kumar 1990; Saalbach et al. 1994; Jelenić et al. 2000), but no transgenic plants have been reported. The first mature transgenic plants were recovered by Böttinger et al. (2001), using a protocol in which transformed shoots obtained from internodal stem segments were rooted or grafted onto young seedlings *in vitro*. Although this protocol has several disadvantages, including the time-consuming (16–24 months) shoot regeneration from callus, the relatively low efficiency and a high percentage (7 out of 11) of morphologically and cytogenetically aberrant plants such as a dwarfed phenotype, inability of flowering, formation of abnormal flowers and pods, or failure in seed setting, it represented a feasible approach for the production of genetically engineered faba beans.

A previous step to transformation is the identification of susceptible faba bean genotypes and/or virulent *Agrobacterium* strains and the development of a protocol for the regeneration of fertile transgenic plants. Three faba bean cultivars, which varied in morphology and geographical origin, were inoculated with nine *Agrobacterium* strains *in vivo* (Jelenić et al. 2000). Cultivars differed in their susceptibility to *Agrobacterium* strains, and plant genotype vs. strain interaction was detected. Although a cultivar (Lobab Lippoi) very susceptible to transformation with *Agrobacterium in vivo* was identified, its *in vitro* transformation was not successful. Possibly the addition of acetosyringone was required for *in vitro* transformation as previously observed in faba bean (Saalbach et al. 1994).

The limitations in regeneration could be cut short by hairy root transformation that is easier to achieve but only allows analysis of gene constructs in root tissue. Transformation protocols using *A. rhizogenes* transformation have been established in faba bean (Ramsay and Kumar 1990; Saalbach et al. 1994; Jelenić et al. 2000). Recently, expression of the chimeric gene *VfLb29* promoter-*gusAint* was analyzed in transgenic hairy roots of *Vicia faba* and other legume and non-legumes plants (Vieweg et al.

Table 9-3 Summary of the *Vicia faba* transformation systems.

Genotype	Method[a]/strain	Explant	Fate of explant	Transferred gene(s)	Reference
V. faba	***R***	Cotyledon, stem	Root clones	*nptII*	Ramsay and Kumar 1990
V. faba	***R*/**RifR 15834	-	Hairy roots	*2S*	Saalbach et al. 1994
Lobab Lippoi, Topolo, Ošlje	***R*/**8196, 15834 ***T*/**A281, B6S3, GV3101(3), C58C1(2)	Third trifoliate stage[b] (*in vivo*), cotyledon, leaf, stem[b] (*in vitro*)	Tumor formation	-	Jelenić et al. 2000
Mythos	***T*/**EHA101, EHA105	Internodal segment	Plant[c]	*uidA, nptII, lysC, sfa8*	Böttinger et al. 2001
Kleine Thueringer	***R*/**Arqual	Seedling in open pots	Hairy roots	VfLb29 promotor-*gusA*int	Vieweg et al. 2004
Mythos, Albatros, and 4 genotypes more	***T*/**EHA101, EHA105	Mature and immature embryo axis slices[d]	Plant[c]	*bar, sfa8, nptII, lysC*	Hanafy et al. 2005
Giza40	Biolistic	Mature embryo axis discs	Plant[c]	*bar, uidA*	Metry et al. 2007
3 genotypes	***T*/**LBA4404	Embryo axis	Plant	*uidA*, defensin gene	Youssef et al. 2007
Otafuku	Biolistic	Leaf	Leaf (guard cell)	*gfp, AtNCED3, AAO3*	Melhorn et al. 2008

[a]*Agrobacterium*-mediated transformation; ***T*/**, with *A. tumefaciens*; ***R***, with *A. rhizogenes*.
[b]Pricked with a needle.
[c]Transformed shoots were grafted onto seedling.
[d]Devoid of the root tip.

2004). The *VfLb29* leghemoglobin gene promoter was found to be specifically active not only in the infected cells of the nitrogen-fixing zone of root nodules but also in arbuscule-containing cells of transgenic *V. faba*, *V. hirsuta* and *M. truncatula* roots colonized by the endomycorrhizal fungus *Glomus intraradices*. These studies are relevant since mycorrhization offers several benefits to the host plant, including improved nutrition, greater drought resistance, and protection from pathogens (Harley and Smith 1983).

Transformation strategies, which minimize the *in vitro* culture period and avoid the callus phase, are advantageous. The transformation system developed by Hanafy et al. (2005) overcomes many of the difficulties previously reported by Böttinger et al. (2001). The use of embryonic axes which were cultivated on media containing high concentrations of cytokinins in combination with low auxin concentrations enhanced the direct shooting without an intermediate callus phase, and the possibility of somaclonal variation was reduced to a minimal level. However transformation frequencies were very low ranging from 0.15 to 2.0% in the cultivars Mythos and Albastros and 0 % in Giza 2, Giza 429, Giza 716 and Giza Blanka (Hanafy et al. 2005). When transgenic faba bean plants were obtained by the gene bombardment delivery system, the transformation percentage was 2% for cultivar Giza 40 (Metry et al. 2007). In both studies the T1 plants were analyzed by testing the expression of the *bar* gene in the greenhouse, and 10 days after BASTA application the treated transgenic plants and leaflets showed complete tolerance, in contrast to non-transformed plants, which showed completely necrotic leaflets (Hanafy et al. 2005; Metry et al. 2007).

The characterization of the number and location of integration sites is very important before transgenic lines can be released to the field. In *V. faba*, using two independent transgenic lines (Vf12-1 and Vf12-2 transformed by *A. tumefaciens*) and the fluorescence *in situ* hybridization technique, the transgene insertion was localized on the long arm of chromosome III (Snowdon et al. 2001). Southern analysis was used to prove T-DNA integration and to estimate the number of integrated copies which ranged between one and three. The high frequency of plants, which did not show the transfer of the introduced genes to the next generations, was possibly due to chimeric events leading to T0 plants and/or escapes (Böttinger et al. 2001; Hanafy et al. 2005). Since short-term selection can increase the possibility for escapes and chimeric plants, it could be recommended that the regenerated shoots should be selected for at least 6 months before recovering the plants by micro-grafting (Hanafy et al. 2005).

Although recently a step forward has been made in the techniques for regeneration of transgenic *V. faba* plants, so far only a few genes useful for breeding have been introduced. Using *A. rhizogenes*, the 2S albumin gene of Brazil nut (*Bertholletia excelsa* H.B.K.) was expressed in hairy roots (Saalbach et al. 1994). Modification of the seed storage protein composition of *V. faba* by means of genetic transformation, to enhance the sulfur containing amino acids methionine and cysteine, was achieved by introducing *lysC* (a mutant aspartate kinase gene) and *sfa8* (a gene encoding a sulfur rich sunflower albumin) (Böttinger et al. 2001; Hanafy et al. 2005). The inheritance and expression of the foreign genes in the seeds of transgenic *Vicia faba* plants was determined by Western blot (Böttinger et al. 2001; Hanafy et al. 2005) and the inheritance of the transgenes was confirmed by PCR until the T3 generation (Hanafy et al. 2005). The low expression or even lack of expression of the *lysC* gene in some lines could be related to a higher number of integrated T-DNA copies (three copies vs. one) (Böttinger et al. 2001). Because transgene expression can be highly dependent on insert location and copy number, information about the number and integration site of inserts is extremely important (Snowdon et al. 2001).

In *Vicia faba*, like in other plants, stomatal closure due to increased levels of abscisic acid (ABA) is important for water stress tolerance. Recently, Melhorn et al. (2008) introduced *AtNCED3* (9-cis-epoxycarotenoid dioxygenase) and *AAO3* (*Arabidopsis* aldehyde oxidase 3) genes directly into guard cells of leaves using biolistic technology. The transient expression of *AtNCED3* and *AAO3* genes in guard cells caused stomatal closure in faba bean. The authors suggested the possibility that these cells might be involved in stress-regulated ABA biosynthesis. In the future, molecular genetic progress will enlarge the list of genes with potential for use in improving abiotic stress tolerance.

9.6 Lentil

Lentil is characterized by poor yields and susceptibility to insect and fungal diseases (Erskine 1984) and is considered as an unexploited crop. The low development of selected cultivars is related to the difficulty of obtaining successful hybridization and the scarce genetic variability that is a characteristic of this species. Modern lentil breeding has a relatively short history since the first major initiative to improve this species began in 1977 (Materne and McNeil 2007). However breeding programs have resulted in substantial success in improving tolerance to drought, insects and diseases leading to high-yielding varieties with greater yield of biomass, and increased tolerance to biotic and abiotic stresses. Breeding for resistance to lodging, pod shattering and a different plant growth habit has increased the yield potential considerably and facilitated mechanized harvesting (Christou 1997).

9.6.1 *In Vitro* Culture and Transformation

The delay in classical breeding in comparison with other legumes is also reflected when biotechnology techniques are considered. Until successful tools have been generalized, application of genetic engineering to lentil breeding cannot be achieved. Pulse crops have long been considered to be recalcitrant to cell and tissue culture, with lentil being among the most difficult legumes for regeneration of whole plants due to problems of root induction. Tissue culture in lentil has been achieved with different explants such as shoot apical meristem, cotyledonary nodes, axilary buds, immature embryos, and embryonic axes. The main difficulty in obtaining whole plantlets was the rooting of the regenerated shoots. Good responses can be obtained from axillary buds in the absence of benzylaminopurine and rooted in inverted orientation (Fratini and Ruiz 2002, 2003).

Compared to other grain legumes, genetic transformation studies in lentil are limited (Eapen 2008; Oktem et al. 2008). Reports on lentil genetic transformation are not frequent, and although *Agrobacterium*-mediated transformation has been the most extended methodology, other alternatives have been tried with different success. Apart from the recalcitrant *in vitro* culture behavior of lentil, other causes may account for the low development of the genetic engineering in lentil. The highly variable response among cultivars and explants hinders the transferability of the methods. For example, it has been reported that explants proven to be good for shoot regeneration were not responsive to genetic transformation. According to different authors shoot regeneration from cotyledonary nodes is very high and largely genotype independent (Warkentin and McHughen 1993; Gulati et al. 2001). However, it was also reported that these explants do not responded to *Agrobacterium*-mediated transformation (Warkentin and McHughen 1993). This result has been attributed to the fact that multiple shoot regeneration occurs from pre-existing meristems in the inner layer of the explants, which are beyond the reach of *Agrobacterium* (Sarker et al. 2003), making it necessary to modify the infection procedure or to use alternative non-*Agrobacterium*-mediated transformation systems (Oktem et al. 1999; Gulati et al. 2002; Mahmoudian et al. 2002b; Akcay et al. 2009).

Alternative methods that have been tried with lentil are summarized in Table 9-4. Among a wide variety of approaches tested, *A. tumefaciens* is the preferred way to introduce foreign genes in the crop. Although different strains are applied, some of them (LBA4404, EHA105, GV2260, C58/C58C1) have been used repeatedly. The constructions employed for the transfer of genes included *uidA* and *nptII* genes.

First reports demonstrated lentil susceptibility to *A. tumefaciens* by *in vivo* inoculation, originating tumor formation with the expression of opine genes (Warkentin and McHughen 1991). Afterwards, protoplasts

Table 9-4 Summary of the *Lens culinaris* transformation systems.

Genotype	Method[a]/strain	Explant	Transferred gene(s)	Transformation evidence	Reference
Laird	***T*/** C58, A281, GV3111, Ach5	Stem (*in vivo*), shoot apex	-	Tumour, opines, Southern blot	Warkentin and McHughen 1991
Laird	***T*/** GV2260	Shoot apex, epicotyl, root	*uidA*	Tumour, histochemical and fluorimetric assays	Warkentin and McHughen 1992
L. culinaris	Lipofection	Leaf protoplast	*uidA, CAT*	Fluorimetric assay and scintillation counting	Maccarrone et al. 1992; 1995
Crimson	Electroporation	Nodal meristems	*uidA*	Transmission T1, histochemical assay	Chowrira et al. 1995
Crimson	Electroporation	Nodal axilary buds	*uidA*	Histochemical assay, transmission T1, T2 (PCR, Southern blot)	Chowrira et al. 1996
11 genotypes	***T*/** A281, A136NC	Stem (*in vivo*)	-	Tumour	Karakaya and Özcan 1998
Eston, Palouse, Brewe, Chilean	***T*/** C58, EHA105	Embryonic axis slices	*uidA*	Histochemical assay	Lurquin et al. 1998
Sultan-1	Biolistic	Cotyledonary node	*uidA*	Histochemical assay	Oktem et al. 1999
Laird, CDC599-23	Biolistic	Cotyledonary node	*ALS*	Transmission T1[b] (PCR, Southern blot), herbicide[r]	Gulati et al. 2002
Sultan-1	***T***[d]/CV2250	Nodal meristem, cotyledonary node (decapitated)	*uidA*	Histochemical assay	Mahmoudian et al. 2002 a, b
21 genotypes	***T***/A281	Leaf, stem	*uidA*	Tumour, histochemical assay	Khawar and Özcan 2002

Table 9-4 contd....

Table 9-4 contd....

Genotype	Method[a]/strain	Explant	Transferred gene(s)	Transformation evidence	Reference
Barimasur 2, Barimasur 4	***T***/LBA4404	Decapitated embryo, cotyledonary node, epicotyl, immature embryo	*uidA*, *nptII*	Kn^r, histochemical assay, PCR	Sarker et al. 2003
ILL6994, ILL5883, ILL7201, ILL7212	***T***/LBA4404, EHA101	Decapitated embryo, half embryo, shoot tip, cotyledonary node, epicotyl	*uidA*	Histochemical assay	Hassan et al. 2007
Verdina	***T***[c]/LBA4404, EHA101	Embryonic axis, decapitated embryo, axylary nodes	*uidA*, *nptII*	Kn^r, histochemical assay	Vaquero et al. 2007
Sultan-1	***T***[d]/EHA105, C58C1, KYRT1	Cotyledonary node	*uidA*, *nptII*	Kn^r, histocchemical assay, transmission T1, T2, T3[b] (PCR, RT-PCR, Southern blot)	Akcay et al. 2009

[a]*Agrobacterium*-mediated transformation; ***T***/, with *A. tumefaciens*; ***R***, with *A. rhizogenes*.
[b]Transformed shoots were grafted onto non-transformed seeding.
[c]SAAT.
[d]Vacuum infiltration, blade wounding.

of lentil were transfected following liposome delivery of DNA by means of electroporation (Maccarrone et al. 1992, 1995). Although this technique seems to work well in genetic transformation, the lack of procedures for regenerating plants has precluded more attempts in this line. Something similar occurred with the very attractive in planta electroporation technique. After the two reports of Chowrira et al. (1995, 1996), no new data on this methodology have been published, apparently because they were difficult to reproduce (Somers et al. 2003). As an alternative, particle bombardment-mediated transformation of cotyledonary nodes was reported to be able to induce genetic transformation (Oktem et al. 1999; Gulati et al. 2002).

However, most research groups have opted for modifications in the *Agrobacterium*-mediated transformation method, by choosing the strain, cultivar and explant displaying the best response and by changing different parameters of the co-culture or the infection procedure. Experiments both *in vivo* (Warkentin and McHughen 1991; Khawar and Özcan 2002) and *in vitro* (Chowrira et al. 1996; Karakaya and Özcan 1998, Lurquin et al. 1998; Sarker et al. 2003; Hassan et al. 2007; Akcay et al. 2009) demonstrated a wide variability in the susceptibility response among different cultivars, and also a high variability among different *Agrobacterium* strains. Unfortunately there is no optimal cultivar or strain that can be chosen for generalized use. Moreover, there seem to be interactions between cultivars and *A. tumefaciens* strains, because the best strain for transformation efficiency in one cultivar can be very inefficient in another. The same rule is observed when different explants are infected with several strains. The strain that is favorable for transformation of a particular explant can be unfavorable for another, and significant differences have been published (Hassan et al. 2007). This is particularly evident when numerous genotypes are tested (Karakaya and Özcan 1998; Khawar and Özcan 2002). Nevertheless, it seems that in order to obtain significant transformation it is better to use some of the so-called hypervirulent strains at high inoculum density (Mahmoudian et al. 2002 a, b; Akcay et al. 2009), and often accessory strategies have been employed which could enhance the infection ability. These adjuvants include: 1) wounding of the explants by performing blade injuries (Warkentin and McHughen 1992; Sarker et al. 2003). Although transformation is detected at the wounding regions, excessive injury must be avoided because it seems to be detrimental for shoot regeneration; 2) vacuum infiltration, which did not affect the regeneration rate, and enhanced the percentage of explants with transient GUS expression (Mahmoudian et al. 2002 a, b); 3) injuries made by blade (better than microprojectile bombardment wounding), jointly with vacuum infiltration (Akcay et al. 2009) and 4) SAAT also was demonstrated to improve the results (Vaquero et al. 2007).

Several groups have reported transient transformation in a variety of lentil tissues including shoot apex, epicotyls, roots (Warkentin and

McHughen 1992), nodal segments (Chowrira et al. 1995, 1996; Mahmoudian et al. 2002a; Vaquero et al. 2007), embryonic axes (Lurquin et al. 1998; Sarker et al. 2003; Hassan et al. 2007; Vaquero et al. 2007) and cotyledonary node (Oktem et al. 1999; Gulati et al. 2002; Mahmoudian et al. 2002b; Sarker et al. 2003; Akcay et al. 2009). Although other explants yielded higher percentages of transformation, cotyledonary nodes were the preferred explants in the majority of studies, likely because of its high shoot regeneration.

Lentil somatic embryogenesis, which was reported in 1987 by Saxena and King, has not been continued although the expected plants would be more uniform and reproducible. Therefore, the way for obtaining transformed plants is always through morphogenetic procedures. Beside this, one feature common in lentil transformation is that the transformation rate is low, with a small number of cells transformed. Consequently most of the obtained transformed shoots are chimeras and it is important that there is a coincidence of regeneration with transformed cells (Sarker et al. 2003; Akcay et al. 2009). During selection these chimeric tissues become etiolated and browned and transformed regions, which are surrounded by them are eliminated, so transformation cannot be accounted for as stable.

It is generally accepted that legumes exhibit a natural resistance to aminoglycoside antibiotics like kanamycin, and this aspect may affect selection efficiency (Christou 1994). Some authors have stated that *npt*II gene-based methods are not suitable for selection of transformants (Puonty-Kaerlas et al. 1990). Others support its use as a selectable marker in lentil, but at a concentration of 200 mg l^{-1} kanamycin (Sarker et al. 2003). In fact, the selectable marker in most cases was *nptII*, but the *ALS* gene, which confers resistance to sulfonylurea, has also been used (Gulati et al. 2002). Due to the particular behavior of lentil cultures, the existence of chimeras and the abundance of escapes, conventional selective procedures are not sufficiently efficient and need to adapt experimental conditions by decreasing (Gulati et al. 2002) or increasing (Akcay et al. 2009) selection pressure. Most of the cited reports showed transient transformation and reports of stable transformation are very scarce. However, there are studies reporting transgene transmission to the progenies in transformed plants obtained by in planta electroporation (Chowrira et al. 1995, 1996), particle bombardment (Gulati et al. 2002) and *Agrobacterium*-mediated transformation (Akcay et al. 2009).

Analysis of progenies detected segregation distortions from Mendelian expectations in T1 and T2 generations (Chowrira et al. 1996; Gulati et al. 2002), and in some of the T1, T2 and T3 families (Akcay et al. 2009). These deviations often show an excess of plants with the integrated genes. PCR assays and Southern hybridization showed that some of the distorted segregations are due to the presence of multiple copies of the analyzed genes (Akcay et al. 2009).

In order to obtain whole transformed plants, rooting has been the main obstacle when shoots have been obtained from highly regenerating explants such as cotyledonary nodes in bencylaminopurine-containing medium. Micrografting of lentil shoots was proposed as an alternative rooting procedure (Gulati et al. 2001). Subsequently the transformed shoots obtained by microprojectile bombardment were micrografted and plantlets successfully transferred to soil (Gulati et al. 2002). Micrografting was also employed in *Agrobacterium*-mediated transformation by Akcay et al. (2009).

Although progress in lentil transformation has begun slowly, recent achievements allow one in expecting a hopeful future. Baum and Madour (2006) related the involvement of ICARDA in genetic engineering of lentil to achieve improved tolerance to fungal diseases, drought and other abiotic stresses, but no results have been yet published. Lentil improvement still needs a reproducible technique for genetic transformation with high efficiency which allows dealing with applied breeding programs based on genetic engineering.

9.7 Grass Pea

Grass pea (*Lathyrus sativus* L.) is the most neglected and underutilized of the cool season grain legumes, and this is also true in biotechnological research, except perhaps in relation to ODAP (Chapter 4). Only a few studies on grass pea genetic transformation have been published and detailed aspects related to grass pea transformation and *in vitro* culture are presented in Section 4.7 of this book.

References

Acharjee S, Sarma BK, Kumar PA, Olsen K, Mahon R, Moar WJ, Moore A, Higgins TJV (2010) Transgenic chickpeas (*Cicer arietinum* L.) expressing a sequence-modified *cry2Aa* gene. Plant Sci 178: 333–339.

Akbulut M, Yücel M, Öktem HA (2008) Analysis and optimization of DNA delivery into chickpea (*Cicer arietinum* L.) seedlings by *Agrobacterium tumefaciens*. Afr J Biotechnol 7: 1011–1017.

Akcay UC, Mahmoudian M, Kamci H, Yucel M, Öktem HA (2009) *Agrobacterium tumefaciens*-mediated genetic transformation of a recalcitrant grain legume, lentil (*Lens culinaris* Medik). Plant Cell Rep 28: 407–417.

Altinkut A, Gozukirmiz N, Bajrovic K, Gozukirmizi N (1997) High percentage of regeneration and transformation in chickpea. In: A Altman, M Ziv (eds) Horticultural Biotechnology *In Vitro* Culture and Breeding. Book Series: Acta Horticulturae, issue 447, pp 319–320.

Anwar F, Sharmila P, Pardha Saradhi P (2008) An ideal protocol for *in vitro* regeneration, efficient rooting and stable transplantation of chickpea. Physiol Mol Biol Plant 14: 329–335.

Anwar F, Sharmila P, Pardha Saradhi P (2009) No more hurdle: *in vitro* chickpea rooting and cent percent transplantation. Aust J Basic Appl Sci 3: 2491–2496.

Anwar F, Sharmila P, Pardha Saradhi P (2010) No more recalcitrant: chickpea regeneration and genetic transformation. Afr J Biotechnol 9: 782–797.

Babaoglu M, Davey MR, Power JB (2000) Genetic engineering of grain legumes: key transformation events. AgBiotechNet 2: 1–8.

Bahgat S, Shabban OA, El-Shihy O, Lightfoot A, El-Shemy HA (2009) Establishment of the regeneration system for *Vicia faba* L. Curr Iss Mol Biol 11: 47–54.

Baum M, Madour M (2006) Development of transgenic crops and their risk assessment. Arab J Plant Prot 24: 178–181.

Bean SJ, Gooding PS, Mullineaux PM, Davies DR (1997) A simple system for pea transformation. Plant Cell Rep 16: 513–519.

Bencheikh M, Gallais A (1996) Study of the variation in the somatic embryogenesis ability of some pea lines (*Pisum sativum* L. and *Pisum arvense* L.). Euphytica 90: 251–256.

Bercetche J, Chriquir D, Adam S, David D (1987) Morphogenetic and cellular reorientation induced by *Agrobacerium rhizogenes* (strains 1855, 2659 and 8196) on carrot, pea and tobacco. Plant Sci 52: 195–210.

Bhatnagar-Mathur P, Vadez V, Devi MJ, Lavanya M, Vani G, Sharma KK (2009) Genetic engineering of chickpea (*Cicer arietinum* L.) with the *P5CSF129A* gene for osmoregulation with implications on drought tolerance. Mol Breed 23: 591–606.

Bhattacharjee B, Mohan M, Nair S (2010) Transformation of chickpea: effect of genotype, explant, *Agrobacterium*-strain and composition of culture medium. Biol Plant 54: 21–32.

Birch RG (1997) Plant transformation: problems and strategies for practical application. Annu Rev Plant Physiol Plant Mol Biol 48: 297–326.

Böttinger P, Steinmetz A, Schieder O, Pickardt T (2001) *Agrobacterium*-mediated transformation of *Vicia faba*. Mol Breed 8: 243–254.

Chandra A, Pental D (2003) Regeneration and genetic transformation of grain legumes: An overview. Curr Sci 84: 381–387.

Chakraborty U, Sarkar B, Chakraborty BN (2003) Protection of soybean rot by *Bradyrhizobium japonicum* and *Trichoderma harzianum* associated with changes in enzyme activities and phytoalexin production. J Mycol Plant Pathol 33: 21–25.

Chiaiese P, Ohkama-Ohtsu N, Molvig L, Godfree R, Dove H, Hocart C, Fujiwara T, Higgins TJV, Tabe LM (2004) Sulphur and nitrogen nutrition influence the response of chickpea seeds to an added, transgenic sink for organic sulphur. J Exp Bot 55: 1889–1901.

Chowrira GM, Akella V, Lurquin PF (1995) Electroporation mediated gene transfer into intact nodal meristems in planta: Generating transgenic plants without *in vitro* tissue culture. Mol Biotechnol 3: 17–23.

Chowrira GM, Akella V, Fuerst PE, Lurquin PF (1996) Transgenic grain legumes obtained by in planta electroporation-mediated gene transfer. Mol Biotechnol 5: 85–95.

Chowrira GM, Cavileer TD, Gupta SK, Lurquin PF, Berger PH (1998) Coat protein-mediated resistance to pea enation mosaic virus in transgenic *Pisum sativum* L. Transgen Res 7: 265–271.

Christou P (1994) The biotechnology of crop legumes. Euphytica 74: 165–185.

Christou P (1997) Biotechnology applied to grain legumes. Field Crop Res 53: 83–97.

Collins CL, Eason PJ, Dunshea FR, Higgins TJV, King RH (2006) Starch but not protein digestibility is altered in pigs fed transgenic peas containing α-amylase inhibitor. J Sci Food Agri 86: 1894–1899.

Constantin GD, Krath BN, MacFarlane SA, Nicolaisen M, Johansen IE, Lund OS (2004) Virus-induced gene silencing as a tool for functional genomics in a legume species. Plant J 40: 622–631.

Crépon K, Marget P, Peyronnet C, Carrouée B, Arese P, Duc G (2010) Nutritional value of faba bean (*Vicia faba* L.) seeds for feed and food. Field Crop Res 115: 329–339.

Davies DR, Hamilton J, Mullinraux P (1993) Transformation of peas. Plant Cell Rep 12: 180–183.

De Kathen A, Jacobsen HJ (1990) *Agrobacterium tumefaciens*-mediated transformation of *Pisum sativum* L. using binary and cointegrate vectors Plant Cell Rep 9: 276–279.

De Kathen A, Jacobsen HJ (1995) Cell competence for *Agrobacterium*-mediated DNA transfer in *Pisum sativum* L. Transgen Res 4: 184–191.

De Sousa-Majer MJ, Turner NC, Hardie DC, Morton RL, Lamont B, Higgins TH (2004) Response to water deficit and high temperature of transgenic peas (*Pisum sativum* L.) containing a seed-specific α-Amylase inhibitor and the subsequent effects on pea weevil (*Bruchus pisorum*) survival. J Exp Bot 55: 497–505.

Dita MA, Rispail N, Prats E, Rubiales D, Singh KB (2006) Biotechnology approaches to overcome biotic and abiotic stress constraints in legumes. Euphytica 147: 1–24.

Eapen S (2008) Advances in development of transgenic pulse crops. Biotechnol Adv 26: 162–168.

Erskine W (1984) Evaluation and utilization of lentil germplasm in an international breeding program. In: JR Witcombe, W Erskine (eds) Genetic Resources and Their Exploitation -Chickpeas, Faba Beans and Lentils. Martinus Nijhoff, Hague, The Netherlands, pp 225–237.

Fontana GS, Santini L, Caretto S, Frugis G, Mariotti D (1993) Genetic transformation in the grain legume *Cicer arietinum* L (chickpea). Plant Cell Rep 12: 194–198.

Fratini R, Ruiz ML (2002) Comparative study of different cytokinins in the induction of morphogenesis in lentil (*Lens culinaris* Medik.). *In Vitro* Cell Dev Biol-Plant 38: 46–51.

Fratini R, Ruiz ML (2003) A rooting procedure for lentil (*Lens culinaris* Medik.) and other hypogeous legumes (pea, chickpea and *Lathyrus*) based on explant polarity. Plant Cell Rep 21: 726–732.

Grant J, Cooper P (2006) Peas (*Pisum sativum* L.) In: K Wang (ed) Methods in Molecular Biology, vol 343: *Agrobacterium* Protocols 2/e, vol 1. Humana Press, Totowa, NJ, USA, pp 337–346.

Grant JE, Cooper PA, McAra AE, Frew TJ (1995) Transformation of peas (*Pisum sativum* L.) using immature cotyledons. Plant Cell Rep 15: 254–258.

Grant JE, Cooper PA, Gilpin BJ, Hoglund SJ, Reader JK, Pither-Joyce MD, Timmerman-Vaughan GM (1998) Kanamycin is effective for selecting transformed peas. Plant Sci 139: 159–164.

Grant JE, Thomson LMJ, Pither-Joyce MD, Dale TM, Cooper PA (2003) Influence of *Agrobacterium tumefaciens* strain on the production of transgenic peas (*Pisum sativum* L.). Plant Cell Rep 21: 1207–1210.

Green BJ, Fujiki M, Mett V, Kaczmarczyk J, Shamloul M, Musiychuk K, Underkoffler S, Yusibov V, Mett V (2009) Transient K protein expression in three *Pisum sativum* (green pea) varieties Biotechnol J 4: 230–237.

Griga M (1998) Direct somatic embryogenesis from shoot apical meristems of pea, and thidiazuron-induced high conversion rate of somatic embryos. Biol Plantarum 41: 481–495.

Griga M (2002) Morphology and anatomy of *Pisum sativum* somatic embryos. Biol Plant 45: 173–182.

Gulati A, Schryer P, McHughen A (2001) Regeneration and micrografting of lentil shoots. *In Vitro* Cell Dev Biol-Plant 37: 798–802.

Gulati A, Schryer P, McHughen A (2002) Production of fertile transgenic lentil (*Lens culinaris* Medik.) plants using particle bombardment. *In Vitro* Cell Dev Biol-Plant 38: 316–324.

Hanafy M, Pickardt T, Kiesecker H, Jacobsen H-J (2005) *Agrobacterium*-mediated transformation of faba bean (*Vicia faba* L.) using embryo axes. Euphytica 142: 227–236.

Hansen G, Chilton MD (1996) 'Agrolistic' transformation of plant cells: integration of T-strands generated in planta. Proc Natl Acad Sci USA 93: 14978–14983.

Harley JL, Smith SE (1983) Mycorrhizal Symbiosis. Academic Press Inc., London, 483 pp.

Hassan F, Hoque MI, Kiesecker H, Jacobsen HJ (2007) Transient GUS expression in decapitated lentil embryos. Plant Tiss Cult Biotechnol 17: 97–102.

Hassan F, Meens J, Jacobsen HJ, Kieseckerd H (2009) A family 19 chitinase (Chit30) from *Streptomyces olivaceoviridis* ATCC 11238 expressed in transgenic pea affects the development of *T. harzianum in vitro*. J Biotechnol 143: 302–308.

Hauggaard-Nielsen H, Jørnsgaard B, Kinane J, Jensen ES (2008) Grain legume–cereal intercropping: the practical application of diversity, competition and facilitation in arable and organic cropping systems. Renewab Agri Food Syst 23: 3–12.

Husnain T, Malik T, Riazuddin S, Gordon MP (1997) Studies on the expression of marker genes in chickpea. Plant Cell Tiss Org cult 49: 7–16.

Husnain T, Fatima T, Ul-Islam R, Riazuddin S (2000) Plant regeneration and expression of beta-glucuronidase gene in hypocotyls tissues of chickpea (*Cicer arietinum* L.). Pak J Biol Sci 3: 842–845.

Ignacimuthu S, Prakash S (2006) *Agrobacterium*-mediated transformation of chickpea with *α*-amylase inhibitor gene for insect resistance. J Biosci 31: 339–345.

Indurker S, Misra HS, Eapen S (2007) Genetic transformation of chickpea (*Cicer arietinum* L.) with insecticidal crystal protein gene using particle gun bombardment. Plant Cell Rep 26: 755–763.

Jaiwal PK, Sonia RS (2001) Chickpea regeneration and transformation. Curr Sci 80: 1368–1369.

Jelenić S, Mitrikeski PT, Papeš D, Jelaska S (2000) *Agrobacterium*-mediated transformation of broad bean *Vicia faba* L. Food Technol Biotechnol 38: 167–172.

Jordan MC, Hobbs SLA (1993) Evaluation of cotyledonary node regeneration system for *Agrobacterium*-medited transformation of pea (*Pisum sativum* L.). *In vitro* Cell Dev Biol 29P: 77–82.

Jones AL, Johansen IE, Bean SJ, Bach I, Maule AJ (1998) Specificity of resistance to pea seed-borne mosaic potyvirus in transgenic peas expressing the viral replicase (*NIb*) gene. J Gen Virol 79: 3129–3137.

Kar S, Johnson TM, Nayak P, Sen SK (1996) Efficient transgenic plant regeneration through *Agrobacterium*-mediated transformation of chickpea (*Cicer arietinum* L.). Plant Cell Rep 16: 32–37.

Kar S, Basu D, Das S, Ramakrishnan NA, Mukherjee P, Nayak P, Sen SK (1997) Expression of *Cry1A(c)* gene of *Bacillus thuringiensis* in transgenic chickpea plants inhibits development of pod-borer (*Heliothis armigera*) larvae. Transgen Res 6: 177–185.

Karakaya A, Özcan S (1998) Susceptibility of different lentil (*Lens culinaris* Medik.) cultivars to *Agrobacterium tumefaciens*. Turk J Bot 22: 313–316.

Khalafalla MM, Hattori K (1999) A combination of thidiazuron and benzyladenine promotes multiple shoot production from cotyledonary node explants of faba bean (*Vicia faba* L.). Plant Growth Regul 27: 145–148.

Khawar KM, Özcan S (2002) *In vitro* induction of crown galls by *Agrobacterium tumefaciens* super virulent strain A281 (pTiBo 542) in lentil (*Lens culinaris* Medik.). Turk J Bot 26: 109–111.

Khawar KM, Özcan S (2004) Hairy root transformation in turkish chickpea (*Cicer arietinum* L.) cultivars. Biotechnol Biotechnol Eq 18: 51–54.

Krishnamurthy KV, Suhasini K, Sagare AP, Meixner M, de Kathen A, Pickardt T, Schieder O (2000) *Agrobacterium* mediated transformation of chickpea (*Cicer arietinum* L.) embryo axes. Plant Cell Rep 19: 235–240.

Kurkdjian A, Manigault P, Beardsley R (1968) Crown gall: effect of temperature on tumorigenesis in pea seedlings. Can J Bot 47: 803–808.

Loiseau J, Michaux-Ferriere N, Le Deunff Y (1998) Histology of somatic embryogenesis in pea. Plant Physiol Biochem 36: 683–687.

Lurquin PF, Cai Z, Stiff CM, Fuerst EP (1998) Half-embryo cocultivation technique for estimating the susceptibility of pea (*Pisum sativum* L.) and lentil (*Lens culinaris* Medik.) cultivars to *Agrobacterium tumefaciens*. Mol Biotechnol 9: 175–179.

Maccarrone M, Dini L, Di Marzio L, Di Julio A, Rossi A, Mossa A, Finazzi-Agró A (1992) Interaction of DNA with cationic liposomes: ability of transfecting lentil protoplats. Biochem Biophys Res Comm 186: 1417–1422.

Maccarrone M, Veldink GA, Finazzi-Agró A, Vliegenthart JFG (1995) Lentil root protoplast—A transient expression system suitable for coelectroporation of monoclonal antibodies and plasmid molecules. Biochim Biophys Acta 1234: 136–142.

Mahmoudian M, Akcay UC, Yucel M, Öktem HA (2002a) Vaccum infiltration based *Agrobacterium* mediated gene transfer to lentil (*Lens culinaris*) tissues. Biotechnol Biotechnol Eq 16: 24–29.

Mahmoudian M, Yucel M, Öktem HA (2002b) Transformation of lentil (*Lens culinaris* M.) cotyledonary nodes via vacuum infiltration of *Agrobacterium tumefaciens*. Plant Mol Biol Rep 20: 251–257.

Manigault P, Kurkdjian A (1967) Influence de la lumiére sur développement des tumeurs de la plantule de pois *Pisum sativum* L., induites par *Agrobacterium tumefaciens* (Smith et Town) Conn. Compt Rend Acad Sci 264: 2304–2306.

Materne M, McNeil D (2007) Breeding methods and achievements. In: SS Yadav, D McNeil, PC Stevenson (eds) Lentil. An Ancient Crop for Modern Times. Spinger, Dordrecht, The Netherlands, pp 241–253.

Melhorn V, Matsumi K, Koiwai H, Ikegami K, Okamoto M, Nambara E, Bittner F, Koshiba T (2008) Transient expression of *AtNCED3* and *AAO3* genes in guard cells causes stomatal closure in *Vicia faba*. J Plant Res 121: 125–131.

Metry EA, Ismail RM, Hussien GM, Nasr El-Din, El-Itriby HA (2007) Regeneration and microprojectile-mediated transformation in *Vicia faba* L. Arab J Biotechnol 10: 23–36.

Morton RL, Schroeder HE, Bateman KS, Chrispeels MJ, Armstrong E, Higgins TJV (2000). Bean α-amylase inhibitor 1 in transgenic peas (*Pisum sativum*) provides complete protection from pea weevil (*Bruchus pisorum*) under field conditions. Proc Natl Acad Sci USA 97: 3820–3825.

Nadolska-Orczyk A, Orczyk W (2000) Study of the factors influencing *Agrobacterium*-mediated transformation of pea (*Pisum sativum* L.). Mol Breed 6: 185–194.

Nicolaisen M, Poulsen GB (1993) Optimization of polyethylene glycol mediated transient gene expression in pea protoplasts. Plant Cell Tiss Organ Cult 35: 93–97.

Ochatt SJ, Pontecaille C, Rancillac M (2000) The growth regulators used for bud regeneration and shoot rooting affect the competence for flowering and seed set in regenerated plants of protein peas. *In Vitro* Cell Dev Biol-Plant 36: 188–193.

Oktem HA, Mahmoudian M, Eyidooan F, Yücel M (1999) GUS gene delivery and expression in lentil cotyledonary nodes using particle bombardment. Lens Newsl 26: 3–6.

Oktem HA, Akcay UC, Bayrac AT, Kamci H, Yucel M (2008) Lentil. In: C Kole, TC Hall (eds) Compendium of Transgenic Crop Plants, vol 3: Transgenic Legumes Grains and Forages. Wiley-Blackwell, Chichester, UK, pp 89–102.

Öz MT, Eyidogan F, Yücel M, Öktem HA (2009) Optimized selection and regeneration conditions for *Agrobacterium*-mediated transformation of chickpea cotyledonary nodes. Pak J Bot 41: 2043–2054.

Pathak MR, Hamzah RY (2008) An effective method of sonication-assisted *Agrobacterium*-mediated transformation of chickpeas. Plant Cell Tiss Organ Cult 93: 65–71.

Patil G, Deokar A, Jain PK, Thengane RJ, Srinivasan R (2009) Development of a phosphomannose isomerase-based *Agrobacterium*-mediated transformation system for chickpea (*Cicer arietinum* L.). Plant Cell Rep 28: 1669–1676.

Pitzschke A, Hirt H (2010) New insights into an old story: *Agrobacterium*-induced tumour formation in plants by plant transformation. EMBO J 29: 1021–1032.

Pniewski T, Kapusta J (2005) Efficiency of transformation of Polish cultivars of pea (*Pisum sativum* L.) with various regeneration capacity by using hypervirulent *Agrobacterium tumefaciens* strains. J Appl Genet 46: 139–147.

Pniewski T, Wachowiak J, Kapusta J, Legocki AB (2003) Organogenesis and long-term micropropagation of Polish pea cultivars. Acta Soc Bot Polon 72: 295–302.

Polowick PL, Quandt J, Mahon JD (2000) The ability of pea transformation technology to transfer genes into peas adapted to western Canadian growing conditions. Plant Sci 153: 161–170.

Polowick PL, Baliski DS, Mahon JD (2004) *Agrobacterium tumefaciens*-mediated transformation of chickpea (*Cicer arietinum* L.): gene integration, expression and inheritance. Plant Cell Rep 23: 485–491.

Popelka JC, Terryn N, Higgins TJV (2004) Gene technology for grain legumes: can it contribute to the food challenge in developing countries? Plant Sci 167: 195–206.
Puonti-Kaerlas J, Stabel P, Eriksson T (1989) Transformation of pea (*Pisum sativum* L.) by *Agrobacterium tumefaciens*. Plant Cell Rep 8: 321–324.
Puonti-Kaerlas J, Eriksson T, Engström P (1990) Production of transgenic pea (*Pisum sativum* L.) plants by *Agrobacterium tumefaciens*- mediated gene transfer. Theor Appl Genet 80: 246–252.
Puonti-Kaerlas J, Eriksson T, Engström P (1992a) Inheritance of a bacterial hygromycin phosphotransferase gene in the progeny of primary transgenic pea plants. Theor Appl Genet 84: 443–450.
Puonti-Kaerlas J, Ottosson A, Ericsson T (1992b) Survival and growth of pea protoplasts after transformation by electroporation. Plant Cell Tiss Organ Cult 30: 141–14.
Ramana RV, Venu C, Jayasree T, Sadanadam A (1996) Direct somatic embryogenesis and transformation in *Cicer arietinum* L. Indian J Exp Biol 34: 716–718.
Ramsay G, Kumar A (1990) Transformation of *Vicia faba* cotyledon and stem tissues *Agrobacterium rhizogenes*: Infectivity and cytological studies. J Exp Bot 41: 841–847.
Reddy AR, Chaitanya KV, Vivekanandan M (2004) Drought induced responses of photosynthesis and antioxidant metabolism in higher plants. J Plant Physiol 161: 1189–1202.
Rekha KT, Thiruvengadam M. (2009) An efficient micropropagation of chickpea (*Cicer arietinum* L.). Phil Agri Sci 92: 320–323.
Richter A, de Kathen A, de Lorenzo G, Briviba K, Hain R, Ramsay G, Jacobsen HJ, Kiesecker H (2006) Transgenic peas (*Pisum sativum* L.) expressing polygalacturonase inhibiting protein from raspberry (*Rubus idaeus*) and stilbene synthase from grape (*Vitis vinifera*). Plant Cell Rep 25: 1166–1173.
Saalbach I, Pickardt T, Machemehl F, Saalbach G, Schieder O, Muntz K (1994) A chimeric gene encoding the methionine-rich 2S albumin of the Brazil nut (*Bertolletia excelsa* H.B.K) is stably expressed and inherited in transgenic grain legumes. Mol Gen Genet 242: 226–236.
Sangwan RS, Bourgeois Y, Brown S, Vasseur G, Sangwan-Norreel BS (1992) Characterization of competent cells and early events of *Agrobacterium*-mediated genetic transformation of *Arabidopsis thaliana*. Planta 188: 439–456.
Sanyal I, Singh AK, Amla DV (2003) *Agrobacterium tumefaciens*-mediated transformation of chickpea (*Cicer arietinum* L.) using mature embryonic axes and cotyledonary nodes. Indian J Biotechnol 2: 524–532.
Sanyal I, Singh AK, Kaushik M, Amla DV (2005) *Agrobacterium*-mediated transformation of chickpea (*Cicer arietinum* L.) with *Bacillus thuringiensis cry1Ac* gene for resistance against pod borer insect *Helicoverpa armigera*. Plant Sci 168: 1135–1146.
Sarker RH, Biswas A, Murtaja B, Mahbub S, Hoque MI (2003) *Agrobacterium*-mediated transformation of lentil (*Lens culinaris* Medik.). Plant Tissue Cult 13: 1–12.
Sarmah BK, Moore A, Tate W, Molvig L, Morton RL, Rees DP, Chiaiese P, Chrispeels MJ, Tabe LM, Higgins TJV (2004) Transgenic chickpea seeds expressing high levels of a bean α-amylase inhibitor. Mol Breed 14: 73–82.
Saxena P K, Kin J (1987) Morphogenesis in lentil *Lens culinaris* Medik.: plant regeneration from callus cultures via somatic embryogenesis. Plant Sci 52: 223–227.
Schaerer S, Pilet PE (1991) Roots, explants and protoplasts from pea transformed with strains of *Agrobacterium tumefaciens* and *rhizogenes*. Plant Sci 78: 247–258.
Schroeder HE, Schotz AH, Wardley-Richardson T, Spencer D, Higgins TVJ (1993) Transformation and regeneration of two cultivars of pea (*Pisum sativum* L.). Plant Physiol 101: 751–757.
Schroeder HE, Gollasch S, Moore A, Tabe LM, Craig S, Hardie DC, Chrispeels MJ, Spencer D, Higgins TVJ (1995) Bean α-Amylase inhibitor confers resistance to the pea weevil (*Bruchus pisorum*) in transgenic peas (*Pisum sativum* L.). Plant Physiol 107: 1233–1239.
Selva E, Stouffes B, Briquet M (1989) *In vitro* propagation of *Vicia faba* L. by micro-cutting and multiple shoot induction. Plant Cell Tiss Org Cult 18: 167–179.

Senthil G, Williamson B, Dinkins RD, Ramsay G (2004) An efficient transformation system for chickpea (*Cicer arietinum* L.). Plant Cell Rep 23: 297–303.

Shrawat AK, Lorz H (2006) *Agrobacterium*-mediated transformation of cereals: a promising approach crossing barriers. J Plant Biotechnol 4: 575–603.

Singh R, Singh NP, Datta S, Yada IS, Singh AP (2009) *Agrobacterium*-mediated transformation of chickpea using shoot meristem. Indian J Biotechnol 8: 78–84.

Snowdon RJ, Böttinger P, Pickardt T, Köhler W, Friedt W (2001) Physical location of transgenes on *Vicia faba* chromosomes. Chrom Res 9: 607–610.

Somers DA, Samac DA, Olhoft PM (2003) Recent advances in legume transformation. Plant Physiol 131: 892–899.

Srinivasan M, Mohapatra T, Sharma RP (1991) *Agrobacterium* mediated genetic transformation of chickpea, *Cicer arietinum* L. Indian J Exp Biol 29: 758–761.

Stachel SE, Messens M, Van Montagu M, Zambryski P (1985) Identification of signal molecules produced by wounded plant cells that activate T-DNA transfer in *Agrobacterium tumefaciens*. Nature 318: 624–629.

Steeves TA, Sussex IM (1989) Patterns in Plant Development. Cambridge Univ Press, New York, USA.

Svabova L, Griga M (2008) The effect of cocultivation treatments on transformation efficiency in pea (*Pisum sativum* L.). Plant Cell Tiss Organ Cult 95: 293–304.

Svabova L, Smykal P, Griga M, Ondrej V (2005) *Agrobacterium*-mediated transformation of *Pisum sativum in vitro* and *in vivo*. Biol Plant 49: 361–370.

Tegeder M, Gebhardt D, Schieder O, Pickardt T (1995) Thidiazuron-induced plant regeneration from protoplast of *Vicia faba* cv. Mythos. Plant Cell Rep 15: 164–169.

Tewari-Singh N, Sen J, Kiesecker H, Reddy VS, Jacobsen HJ, Guha-Mukherjee S (2004) Use of a herbicide or lysine plus threonine for non-antibiotic selection of transgenic chickpea. Plant Cell Rep 22: 576–583.

Timchenko T, Katul L, Aronson M, Vega-Arreguín JC, Ramirez BC, Vetten HJ, Gronenborn B (2006) Infectivity of nanovirus DNAs: induction of disease by cloned genome components of *Faba bean necrotic yellows virus*. J Gen Virol 87: 1735–1743.

Timmerman-Vaughan GM, Pither-Joyce MD, Cooper PA, Russell AC, Goulden DS, Butler R, Grant JE (2001) Partial resistance of transgenic peas to alfalfa mosaic virus under greenhouse and field conditions. Crop Sci 41: 846–853.

Tzitzikas EN, Bergervoet M, Raemakers K, Vincken JP, van Lammeren A, Visser RGF (2004) Regeneration of Pea (*Pisum sativum* L.) by a cyclic organogenic system. Plant Cell Rep 23: 453–460.

Vaquero F, Martínez C, Llorente A, Vences FJ (2007) Evaluación de la eficacia de diferentes métodos de transformación en lenteja (*Lens culinaris* Medik.) por *Agrobacterium tumefaciens*. XXXIV Congreso de la Sociedad Española de Genética. Universidad de León, p 191. ISBN 978-84-9773-354-0.

Vieweg MF, Frühling M, Quandt H-J, Heim U, Bäumlein H, Pühler A, Küster H, Perlick AM (2004) The promoter of the *Vicia faba* L. leghemoglobin gene *VfLb29* is specifically activated in the infected cells of root nodules and in the arbuscule-containing cells of mycorrhizal roots from different legume and nonlegume plants. Mol Plant-Microbe Interact 17: 62–69.

Vignesh M, Shanmugavadivel PS, Prabha M, Kokiladevi E (2010) Transformation studies in pea. A review. Agri Rev 31: 68–72.

Wang Z, Luo Y, Li X, Wang L, Xu S, Yang J, Weng L, Sato S, Tabata S, Ambrose M, Rameau C, Feng X, Hu X, Luo D (2008) Genetic control of floral zygomorphy in pea (*Pisum sativum* L.). Proc Natl Acad Sci USA 105: 10414–10419.

Warkentin TD, McHugen A (1991) Crown gall transformation of lentil (*Lens culinaris* Medik.) with virulent strains of *Agrobacterium tumefaciens*. Plant Cell Rep 10: 489–493.

Warkentin TD, McHughen A (1992) *Agrobacterium tumefaciens* mediated beta-glucuronidase (GUS) gene expression in lentil (*Lens culinaris* Medik.) tissues. Plant Cell Rep 11: 274–278.

Warkentin TD, McHugen A (1993) Regeneration from lentil cotyledonary nodes and potential of this explant for transformation by *Agrobacterium* tumefaciens. Lens Newsl 20: 26–28.

Warkentin TD, Jordan MC, Hobbs LA (1992) Effect of promoter-leader sequences on transient reporter gene expression in particle bombardment pea (*Pisum sativum* L.) tissues. Plant Sci 87: 171–177.

Welham T, Domoney C (2000) Temporal and spatial activity of a promoter from a pea enzyme inhibitor gene and its exploitation for seed quality improvement. Plant Sci 159: 289–299.

Wisniewski JP, Brewin NJ (2000) Construction of transgenic pea lines with modified expression of diamine oxidase and modified nodulation responses with exogenous putrescine. Mol Plant-Microbe Interact 13: 922–928.

Youssef SS, Moghaieb REA, Saker MM, El Awaidy M, El Sharkawy A (2007) Transformation of faba bean (*Vicia faba* L.): a non-tissue culture based approach for generating transgenic plants. *In Vitro* Cell Dev Biol 43: S24–S32.

10

Nitrogen Fixation in Legumes

José Olivares,[1,a,*] *María J. Soto,*[1] *Cesar Arrese-Igor*[2] and *Juan Sanjuán*[1]

ABSTRACT

Legume nitrogen fixation plays a key role in the maintenance of crop production. However, this property is often not considered in new cultivar development programs. The chapter reviews first the symbiotic nitrogen fixation process, its effectiveness and efficiency, nodule functioning and nitrogen assimilation. The second part reviews both the structural and functional genomics studies on the plant-bacteria interactions. The availability of several rhizobia genomes has opened new possibilities to identify important symbiotic functions through the use of DNA arrays. The identification of symbiotic genes in legumes has been speeded up due to the use of model legume species and functional genomics technologies such as map-based gene cloning, transcriptomics, proteomics and metabolomics. The chapter ends stressing on the importance of breeding for enhancing nitrogen fixation. Breeders often do not evaluate the symbiotic performance of cultivars. In legume breeding programs, more attention should be directed towards understanding and evaluating plant mechanisms that control root nodule formation and activity, otherwise biological nitrogen fixation may be compromised.

Keywords: nitrogen fixation, *Rhizobium*-legume symbiosis, plant-bacteria interactions, microarrays, signal transduction, structural and functional genomics

[1]Departamento de Microbiología del Suelo y Sistemas Simbióticos. Estación Experimental del Zaidín. CSIC. Prof. Albareda, 1. 18008 Granada, Spain.
[a]e-mail: *Olivares@eez.csic.es*
[2]Departamento de Ciencias del Medio Natural. Universidad Pública de Navarra. Campus de Arrosadía, s/n. 31006 Pamplona, Spain.
*Corresponding author

10.1 Introduction

Nitrogen fixation in legumes plays a key role in the maintenance of crop production. The most important legume crops including cool season grain legumes (pea, faba bean, lentils, grass pea, chickpea, lupins) are able to use atmospheric nitrogen in symbiotic association with rhizobia, that renders them partially or fully independent of nitrogen fertilizer. However, this property is often not considered in new cultivar development programs, and worse, non-reasonable amounts of chemically synthesized nitrogen are many times applied to legume crops that are otherwise capable of nitrogen fixation. General aspects of the symbiotic nitrogen fixation process, together with a review of the knowledge accumulated on the legume-rhizobia interaction will be the subject of this chapter. In addition, the importance and possibilities of breeding for enhanced nitrogen fixation in crop legumes are discussed.

10.2 Nitrogen Fixation

Molecular nitrogen or dinitrogen (N_2) is the major component of the atmosphere but can not be directly used by most living organisms. The term nitrogen fixation is applied to the combination of N_2 with oxygen or hydrogen to give oxides or ammonia. While the oxidation occurs by abiotic processes, ammonia is the result of the reduction of N_2 by bacteria, either free-living or in association with plants (biological fixation), or produced by chemical synthesis following the Haber-Bosch process with a high consumption of energy (chemical fixation). The oxides are formed in natural conditions as a consequence of lightning (electric sparks) or combustion processes and are swept down into the soil by the rain.

Nitrogen fixation is globally and quantitatively the most important way of incorporating N_2 in the biosphere, being estimated as 250 millions tons per year of which more than 50% correspond to biological fixation. Biological N_2 fixation (BNF) is an exclusive property of certain prokaryota, such as proteobacteria, actinomycetes, cyanobacteria; and even archaea. It is highly energy consuming and proceeds according to the equation:

$$N_2 + 8H^+ + 8e + 16ATP \rightarrow 2NH_3 + H_2 + 16ADP + 16Pi$$

For BNF the enzyme nitrogenase is always required. It is constituted by two components, one called ferroprotein or nitrogenase reductase, which contains Fe, and the second, nitrogenase, also containing Fe, and with Mo as cofactor. Nitrogenase is well conserved in all N_2-fixing microorganisms. It is very sensitive to oxygen and its synthesis is subjected to a strict regulation so that it is not formed in the presence of this element and/or combined nitrogen. The nitrogenase presents a wide range of activity, reducing other molecules containing triple bonds, what is the basis of an easy and rapid assay method to measure activity using acetylene as substrate, which is reduced to ethylene and detected by gas chromatography.

Due to the energy requirements, BNF carried out by free-living bacteria (e.g., *Azotobacter*, *Clostridium*), has quantitatively low significance, in contrast with BNF by plant-associated bacteria, particularly in symbiosis. Free-living fixers can not find in the soil available energy sources enough to support significant rates of nitrogen fixation, while in mutualistic symbioses the host plant can supply the high nutrient requirements of the process to the associated bacteria. Symbiotic nitrogen fixation is restricted to only four related orders within the Eurosid clade of angiosperms. Among the families able to fix, the Leguminosae (Fabaceae) is the most important from the quantitative point of view, as well as for its agricultural, economical and ecological significance. There are other plant species, shrubs or trees with forestry or environmental interest like *Alnus*, *Casuarina*, *Myrica*, etc. belonging to different families, which are able to fix nitrogen in symbiosis with the actynomycete *Frankia*, constituting the so-called actinorhizal symbioses.

10.3 *Rhizobium*-Legume Symbiosis

10.3.1 The Legume

Since Roman times it has been known that legumes enrich the soil and benefit subsequent crops, but it was not until 1888 that these plants were discovered to have the ability to fix N_2 in symbiosis with bacteria, which were initially called *Bacillus radicicola* and today are known as *rhizobia* in a broad sense. These bacteria were inside "small tumors" formed in the roots, which were already described by Malpighi in the XVII century.

Not all genera and species belonging to this plant family (Leguminosae) are able to establish symbiosis with *Rhizobium*. Practically all Papilionoideae and many Mimosoideae do form nodules, while this association is rare in Caesalpinioideae. Most of the plants of economical or ecological interest are nodulated when grown in natural conditions, even in not very adverse environments such as those affected by drought or salinity.

An important feature of symbiotic legumes is the highly diverse behavior when growing at the expense of N_2 fixation or under fertilization with nitrate or ammonia. The production of biomass is higher in plants treated with nitrate, than ammonia and followed by N_2, but the crop index (percentage of pods or seeds on biomass) is on the contrary, higher for N_2, followed by ammonia and finally by nitrate. In addition, nitrogen-fixing plants flower earlier than those which received nitrate or ammonia fertilizers, which is an agricultural advantage under most plant growth and environmental conditions. These results evidence the importance of nitrogen fixation in legume productivity. Due to the involvement of both plants and bacteria in the process, agronomists and microbiologists must engage collaboratively in the study of this symbiosis to improve legume crop performance.

10.3.2 The Bacteria

Bacteria able to fix N_2 in symbiosis with legumes, are collectively known as the rhizobia, which may belong to the α or ß subdivisions of proteobacteria. Those α-proteobacteria, considered in the strictest sense as *Rhizobium*, are heterotrophic gram-negative bacilli soil inhabitants (Kuykendall 2005) distributed within the Order Rhizobiales in the families Rhizobiaceae, Phyllobacteriaceae, Bradyrhizobiaceae, Hyphomicrobiaceae and Methyllobacteriace. The most important genera from the agricultural point of view are *Rhizobium*, *Sinorhizobium*, *Mesorhizobium* and *Bradyrhizobium* and all of them, except *Sinorhizobium*, are generally able to infect several cool season grain legumes. Other α-proteobacteria of the genera *Azorhizobium*, *Methylobacterium*, *Devosia*, *Ochrobactrum* and *Phyllobacterium*, together with some β-rhizobia belonging to the genera *Burkholderia*, *Ralstonia* and *Cupriavidus*, are generally associated with tropical legumes (Willems 2006). All of them are able to infect the roots, in some cases also stems, of their corresponding hosts, producing a special organ called nodule, where N_2 fixation takes place.

10.3.3 Root Infection and Nodulation

Concerning the interaction with legumes, three characteristics have been classically recognized in rhizobia: specificity, infectiveness and effectiveness. The fourth one that could be added is competitiveness, which is closely associated to infectiveness. The first two characteristics refer to the ability of a given rhizobial species to recognize, infect and nodulate in the roots of a limited number of related legume species. For example, *R. leguminosarum* bv. *viciae*, *S. meliloti*, *B. japonicum* and *M. ciceri*, can infect *Vicia*, *Medicago*, *Glycine* and *Cicer*, respectively. However, there are certain strains, although not common, which are less specific and are able to interact with a broad range of legume species. In case of the main cool season grain legumes, *Pisum sativum*, *Vicia faba*, *Lens culinaris* and *Lathyrus sativus* are infected by different strains of *R. leguminosarum* bv. *viciae*. *Cicer arietinum* is infected by the strains of *M. ciceri* and *M. mediterraneun*, while *Lupinus* species are associated with yet not well defined species of the genus *Bradyrhizobium* and even other α-rhizobia like *Ochrobactrum* (Wang and Martínez-Romero 2000; Trujillo et al. 2005).

Competitiveness indicates the ability of a given strain of rhizobia to infect roots of a plant species or genotype in the presence of other rhizobial strains (see Dowling and Broughton 1986). This character is of great importance since it is frequently the cause of failure of inoculation of legumes with selected rhizobial strains, particularly in soils where indigenous rhizobia are present. Rhizobial competitiveness is determined by different factors depending either on the bacteria, such as survival and rate of multiplication in the rhizosphere, the production of substances that inhibit the growth of other strains (Triplett and Barta 1987), etc., or on the complementary genotype of both partners involved in the association. For instance, cv. Afghanistan peas are not nodulated by most European strains of *R. leguminosarum* bv. *viciae*, and further, these non-nodulating strains can competitively block nodulation of this cultivar by nodulating strains. The bacterial gene *nodX* and the pea gene *Sym2* are involved in this particular phenomenon (Hogg et al. 2002). Rhizobial competition is being managed by different approaches in order to solve this problem that often limits the success of inoculants (Sessitsch et al. 2002).

When a legume species is going to be cultivated in a soil where this plant or any related species have never grown, it is necessary to inoculate with the adequate rhizobial strain. This is the case, for example, of soybean in Europe. In such cases the selection of the most adequate bacterial strain for the plant cultivar to be sown will lead to the highest productivity.

Once the legume seed germinates, *Rhizobium* present in the soil colonizes the rhizosphere attracted by root exudates containing many compounds including flavonoids (2-phenyl-1,4-benzopyrone derivatives), which are specifically recognized by the bacteria. As a previous step in root infection, bacteria adhere to the root surface, particularly on the tip of the root hairs, beginning then a specific talk between both partners. Some of the plant flavonoids exuded are able to induce the transcription of certain bacterial genes, generically called nodulation genes (*nod*, *nol*, *noe*), which codify for the synthesis and transport out of the cell of a compound, named Nod factor, which induces important responses on root hairs and root cortex cells. The expression of these *nod* genes is induced by the product of the *nodD* gene in response to specific flavonoids produced by the compatible host. For instance, the major inducer flavonoid for *S. meliloti* (symbiont of alfalfa) is luteolin, while naringenin and genistein are preferred by *R. leguminosarum* bv. viciae and *B. japonicum*, respectively.

These Nod factors, also named lipochitooligosaccharides (LCO) due to their chemical nature, are compounds constituted by a short chain (3 to 5 residues) of N-acetyl glucosamine. The chitin backbone is modified on the nonreducing glucosamine by a fatty acid of variable size and saturation state. Nod factors can be further decorated with a variety of chemical substituents. The chemical structure of the Nod factor which is species-specific determines whether the host allows bacterial invasion to proceed. Figure 10-1 shows the structure of the molecule synthesized by *R. leguminosarum* bv. *viciae*. At very low concentrations (10^{-8}–10^{-10} molar), these LCOs are able to trigger a plant response that will determine the infection of root cells and the formation of nodules.

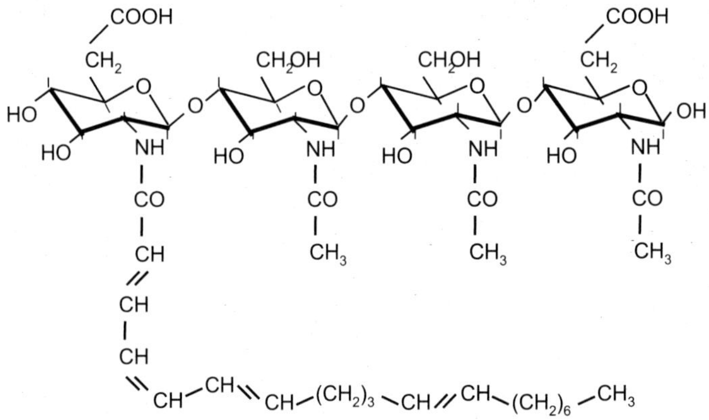

Figure 10-1 Chemical structure of the Nod factor synthesized by *Rhizobium leguminosarum* bv. *Viciae*.

Nod factors are not the only bacterial effectors required for bacteria to successfully infect root cells. Polysaccharides such as cyclic β-glucans, extracellular polysaccharides (EPS) as well as lipopolysaccharides (LPS) play important roles in the process and in the control of the plant defense response. Other additional molecular signals, such as secreted proteins (known as nodulation outer proteins or Nops), seem to be specific for some rhizobia with homologs in bacterial pathogens (Deakin and Broughton 2009). Despite plant defense responses are meant to limit bacterial infection, it is thought that a certain degree of response is necessary for the progression of the infection. The production of reactive oxygen species (ROS) by the plant may act as facilitator of the infection of the root cells when they are in a correct balance. ROS, which are liberated by the plant in defense response against invading pathogens, seem to be necessary within certain limits for infection thread progression. An excess of ROS is detrimental so that anti-ROS activities expressed by rhizobia are essential to control the level of the oxidative pulse generated in the plant-bacteria interaction (see Jones et al. 2007). This fact is an example of the close relation existing between symbiosis and pathogenesis as has been recently reviewed by Soto et al. (2009).

A Nod factor-independent infection strategy has been reported in photosynthetic *Bradyrhizobium* strains that elicit nodules on a particular group of *Aeschynomene* plants. These bacteria, which lack genes required for Nod factor synthesis, can infect and nodulate by using cytokinins to activate the nodule organogenesis program. This is not the case of cool season legumes, which are unavoidably required of the Nod factor-plant receptor cross talk. However, hormones like cytokinins and auxins are always involved in nodulation of all legumes.

The presence of the corresponding Nod factor is recognized by the receptors located in the root hair cell membrane and the signal is transduced through a cascade of events (membrane depolarization, calcium spiking, etc.) that involves the expression of many plant genes (see Section 10.4).

After recognition, root hairs curl trapping intimately associated bacteria which are then able to invade the plant root through a tubular structure known as infection thread. The infection thread carries bacteria from the root surface towards the dividing cells of the root cortex where they are released surrounded by a membrane, constituting the so-called symbiosomes. Bacteria are there transformed into bacteroids, morphologically different from the free-living bacteria, which are able to fix nitrogen. This organized growth of root cells in the site of infection causes the formation of a "small tumor", called nodule (Fig. 10-2), with most of the central zone cells filled with bacteroids. A vascular system formed in the nodule allows, on one side, the arrival of photosynthate as a source for the high energy-consuming process of nitrogen fixation, as well as for the provision of reduction power and carbonate skeleton to incorporate the ammonia produced, and, on the other, the transfer of the combined nitrogen to the plant.

Considering morphological and developmental criteria, three types of nodules can be recognized: indeterminate, determinate and lupinoid nodules. Indeterminate nodules, formed in most cool season legumes,

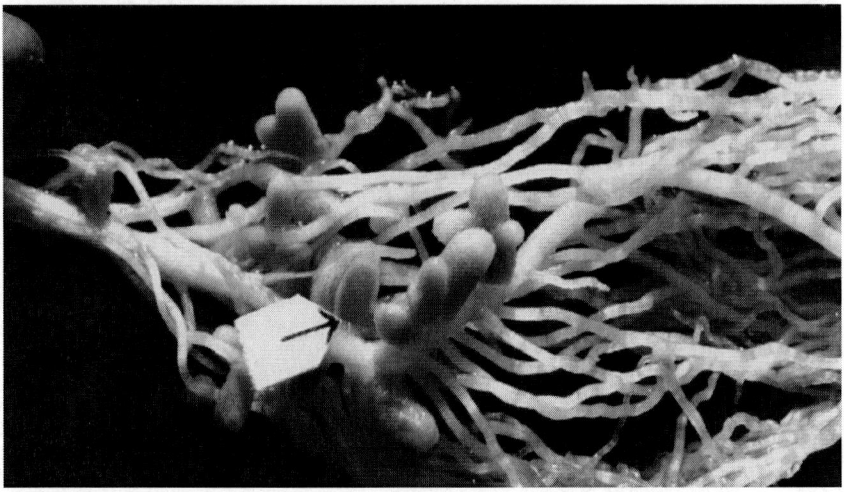

Figure 10-2 Nodulated root of *Pisum sativum*. The arrow indicates a sectioned nodule.
Color image of this figure appears in the color plate section at the end of the book.

have a cylindrical form and contain an apical meristem continually active, producing new cells throughout the live of the nodules. In determinate nodules, formed in tropical plants such as *Glycine*, *Phaseolus* or *Lotus*, the meristem ceases its activity shortly after nodule initiation, conducing to nodules of spherical shape that grow by cell expansion. The lupinoid type appears on *Lupinus* roots where *Rhizobium* enters intercellularly in the base of root hairs, invading the subjacent cells that at dividing form the mature nodule.

10.3.4 Effectiveness and Efficiency

By effectiveness we understand the capacity of a given rhizobial strain to fix N_2 in symbiosis with the corresponding plant inside a mature nodule. Assumed to be good receptivity of the host for bacteria and a convenient control of the defense response, an adequate provision of nutrients to the nodule and an acceptable drain of the nitrogenous compounds synthesized, the level of nitrogen fixed depends, among other factors, on the optimal use by the bacteroids of the available photosynthate.

Some attempts have been reported about the manipulation of rhizobial strains to raise the rate of nitrogen fixed by increasing the dosage of some *nif* genes. Other strategies are related with elevating the efficiency of the process by optimizing energy utilization. For instance, an effective *Rhizobium* strain may be more or less efficient according to its ability to recycle the molecular hydrogen released by the nitrogenase activity concomitantly with ammonia derived from reduction of protons. Up to 25% of the available energy may be lost if bacteria do not harbor a functional hydrogenase, the enzyme involved in H_2 recycling (Evans et al. 1988). Genetic manipulation of this character can lead to more efficient rhizobial inoculants (Brito et al. 2002).

Agricultural practices can increase the contribution of the macrosymbiont to the symbiosis. For example, an appropriate plant density can increase C and energy supply to the nodule. In this sense, plants with wider foliar surface will contribute to improve the nitrogen fixing rate by providing more photosyntate to the nodule, which is an important factor limiting symbiotic nitrogen fixation.

10.3.5 Nodule Functioning

Nitrogenase activity requires the occurrence of an energy-yielding substrate, an adequate O_2 balance and the maintenance of a satisfactory N status. Nodules provide a special environment that allows rhizobia, which are not able to fix N_2 as free-living cells, to become N_2-fixing bacteroids. The main features of nodule functioning are described below.

10.3.5.1 Carbon Supply

The energy required for BNF is ultimately supplied from shoot photosynthesis. Photoassimilates are transported from the aerial parts of the plant to nodules, via the phloem, largely in the form of sucrose (Avigad 1982). In nodules, sucrose is potentially cleaved to enter the glycolytic pathway by two different enzymes: alkaline invertase, which produces glucose and fructose, and sucrose synthase (SS), which catalyzes the reversible reaction UDP + sucrose + $H^+ \leftrightarrow$ UDP-glucose + fructose, although the enzyme is thought to function primarily in the direction of sucrose cleavage in nodules (Morell and Copeland 1985). SS, which is thought to play a predominant role, is known to be reversibly phosphorylated, although its physiological significance is still under debate (Duncan and Huber 2007). It has been recently shown that SS can be transcriptional and post-translational regulated by the cellular redox state (Marino et al. 2008). This makes SS a crucial step of the regulation of metabolic fluxes in nodules. In conditions where plants face survival rather than maximize growth, preventing nodules from becoming an excessive burden on a diminishing plant carbohydrate pool, can be a mechanism that may have been subjected to evolutionary selection (Arrese-Igor et al. 1999).

Products of sucrose cleavage may enter the glycolytic pathway, whose enzyme activities are highly expressed in nodules. Phosphoenolpyruvate is irreversibly carboxylated to oxaloacetate by phosphoenolpyruvate carboxylase (PEPC). PEPC requires HCO_3^-, which is provided by the occurrence of a very active carbonic anhydrase (CA). The CA-PEPC activity is thought to provide ca. 30% of the total C required for BNF (Nomura et al. 2006). Oxaloacetate is readily reduced to malate by malate dehydrogenase, which shows a very high, non-limiting activity in order to provide an adequate supply of malate to bacteroids. This is important as malate acts as a C-source both for energy production in bacteroids and also as C-backbone for the synthesis of new compounds. The symbiosome membrane is highly permeable to malate, but not to sugars (Ou-Yang et al. 1990). Bacteroids have their own dicarboxylate transport system, called Dct, which is essential for BNF (Yurgel and Kahn 2004).

Once incorporated, dicarboxylic acids are oxidized during the intense bacteroid respiration that provides ATP and reducing power for nitrogenase activity. This oxidation is carried out via the tricarboxylic acid cycle in bacteroids from fast-growing rhizobia, whereas it has been suggested that the full tricarboxylic acid cycle may not be essential for those derived from slow-growing rhizobia.

10.3.5.2 Assimilation of Fixed N

Most of the N assimilatory enzymes, other than the nitrogenase complex, are expressed at very low or negligible levels in bacteroids (in contrast to free-living bacteria) and, therefore, diffusion of NH_4^+ occurs from bacteroids to the symbiosome space for further transport across the symbiosome membrane to the host plant cytosol. It is debatable whether this movement occurs only via diffusion because of a large concentration gradient or this is facilitated by the occurrence of a nonselective cation channel, capable of NH_4^+ transport, or NH_3 through another channel type (White et al. 2007).

Once in the host plant cell, NH_4^+ is assimilated into glutamine (Gln) via the glutamine synthetase (GS)/Glutamic synthase pathway (Patriarca et al. 2002). In brief, GS catalyses the addition of NH_4^+ to glutamic acid (Glu) with the consumption of ATP, and the NADH-dependent GS transfers the amide group of Gln to α-ketoglutarate with formation of two molecules of Glu. In addition to Glu, the amide group of Gln can be transferred to aspartic acid in order to render asparagine (Asn). This reaction is catalyzed by Asn synthetase and it requires the hydrolysis of ATP to AMP and pyrophosphate (Vance 2008).

Together with this classical view of N-assimilation in nodules, a model of amino-acid cycling between both partners operating further to the simple malate/NH_4^+ exchange has been recently proposed (Lodwig et al. 2003). According to this model, the interaction between the two symbionts would be more complex than previously envisaged, and both partners would have apparently developed a total dependence on each other.

Nitrogen assimilated in nodules is exported to the aerial parts via the xylem in the form of amides or ureides. Most legumes, particularly those of temperate origin, export fixed N in the form of amides, such as Asn, Gln and, to a much lesser extent, 4-methylen-Gln. It is generally described that legumes of tropical and subtropical origin export fixed N in the form of ureides, such as allantoin, allantoic acid and citrulline. Indeed, ureides, as the main export product, are restricted to some species from the tribes Phaseoleae, Desmodieae and Psoraleae, although some of them are really relevant in terms of economic value. Ureides are synthesized from amides, involving the de novo synthesis of purines and its subsequent oxidation to form ureides. Plant enzymes of both infected (synthesis of purines) and interstitial (purines oxidation) cells are involved. In brief, xanthine is synthesized from Asp, a process that requires 10 different enzyme activities, and further metabolized to uric acid in infected cells (Vance 2008). Uric acid is then transported to uninfected cells and is metabolized to allantoin by peroxisomal uricase. Subsequent conversion of allantoin to allantoic acid is associated with the endoplasmic reticulum.

Long-distance transport of fixed N in the form of ureides represents C savings in terms of transport efficiency, as they have a C:N ratio of 1, compared to 2–2.5 in amides. However, ureides are considerably less soluble that amides, which may explain why they are mostly exported in tropical legumes. Tropical and subtropical legumes have determinate nodules with a closed vascular system, which allows a higher hydrostatic pressure. In contrast, indeterminate nodules, typical of legumes of temperate origin, have abundant ramifications and, therefore, greater resistance to water flow, which probably would prevent to maintain ureides in a soluble form.

10.3.5.3 The Oxygen Balance

Bacteroids need O_2 to synthesize the ATP required for metabolic processes and, particularly, to fix N_2. However, O_2 inhibits the synthesis of nitrogenase and irreversibly inactive the nitrogenase complex (Hill 1988). The O_2 balance to keep a low O_2 concentration while allowing a high O_2 flux is achieved by the combination of three different mechanisms.

The occurrence of leghemoglobin (Lb), which is a monomeric protein of 16 kDa, is located exclusively in the cytosol of infected cells, with a prosthetic protohemo IX-type group that reversibly binds O_2. In physiological conditions, the Fe present in the heme group is in its reduced state (ferrous form). Lb has a high affinity for O_2 (Km = 48–60 nM). It can be calculated that the O_2 concentration bound to Lb is 50,000-times greater than the concentration of free O_2. Under these conditions, Lb allows a high O_2 delivery from the plasma membrane of infected cells, adjacent to intercellular spaces, to the symbiosome membranes. Then, free O_2 diffuses through the peribacteroid space, which lacks Lb, to reach the high affinity terminal oxidases of bacteroids. The diffusion of O_2 provided by the Lb and high rates of bacteroid respiration ensure that the O_2 concentration in the environment of nitrogenase can be maintained around 20–100 nM avoiding any damage to the nitrogenase complex.

However, under such a low O_2 concentration bacteroid respiration might be impaired. Indeed, free-living rhizobia express the "typical" aerobic cytochromes aa_3 and *o*, but bacteroids are able to synthesize cytochrome bcb_3, a product of the genes *fixNOPQ*, whose Km for O_2 is as low as 7 nM. In this way, bacteroid respiration and nitrogenase activity can take place simultaneously.

In addition to these two mechanisms, a variable O_2 diffusion barrier (ODB) in nodules was described (Witty et al. 1987). It has been suggested that changes in water content of intercellular spaces may alter the O_2 path from the soil to the infected cells (since resistance to O_2 diffusion is 10,000 times higher in water than in air). These alterations may be related to changes in the conformation of glycoproteins located in intercellular spaces, that alter their level of avidity for water, or changes in the content of sucrose, which, in turn, determine water fluxes from adjacent cells.

10.4 Genomic Studies of the *Rhizobium*-Plant Interaction

The identification of the molecular bases of the establishment of nitrogen-fixing *Rhizobium*-legume symbiosis has been the subject of an intense study during the last decades. The use of classical biochemical, physiological, and genetic approaches has allowed obtaining important insights into microbial components essential for the symbiosis. For example, genes essential for nodulation (*nod, nol, noe*), plant invasion (surface polysaccharide biosynthetic genes and their regulatory networks), and nitrogen fixation (*nif, fix*) have been identified and the chemical structure of the Nod factor signal has been determined for several rhizobial species. On the contrary, before the era of genomics, the information available from the macrosymbiont was less abundant due to the inherent complexity of the plant. The use of approaches, such as differential display or differential hybridization, resulted in the identification of nodule-specific or nodule-enhanced genes (nodulins) in several legumes. The advances in genome sequencing, plant genetics and the application of modern genomic tools has allowed the identification of new key determinants in the interaction between rhizobia and their host plants. We will describe some of the more recent insights into the *Rhizobium*-plant interaction in general, focusing on advances related to cool season grain legumes.

10.4.1 Genomic Insights into the Microsymbiont

Although it has been described that β-proteobacterial strains of the genus *Burkholderia* and *Ralstonia* are able to nodulate legumes like *Aspalathus* and *Mimosa* spp. (Chen et al. 2003), the majority of rhizobia are α-proteobacterial belonging to the Rhizobiales order, also known as α-rhizobia. The wide biodiversity of rhizobia can be explained by horizontal transfer events of symbiotic plasmids and islands together with subsequent genome adaptation under selection pressure by the plant as recently proved (Marchetti et al. 2010).

10.4.1.1 Structural Genomics

Up to April 2009, the complete genome sequences of 24 α-rhizobia was available in databases. These include the genome sequences of the most intensively studied rhizobia: *S. meliloti* (alfalfa symbiont), *M. loti* (*Lotus* spp.

symbiont), *R. etli* (common bean symbiont), *B. japonicum* (soybean symbiont) and *R. leguminosarum* (which nodulates *Vicia, Pisum, Lathyrus* and *Lens*). These organisms contain complex large genomes. With the exception of *Bradyrhizobium,* which only contains a large circular chromosome of more than 9 Mb, rhizobial genomes usually contain several large plasmids ranging in size from about 100 kilobases to more than two megabases. Moreover, these genomes show high variability with a significant percentage of genes lacking orthologs in other rhizobia. These characteristics, together with the presence of a large number of genes involved in transport and catabolic functions, probably confer to these bacteria the versatility needed to grow and survive in a heterogeneous and changing environment as the soil.

A common feature found in rhizobial genomes is that genes involved in the symbiotic process are located in specific symbiotic genome compartments (SGCs), either in symbiotic plasmids (pSyms), in specific regions within the chromosome or in symbiotic islands. It is frequent that these symbiotic compartments contain reiterated genes, pseudogenes and a large number of insertion sequence-related elements, suggesting that SGC are mosaic structures that have evolved from recombination, horizontal transfer and transposition events.

Genome sequencing of rhizobial genomes has led to somewhat unexpected findings. One of these is the presence of hundreds of genes homologous to the virulence factors of pathogens (van Sluys et al. 2002). This is the case of genes coding for very specialized protein secretion systems like the type III and type IV secretion systems, or genes coding for two-component regulatory systems. Interestingly, the functional characterization of some of them, indicate a likely role in the *Rhizobium*-legume symbiosis, suggesting that rhizobia and pathogenic bacteria might use similar strategies to invade their corresponding hosts (Soto et al. 2006; Deakin and Broughton 2009).

Not less surprising was finding that canonical nodulation genes and typical Nod factors are not required for symbiosis in some legumes, discarding a long-term held dogma of the *Rhizobium*-legume interaction. This discovery has ocurred after completion of genome sequencing of two symbiotic, photosynthetic, *Bradyrhizobium* strains, which induce nodules on both the root and the stem of the aquatic legume *Aeschynomene* (Giraud et al. 2007). Genetic studies suggest that these rhizobia use an alternative pathway to trigger nodule formation in which a purine derivative could be involved instead of LCOs.

10.4.1.2 Functional Genomics: Transcriptomics and Proteomics

The availability of the entire sequence of several rhizobia genomes has opened new possibilities to identify important symbiotic functions through the use of DNA arrays. This technology, available up to now as microarrays for *S. meliloti*, *B. japonicum*, *R. leguminosarum* and *M. loti*, allows large-scale expression analysis of the mRNA under specific cellular conditions. A significant number of transcription profiling studies have been already performed, mostly in *S. meliloti*, permitting the identification of bacterial genes induced in symbiotic conditions or in free-living cultures under a variety of situations, which could have a connection with symbiosis, including microaerobic conditions, oxidative and osmotic stress, phosphate starvation, nitrogen metabolism, quorum sensing, and response to plant flavonoids. Likewise, this technology is being applied to rhizobial mutants affected in key regulatory genes or strains that overexpress such genes allowing the discovery of regulatory networks. A recent transcriptomic analysis of *R. leguminosarum* in symbiosis with two different hosts (pea and vetch) has confirmed results previously obtained in other rhizobia and added new knowledge about the complex biochemical and developmental changes undergone during the symbiotic interaction (Karunakaran et al. 2009). Especially interesting are dual genome symbiosis chips, which allow the study of gene expression in both partners simultaneously (Barnett et al. 2004). Using this tool and comparing transcription profiles in wild type and mutant nodules revealed striking differences in gene expression in both bacteria and plant.

Some relevant conclusions have been derived from transcriptomic studies in rhizobia. It has been shown that *Rhizobium* suffers a profound modification of its gene expression during bacteroid differentiation, with a general slow down of cell metabolism and induction of nitrogen fixation functions. Although the presence of a low oxygen concentration plays an important role in the control of gene expression in endosymbiotic bacteria, there is evidence that suggests that additional biochemical and/or physiological factors might also be involved. In other studies, the great influence of the host plant genotype on the profile of bacterial regulated genes, the expression of rhizobial virulence genes during the infection process or the role played in symbiosis by global regulators, have been revealed. Another important finding highlighted by several transcriptomic studies in rhizobia is the coordinated regulation of motility and expression of important genes for plant infection. This characteristic, which has also been observed in some phytopathogenic bacteria, could help these microbes to make the transition from saprophytic lifestyle to that of a plant-interacting bacterium. Similar chemical signals and regulatory genes have been involved in turning off behavior like motility aimed to free-living survival, and tuning on those required for host colonization and invasion.

Proteomics approaches have been used to study protein patterns in several rhizobial species including *R. leguminosarum*, *R. etli* and *S. meliloti*. Studies have been performed to identify rhizobial proteins differentially expressed in the exponential or stationary phase, induced by plant flavonoids, anaerobiosis or in bacteroids. This approach has turned out to be a complementary tool to transcriptomic studies. Besides confirming genetic and biochemical data, it offers important advantages: allows the identification of proteins not annotated in the genome and reveals putative posttranscriptional regulatory mechanisms.

Amongst new relevant results found with this technique is the important role played by quorum sensing signals in bacteroid differentiation and/or functioning, the evidence that nodule bacteria cope with several stresses (osmotic and oxidative stresses), and the identification of additional genes that may not be amenable to single gene mutagenesis but that may play important roles in the establishment of symbiosis (Djordjevic 2004; Gao et al. 2005).

10.4.2 Important Symbiotic Functions in Legumes Unveiled by Genetics and Functional Genomics

Plant mutants in legumes like pea or soybean obtained by chemical mutagenesis and defective in the *Rhizobium*-legume symbiosis have been available for many years. Mutant phenotypes ranged from defects in early steps as root hair curling and infection thread formation to deficiencies in nodule organogenesis, autoregulation (control of nodule number), symbiosome differentiation and nitrogen fixation. The existence of such mutants clearly demonstrated that plant genes were essential for the establishment of the nitrogen-fixing symbiosis. The observation in the late 1980s was interesting that some non-nodulating pea mutants were not colonized by arbuscular mycorrhizal fungi, suggesting common steps in the two symbiotic pathways. The overlapping of the symbiotic pathways was later confirmed in model legume species in which several non-nodulating mutants were also impaired in mycorrhization. Mutants were initially classified into different classes on the basis of their phenotypes regarding calcium flux, root hair deformation, calcium spiking, expression of early nodulin genes, and cortical cell division.

Ultrastructural analysis of pea mutants forming ineffective nodules has genetically dissected some of the late stages of nodule organogenesis. Some phenotypic alterations associated to these mutants affect endocytosis, symbiosome differentiation or nodule persistence. Unfortunately, the genes altered have not been isolated yet. On the contrary, significant advances have been made during the last decade on the characterization of plant genes essential in Nod factor perception and signal transduction.

In general, the identification of symbiotic genes in legumes has been sped up, mainly due to two general developments: i) the use of two model legumes: *Medicago truncatula* and *Lotus japonicus,* which overcome some of the limitations of crop legumes like their large genomes and long life cycles, and ii) functional genomics technologies such as map-based gene cloning, transcriptomics, proteomics and metabolomics.

10.4.2.1 Plant Genes with a Role in Nod Factor Perception and Signal Transduction

The first plant symbiotic gene to be characterized was isolated from the non-nodulated mutant *nin-1* of *L. japonicus* (Schauser et al. 1999). The gene affected encoded a transcription factor that is not required for Nod factor perception but facilitates early symbiosis at the infection-thread stage and cortical cell division. Interestingly, a *nin*-like gene is involved in nitrate sensing and metabolism in the non-nodulating plant *Arabidopsis* (Castaings et al. 2009). More recently, the characterization of several defective in nodulation has begun to reveal the pathway for NF perception and signal transduction (for review see Oldroyd and Downie 2008). The *NFR1* and *NFR5* genes were isolated from *L. japonicus* mutants devoid of early Nod signal perception events. These genes encode receptor-like kinases (RLK) with N-acetyl-glucosamine-binding lysine motifs (LysM) in the extracellular domain. The specificity of legume/rhizobial symbiosis seems to be defined by the LysM domains of these RLK. This fact supports the hypothesis that NFR1 and NFR5 form part of a putative heteromeric Nod-signal receptor, although more proof is needed for complete confirmation. A pea mutant also blocked in NF signaling has been identified and the corresponding gene *sym10* has shown to encode a LysM RLK (Madsen et al. 2003). The pea *sym2* allele from cv. Afghanistan has been hypothesized to encode an entry receptor (Geurts et al. 1997). Infection of this cultivar is arrested if the *R. leguminosarum* bv. *viciae* strain does not produce a Nod signal with a specific chemical modification (i.e. acetylation, catalyzed by the product of the *nodX* gene). Unlike *sym10*, the *sym2* mutant shows some early Nod-signal-induced responses like calcium spiking. Identification of a cluster of genes in *M. truncatula* at a syntenic location with the *sym2* locus in pea indicates that the gene could code for another LysM RLK. It has been suggested that this locus may function in Nod signal induction of infection thread formation.

Several plant components with additional roles in NF perception have been identified after the characterization of mutants devoid of calcium spiking in response to the Nod signal, but showing root hair deformation. These include an RLK containing extracellular leucine-rich repeat domains (DMI2 in *M. truncatula*, Sym19 in pea), two nucleoporins, and two putative cation channels in the nuclear membrane (one of them DMI1 in *M. truncatula*, Sym8 in pea). All these genes form part of the common pathway used by both rhizobia and arbuscular mycorrhizal fungi. In the case of the *Rhizobium*-

legume symbiosis, the following working hypothesis is accepted: NF perception by multiple RLKs probably causes the generation of a second messenger, which links NF perception at the plasma membrane with calcium changes in the nucleus by activating several calcium channels.

The calcium-spiking signal is perceived by a calcium- and calmodulin-dependent protein kinase (CCaMK) (DMI3 in *M. truncatula*, Sym9 in pea) that can bind Ca directly and in a complex with calmodulin. The calcium-spiking signal finally is transduced into changes in gene expression due to the interaction between CCaMK and several transcriptional regulators. How the calcium signal is perceived and how this activates transcription are important areas of investigation at the moment.

10.4.2.2 Nodulation in the Absence of NF or Rhizobia

As it has been described above, CCaMK plays an essential role in the signal transduction triggered after NF perception. Interestingly, gain-of-function mutations of CCaMK lead to spontaneous nodulation and expression of nodulin genes in the epidermis but without root hair deformation or formation of preinfection thread structures. Therefore, activation of this gene leads to coordinated responses in both epidermis and cortex, although additional components are necessary for the activation of processes related with bacterial invasion.

Characterization of a *L. japonicus* mutant showing spontaneous mutation led to the identification of LHK1, a histidin kinase that functions as a cytokinin receptor (Trichine et al. 2007). Whereas gain-of-function mutation in this gene leads to nodule organogenesis in the absence of NF or rhizobia, loss-of-function mutants lose the ability to form nodule primordia, although they are not affected in the infection process. These results demonstrate that infection of legumes by rhizobia is possible without the need of nodule organogenesis and corroborate the essential role of cytokinins in nodule formation but not for bacterial infection in the epidermis.

10.4.2.3 Autoregulation or Plant Control of Nodule Number

Nodule formation is an expensive process for plants. To avoid overproduction of nodules, which could have detrimental effects on optimal plant growth, legumes have developed negative regulatory mechanisms to repress nodulation. It is known that formation of the first nodule primordia leads to a rapid and systemic repression of further rhizobial infection on other parts of the root. This mechanism is known as autoregulation of nodulation. Mutants defective in this control have been isolated in several legumes like *har1* and *klavier* in *L. japonicus*, *sunn* in *M. truncatula*, *nark* in soybean or *sym29* in pea. These mutants show hypernodulation after *Rhizobium* inoculation coupled with reduced root growth. Grafting experiments using these autoregulation mutants and wild-type plants showed that the hypernodulation phenotype was determined by the shoot genotype, implying a long-distance communication between shoots and roots. After the first nodule primordia are formed, a molecule is synthesized in the shoots and transported down to the roots where it inhibits further nodule development. The molecular cloning of *har1*, *GmNARK*, *sym29* and *sunn* revealed that the corresponding mutants were affected in a CLAVATA1 (CLV1)-like receptor kinase, which is expressed mainly in the phloem tissues of different organs. The nature of the root-derived and shoot-derived signals is not known yet. Recently, the locus *tml* of *L. japonicus* has been described as another component of the autoregulatory mechanism. *Tml* seems to be a root regulator participating in the same genetic pathway than *Har1* acting as a receptor or mediator of the shoot-derived signal (Magori et al. 2009).

10.4.2.4 Essential Plant Genes in Nodule Functioning

The numerous advances in the identification and characterization of plant genes with a role in nodule development contrast with the rather scarce information related to plant cell differentiation during symbiosis, especially genes required for metabolic differentiation. Map-based cloning of genes in legume mutants that develop nodules, which cannot fix nitrogen efficiently and the use of the antisense and RNA-interference technologies has revealed some of them. In this case the gene that encodes nodule-induced sucrose synthase, which has been shown to be important for carbon and energy metabolism in pea nodules and for symbiotic nitrogen fixation (Gordon et al. 1999). Using reverse rather than forward genetics, genes coding for glutamate synthase and leghemoglobin have been identified in alfalfa and *L. japonicus*, which are required for nitrogen assimilation and oxygen transport/energy metabolism in nodules, respectively (Cordoba et al. 2003; Ott et al. 2005). Udvardi's group, using map-based cloning, identified the gene *Sst1* (for symbiotic sulfate transporter) that encodes a nodule-specific sulfate transporter located in the symbiosome membrane, which is essential for nitrogen fixation in *Lotus* (Krusell et al. 2005). By reverse genetics approach, it has been proved that another transporter, the ammonium transporter LjAMT1 plays a role in retaining ammonium derived from symbiotic nitrogen fixation in plant cells prior to its assimilation, contributing to the efficiency of the symbiosis (Rogato et al. 2008).

10.4.2.5 Transcriptomics, Proteomics and Metabolomics

Transcriptional profiling studies have been performed in *M. truncatula*, *L. japonicus* and soybean using custom-designed oligonucleotide arrays, which led to the identification of numerous genes whose expression is either up- or down-regulated at different stages of nodule development and in different plant genetic backgrounds (Colebatch et al. 2002; Lohar et al. 2006; Brechenmacher et al. 2008). Among the genes identified are genes functioning in defense, metabolism, transcriptional regulation, oxidation or iron storage. Although it is difficult to make comparisons between the different studies, some conclusions can be drawn. The numerous genes coding for transporters, which are up-regulated in response to rhizobia inoculation, suggest the important role of the exchange of carbon sources and nitrogen compounds between plant and bacteria. Defense-related genes are up-regulated during the early steps of the symbiosis and then decrease in nodules, suggesting that the invading rhizobia suppress the host's defense responses to successfully establish a chronic infection. Moreover, there is evidence to suggest that like bacteroids, nodule cells experience P-limitation and osmotic stress. Additional transcriptomic studies have compared the differential responses of legume plants to wild-type and rhizobial strains defective in the production of surface polysaccharides (D'Antuono et al. 2008; Jones et al. 2008). The results indicate that appropriate symbiotically active exopolysaccharides act as signals to plant hosts and in its absence plants induce defense responses that could abort the infection process.

From the bacterial side, proteomic investigations into nodule biology complement the transcriptomic studies. Two-dimensional polyacrylamide gel electrophoresis and gel-based liquid chromatography-mass spectrometry approaches have been used. Soybean, *Lotus* and *Medicago* have been the subject of this kind of studies, examining changes in root hairs after inoculation with rhizobia and during nodule development. One of the focal points for proteomics in legumes has been the peribacteroid membrane that separates the microsymbiont from the plant cytoplasm. Many of the proteins identified with this approach are involved in the control of nutrient and signal exchange between the symbionts that were identified in biochemical genetics and transcriptomics studies.

Metabolic changes in both the plant host and the bacteria occur during symbiosis. Metabolite profiling helps to better understand how changes in transcription and translation affect cellular functions. A metabolomic study has been performed in *L. japonicus* using gas-chromatography-coupled with mass spectrometry, identifying several nodule-enriched metabolites like specific amino and organic acids, polyols, phosphates and nitrogen-containing compounds (Desbrosses et al. 2005). Data confirmed results obtained in different approaches, like the activation of glycolysis in nodules compared to roots, or the existence of osmotic stress in nodule cells. The integration of "omics" data from different levels, like transcriptomics and metabolomics, will enrich our understanding on the *Rhizobium*-legume symbiosis.

10.5 Breeding for Enhanced Nitrogen Fixation

Although there is no doubt of the economical and ecological importance of BNF in legumes, breeders often do not evaluate the symbiotic performance of newly released cultivars, but pay greater attention to disease resistance and yield. In legume breeding programs, more attention should be directed towards understanding and evaluating plant mechanisms that control root nodule formation and activity, otherwise BNF may be compromised. For instance, van Kessel and Hartley (2000) verified a trend of declining N-fixation in more recent cultivars of soybeans. Similarly, Bohrer and Hungria (1998) found that in new Brazilian genotypes, losses in BNF capacity may have resulted from the introduction of less efficient cultivars. In contrast, new high-yielding cultivars and new intensive crop management result in even higher plant demands for nitrogen. This creates a conflicting situation with higher yielding and N-demanding cultivars but lower nitrogen fixation capacity, which is usually balanced with higher applications of fertilizer N, a clearly non-sustainable approach. Therefore, increasing nitrogen fixation efficiency in high-yielding cultivars should be a priority to maintain legume properties as soil-fertility-enriching crops and to maximize their contribution to sustainable agriculture and human needs. For this, legume interactions with beneficial microbes, in general, and symbiotic nitrogen fixation in particular, must be considered as important selective criteria. For all crop legumes, there is great potential to increase the percentage of legume N derived from N_2 fixation as well to enhance the total N_2 fixed through improved management and genetic modification of the plants. Future programs may benefit from increased integration into mainstream legume breeding programs focused on diverse traits, and in which all material is evaluated in low N soils to provide the most appropriate selection pressure for highly effective nodulation and N_2 fixation (Herridge and Rose 2000). A number of strategies for enhancing legume nitrogen fixation through breeding have been suggested, such as to maximize plant biomass and seed yield within the constraints imposed by agronomic management and the environment, to enhance the ability of the legume to nodulate and fix nitrogen in presence of soil nitrate, and optimizing nodulation through specific nodulation traits, including the continuous matching of plant and rhizobial genotypes, particularly for specific environmental niches (Herridge and Rose 2000).

Classical breeding may provide useful results. For instance, a new pea cultivar has been recently developed in Russia by crossing two previously existing genotypes, a commercial cultivar providing desirable agricultural traits, and a non-commercial genotype specifically selected for symbiotic effectiveness of interaction with beneficial microbes (mainly *Rhizobium* and arbuscular mycorrhiza). Individual hybrids were selected for high productivity in low fertility soils and responsiveness to inoculation with beneficial microbes. The selected genotype (named Triumph) has been already tested in field trials showing productivity similar to commercial cultivars that require high fertilizer inputs (Alexey Borisov, ARRIAM, St. Petersburg, Russia; pers. comm.). This is an example of how traditional breeding can help to improve legume productivity and nitrogen fixation simultaneously. Several legumes have emerged as models for genetic and genomics research, mainly *M. truncatula* and *L. japonicus*, and to a lesser extent *Glycine max* and *Pisum sativum*. To facilitate the study of plant-microbe symbiotic interactions in legumes, different tools for classical, molecular and reverse genetics, along with functional genomics have been developed in model legumes, particularly *M. truncatula* and *L. japonicus*, including marker saturated genome maps, EST libraries and a great progression towards complete genome sequencing, along with transcriptomic, proteomic and metabolomic tools (Schauser et al. 2008; Rispail et al. 2010). Altogether, the resources developed in the model plant species make them ideal candidates to study legume physiology. In addition, the model legume species are also affected by most biotic and abiotic stresses limiting crops yield, so that the different resources developed on them provide a great advantage to improve the breeding for the specific needs of crops such as cool season grain legumes. The high level of synteny existing between legume genomes makes possible the transfer of knowledge from model legumes to agriculturally important species (Choi et al. 2004). As molecular markers are usually found in similar genome blocks, the discovery of a marker associated to a phenotype in a model legume provides a tool to identify the same characteristic in unexplored legume genomes (Schauser et al. 2008; Rispail et al. 2010). QTLs related to nodulation and nitrogen fixation efficiency have been identified for common bean and soybean (Nodari et al. 1993; Nicolás et al. 2006), and it should be feasible to successfully carry out marker-assisted selection programs for enhancing nitrogen fixation in cool season legume species. Although the root nodule nitrogen fixing symbiosis is an efficient system for incorporating nitrogen into biomass it is restricted to only a few plants of agronomical importance. The transfer of this characteristic to other plants will be a challenge that will be accomplished by accumulating knowledge derived from the study of current microbe-plant associations, particularly the *Rhizobium*-legume symbiosis. The fact that many of the genetic programs involved in the establishment of this type of symbiosis are well conserved in other plants will make the task easier (see Markmann and Parniske 2009).

References

Arrese-Igor C, González EM, Gordon AJ, Minchin FR, Gálvez L, Royuela M, Cabrerizo PM, Aparicio-Tejo PM (1999) Sucrose synthase and nodule nitrogen fixation under drought and other environmental stresses. Symbiosis: 27: 189–212.

Avigad G (1982) Sucrose and other dissacarides. In: FA Loewus, W Tanner (eds) Encyclopedia of Plant Physiology (New Series), vol 13A.Springer, Berlin, Germany, pp 216–247.

Barnett MJ, Toman CJ, Fisher RF, Long SR (2004) A dual-genome symbiosis chip for coordinate study of signal exchange and development in a prokaryote-host interaction. Proc Natl Acad Sci USA 101: 16636–16641.

Bohrer TRJ, Hungria M (1998) Avaliação de cultivares de soja quanto à fixação biológica do nitrogênio. Pesquisa Agropecu Bras 33: 937–953.

Brechenmacher L, Kim MY, Benitez M, Li M, Joshi T, Calla B, Lee MP, Libault M, Vodkin LO, Xu D, Lee SH, Clough SJ, Stacey G (2008) Transcription profiling of soybean nodulation by *Bradyrhizobium japonicum*. Mol Plant-Microbe Interact 21: 631–645.

Brito B, Palacios JM, Imperial J, Ruiz-Argüeso T (2002) Engineering the *Rhizobium leguminosarum* bv. *viciae* hydrogenase system for expression free-living microaerobic cells and increased symbiotic hydrogenase activity. Appl Environ Microbiol. 68: 2461–2467.

Castaings L, Camargo A, Pocholle D, Gaudon V, Texier Y, Boutet-Mercey S, Taconnat L, Renou JP, Daniel-Vedele F, Fernandez E, Meyer C, Krapp A (2009) The nodule inception-like protein 7 modulates nitrate sensing and metabolism in *Arabidopsis*. Plant J 57: 426–435.

Chen WM, Moulin L, Bontemps C, Vandamme P, Béna G, Boivin-Masson C (2003) Legume symbiotic nitrogen fixation by β-proteobacteria is widespread in nature. J Bacteriol 185: 7266–7272.

Choi HK, Mun JH, Kim DJ, Zhu H, Baek JM, Mudge J, Roe BA, Ellis N, Doyle J, Kiss GB, Young ND, Cook DR (2004) Estimating genome conservation between crop and model legume species. Proc Natl Acad Sci USA 101: 15289–15294.

Colebatch G, Kloska S, Trevaskis B, Freund S, Altmann T, Udvardi MK (2002) Novel aspects of symbiotic nitrogen fixation uncovered by transcript profiling with cDNA arrays. Mol Plant-Microbe Interact 15: 411–420.

Cordoba E, Shishkova S, Vance CP, Hernández G (2003) Antisense inhibition of NADH glutamate synthase impairs carbon/nitrogen assimilation in nodules of alfalfa (*Medicago sativa* L.). Plant J 33: 1037–1049.

D'Antuono AL, Ott T, Krusell L, Voroshilova V, Ugalde RA, Udvardi M, Lepek VC (2008) Defects in rhizobial cyclic glucan and lipopolysaccharide synthesis alter legume gene expression during nodule development. Mol Plant-Microbe Interact 21: 50–60.

Deakin WJ, Broughton WJ (2009) Symbiotic use of pathogenic strategies: rhizobial protein secretion systems. Nat Rev Microbiol 7: 312–320.

Desbrosses GG, Kopka J, Udvardi MK (2005) *Lotus japonicus* metabolic profiling. Development of gas chromatography-mass sspectrometry resources for the study of plant-microbe interactions. Plant Physiol 137: 1302–1318.

Djordjevic MA (2004) *Sinorhizobium meliloti* metabolism in the root nodule: a proteomic perspective. Proteomics 4: 1859–1872.

Dowling DN, Broughton WJ (1986) Competition for nodulation of legumes. Annu Rev Microbiol 40: 131–157.

Duncan KA, Huber SC (2007) Sucrose synthase oligomerization and F-actin association is regulated by sucrose concentration and phosphorylation. Plant Cell Physiol 48: 1612–1623.

Evans HJ, Russwell SA, Hanus FJ, Ruiz-Argüeso T (1988) The importance of hydrogen recycling in nitrogen fixation by legumes. In: RJ Summerfield (ed) World Crops: Cold Season Food Legumes. Kluwer Academic Press, Boston, MA, USA, pp 777–791.

Geurts R, Heidstra R, Hadri AE, Downie JA, Franssen H, Van Kammen A, Bisseling T (1997) *Sym2* of pea is involved in a nodulation factor-perception mechanism that controls the infection process in the epidermis. Plant Physiol 115: 351–359.

Giraud E, Moulin L, Vallenet D, Brabe V, Cytryn E, Avarre JC, Jaubert M, Simon D, Cartreaux F, Prin Y, Bena G, Hannibal L, Fardoux J, Kojadinovic M, Vuillet L, Lajus A, Cruveiller S, Rouy Z, Margenot S, Sequrens B, Dossat C, Franck WL, Chang WS, Sannders E, Bruce D, Richardson P, Normand P, Dreyfus B, Pignol D, Stacey G, Emerich D, Vermeglio A, Mèdique C, Sadowsky M (2007) Legumes symbioses: Absence of *Nod* genes in photosynthetic bradyrhizobia. Science 316: 1307–1312.

Gordon AJ, Minchin FR, James CL, Komina O (1999) Sucrose synthase in legume nodules is essential for nitrogen fixation. Plant Physiol 120: 867–78.

Herridge D, Rose I (2000) Breeding for enhanced nitrogen fixation in crop legumes. Field Crop Res 65: 229–248.

Hill S (1988) How is nitrogenase regulated by oxygen? FEMS Microbiol Rev. 4: 111–129.

Hogg B, Davies ED, Wilson KE, Bisseling T, Downie JA (2002) Competitive nodulation blocking of cv. Afghanistan pea is related to high levels of nodulation factors by some strains of *Rhizobium leguminosarum* bv. *viciae*. Mol Plant-Microbe Interact 15: 60–68.

Jones KM, Kobayashi H, Davies BW, Taga ME, Walker GC (2007) How rhizobial symbionts invade plants: The Sinorhizobium-Medicago model. Nat Rev Microbiol 5: 619–633.

Jones KM, Sharopova N, Lohar DP, Zhang JQ, VandenBosch KA, Walker GC (2008) Differential response of the plant *Medicago truncatula* to its symbiont *Sinorhizobium meliloti* or an exopolysaccharide-deficient mutant. Proc Natl Acad Sci USA 105: 704–709.

Karunakaran R, Ramachandran VK, Seaman JC, East AK, Mouhsine B, Mauchline TH, Prell J, Skeffington A, Poole PS (2009) Transcriptomic analysis of *Rhizobium leguminosarum* biovar viciae in symbiosis with host plants *Pisum sativum* and *Vicia cracca*. J Bacteriol 191: 4002–4014.

Kessel C van, Hartley C (2000) Agricultural management of grain legumes: has it led to an increase in nitrogen fixation? Field Crop Res 65: 165–181.

Krusell L, Krause K, Ott T, Desbrosses G, Krämer U, Sato S, Nakamura Y, Tabata S, James EK, Sandal N, Stougaard J, Kawaguchi M, Miyamoto A, Suganuma N, Udvardi MK (2005) The sulfate transporter SST1 is crucial for symbiotic nitrogen fixation in *Lotus japonicus* root nodules. Plant Cell 17: 1625–1636.

Kuykendall LD (2005) Order VI. Rhizobiales ord. nov. In: DJ Brenner, NR Krieg, JT Stanley (eds) Bergey's Manual of Systematic Bacteriology, vol 2, part C. Springer, New York, USA, pp 324–574.

Lodwig EM, Hosie AHF, Bordes A, Findlay K, Allaway D, Karunakaran R, Downie JA, Poole PS (2003) Amino-acid cycling drives nitrogen fixation in the legume-*Rhizobium* symbiosis. Nature 422: 722–726.

Lohar DP, Sharopova N, Endre G, Peñuela S, Samac D, Town C, Silverstein KA, VandenBosch KA (2006) Transcript analysis of early nodulation events in *Medicago truncatula*. Plant Physiol 140: 221–234.

Madsen EB, Madsen LH, Radutoiu S, Olbryt M, Rakwalska M, Szczyglowski K, Sato S, Kaneko T, Tabata S, Sanfal N, Stougaard J (2003) A receptor kinase gene of the LysM type is involved in legume perception of rhizobial signals. Nature 425: 637–640.

Magori S, Oka-Kira E, Shibata S, Umehara Y, Kouchi H, Hase Y, Tanaka A, Sato S, Tabata S, Kawaguchi M (2009) Too much love, a root regulator associated with the long-distance control of nodulation in *Lotus japonicus*. Mol Plant-Microbe Interact 22: 259–268.

Marchetti M, Capela D, Glew M, Cruveiller S, Chane-Woon-Ming B, Gris C, Timmers T, Poinsot V, Gilbert LB, Heeb P, Médigue C, Batut J, Masson-Boivin C (2010) Experimental evolution of a plant pathogen into a legume symbiont. PLoS Biol 8: e1000280: doi: 10.1371/journal.pbio.1000280.

Marino D, Hohnjec N, Küster H, Moran JF, González EM, Arrese-Igor C (2008) Evidence for transcriptional and post-translational regulation of sucrose synthase in pea nodules by the cellular redox state. Molec Plant-Microbe Interact 21: 622–630.

Markmann K, Parniske M (2009) Evolution of root endosymbiosis with bacteria: how novel are nodules? Trends Plant Sci 14: 77–86.

Morell M, Copeland L (1985) Sucrose synthase in soybean nodules. Plant Physiol 78: 149–154.

Nicolás M, Hungria M, Arias CAA (2006). Identification of quantitative trait loci controlling nodulation and shoot biomass in progenies from two brazilian soybean cultivars. Field Crop Res 95: 355–366.

Nodari RO, Tsai SM, Guzmán P, Gilbertson RL, Gepts P (1993) Towards an integrated linkage map of common bean. III. Genetic factors controlling host-bacteria interactions. Genetics 134: 341–350.

Nomura M, Mai HT, Fujii M, Hata S, Izui K, Tajima S (2006) Phosphoenolpyruvate carboxylase plays a crucial role in limiting nitrogen fixation in *Lotus japonicus* nodules. Plant Cell Physiol 47: 613–621.

Oldroyd GED, Downie JA (2008) Coordinating nodule morphogenesis with rhizobial infection in legumes. Annu Rev Plant Biol 59: 519–546.

Ott T, van Dongen JT, Günther C, Krusell L, Desbrosses G, Vigeolas H, Bock V, Czechowski T, Geigenberger P, Udvardi MK (2005) Symbiotic leghemoglobins are crucial for nitrogen fixation in legume root nodules but not for general plant growth and development. Curr Biol 15: 531–535.

Ou-yang L, Udvardi MK, Day DA (1990) Specificity and regulation of the dicarboxylate carrier on the peribacteroid membrane of soybean nodules. Planta 182: 437–444.

Patriarca EJ, Tate R, Iaccarino M (2002) Key role of bacterial NH_4^+ metabolism in *Rhizobium*-plant symbiosis. Microbiol Mol Biol Rev 66: 203–222.

Rispail N, Kaló P, Kiss GB, Ellis THN, Gallardo K, Thompson RD, Prats E, Larrainzar E, Ladrera R, González EM, Arrese-Igor C, Ferguson BJ, Gresshoff PM, Rubiales D (2010) Model legumes contribute to faba bean breeding. Field Crop Res 115: 253–269.

Rogato A, D'Apuzzo E, Barbulova A, Omrane S, Stedel C, Simon-Rosin U, Katinakis P, Flemetakis M, Udvardi M, Chiurazzi M (2008) Tissue-specific down-regulation of LjAMT1;1 compromises nodule function and enhances nodulation in *Lotus japonicus*. Plant Mol Biol 68: 585–595.

Schauser L, Roussis A, Stiller J, Stougaard J (1999) A plant regulator controlling development of symbiotic root nodules. Nature 402: 191–195.

Schauser L, Udvardi M, Tabata S, Stougaard J (2008) Legume genomics relevant to nitrogen fixation. In: MJ Dilworth, EK James, JI Sprent, WE Newton (eds) Nitrogen-fixing Leguminous Symbioses, vol 7. Springer, Berlin, Heidelberg, Germany; New York, USA, pp 211–239.

Sessitsch A, Howieson JG, Perret X, Antoun H, Martínez-Romero E (2002) Advances in *Rhizobium* research. Crit Rev Plant Sci 21: 323–378.

Soto MJ, Sanjuán J, Olivares J (2006) Rhizobia and plant-pathogenic bacteria: common infection weapons. Microbiology 152: 3167–3174.

Soto MJ, Domínguez-Ferreras A, Pérez-Mendoza D, Sanjuan J, Olivares J (2009) Mutualism versus pathogenesis: the give and take in plant-bacteria interactions. Cell Microbiol 11: 381–388.

Trichine L, Sandal N, Madsen LH, Radutoiu S, Albrektsen AS, Sato S, Asamizu E, Tabata S, Stougaard J (2007) A gain-of-function mutation in a cytokinin receptor triggers spontaneous root nodule organogenesis. Science 315: 104–107.

Triplett EW, Barta TM (1987) Trifolitoxin Production and Nodulation Are Necessary for the Expression of Superior Nodulation Competitiveness by *Rhizobium leguminosarum* bv. *trifolii* Strain T24 on Clover. Plant Physiol 85: 335–342.

Trujillo ME, Willems A, Abril A, Planchuelo A-M, Rivas R, Lueña D, Mateos PF, Martínez-Molina E, Velázquez E (2005) Nodulation of *Lupinus* by strains of *Ochrobactrum lupini* sp nov. Appl Environ Microbiol 71: 1318–1327.

Vance CP (2008) Carbon and nitrogen metabolism in legume nodules. In: MJ Dilworth, EK James, JI Sprent, WE Newton (eds) Nitrogen-fixing Leguminous Symbioses. Springer, Dordrecht, The Netherlands, pp 293–320.

Van Sluys MA, Monteiro-Vitorello CB, Camargo LEA, Menck CFM, da Silva ACR, Ferro JA, Oliveira MC, Setubal JC, Kitajima JP, Simpson AJ (2002) Comparative genomic analysis of plant-associated bacteria. Annu Rev Phytopathol 40: 169–189.

Wang ET, Martínez-Romero E (2000) Phylogeny of root- and stem nodule bacteria associated with legumes. In: EW Triplett (ed) Prokaryotic Nitrogen Fixation: A Model System for Analysis of a Biological Process. Horizon Scien Press, Wymondham, UK, pp 177–186.

White J, Prell J, James EK, Poole PS (2007) Nutrient sharing between symbionts. Plant Physiol 144: 604–614.

Willems A (2006) The taxonomy of rhizobia: an overview. Plant Soil 287: 3–14.

Witty JF, Skot L, Revsbech NP (1987) Direct evidence for changes in the resistance of legume root-nodules to O_2 diffusion. J Exp Bot 38: 1129–1140.

Yurgel SN, Kahn ML (2004) Dicarboxylate transport by rhizobia. FEMS Microbiol Rev 28: 489–501.

11

Broomrape (*Orobanche* spp.)

Belén Román,[1,a,*] *José Vicente Die,*[1] *Salvador Nadal,*[2] and *Clara Isabel González-Verdejo*[1]

ABSTRACT

Some parasitic plant species have become weeds, posing severe constraints to major crops including cool season grain legumes. In this chapter, we focus specifically on the genus *Orobanche*, a major threat to cool season legumes and other crops in many temperate areas. Broomrapes (*Orobanche* spp.) are annual or perennial obligate parasites without chlorophyll that infect roots of dicotyledonous plants. Broomrape control is quite difficult due to the nature of this parasite, as it produces a vast number of seeds that are able to remain dormant in the soil for many years. The high complexity of the plant-parasite interaction determines a great lack in our knowledge on the genetic basis of resistance that obviously hampers the development of resistant cultivars. Traditional plant breeding has failed to produce stable resistance. However, some sources of genetic resistance exit and have been used in cool season legume breeding. Molecular mapping of broomrape resistance QTLs together with the development of MAS techniques are promising approaches to rapidly improve crop resistance. With the emergence of large-scale genomic tools, the combination of genetic mapping with gene expression studies, can offer an integrated approach to study resistance to parasitic plants. In this sense, testing the role of candidate genes selected from expression experiments, can simplify the search of sequence polymorphisms for a more efficient and rapid MAS strategy.

Keywords: Broomrape, parasitic plants, legumes, resistance, breeding for resistance, host-parasite interactions, gene expression

[1]IFAPA Alameda del Obispo, Área de Mejora y Biotecnología, Apdo. 3092, 14080 Córdoba, Spain.
[a]e-mail: *mariab.roman@juntadeandalucia.es*
[2]IFAPA Alameda del Obispo, Área de Producción Agraria, Apdo. 3092, 14080 Córdoba, Spain.
*Corresponding author

11.1 Introduction

Parasitic plant species have lost their autotrophic way of life during evolution and have adapted to parasitize other plants in order to supply themselves with water and nutrients. Unfortunately for farmers, a small number of these species have become weeds, posing severe constraints to major crops including grain legumes. This is the case of the witchweeds (*Striga* spp.), dodders (*Cuscuta* spp.) and broomrapes (*Orobanche* spp.). In this chapter we focus on parasitic plants infecting legumes, specifically the genus *Orobanche*. Other genera of weedy root parasites have been already reviewed (Parker and Riches 1993; Press and Graves 1995).

Broomrapes (*Orobanche* spp.) are obligate parasites that infect roots of dicotyledonous plants. They have the ability to attach themselves to the roots of other plants, from which they then draw all their water and nutritional needs. Although most of the species within the genus are found in natural vegetation, however some have abandoned their natural hosts and become very noxious parasitic weeds causing serious losses in yields of economically important crops. This can have an enormous detrimental impact on the host plant, resulting in dramatic crop losses in affected regions, and placing *Orobanche* among the world's most destructive weeds.

Orobanche species are annual or perennial plants without chlorophyll, parasitic on roots of other plants and emerging from the soil in an asparagus-like fashion. The roots of broomrapes are partly suckers, partly adventitious without root hairs. Stems are erect, single and thicker below. Chlorophyll is completely absent, resulting in a total inability to produce organic substances through photosynthesis. Leaves are reduced to scale leaves. The flowers grow in single spikes from the axils of scale-like bracts and produce seeds. *Orobanche* seeds are very small and light and easily dispersed by the wind. Since the endosperm contains oil, the germination capacity of many species is maintained for 10–12 years increasing the virulent capacity of the seedbank in the fields.

The life cycle of the root parasite *Orobanche* is composed of several key steps: preconditioning of seeds, germination (which only occurs when the broomrape seed contacts with the host plant roots, triggered by certain chemical signals released by the roots of the host plant), development of a small seedling that attaches to the host root and differentiates an attachment organ (appressorium), penetration through host tissues, formation of the haustorium (a specialized transfer structure connected to the host vascular tissues) and connection to the vascular tissues, development of a tubercle and apex, stem growth and emergence, and flowering and setting of seeds (see Parker and Riches 1993; Press and Graves 1995; Joel et al. 2007 for review). The haustorium allows an intimate contact between host and parasite, with the consequent exchange of water and nutrients. The formation of a such structure involves a complex process in which the parasites finishes its short independent cycle that is facilitated by the consumption of material stored in the seed and begin its parasitic life stage (Joel 2000).

11.2 Broomrape Species Distribution, Description and Range of Legume Hosts

The species of the genus *Orobanche* are found largely in the Mediterranean and warm temperate areas of Europe, North Africa and the Middle East. *Orobanche* parasitize a large number of crops such as legumes, crucifers, tomato, sunflower or tobacco among others and constitute one of the most important biotic constraints to the food crop production in southern and eastern Europe, North and East Africa, the Middle East and the Indian subcontinent (Joel et al. 2007). In the case of legumes, broomrapes are among the major biotic constraints to the final production in different parts of the world, and unfortunately the efficacy of available means of control is limited. In particular, crenate broomrape (*Orobanche crenata*) is the most economically damaging parasitic plant species on legumes, but some other broomrapes such as *O. aegyptiaca*, *O. foetida* and *O. minor* can be of increasing importance.

Many *Orobanche* species infect only a single type of host plant, others parasitize a range of hosts; some on at least 50 different hosts. On the other hand, some plants host a single *Orobanche* species, while others may have several *Orobanche* species as parasites. For example, *O. cumana* or *O. crenata* parasitize on a small group of genera or species, while others have a much broader spectrum of hosts as is the case of *O. aegyptiaca* and *O. ramosa* (Parker and Riches 1993; Press and Graves 1995).

Crenate broomrape (*Orobanche crenata* Forsk.) has been known to threaten legumes crops since antiquity (Cubero and Moreno 1996). It is an important parasite to cool season legumes such as faba bean (*Vicia faba*), pea (*Pisum sativum*), lentil (*Lens culinaris*), vetches (*Vicia* spp.), grass and chickling pea (*Lathyrus sativus* and *L. cicera*), narbon bean (*Vicia narbonensis*) but also to other grain and forage legumes in the Mediterranean basin and Middle East (Sauerborn 1991). *Orobanche crenata* has its greatest economic impact on faba bean. In 1991, some 180,000 ha were estimated to be infested in Morocco, Portugal, Spain and Syria, representing 50–70% of the areas grown in those countries. In Morocco, the losses were estimated at 12% in the infested areas, or 7.8% of the national yield. Some estimates of yield loss were even higher: in Egypt (33%), Malta (50–100%) and Turkey (30–70%). Other countries in which *O. crenata* is a significant problem on legumes include Algeria, Cyprus, Iraq, Italy, Jordan, Lebanon and Tunisia. It has been estimated that over 4 million ha of legumes could be at risk from *Orobanche* (mainly *O. crenata*) across the Mediterranean and West Asia (reviewed by Parker 2009).

Although field pea is the most widely grown grain legume in Europe and the fourth most widely grown in the world, its cultivation is strongly hampered in Mediterranean and Middle East farming systems by the occurrence of crenate broomrape causing important yield losses. Severe damages on pea have been reported in southern Spain (García-Torres et al. 1996; Rubiales et al. 2003a), Morocco (Mabsoute and Saadaoui 1996), Egypt (Korashi et al. 1996) and Israel (Bernhard et al. 1998). Yield loss can be huge, as high as 80% (Korashi et al. 1996) or even 100% (Bernhard et al. 1998). Early reports of severe infestations of crenate broomrape on pea point to the fact that pea is more sensitive to broomrape damage than is faba bean (Moreno 1944). Interestingly, pea is only infected by *O. crenata*, and not by other broomrapes such as *O. foetida* (Poir) or *O. aegyptiaca* (Pers.) like other legumes (Kharrat et al. 1992; Fernández-Aparicio et al. 2009a; Rubiales et al. 2005a). Narbon vetch is very seriously infested by crenata broomrape and thus has been grouped among the very susceptible species (Linke et al. 1993), supporting in some cases up to 200 *O. crenata* shoots per m^2. This pest could be a major problem for *V. narbonensis* cultivation in southern Spain and other places where broomrape is present.

On the other hand, in many Mediterranean areas, chickpea (*Cicer arietinum*), a well established cool season food legume, is attacked by broomrape but the level of susceptibility seems to be much lower than that in the Vicieae (Cubero 1991) and resistance, although scarce in many legumes, is common in chickpea germplasm.

Egyptian broomrape (*Orobanche aegyptiaca* Pers.) (syn. *Phelipanche aegyptiaca* Pers.) attacks faba bean, common vetch, chickpea and lentil, but in addition it can also attack peanut (*Arachis hypogea*), cruciferous crops (particularly cabbage and oilseed rape) and several members of the families Solanaceae, Apiaceae and Asteraceae (Parker and Riches 1993). It is widely distributed in the eastern parts of the Mediterranean, in the Middle East and in parts of Asia. The main range in the Eurasia context is from Greece through Turkey to the Black Sea and down through Syria and Israel to Egypt (Schneeweiss 2009) and from the eastern Mediterranean through to Afghanistan, Pakistan, India and Nepal.

Foetida broomrape (*Orobanche foetida* Poir.) is common in native habitats in the western Mediterranean area (Portugal, Spain, Morocco, Algeria, Tunisia) parasitizing wild herbaceous leguminous plants, but recently it has become an important threat as an agricultural parasite in north-central Tunisia, on faba beans, chickpea and common vetch (Kharrat et al. 1992). More recently it also has been found on common vetch in Taounate region in Morocco (Rubiales et al. 2005b). In the Beja region of Tunisia it has become a huge problem in faba bean fields with yield losses ranging 66–83% (Kharrat 1999). This species seems to be more aggressive on faba bean and common vetch than on other legumes (Kharrat 2002). Of the cool season grain legumes only pea (*Pisum sativum*) escapes its attack (Kharrat 1999). Infection by this broomrape species has also been reported in Tunisia on *Lathyrus odoratus, L. sativus, Trifolium alexandrinum, Medicago truncatula* and *V. sativa* ssp. *amphicarpa* (Kharrat 2002). In Morocco, *O. foetida* is abundant on weeds in close vicinity to crop fields, or even within them, and recently the species has been described on crop plants themselves (Rubiales et al. 2005b), thus, the potential of this species to become a threat to agriculture exists and has to be taken into account.

Small broomrape or clover broomrape (*Orobanche minor* Sm.) is a holoparasitic weed native to Europe that has spread to Africa, Australia, New Zealand, Japan and the US (James and Frater 1977; Parker and Riches 1993; Hassan 1998). It has a wide host range among forage legumes in temperate climates and is mainly a problem in clovers and lucerne crops grown for seed, where it has the opportunity to shed seed and build up to high populations. New infestations continue to appear. In Oregon, USA, *O. minor* was first identified in a single field in 1998. By 2001 it was found from 15 additional sites becoming a problem on red clover in Oregon (Osterbauer and Rehms 2002; Eizenberg et al. 2004). In South and West Australia it has also been introduced and the species can be seen in gardens and in crop and pasture paddocks (Hussey et al. 1997) but is reported as a potential problem only for *Vicia ervilia* (Carter et al. 1996). In pot experiments, up to 50% reduction in clover biomass was measured, while inflorescence mass could be reduced by 80% (Lins et al. 2007).

Other parasitic plants, not belonging to the genus *Orobanche,* can also attack grain legumes and cause considerable yield reduction in semi-arid areas of sub-Saharan Africa such as, *Striga gesnerioides* and *Alectra vogelii* (Parker and Riches 1993; Rubiales et al. 2006). Cowpea is the most affected crop, yield losses range from 30–50% to the total crop loss (Aggarwal and Ouedraogo 1989; Riches 1989; Emerchebe et al. 1991).

11.3 Broomrape Control

Broomrape control is quite difficult due to the nature of this parasite, as it produces a vast number of seeds that are able to remain dormant in the soil for many years waiting for the germination stimulants exuded from plant roots. Different control methods have been proposed to control *Orobanche* species including chemical control, cultural practices, biological control or genetic resistance, but none with complete success.

11.3.1 Chemical Control

Regarding chemical control, soil fumigants such as methyl bromide, ethylene dibromide, methamsodium or formalin are effective for the control of broomrape (Jacobsohn et al. 1988), but the high cost together with the negative environmental effects seldom justifies their use in the control of broomrapes in legume production.

Glyphosate is a non-selective broad spectrum, post-emergence, foliar applied herbicide that is absorbed through the leaves and is translocated primarily in the symplast from the point of application to all metabolic active parts of the plant ("metabolic sinks"). The mode of action of glyphosate is the inhibition of the enzyme EPSP synthase of the shikimic acid pathway, inhibiting aromatic amino acid synthesis and thus protein synthesis and growth (Amrhein et al. 1980). The attached root parasite acts as a strong metabolic active part of the host plant, rapidly absorbing and accumulating the herbicide through its direct connection with the host root vascular system. Glyphosate has been effective in controlling *O. crenata* in faba bean fields, but problems of lack of effectiveness or phytotoxicity can arise when it is not properly applied. Standard recommendations are: application as a post-emergence foliar spray (64 g a.i. ha^{-1}) at the beginning of flowering (2–3 sprays at 2–3 weeks intervals), which results in good control (García-Torres et al. 1998). Moreover, faba bean cultivars with determinate growth habit seem to tolerate higher rates of glyphosate to ensure broomrape control (Nadal et al. 2001). The tolerance of narbon bean (*V. narbonensis* L.) to glyphosate was also studied under field conditions. Its effectiveness was measured both in terms of broomrape shoot control and seed yield performance. When broomrape was sprayed twice with glyphosate (when the broomrape attachment to the narbon bean roots was at the tubercle stage of development and 14 days later), there was an increased seed yield in the crop as compared to the control treatment. The best dose range of application was 35–67 g a.i. ha^{-1}, depending on weather conditions (Nadal et al. 2008).

The "sulfonylureas" are systemic herbicides absorbed through roots, shoots and foliage of plants with rapid acropetal and basipetal translocation trough the xylem and phloem. The mode of action of these herbicides is the inhibition of acetolactate synthase (ALS), a key unique plant enzyme in the biosynthesis of the branched-chain amino acids isoleucine, leucine and valine (Schloss 1995). Various legumes are resistant to some of the imidazolinone herbicides and this resistance has led to selective use of these herbicides in certain legume crops. Legumes are tolerant to "imazapyr" because they can metabolize it to an inactive form (Shaner 1989). García-Torres et al. (1998) reported selective *O. crenata* control in faba bean in pre-emergence and post-emergence applications of imazethapyr, imazapyr and imazaquin. A post emergence application of 20 g ha^{-1} of imazethapyr on garden and field pea (*P. sativum* and *P. arvense*, respectively) one month after planting, and an additional treatment of 20–40 g ha^{-1} two weeks later, was selective for pea and efficient in *Orobanche* control (Jacobsohn 1994).

There are also reports of some promising results of *O. crenata* control by faba bean, lentil and pea seed treatments with imazethapyr (Jurado-Expósito et al. 1997, 1999). Moreover, an international seed company has released the variety RH44 of lentil (*Lens culinaris*) derived through mutation breeding which is tolerant to imidazolinone herbicides (Slinkard et al. 2007).

In the case of red clover, imidazolinone herbicides such as imazamox control small broomrape (*O. minor*). Studies were conducted to determine the imazamox route of uptake by small broomrape on red clover, and to determine the potential for imazamox to be exuded from red clover and the subsequent effect on small broomrape. Small broomrape control was best at 90% when imazamox was foliar-applied, and worst at 42% or less when imazamox was soil-applied (Colquhoun et al. 2006).

Herbicides have to be applied either at crop pre-emergence or at late crop post-emergence, but always before broomrape emergence. For this reason control decision making is very difficult because farmers have to predict the level of infestation they need to control, which depends on the infestation level of previous years, climatic conditions, sowing date, etc.

11.3.2 Cultural Practices

Regarding cultural practices, different methods have been also proposed to control *Orobanche* spp. in legumes. The **sowing date** is important for the intensity of *Orobanche* infection. Thus, delayed sowing is recommended in order to prevent severe attacks of the parasite (Moreno-Márquez 1947; Mesa-García and García-Torres 1991; Zaitoun and Ibrahim 1998). However, some legumes have a long life cycle and delayed sowing can decrease yield because of the shortened cropping season (Saxena et al. 1994; Zaitoun and Ibrahim 1998). This reduction in the potential crop yield could be equilibrated by a compromise of a few weeks, so the decrease in crop biomass is compensated by the reduction of the infection. Differences in establishment, development and shoot emergence of the parasite were compared in resistant and susceptible cultivars of faba bean and vetch cultivars (Pérez-de-Luque et al. 2004). Susceptible cultivars were more severely affected by the parasite with early sowing and *O. crenata* development was also delayed in resistant cultivars. Nevertheless, crop yield, estimated by the number of pods per plant, decreased with late sowings. Although chickpea is a host of crenate broomrape that suffers little damage in the traditional spring sowing, infection in this species is also favored by very early sowing dates (Rubiales et al. 2003b). The combined use of resistant cultivars and early sowing is a useful tool as part of an integrated control strategy. In this sense, resistant cultivars allow early sowing (with low *O. crenata* attack), thus avoiding yield losses due to the short crop cycle with late sowing.

Hand-weeding can only be recommended in cases of limited infestation to prevent any further increase in the parasite population and to reduce the seed bank in the soil. However, this practice is still commonly used in some countries where no other feasible means of control are available and the wages for labor are cheap.

Intercropping is the growing of two or more crops in proximity to promote beneficial interaction between them. This method is regarded as an ecological method to manage pests, diseases and weeds via natural competitive principles that allow for more efficient resource utilization (Liebman and Dyck 1993). Field experiments have shown that *O. crenata* infection on faba bean and pea is reduced when these host crops are intercropped with oat (Fernández-Aparicio et al. 2007). The number of *O. crenata* plants per host plant decreased as the proportion of oats increased in the intercrop. The finding that germination of seeds exposed to the synthetic germination stimulant is inhibited in presence of cereal roots suggest that cereal roots might be exuding substances that inhibit *O. crenata* seed germination. Thus, the suggested mechanism for reduction of *O. crenata* infection has been the release of allelochemicals by cereal roots (Fernández-Aparicio et al. 2007).

Rotation into non-host crops is seldom the only valid recommendation for farmers, but this option is not always acceptable due to the importance of the host crops for the economy and income of subsistence farmers. The use of trap crops offers the advantage of stimulating broomrape germination without themselves being parasitized. Flax (*Linum usitatissimum* L.), fenugreek (*Trigonella foenum-graecum* L.) and Egyptian clover (*Trifolium alexandrinum* L.) are known trap crops for *O. crenata* (Al-Menoufi 1991).

The **solarization** method is based on the trapping of radiation in moist soil by polyethylene mulch for at least 5–6 weeks in the summer. The humidity is achieved by irrigation prior to mulching or during the solarization process by drip irrigation. *Orobanche* seeds and soil borne disease control is achieved in the upper 15–20 cm. where the temperature exceeds 50–55°C. Solarization is used in limited areas mainly integrated with lower rates of soil fumigants for soil borne pest control including *O. crenata*, especially in very hot regions for winter legumes (Kleifeld et al. 1999).

11.3.3 Biological Control

Biological control of *Orobanche* and other weeds has been advocated for integrated management, which includes measures such as the use of insect predators on *Orobanche* (Kroschel and Klein 1999) and pathogenic fungi (Thomas et al. 1998). The young attached parasite acts as a strong metabolic sink on the host plant, rapidly accumulating water and nutrients for its latter growth stages. Such a nutrient–rich plant organ is a good target for attacking pests that could be used for biological control. The adult insects of *Phytpomiza orobanchia*, a Diptera, suborder Brachycera (flies) of the family Agromyzidae are attracted by *Orobanche* flowers and feed on its nectar. Eggs are laid under the epidermis of shoots and flowers and the larvae penetrate into the plant tissue and feed on the shoots and capsules. As a consequence, a natural reduction of *Orobanche* seed production is achieved. However, this reduction might still have little epidemiological impact in heavily infested areas due to the huge seed production of broomrape and its longevity in the soil. Nevertheless, biocontrol with *P. orobanchia* can be helpful to slow down further dissemination and infestation in weakly infested areas and can be part of an integrated control approach to reduce the seed bank in heavily infested soils. In many temperate countries *Orobanche* (and specially *O. crenata*) flowering in early spring escapes from heavy *Phytomyza* damage, since the low soil winter temperatures kill a high amount of pupa (Kurbanov 1970).

Pathogenicity tests showed that fungi, such as *Fusarium* spp. and *Alternaria alternata*, attacked healthy living tissue of *Orobanche* spikes. These fungi caused lesions of black soft rot and complete deterioration within 5–7 days. They also attacked *Orobanche* seeds, arresting their germination and causing maceration of non-germinated and germinated seeds after 5–7 days of incubation. Meanwhile, *Dendrophora* spp. and *Chaetomium* spp. caused limited lesion at first, but were able to colonize the tissue as it aged and senesced. The potential endogenous pathogenic fungi existing in Jordan, have been proposed as a biological control for *Orobanche* (Hameed et al. 2001). In other studies different fungi have been isolated from broomrape species, some of which (*Fusarium solani, F. oxysporum, Macrophomina phaseolina*, and *Uloclaudium atrum* among others) were effective in controlling broomrape under laboratory conditions (Linke et al. 1992; Amsellem et al. 1999; Dor et al. 2001). However the technology for their application in biocontrol under field conditions is not ready yet.

11.3.4 Genetic Resistance

In addition to the control methods described above, one of the most desirable approaches, which is also environmentally friendly and can be considered a long-term strategy for limiting the damage by the parasite, is the development of genetic resistant cultivars. Traditional plant breeding has failed to produce stable and lasting resistance or under variations of location and parasite pressure (Cubero 1991). The main reasons for the lack of good results are the existence of several associated difficulties: the unknown interaction between the host legume plant and the parasitic plant, the complex nature of resistance and the lack of adequate resistance evaluation methods.

The high complexity of the interaction and the nature of the parasite determine a great lack in our knowledge on the genetic basis of resistance that obviously hampers the development of resistant cultivars. Moreover, resistance against most parasitic weeds is in general difficult to find, scarce, complex in genetic nature and of low heritability, making breeding for resistance a quite difficult task. Although qualitative resistance to *Orobanche* has been found in other species allowing, for instance, rapid progress in sunflower breeding against *O. cumana* (Wallr.) (Fernández- Martínez et al. 2000), in legumes only moderate to low levels of incomplete resistance of complex inheritance have been identified against *Orobanche*. Nevertheless, the existence of monogenic resistance against a parasitic plant included in a different genus (*Striga gesnerioides*) has been reported in the leguminous crop cowpea (*Vigna ungiculata)* (Lane et al. 2006). The development of improved cultivars with monogenic resistance is often straightforward if a good source of resistance is available and an efficient, easily controlled and practical screening procedure exists to provide sufficient selection pressure. However, breeding programs based on only a few dominant genes are in serious risk of resistance breakdown. On the other hand, the multigenic-quantitative systems controlling resistance dramatically slow down the breeding process. Finally, breeding programs for broomrape resistance rely primarily on field evaluations where natural parasite populations occur and homogeneity of seed distribution in the soil is difficult to reach. The environment can also affect the infectious ability of the parasite and although evaluations in controlled environments could avoid these difficulties, they are expensive, difficult to manage and frequently do not correlate with field data. In this sense *in vitro* or greenhouse screening methods based on the reaction of root and/or foliar tissues are usually superior to field screenings, being necessary to demonstrate a strong correlation between *in vitro* and field data to utilize them in plant breeding (Cubero et al. 1994a). The correlation should moreover be calculated for every environment in which selection is practiced. These deficiencies in the evaluation method also make broomrape resistance a difficult trait to manipulate in legume breeding programs.

In spite of these difficulties, which clearly make broomrape resistance a difficult trait to analyze and have slowed down the breeding process, the detection of partial quantitative resistance within genotypes of crop germplasm collections allowed further development of genetic crop improvement. Although the introduction of quantitative resistance needs of tedious selection procedures, in the last decades it has resulted in the release of legume cultivars with useful levels of incomplete resistance combined with a degree of tolerance. The resulting resistance, which might be based on a combination of resistance mechanisms, is more likely to last longer than resistances based on a single gene.

No single method would alone lead to a long term solution of the problem posed on agricultural crops by root parasitic weeds. A comprehensive approach, employing a variety of methods, with an emphasis on methods that are specifically developed to neutralize key developmental processes in the parasite, should lead to a reasonable management strategy that will reduce crop damage and suppress the development of virulent seed bank in the fields (Joel 2000).

11.4 Sources of Resistance and Breeding for Resistance

A faba bean line showing a high degree of resistance to *O. crenata* was identified by Egyptian scientists early in the 1970s (Nasib et al. 1982) and subsequently confirmed in laboratory and field experiments. The resistant line F402 showed a high level of field resistance as well as agronomically favorable characteristics. It was also possible to perform crosses to study the genetics of resistance by using a wider variation range. F402 proved to be resistant under both greenhouse and field conditions and under very severe conditions of both natural and artificial infestations. Subsequently, F402 gave origin to the cultivar "Giza 402". The Line VF1071, selected from the Egyptian cultivar Giza 402 under field conditions in southern Spain, was crossed with the Spanish cultivars Alameda and Brocal to produce the cultivar Baraca, a well-adapted, high yielding cultivar that was registered (Cubero et al. 1992). The strategy followed combined recurrent and pedigree selection methods, the latter under insect-proof cages (Cubero 1994b). Baraca shows a higher degree of field resistance than VF1071, very likely because some genes from Alameda, known to be a tolerant cultivar, were transferred during the selection process in heavily infested plots. Experimental lines leading to Baraca proved to be stable under different conditions in southern Spain, Syria and Morocco. The fact that Baraca showed a somewhat higher degree of resistance than the resistant parent line highlight the importance of incorporating any kind of resistant and/or tolerant material in a breeding program. Even lines showing a lower degree of *in vitro* parasitic seed germination should be included in a crossing program as they may carry genes controlling certain steps in a long chain leading to field resistance. Lines such as ICARDA BPL 2210 (Aalders and Pieters 1987), VF172 (Cubero 1973) or landraces such as Locale di Castellano (Perrino et al. 1988) should be regarded as sources of useful, although minor genes.

Resistance against *O. foetida* has been identified in faba bean germplasm in Tunisia and a resistant cultivar (Bader) is in the process of registration by INRAT in Tunisia. One of the parents of this cultivar was a selection from an advanced breeding line provided by ICARDA with resistance derived from Giza 402. Some lines selected against *O. crenata* had also shown a high level of resistance to *O. foetida* and showed a high yield potential (Kharrat and Halila 1994).

In the case of pea varieties, no resistance against *O. crenata* is available and only incomplete resistance was found in pea germplasm (Rubiales et al. 2003a, 2005a). The low levels of broomrape resistance present in these lines would be enough to retard broomrape attachment and/or development in normal years, but not in favorable years and in heavily infested fields. Various levels of resistance have been reported in accessions of the subspecies of *P. sativum* namely: *sativum*, *abyssinicum*, *arvense* and *elatius* and in *P. fulvum* (Rubiales et al. 2005a). All these accessions have been successfully crossed with pea cultivars and included in a breeding program (Rubiales et al. 2009).

Orobanche crenata also attacks the common vetch (*Vicia sativa* L.). In contrast to the situation with faba bean, broomrape was never a serious problem in this crop, although there were obvious differences among landraces. However a search for resistance was undertaken to fulfill a double aim: to develop resistant cultivars and to study the genetics of resistance to the parasite for its use as a model in faba bean. This latter objective is based both on the phylogenetic proximity of *V. faba* to *V. sativa* and on the fact that *V. sativa* is a strict self-pollinating species, whereas *V. faba* is a partial outcrossing species and hence more difficult to handle. Some strongly resistant lines have been selected in common vetch following an individual plant selection method after screening of a wide germplasm collection (Martín et al. 1982; Gil et al. 1987). Lines derived from many accessions showed complete resistance under field conditions. The resistance was also tested in a greenhouse with artificial inoculation and some lines had stable resistance under a wide range of inoculum dose. Three common vetch cultivars adapted to southern Spain conditions with good forage and grain production and resistance to broomrape were obtained and registered (Martín et al. 1982). Resistance to *O. crenata* has also been reported in local cultivars of common vetch in Syria (Linke et al. 1993).

Whereas common vetch is susceptible to *O. aegyptiaca* and *O. crenata*, purple vetch (*V. atropupurea*) and hairy vetch (*V. villosa*) are resistant. The resistance of *V. atropupurea* varieties "Sadot" and "Popany" to *Orobanche* is evident in spite of the fact that both induce a high rate of *Orobanche* seed germination. The resistance results from the inability of the parasite haustorium to penetrate into the vascular cylinder of the host root and connect with the host vascular system. A chemical and/or mechanical barrier active at the endodermis cell layer seem to obstruct the intruding haustorium (Goldwasser et al. 1999).

Little is known about the levels of resistance to *O. crenata* available in narbon vetch (*V. narbonensis* L.). A germplasm collection of 200 accessions from different countries was screened for resistance to crenate broomrape under field conditions. Thirteen accessions were selected for more detailed screening under controlled conditions and for additional field testing.

Resistance to *O. crenata* was manifested by lower germination of broomrape seeds, reduced emergence of *Orobanche* shoots and fewer root attachments per narbon vetch plant. Differences in pod yield among narbon vetch accessions were also found in response to *Orobanche* infestation. Finally, eight accessions were selected and identified as new sources of narbon vetch resistant to broomrape (Nadal et al. 2007).

Resistance has not been identified in cultivated grass pea (*Lathyrus sativus* L.) and chickling pea (*Lathyrus cicera* L.), both susceptible to broomrape (*O. crenata*). However, high levels of resistance to *O. crenata* have been found in other *Lathyrus* species, particularly *L. ochrus* and *L. clymenum* (Linke et al. 1993; Sillero et al. 2005). In both species, the main mechanism of resistance seems to be an early barrier to the establishment of the broomrape as none or few broomrape tubercles were recorded under field conditions (Sillero et al. 2005).

Although resistance to this parasitic weed is scarce in many legumes it is common in chickpea (*Cicer arietinum* L.) germplasm. A detailed *in vitro* study has shown that resistance in chickpea is the result of the combination of at least two mechanisms (Rubiales et al. 2003c). First and most important, the two chickpea lines studied presented low rates of stimulant production. Once germination is induced by exogenous applications of the synthetic germination stimulant GR24, thus overcoming the primary resistance mechanism in these lines, a second resistance mechanism is apparent. This is evidenced by a darkening of host cell tissue in contact with the broomrape radicle, leading to failure of establishment, which was frequently observed in the chickpea accessions. Anatomical studies have shown that this apparently "hypersensitive" response does not correspond with the death of host cells in contact with the parasite cells but corresponds to blocking and death of the penetration structures of the parasite.

Although no resistance against *O. crenata* was reported in lentils (Sauerborn et al. 1987; Khalil and Erskine 1999), recently, complete resistance has been observed in wild lentils (*Lens ervoides*, *L. odemensis*, and *L. orientalis*) (Fernández-Aparicio et al. 2009b). Nevertheless, only partial resistance against *O. crenata* was detected in a Spanish germplasm collection of *L. culinaris* (Fernández-Aparicio et al. 2008).

In germplasm collections there is variation for broomrape resistance. Landraces have been widely replaced by homogeneous cultivars that may not possess the combined tolerance and resistance already accumulated by natural selection in land races or wild accessions. Genetic resources remain highly unexplored and underused and might contain very valuable additional sources of resistance. Understanding the escape and resistance factors will help to detect existing genetic diversity for mechanisms that hamper infection.

Likewise, the combination of different resistance mechanisms into a single cultivar will provide a durable outcome. This can be facilitated by the use of *in vitro* screening methods that allow the identification of highly heritable resistance components. In general, the parasite can be stopped at three different levels after attempted penetration in resistant host roots: first, in the root cortex observed in resistant faba bean, pea and chickpea (Rubiales et al. 2003c; Pérez-de-Luque et al. 2005); second, at the endodermis observed in faba bean and vetch (Pérez-de-Luque et al. 2005, 2007) and finally, inside the central cylinder observed in *Medicago truncatula* (Lozano-Baena et al. 2007).

11.5 Molecular Analyses of the Host-Parasite Interactions

Molecular mapping of broomrape resistance QTLs together with the development of MAS techniques (see Section 11.2.3) are promising approaches to rapidly improve crop resistance, since screening for this trait is difficult, expensive and sometimes unreliable. With the emergence of large-scale genomic tools, the combination of genetic mapping with gene expression studies can offer an integrated approach to study resistance to parasitic plants. In this sense, testing the role of candidate genes selected from expression experiments, can simplify the search of sequence polymorphisms for a more efficient and rapid MAS strategy:

As suggested previously (see Section 11.4), the best long-term strategy to control parasitic plants is the development of resistant (or to some extent partially resistant) varieties by pyramiding of the minor resistance genes (pyramiding of multiple resistance mechanisms) into locally adapted cultivars. To achieve this, it is crucial identifying key genes that govern resistance to parasitic plants. Thus, in recent years information on host and non-host responses at the molecular level to *Orobanche* spp. has been accumulated.

The recent advances on identification of host plant genes are based on the suitability of using model plants that have the potential to establish the parasitic interaction like *Nicotiana tabacum* (Foy et al. 1989), or *Arabidopsis thaliana* (Goldwasser et al. 2000; Westwood 2000). Rodríguez-Conde et al. (2004) have reported a resistant *M. truncatula* collection available for the study of parasitism with different mechanisms of resistance against *O. crenata* (essentially identification of genotypes that allow and do not allow the establishment of the parasite, or in the same way, incompatible interactions appearing before and after the development of the parasite haustorium) increasing the *M. truncatula* value as a resource for understanding legume resistance against *O. crenata* (Lozano-Baena et al. 2007). The suitability of *Lotus japonicus* as a model to study *O. aegyptyaca* parasitism has been also recently reported (Kubo et al. 2009).

11.5.1 *Targeted-Gene Expression Approaches*

Most of the gene expression analysis performed in host plant-parasite interaction has consisted of the study of single or few genes, mainly based on the knowledge already gained from other plant stress studies. Although the biosynthesis pathways involved in defense mechanisms or how these pathways are regulated remain still unclear, and molecular studies of host plant-*Orobanche* interactions are still limited to a few reports, it is probable that the dissection of the molecular response in the host will be increased rapidly and data gained on model systems will accelerate our understanding of the legume-*Orobanche* interactions.

The expression of a β-glucuronidase (GUS) reporter gene fused to the promoter of the PR gene *PR-1b* was observed in tobacco roots infected with *O. aegyptiaca* at the infection site indicating that *Orobanche* releases appropriate elicitors and that tobacco plants do sense *Orobanche* invasion, in spite of the fact that they do not resist haustorial invasion (Joel and Portnoy 1998). Moreover, Westwood et al. (1998) demonstrated that parasitization by *Orobanche* induced the expression of *hmg2*, a defense-related isogene of 3-hydroxy-3methylglutaryl CoA reductase (HGMR) in transgenic tobacco as early as 1 day after root penetration. This protein is involved in isopropenoid biosynthesis pathway and is activated specifically during defense responses associated with phytoalexins and the production of sesquiterpenes. The enhanced expression was spread in the cortical and vascular tissues of the host, around the penetration site and persisted on infested roots for 4 weeks. The demonstrated induction of a pathogenesis-related protein PR1 gene in response to *Orobanche* attack (Joel and Portnoy 1998) together with the activation of a *hmg2* promoter (Westwood et al. 1998), indicates that *Orobanche* parasitization may trigger a broad array of defense responses that are common to pathogen attack such as increased production of phytoalexins, lignin precursors, and lytic enzymes (Lamb et al. 1989). The expression patterns of *hmg2* and a second isogene encoding HMGR, *hmg1*, has been analyzed in tomato roots parasitized with *O. aegyptiaca* showing a coordination of metabolic events in the isoprenoid pathway promoting the alteration of gene expression in the host plant by activating *hmg2* and repressing the squalene synthase involved in the general metabolism, thus mobilizing the host energy toward the production of defense compounds (Griffitts et al. 2004).

An accurate kinetic study of the gene expression pattern in host roots from the earliest contact with *Orobanche* seed to the installation of a functional haustoria, could also contribute to better understand the mechanisms involved in the host defense against the parasite. In this sense, Vieira Dos Santos et al. (2003a) studied the effect of *O. ramosa* on the expression patterns of *A. thaliana* genes that were known to be involved in metabolic pathways and resistance mechanisms. Along different time points, several defense genes involved in the jasmonate (JA) and Ethylene signaling pathways were up-regulated in response to germinating *O. ramosa* seeds. This response was detected even before the attachment of the parasite indicating that signaling from the *Orobanche*-germinated seed is recognized by the host and this is able to activate several defense reactions against its aggressor. However, genes mediated by a salicylic acid (SA)-dependent pathway were not activated. The issue concerning plant responses pathways involved in the activation of defenses against *Orobanche* spp. is still open and will need further research. Histological studies aimed to elucidate the effects of SA and JA on resistance to *Orobanche* have been recently reported. The few studies that have addressed host responses to parasitic plants suggest that both salycilates and jasmonates can mediate effective defenses even though the roles that plant hormones play in complex signaling pathways require further characterization (Smith et al. 2009). For example, Kusumuto et al. (2007) have shown that the pretreatment of red clover plants with chemical defense-inducers leading to the activation of SA-dependent responses inhibited *Orobanche minor* parasitism, whereas exogenous induction of JA-dependent responses had little effect. However with a different experimental system, Bar-Nun and Mayer (2008) have reported a completely different result suggesting that methyl jasmonate could induce almost full resistance in *A. thaliana* to the infection by *O. aegyptiaca*. The reason for this discrepancy is not clear, but seems indicate that both experimental pathosystem and the way of applying the compounds might cause differences in the response. However, these observations highlight that if the right set of defense mechanisms is activated in the host, the infection process may be prevented. Additional studies are needed since the identification of key regulator genes leading to the full signaling pathway induction may be useful in engineering appropriate resistance mechanism in economically important crops.

11.5.2 Non-Targeted Gene Expression Approaches

High-throughput expression profiling technologies are significantly transforming molecular genetics approaches in the field of plant-parasitic plant interactions. The emergence of the "omic" technologies is providing a more integrated view of biological processes. One prerequisite for a successful application of genomic research based on high-throughput technologies is the availability of model systems and their vast available information in data bases. During this decade considerable advance in our knowledge about the molecular changes in the host during *Orobanche* parasitism has been gained through the use of genomic approaches for the model species *A. thaliana*, and especially for legumes, thanks to *M. truncatula* and *L. japonicus*.

Strategies such as suppression subtractive hybridization (SSH) cDNA libraries require no prior knowledge of the sequences of the transcripts and can discover previously unknown transcripts. The technique combines normalization and subtraction in a single procedure (Diatchenko et al. 1999). There are several examples demonstrating the power of suppressive subtractive approaches to identify differentially expressed genes (Doll et al. 2003; Brechenmacher et al. 2004; Frenzel et al. 2005).

Owing to the potential of identifying rare transcripts or genes expressed locally, SSH is usually not redundant but quite complementary to cDNA microarray-based transcriptome profiling. Combining both, SSH-cDNA libraries construction and array-based transcriptome technologies represent a comprehensive basis for elucidating the global genetic program activated during *Orobanche* parasitism. In this sense, several genes were isolated from *A. thaliana* roots inoculated with *O. ramosa* (Vieira-Dos-Santos et al. 2003b). The molecular analysis of resistance mechanisms to *O. cumana* in sunflower has also been studied using this strategy and time-course expression of the cDNAs identified revealed qualitative differences between the resistant and susceptible genotypes analyzed (Letousey et al. 2007). Mechanical barriers to infection are common resistance factors to microbial pathogens. The authors have correlated callose depositions and an encapsulation layer preventing entry of the parasite in the sunflower resistant genotype LR1 with the regulation of a defensin gene using the SSH approach. Moreover, they have observed little evidence for induction expression of genes associated with the development of a hypersensitive response (HR) such as reactive oxygen species, NADHP oxidases and peroxidases. Although further characterization of this response at the molecular level is necessary, this work and some other initial reports (Perez-de-Luque et al. 2005) seem to show a questionable HR-like response as member of the defense response of host to *Orobanche* spp. infection.

More recently, this approach has been used in one model legume *L. japonicus*. Since Kubo et al. (2009) reported that *L. japonicus* is a suitable host for the study of the parasitism in plants, research for understanding host plant resistance at molecular level has been initiated. Hiraoka et al. (2009) have shown through a SSH-cDNA library construction the expression of genes involved in JA biosynthesis, some of which were systemically expressed. Moreover, the fact that genes involved in nodulation were interestingly induced might suggest that *L. japonicus* is recognizing the parasite as a symbiont similar to *Rhizobium*.

Our group has focused its research on the elucidation of the molecular mechanisms underlying resistance to *O. crenata* in legumes. In order to obtain information about specific mechanisms mediating the defense response, a cDNA-SSH library was developed using the model legume *M. truncatula* (Die et al. 2007). The library has allowed the identification of a collection of EST-sequences, most of them involved in different plant response pathways, activated during several plant-pathogen interactions. However, 18% of the sequences seem to be novel sequences and are therefore good candidates for further research in order to gain insight into novel activated defense pathways. The determination of the specific role of candidate genes selected by their transcriptomic patterns is possible using functional approaches like TILLING collections. We are examining, through phenotypic analysis combined with transcriptomic characterization of different TILLING mutants, the function of a differentially expressed gene identified in the SSH library in the phenylpropanoid pathway-mediated associated with the lignin biosynthesis.

In parallel, candidate genes for *Orobanche* resistance provided by the SSH approach have been analyzed in key stages of the parasitism process in both, *M. truncatula* and *Pisum sativum*. Results reflected the usefulness of gene expression studies in dissecting the molecular mechanism governing resistance to parasitic plants (Die et al. 2009a, b). Our EST collection combined with the results of the transcriptome profiling based on microarray from Dita et al. (2009) represent a comprehensive knowledge on host legumes responses to *O. crenata* at the molecular level and can provide support to conducting breeding strategies for resistance in the future. In order to understand the regulation of the defense response, a genome-wide identification assay, testing more than 1,000 transcription factors (TFs) through high-throughput quantitative RT-PCR, has been also carried out by our group using an available platform for massive analysis of *M. truncatula* TF's expression (Kakar et al. 2008). Analysis of data is currently underway but preliminary results show quantitative and qualitative differences in two *Medicago* genotypes differing in their sensitivity to *O. crenata*.

The combination of different approaches like genomics and functional analysis in model legumes will accelerate the characterization of molecular mechanisms involved in the resistance to *Orobanche*. The high synteny reported between *Medicago* and important legume crops should facilitate positional cloning providing new markers and targets for legume breeding programs (VanderBosch and Stacey 2003; Young et al. 2005; Varshney et al. 2009).

11.6 Diversity Studies

Understanding the variability within and among weedy broomrape populations is particularly important in any attempt to develop resistance breeding strategies in relevant host crops against these parasites. Similarly, the existence of host specialization and parasite races is also of vital importance for the development of resistant cultivars. Information on population genetics in broomrape species is limited, especially with regard to molecular analysis, mainly because of the lack of suitable co-dominant markers such as single sequence repeats (SSRs, i.e., microsatellites), limited geographical scope of sampling and rather low sample sizes of the analyzed populations. Variability within and among populations of some species of *Orobanche* has been observed on the morphological level (Musselman and Parker 1982; Musselman 1994), and also on the host-range and host-preference levels. Molecular analyses on the genetic relationships among species of *Orobanche* have been carried out, first by using RAPD markers (Román et al. 2003) and later by sequencing the internal transcribed spacer (ITS) region of the nuclear rDNA (Schneeweiss et al. 2004; Wolfe et al. 2005) and specific genes of the plastid genome (Manen et al. 2004). The amount and structure of population genetic diversity as a consequence of the mating system, migration, genetic drift and host differentiation, has been recently reviewed by Satovic et al. (2009).

Molecular marker analyses carried out among *O. crenata* populations from Egypt (Zeid et al. 1997), Israel (Paran et al. 1997; Román et al. 2002), Syria (Verkleij et al. 1991a) and Spain (Verkleij et al. 1991b; Román et al. 2001, 2002) have generally revealed low inter-population differentiation within countries. Román et al. (2001), in their analysis of individual *O. crenata* plants collected *in situ* from six faba-bean fields in Andalusia (South of Spain), detected that as high as 95% of the genetic variability was attributable to differences among individuals within a population. The obvious conclusion was that in spite of geographical distances between populations, the gene flow was continuous and strongly favored by an efficient dispersal of the parasite seeds by humans, machinery, animals or wind, as well as on host seeds. However, the genetic differentiation among *Orobanche* populations from distant regions such as Spain and Israel, estimated by using ISSR markers (Román et al. 2002), revealed significant genetic differentiation and considerably high variation within populations in both countries, indicating that geographic distance basically provides a substantial barrier to gene flow as long as there is no commercial exchange of host seeds between the regions. As increasing global crop-seed exchange and transport play an important role in *Orobanche* seed migration, geographical differentiation would be difficult to discern because the population substructure may not depend on spatial distances or barriers for gene flow, that are common in natural ecosystems.

Considering molecular variation according to the parasitized host, when analyzing populations of *O. aegyptiaca* attacking five different crops (eggplant, vetch, tomato, chickpea and faba bean) by using RAPD markers, Paran et al. (1997) found no evidence for differentiation due to the host species, as shown by the high similarity between the mean Jaccard's distance of plants collected from the same host and those collected from different hosts. The same was true for populations of *O. crenata* collected from faba bean, vetch and carrot fields. Hence, as both parasites are outcrossing species, most of the variations were detected among individuals within populations. Nevertheless, host-induced selection could play an important role in population dynamics of these species without even being noticed by overall genome scans. Theoretically, the selection could act upon just a single gene or very small portions of the genome while the rest of the genome is predominantly shaped by other evolutionary forces.

Molecular markers can also assist diagnostic studies in the *Orobanche* genus. Broomrape seeds are very small in size and therefore cannot be easily identified in infested fields or in seed lots. Early detection of the parasite prior to its germination is important for agriculture because *Orobanche* seeds germinate only in the presence of a susceptible host. Because of its miniscule seed size, contamination of fields and legume seed lots by broomrape seeds is difficult to detect and confirm via conventional methods. In addition, because differences in host preferences were noted among *Orobanche* species, early detection of seeds of a certain *Orobanche* species in infested soils can be important in directing the choice of crop to be cultivated. In the case of contaminated seed lots, the identification of parasitic seeds could avoid their further spread to parasite-free areas. However, practical diagnosis of the species that are serious weeds in agricultural fields has been problematic, particularly the analyses of soil seed banks, not only because the seed-coat texture of some species does not show a sufficient number of differential features, but also because the seeds tend to lose diagnostic surface features when buried in soil for a long period of time, resulting from biotic activities (Joel et al. 1998a). RAPD (Katzir et al. 1996) and SCAR markers (Joel et al. 1998b) can be used for the identification of single tiny seeds of the various *Orobanche* species, and Portnoy et al. (1997) applied these methods for the identification of soil-borne *Orobanche* seeds. Single seeds of *O. minor* infecting alfalfa, red and white clover, can similarly be detected by a rapid and reliable PCR-assay consisting of the development of primers based upon unique sequences in the internal transcribed spacers (ITS) regions of the nuclear ribosomal DNA (Osterbauer and Rehms 2002). Román et al. (2007) proposed a molecular protocol to distinguish the main parasitic plants in crop fields of southern Spain by amplifying a non-coding region from the plastid genome and detecting sequence differences among the amplified fragments.

As mentioned before, unfortunately, in many crops no resistant varieties have been produced to date. Control strategies based on a single treatment are often only partially or inconsistently effective and are affected by environmental conditions. Overall, control strategies have not proven to be as effective, economical and applicable as desired. Current means for controlling parasitic weeds are focusing on reducing the soil seed bank, preventing seed set and inhibiting seed movement from infested to non-infested areas. In this sense, the diagnosis and quarantine means and the assistance of all those described techniques could be of utmost importance.

References

Aalders AJG, Pieters R (1987) Resistance in *Vicia faba* to *Orobanche crenata*: true resistance versus hidden susceptibility. Euphytica 36: 227–236.

Aggarwal VD, Ouedraogo JT (1989) Estimation of cowpea yield loss from *Striga* infestation. Trop Agri 66: 91–92.

Al-Menoufi OA (1991) Crop rotation as a control measure of *Orobanche crenata* in *Vicia faba* fields. In: K Wegman, LJ Musselman (eds) Progress in *Orobanche* Research. Publ Ebenhard-Karts-Universitat, Tubengen, Germany, pp 241–247.

Amrhein N, Deus B, Gehrke P, Steinrucken HC (1980) The site of the inhibition of the shikimate pathway by glyphosate. II. Interference of glyphosate with chorismate formation *in vivo* and *in vitro*. Plant Physiol 66: 830–834.

Amsellem Z, Zidack NK, Quimby PC, Gressel J (1999) Long term dry preservation of active mycelia of two mycoherbicidal organism. Crop Protec 18: 643–649.

Bar-Nun N, Mayer AM (2008) Methyl jasmonate and methyl salicylate, but not cis-jasmone, evoke defenses against infection of *Arabidopsis thaliana* by *Orobanche aegyptiaca*. Weed Biol Manag 8: 91–96.

Bernhard RH, Jensen JE, Andreasen C (1998) Prediction of yield loss caused by *Orobanche* spp. in carrot and pea crops based on the soil seedbank. Weed Res 38: 191–197.

Carter RJ, Cooke DA, Barker WR, Csurher SM (1996) Occurrence and control of *Striga*, *Orobanche* and *Cuscuta* in Australia. In: MT Moreno, JI Cubero, D Berner, D Joel, LJ Musselman, C Parker (eds) Advances in Parasitic Plant Research. Consejería de Agricultura y Pesca, Junta de Andalucía, Sevilla, Spain, pp 801–808.

Colquhoun JB, Eizenberg H, Mallory Smith CA (2006) Herbicide placement site affects small broomrape (*Orobanche minor*) control in red clover (*Trifolium pratense*). Weed Tecnol 20: 356–360.

Cubero JI (1973) Resistance to *Orobanche crenata* Forsk in *Vicia faba*. In: Parasitic Weeds Research Group (eds) Proc Eur Weed Research Council Symposium on Parasitic Weeds. Malta University Press, Malta, pp 205–217.

Cubero JI (1991) Breeding for resistance to *Orobanche* species: a review. In: K Wegmann, LJ Musselman (eds) Progress in *Orobanche* Research. Eberhard-Karls-Universität, Tübingen, Germany, pp 257–277.

Cubero JI (1994b) Breeding work in Spain for *Orobanche* resistance in faba bean and sunflower. In: AH Pieterse, JAC Verkleij, SJ ter Borg (eds) Biology and Management of *Orobanche*. Proc 3rd Int Workshop on *Orobanche* and Related *Striga* Research. Royal Tropical Institute, Amsterdam, The Netherlands, pp 465–473.

Cubero JI, Moreno MT (1996) Parasitic plant science: a quarter century. In: MT Moreno, JI Cubero, D Berner, DM Joel, LJ Musselman, C Parker (eds) Advances in Parasitic Plant Research. Consejería de Agricultura y Pesca, Junta de Andalucía, Sevilla, Spain, pp 15–23.

Cubero JI, Moreno MT, Hernández L (1992) A faba bean (*Vicia faba* L.) cultivar resistant to broomrape (*Orobanche crenata* Forsk.). In: Proc 1st Eur Conf on Grain Legumes. Association Européenne des Protéagineux, Angers, France, pp 41–42.

Cubero JI, Pieterse AH, Khalil SA, Sauerborn J (1994a) Screening techniques and sources of resistance to parasitic angiosperms. Euphytica 73: 51–58.

Diatchenko L, Lukyanov S, Lau YF, Siebert PD (1999) Suppression subtractive hybridization: a versatile method for identifying differentially expressed genes. Meth Enzymol 303: 349–380.

Die JV, Dita MA, Krajinski F, González-Verdejo CI, Rubiales D, Moreno MT, Román B (2007) Identification by suppression hybridization and expression analysis of *Medicago truncatula* putative defence genes in response to *Orobanche* parasitization. Physiol Mol Plant Pathol 70: 49–59.

Die JV, González-Verdejo CI, Dita MA, Nadal S, Román B (2009a) Gene expression analysis of molecular mechanisms of defense induced in *Medicago truncatula* parasitized by *Orobanche crenata*. Plant Physiol Biochem 47: 635–641.

Die JV, Román B, Nadal S, Dita MA, González-Verdejo CI (2009b) Expression analysis of *Pisum sativum* putative defence genes during *Orobanche crenata* infection. Crop Pasture Sci 60: 490–498.

Dita MA, Die JV, Román B, Krajinski F, Küster H, Moreno MT, Cubero JI, Rubiales D (2009) Gene expression profiling of *Medicago truncatula* roots in response to the parasitic plant *Orobanche crenata*. Weed Res 49: 66–80.

Dor E, Plakhine D, Herschenhorn J (2001) Phytopathogenic fungi of the parasitic plant Egyptian broomrape (*Orobanche aegyptiaca*). Phytoparasitica 29: 272–273.

Eizenberg H, Colquhoun J, Mallory-Smith CA (2004) The relationship between temperature and small broomrape (*Orobanche minor*) parasitism in red clover (*Trifolium pratense*). Weed Sci 52: 735–741.

Emerchebe AM, Singh BB, Leleji OI, Atokple IDK, Adu JK (1991) Cowpea-*Striga* problems and research in Nigeria. In: SK Kim (ed) Combating *Striga* in Africa. IITA, Ibadan, Nigeria, pp 18–28.

Fernández-Aparicio M, Sillero JC, Rubiales D (2007) Intercropping with cereals reduces infection by *Orobanche crenata* in legumes. Crop Protec 26: 1166–1172.

Fernández-Aparicio M, Sillero JC, Pérez de Luque A, Rubiales D (2008) Identification of sources of resistance to crenate broomrape (*Orobanche crenata*) in Spanish lentil (*Lens culinaris*) germplasm. Weed Res 48: 85–94.

Fernández-Aparicio M, Sillero JC, Rubiales D (2009a) Resistance to broomrape species (*Orobanche* spp.) in common vetch (*Vicia sativa* L.) Crop Protec 28: 7–12.

Fernández-Aparicio M, Sillero JC, Rubiales D (2009b) Resistance to broomrape in wild lentils (*Lens* spp.). Plant Breed 128: 266–270.

Fernández-Martínez JM, Melero-Vara J, Muñoz-Ruz J, Ruso J, Domínguez J (2000) Selection of wild and cultivated sunflowers for resistance to a new race of broomrape which overcomes resistance of the *Or5* gene. Crop Sci 40: 550–555.

Foy CL, Jain R, Jacobsohn R (1989) Recent approaches for chemical control of broomrape (*Orobanche* spp.). Rev Weed Sci 4: 123–152.

García-Torres L, Jurado-Expósito M, Díaz-Sánchez J, Castejón-Muñoz M, López-Granados F (1996) Cultive bien guisantes. Agricultura 65: 755–759.

García-Torres L, López-Garnados F, Jurado-Expósito M, Díaz-Sánchez J (1998) The present state of *Orobanche* spp. Infestations in Andalusia and the prospects for its management. In: J Maillet (ed) Comptes rendu 6ème Symp Méditerranéen EWRS. ENSA, Montpellier, France, pp 141–145.

Gil J, Martín LM, Cubero JI (1987) Genetics of resistance in *V. sativa* to *O. crenata* Forsk. Plant Breed 99: 134–143.

Goldwasser Y, Hershenhorn J, Plakhine D, Kleifeld Y, Rubin B (1999) Biochemical factors involved in vetch resistance to *Orobanche aegyptiaca*. Physiol Mol Plant Pathol 54: 87–96.

Goldwasser Y, Plakhine D, Yoder JI (2000) *Arabidopsis thaliana* susceptibility to *Orobanche* spp. Weed Sci 48: 342–346.

Griffitts A, Cramer CL, Westwood JH (2004) Host gene expression in response to Egyptian broomrape (*Orobanche aegyptiaca*). Weed Sci 52: 697–703.

Hameed KM, Saadoun IM, Al-Shyab Z (2001) Potential biological control of *Orobanche* by fungi isolated from diseased specimens in Jordan. Plant Pathol J 17: 257–263.

Hassan EA (1998) Broomrape species in Egypt, a recent survey in relation to geographical distribution. In: J Maillet (ed) Comptes rendu 6ème Symposium Méditerranéen EWRS. ENSA, Montpellier, France, pp 13–15.

Hiraoka Y, Ueda H, Sugimoto Y (2009) Molecular responses of *Lotus japonicus* to parasitism by the compatible species *Orobanche aegyptiaca* and the incompatible species *Striga hermonthica*. J Exp Bot 60: 641–650.

Hussey BMJ, Kieghery GJ, Cousens RD, Dodd J, Lloyd SG (1997) Western weeds, a guide to the weeds of Western Australia. The Plant Protection Society of Western Australia, Perth, Australia, 188 pp.

Jacobsohn R (1994) The broomrape problem in Israel and an integrated approach to its control. In: AH Pieterse, JAC Verkleij, SJ ter Borg (eds) Biology and Management of *Orobanche*. Proc 3rd Int Workshop on *Orobanche* and Related *Striga* Research. Royal Tropical Institute. Amsterdam, The Netherlands, pp 652–658.

Jacobsohn R, Kelman Y, Shaked R, Klein L (1988) Broomrape (*Orobanche* spp.) control with ethylene dibromide and chloropicrin. Weed Res 28: 151–157.

James RW, Frater KC (1977) The control of broomrape (*Orobanche minor*) in flue-cured tobacco. New Zealand Tobacco Growers Journal pp 10–13.

Joel DM (2000) The long term approach to parasitic weeds control: manipulation of specific developmental mechanisms of the parasite. Crop Protec 19: 753–758.

Joel DM, Portnoy VH (1998) The angiospermous root parasite *Orobanche* L. (Orobanchaceae) induces expression of a pathogenesis related (PR) gene in susceptible tobacco roots. Ann Bot 81: 779–781.

Joel DM, Portnoy V, Tzuri G, Katzir N (1998a) Use of DNA fingerprinting for soil-borne seed identification. Aspects Appl Biol 51: 23–27.

Joel DM, Benharrat H, Portnoy VH, Thalouarn P (1998b) Molecular markers for *Orobanche* species—New approach and their potential uses. In: K Wegmann, LJ Musselman, DM Joel (eds) Current problems of *Orobanche* researches, Proc 4th Int Workshop on *Orobanche*. Institute for Wheat and Sunflower, Albena, Bulgaria, pp 115–124.

Joel DM, Hershenhorn Y, Eizenberg H, Aly R, Ejeta G, Rich PJ, Ransom JK, Sauerborn J, Rubiales D (2007) Biology and management of weedy root parasites. Hort Rev 33: 267–349.

Jurado-Expósito M, García-Torres L, Castejón-Muñoz M (1997) Broad bean and lentil seed treatments with imidazolinones for the control of broomrape (*Orobanche crenata*) J Agri Sci 129: 307–314.

Jurado-Expósito M, Castejón-Muñoz M, García-Torres L (1999) Uptake and translocation of imazethapyr in peas as affected by parasitism of *Orobanche crenata* and herbicide application methods. Weed Res 39: 129–136.

Kakar K, Wandrey M, Czechowski T, Gaertner T, Scheible WR, Stitt M, Torres-Jerez I, Xiao Y, Redman JC, Wu HC, Cheung F, Town CD, Udvardi MK (2008) A community resource for high-throughput quantitative RT-PCR analysis of transcription factor gene expression in *Medicago truncatula*. Plant Meth 4: 18.

Katzir N, Portnoy V, Tzuri G, Castejón-Muñoz M, Joel DM (1996) Use of random amplified polymorphic DNA (RAPD) markers in the study of the parasitic weed *Orobanche*. Theor Appl Genet 93: 367–372.

Khalil S, Erskine W (1999) Breeding for *Orobanche* resistance in faba bean and lentil. In: JI Cubero, MT Moreno, D Rubiales, J Sillero (eds) Resistance to *Orobanche:* The State of the Art. Junta de Andalucía, Spain, pp 63–76.

Kharrat M (1999) *Orobanche* research activities on faba bean in Tunisia. In: JI Cubero, MT Moreno, D Rubiales, J Sillero (eds) Resistance to Orobanche: The State of the Art. Junta de Andalucía, Spain, pp 77–81.

Kharrat M (2002) Etude de la virulence de l'ecotype de Béja d'Orobanche foetida sur différentes espèces de légumineuses. In: M Kharrat, F Abbad Andaloussi, MEH Maatougui, M Sadiki, W Bertenbreiter (eds) Devenir des l´egumineuses alimentaires dans le Maghreb. GTZ, Tunis, Tunisia, p 89.

Kharrat M, Halila MH (1994) *Orobanche* species on faba beans (*Vicia faba* L.) in Tunisia: Problems and management. In: AM Pieterse, JAC Verkleij, SJ ter Borg (eds) Biology and Management of *Orobanche*. Proc 3rd Int Workshop on *Orobanche* and related *Striga*. Royal Tropical Institute, Amsterdam, The Netherlands, pp 639–643.

Kharrat M, Halila MH, Linke KH, Haddar T (1992) First report of *Orobanche foetida* Poiret on faba bean in Tunisia. FABIS Newsl 30: 46–47.

Kleifeld Y, Goldwasser Y, Plakhine D, Eizenberg H, Herzlinger G, Golan S (1999) Selective control of *Orobanche* spp. in various crops with sulfonylurea and imidazolinones herbicides. In: J Kroschel, M Abderabihi, H Betz (eds) Advances in parasitic weed control at on-farm level, vol 2. Joint Action to Control *Orobanche* in the WANA Region. Margraf Verlag, Weikersheim, Germany, pp 251–257.

Korashi AA, El-Borollosy MM, Hassa EA, Abo El-Suoud MR, Zain El-Deen, Koraim A (1996) Hosts of *Orobanche* spp. and yield losses in Delta and Upper Egypt. In: MT Moreno, JI Cubero, D Berner, DM Joel, LJ Musselman, C Parker (eds) Advances in Parasitic Plant Research. Consejería de Agricultura y Pesca, Junta de Andalucía, Spain, pp 487–493.

Kroschel J, Klein O (1999) Biological control of *Orobanche* spp. with *Phytomyza orobanchia* Kalt.: a review. In: J Kroschel, M Abderabihi, H Betz (eds) Advances in Parasitic Weed Control at On-farm Level, vol 2: Joint Action to Control *Orobanche* in the WANA Region. Margrat Verlag, Weikersheim, Germany, pp 135–159.

Kubo M, Ueda H, Park P, Kawaguchi M, Sugimoto Y (2009) Reactions of *Lotus japonicus* ecotypes and mutants to root parasitic plants. J Plant Physiol 166: 353–362.

Kurbanov T (1970) The use of Phytomyza on the fields of Solchos. Biological control of the Egyptian broomrape in vegetables and melons. Sel'skoe Chozjajstvo Usbekistana 9: 50.

Kusumoto D, Goldwasser Y, Xie X, Yoneyama K, Takeuchi Y, Yoneyama K (2007) Resistance of red clover (*Trifolium pratense*) to the root parasitic plant *Orobanche minor* is activated by salicylate but not by jasmonate. Ann Bot 100: 537–544.

Lamb CJ, Lawton MA, Dron M, Dixon RA (1989) Signals and transduction mechanisms for activation of plant defenses against microbial attack. Cell 56: 215–224.

Lane JA, Bailey JA, Butler RC, Terry PJ (2006) Resistance of cowpea [*Vigna unguiculata* (L.) Walp.] to *Striga gesnerioides* (Willd.) Vatke, a parasitic angiosperm. New Phytol 125: 405–412.

Letousey P, De Zélicourt A, Vieira Dos Santos C, Thoiron S, Monteau F, Simier P, Thalouarn P, Delavault P (2007) Molecular analysis of resistance mechanisms to *Orobanche cumana* in sunflower. Plant Pathol 56: 536–546.

Liebman M, Dyck E (1993) Crop rotation and intercropping strategies for weed management. Ecol Appl 3: 92–122.

Linke KH, Scheibel C, Saxena MC, Sauerborn J (1992) Fungi occurring on *Orobanche* spp. and their preliminary evaluation for *Orobanche* control. Trop Pest Manag 38: 127–130.

Linke KH, Abd El-Moneim AM, Saxena MC (1993) Variation in resistance of some forage legumes species to *Orobanche crenata* Forsk. Field Crops Res 32: 277–285.

Lins RD, Colquhoun JB, Mallory-Smith CA (2007) Effect of small broomrape (*Orobanche minor*) on red clover growth and dry matter partitioning. Weed Sci 55: 517–520.

Lozano-Baena MD, Moreno MT, Rubiales D, Pérez-de-Luque A (2007) *Medicago truncatula* as a model for non-host resistance in legume-parasitic plant interactions. Plant Physiol 145: 437–449.

Mabsoute L, Saadaoui EM (1996) Acquis de recherche sur le parasitisme des legumineuses alimentaires au Maroc: synthese bibliographique. Al Awamia 92: 55–67.

Manen JF, Habashi C, Jeanmonod D, Park J-M, Schneeweis GM (2004) Phylogeny and intraspecific variability of holoparasitic *Orobanche* (Orobanchaceae) inferred from plastid *rbcL* sequences. Mol Phylogenet Evol 3: 482–500.

Martín LM, Moreni MT, Gil J (1982) Tres nuevos cultivares de veza común (*Vicia sativa* L.). Anales Inst Nacional Investigaciones Agrarias 21: 3.

Mesa-García J, García-Torres L (1991) Status of *Orobanche crenata* in faba bean in the Mediterranean region and its control. Options Méditerranéennes 10: 75–78.

Moreno V (1944) Apuntes biológicos sobre el jopo de las habas (*Orobanche crenata* Forsk.). Bol Patol Veg Entomol Agrí 13: 335–376.

Moreno-Márquez V (1947) El jopo (*Orobanche crenata* Forsk.) y las siembras tardías de habas. Bol Patol Veg Entomol Agrí 15: 97–108.

Musselman LJ, Parker C (1982) Preliminary host ranges of some strains of economically important broomrapes (*Orobanche*). Econ Bot 36: 270–273.

Musselman LJ (1994) Taxonomy and spread of *Orobanche*. In: AH Pieterse, JAC Verkleij, SJ ter Borg (eds) Biology and Management of *Orobanche*. Proc 3rd Int Workshop of *Orobanche* and Related *Striga* Research. Royal Tropical Institute: Amsterdam, The Netherlands, pp 27–35.

Nadal S, Rubiales D, Moreno MT, Cubero JI (2001) "Retaca" a faba bean cultivar for green pod consumption of determinate growth habit that escape from broomrape attack and tolerate higher glyphosate doses. In: P Fer A Thalouarn, DM Joel, LJ Musselman, C Parker, JAC Verkleij (eds) Proc 7th Int Parasitic Weed Symp, Nantes, France, p 292.

Nadal S, Cubero JI, Moreno MT (2007) Sources of resistance to broomrape (*Orobanche crenata* Forsk.) in narbon vetch. Plant Breed 126: 110–112.

Nadal S, Moreno MT, Román B (2008) Control of *Orobanche crenata* in *Vicia narbonensis* by glyphosate. Crop Protec 27: 873–876.

Nasib AM, Ibrahim AA, Khalil SA (1982) Breeding for resistance to *Orobanche*. In: G Hawtin, C Webb (eds) Faba Bean Improvement. Martinus Nijhoff, Hague, The Netherlands, pp 199–206.

Osterbauer NK, Rehms L (2002) Detecting single seeds of small broomrape (*Orobanche minor*) with a polymerase chain reaction. Plant Health Prog doi: 10.1094/PHP-2002-1111-01-RS.

Paran I, Gidoni D, Jacobsohn R (1997) Variation between and within *Orobanche* (*Orobanche*) species revealed by RAPD markers. Heredity 78: 68–74.

Parker C (2009) Observations on the current status of *Orobanche* and *Striga* problems worldwide. Pest Manag Sci 65: 453–459.

Parker C, Riches CR (1993) *Orobanche* species: the Broomrapes. In: Parasitic weeds of the World: Biology and control. CABI Publ, Wallingford, UK, pp 111–164.

Pérez-de-Luque A, Sillero JC, Moral A, Cubero JI, Rubiales D (2004) Effect of sowing date and host resistance on the establishment and development of *Orobanche crenata* in faba bean and common vetch. Weed Res 44: 282–288.

Pérez-de-Luque A, Rubiales D, Cubero JI, Press MC, Acholes J, Yoneyama K, Takeuchi Y, Plakhine D, Joel DM (2005) Interaction between *Orobanche crenata* and its host legumes: unsuccessful haustorial penetration and necrosis of the developing parasite. Ann Bot 95: 935–942.

Pérez-de-Luque A, Lozano MD, Moreno MT, Testillano PS, Rubiales D (2007) Resistance to broomrape (*Orobanche crenata*) in faba bean (*Vicia faba*): cell wall changes associated with pre-haustorial defensive mechanisms. Ann Appl Biol 151: 89–98.

Perrino P, Pace MS, Polignanao GB (1988) Evaluation for tolerance to broomrape in a germplasm collection of *Vicia faba*. FABIS Newsl 20: 40–44.

Portnoy VH, Katzir N, Joel DM (1997) Species identification of soil-borne *Orobanche* seeds by DNA fingerprinting. Pest Biochem Physiol 58: 49–54.

Press MC, Graves JD (eds) (1995) Parasitic Plants. Chapman & Hall, London, UK, p 292.

Riches CR (1989) The biology and control of *Alectra vogelii* Benth. (Scrophulariaceae) in Botswana. PhD Thesis, University of Reading, Reading, UK, p 208.

Rodríguez-Conde MF, Moreno MT, Cubero JI, Rubiales D (2004) Characterization of the *Orobanche-Medicago truncatula* association for studying early stages of the parasite-host interaction. Weed Res 44: 218–223.

Román B, Rubiales D, Torres AM, Cubero JI, Satovic Z (2001) Genetic diversity in *Orobanche crenata* populations from southern Spain. Theor Appl Genet 103: 1108–1114.

Román B, Satovic Z, Rubiales D, Torres AM, Cubero JI, Katzir N, Joel DM (2002) Variation among and within populations of the parasitic weed *Orobanche crenata* from Spain and Israel revealed by inter simple sequence repeat markers. Phytopathology 92: 1262–1266.

Román B, Alfaro C, Torres AM, Moreno MT, Satovic Z, Pujadas A, Rubiales D (2003) Genetic relationships among *Orobanche* species as revealed by RAPD analysis. Ann Bot 91: 637–642.

Román B, González Verdejo CI, Satovic Z, Madrid MD, Cubero JI, Nadal S (2007). Detecting *Orobanche* species by using cpDNA diagnostic markers. Phytoparasitica 35: 129–135.

Rubiales D, Pérez-de-Luque A, Cubero JI and Sillero JC (2003a) Crenate broomrape (*Orobanche crenata*) infection in field pea cultivars. Crop Protec 22: 865–872.

Rubiales D, Alcántara C, Pérez de Luque A, Gil J, Sillero JC (2003b) Infection of chickpea (*Cicer arietinum*) by crenate broomrape (*Orobanche crenata*) as influenced by sowing date and weather conditions. Agronomie 23: 359–362.

Rubiales D, Pérez-de-Luque A, Joel DM, Alcántara C, Sillero JC (2003c) Characterization of resistance in chickpea to crenate broomrape (*Orobanche crenata*). Weed Sci 51: 702–707.

Rubiales D, Moreno MT, Sillero JC (2005a) Search for resistance to crenate broomrape (*Orobanche crenata*) in pea germplasm. Genet Resour Crop Evol 52: 853–861.

Rubiales D, Sadiki M, Román B (2005b) First report of *Orobanche foetida* on common vetch (*Vicia sativa*) in Morocco. Plant Dis 89: 528.

Rubiales D, Pérez-de-Luque A, Fernández-Aparicio M, Sillero JC, Román B, Kharrat M, Khalil S, Joel DM, Riches C (2006) Screening techniques and sources of resistance against parasitic weeds in grain legumes. Euphytica 147: 187–199.

Rubiales D, Fernández-Aparicio M, Pérez de Luque A, Castillejo MA, Prats E, Sillero JC, Rispail N, Fondevilla S (2009) Breeding approaches for crenate broomrape (*Orobanche crenata* Forsk.) management in pea (*Pisum sativum* L.) Pest Manag Sci 65: 553–559.

Sauerborn J (1991) Parasitic Flowering Plants: Ecology and Management. Margraf Scientific Books, Weikersheim, Germany.

Sauerborn J, Masri H, Saxena MC, Erskine W (1987) A rapid test to screen lentil under laboratory conditions for susceptibility to *Orobanche*. Lens Newsl 14: 15–16.

Satovic Z, Joel DM, Rubiales D, Cubero JI, Román B (2009) Population genetics in weedy species of *Orobanche*. Australas Plant Pathol 38: 228–234.

Saxena MC, Linke K-H, Sauerborn J (1994) Integrated control of *Orobanche* in cool-season food legumes. In: AH Pieterse, JAC Verkleij, SJ ter Borg (eds) Biology and Management of *Orobanche*. Proc 3rd Int Workshop on *Orobanche* and Related *Striga* Research, Amsterdam, The Netherlands, pp 419–431.

Schloss JV (1995) Recent advances in understanding the mechanism and inhibition of acetolactate synthase. In: J Setter (ed) Herbicides Inhibiting Branch Chain Amino Acid Biosynthesis. Springer, New York, USA, pp 4–11.

Schneeweiss GM (2009) List of actual and potential broomrape pest species: http://www.botanik.univie.ac.at/plant chorology/PestSpeciesOrobanche.pdf (Cited Jan 2009).

Schneeweiss GM, Colwell A, Park J-M, Jang CG, Stuessy TF (2004) Phylogeny of holoparasitc *Orobanche* (Orobanchaceae) inferred form nuclear ITS sequences. Mol Phylogenet Evol 30: 465–478.

Shaner DL (1989) Factors affecting soil and foliar bioavailability of imidazolinones. Technical Information Rep Agri Res Division. American Cyanamid Company, Princeton, NJ, USA.

Sillero JC, Cubero JI, Fernández Aparicio M, Rubiales D (2005) Search for resistance to crenate broomrape (*Orobanche crenata*) in *Lathyrus*. Lathyrus Lathyrism Newsl 4: 7–9.

Slinkard AE, Vandenberg A, Holm A (2007) Lentil plants having increased resistance to imidazolinone herbicides. Patent number 7232942: http://www.freepatentsonline.com/7232942.html (Cited Sept 2010).

Smith JL, De Moraes CM, Mescher MC (2009) Jasmonate- and salicylate-mediated plant defense responses to insect herbivores, pathogens and parasitic plants. Pest Manag Sci 65: 497–503.

Thomas H, Saurborn J, Muller-Stover D, Zeigler A, Badi JS, Kroschel J (1998) The potential of *Fusarium oxysporum* f. sp. *orthoceras* as a biological control agent for *Orobanche cumana* in sunflower. Biol Control 13: 41–48.

VanderBosch KA, Stacey G (2003) Summaries of legume genomics projects from around the globe. Community resources for crops and models. Plant Physiol 131: 840–865.

Varshney RK, Close TJ, Singh NK, Hoisington D, Cook DR (2009) Orphan legume crops enter the genomic era! Curr Opin Plant Biol 12: 202–210.

Verkleij JAC, Egbers WS, Pieterse AH (1991a) Allozyme variations in populations of *Orobanche crenata* from Syria. In: K Wegmann, LJ Musselman (eds) Progress in *Orobanche* Research. Proc Int Workshop on *Orobanche* Research. Eberhard-Karls University, Tübingen, Germany, pp 304–317.

Verkleij JAC, Koevoets P, López-Granados F, Egbers WS, García-Torres L, Pieterse AH (1991b) Genetic variability in populations of *Orobanche crenata* from Spain. In: JK Ransom, LJ Musselman, AD Worsham, C Parker (eds) Proc 5th Int Symp on Parasitic Weeds. CIMMYT, Nairobi, Kenya, pp 462–469.

Vieira Dos Santos C, Letousey P, Delavault P, Thalouarn P (2003a) Defense gene expression analysis of *Arabidopsis thaliana* parasitized by *Orobanche ramosa*. Phytopathology 93: 451–457.

Vieira Dos Santos C, Delavault P, Letousey P, Thalouarn P (2003b) Identification by supresión subtractive hybridization and expression analysis of *Arabidopsis thaliana* putative defence genes during *Orobanche ramosa* infection. Physiol Mol Plant Pathol 62: 297–303.

Westwood JH (2000) Characterization of the *Orobanche-Arabidopsis* system for studying parasite-host interactions. Weed Sci 48: 742–748.

Westwood JH, Yu X, Foy CL, Cramer CL (1998) Expression of a defense-related 3-hydroxy-3-methylglutaryl CoA reductase gene in response to parasitization by *Orobanche* spp. Mol Plant-Microbe Interact 11: 530–536.

Wolfe AD, Randle CP, Liang L, Steiner KE (2005) Phylogeny and biogeography of Orobanchaceae. Folia Geobot 40: 115–134.

Young ND, Cannon SB, Sato S, Kim D, Cook DR, Town CD, Roe BA, Tabata S (2005) Sequencing the genespaces of *Medicago truncatula* and *Lotus japonicus*. Plant Physiol 137: 1174–1181.

Zaitoun FMF, Ibrahim HM (1998) Effect of planting date and faba bean genotypes on *Orobanche crenata* growth. In: K Wegmann, LJ Musselman, DM Joel (eds) Current problems of *Orobanche* researches. Proc 4th Int Workshop on *Orobanche*, Albania, Republic of Albania, pp 439–446.

Zeid M, Madkour M, Koraiem Y, Nawar A, Soliman M, Zaitoun F (1997) Molecular studies on *Orobanche*. J Phytopathol 145: 351–355.

12

Future Prospects

Jose I. Cubero,[1,*] *Ana M. Torres*[2] and *Marcelino Pérez de la Vega*[3]

ABSTRACT

In a world in which food demand is steadily increasing, cool season grain legumes have to play a key role due to their high nutritive value and protein content, their contribution to a healthier nutrition, their ability to improve soil nutrients, their role in crop rotation increasing yield of subsequent crops, etc. Many of these characteristics are a consequence of the legume's ability of symbiotically fixing atmospheric nitrogen. But this characteristic increases breeding difficulties in these species since this ability depends not only of the plant genome but also of the bacterium genome and their interactions. Yield and yield stability were, and still are, the main objectives of traditional breeding in these species, but breeding for stress resistance and quality are contributing to a new breeding era in which molecular markers and genomic technologies are incorporating with different speed depending upon the species. Modern technologies, such as genomics and transformation, are less developed, and thus have been less applied to breeding, than in other crops such as cereals. To increase the role of cool season grain legumes, and other grain legume species, in food security the first step is more research.

Keywords: Chickpea, faba bean, grass pea, lentil, pea, breeding, future prospects.

12.1 The Past and the Present

Plant production has to meet considerably mounting demands for the future. By 2050 we will need to double our agricultural production to

[1]Departamento de Genética, Universidad de Córdoba, Campus de Rabanales. Edificio C5, 2nd planta. 14071 Córdoba, Spain; e-mail: *ge1cusaj@uco.es*

[2]IFAPA, Centro Alameda del Obispo, Área de Mejora y Biotecnología, Apdo. 3092, E-14080 Córdoba, Spain.

[3]Area de Genética, Universidad de León, 24071 León, Spain.

*Corresponding author

guarantee food for a world population of about 9,000 million. In a world demanding more food from a limited amount of land, improved resistance to biotic and abiotic stresses is a priority. How to produce the required amount of food is itself the main problem, but the challenge will be to do it while preserving the environment (to talk of improving it sounds, in principle, unrealistic) allowing for clean water, pure air and fertile soils. The recognition that biodiversity and the quality of the environment needs to be preserved for future generations also leads to competition for land use and can also negatively affect crop yields through reduced use, or total exclusion of inorganic fertilizers and pesticides (Jellis 2009; Den Herder et al. 2010).

At the same time the demand for food is intensifying, the climate is changing, with inevitable consequences for agriculture and the world's food supply (Rosenzweig and Hillel 1995). The impact of climate change extends to all economic activities in rural areas. This climatic reality, in addition to the greater specialization required for the correct use of land, the rising complexity and cost of the food chain and simultaneously increasing concern for food safety are associated challenges, which need feasible solutions in a short period of time: 40 years from now, indeed, seem not to leave much room to hope.

One of the solutions might be to rely on the domestication of new crops, and indeed some efforts are being devoted to domesticate wild plants , and at the present moment, looking for new molecules of industrial interest (Hajjar and Hodgkin 2007). Revisiting well known crops (e.g., maize, wheat and sugarcane, etc.) to produce specialized plant products for processing or for new uses such as for biofuel production and medical-pharmaceutical uses is also a current practice in our growing markets (Heslop-Harrison 2002; Davies 2010).

Food legumes can meet the major challenges for sustainable agriculture. Given their unique ability to fix atmospheric N, legumes contribute to: 1) improving soil fertility and reduce the use of fossil energy and greenhouse gas emission through a decrease use of nitrogen fertilizers; 2) diversifying the cropping systems by crop rotation with cereals; 3) reducing the need for external inputs such as pesticides (as legumes break the disease cycle and reduce levels of the pathogens in the soil); 4) avoiding protein import dependency by producing home-grown proteins with high nutritional value. Finally, legumes are crops well known by farmers around the world as they have been domesticated and grown in all agricultural regions.

Moreover, their uses have been revitalized due to the potential health benefits of specific legume phytochemicals (e.g., isoflavones, phytosterols, protease inhibitors, saponins, etc.). Studies have revealed the beneficial or protective effects of the consumption of legumes with regard to hypercholesterolaemia and coronary heart disease, obesity, diabetes mellitus, menopause and, are strong candidates for a role as natural cancer-protective food (Salunkhe et al. 1982; Messina 1999).

In spite of these advantages, food legumes still show some limitations: 1) yields are relatively low; averages do not reach, in general, 1 t/ha, values from 0.6–0.8 t/ha being very common, 1) biotic and abiotic stresses still reduce significantly their yield potential in specific areas, 3) genotype-environmental (GxE) interaction are usually large thus, limiting breeding strategies and, 4) the level of mechanization is rather poor.

Low yields can be explained by a set of inter-related facts. Legumes have a much higher protein content than that of cereals, proteins requiring in their synthesis much more energy than both cereals and oil rich crops. Because of that requirement of energy, a legume will probably never reach the yield level of a cereal. Moreover, legumes obtain the nitrogen needed for protein synthesis through the symbiosis with nitrogen-fixing bacteria (rhizobia); although this association is referred to as **symbiosis**, the legume plant is in fact the host of the bacteria, which draws carbohydrates from it before they can fix atmospheric nitrogen and benefit the host. This disease-like symbiosis is also a cause of energy loss with impact on the yield.

In order to overcome these restrictions, and instead of improving legumes themselves, one of the objectives in modern plant breeding has been to look for cereals with the ability of fixing atmospheric nitrogen, although a practical solution seems to be far as yet (Bruijn et al. 1995; Mehbood et al. 2009; Montanez et al. 2009). Transferring the whole fixing systems by genetic engineering to non-legume crops has been proposed (Dixon et al. 1997), but it seems to be a daunting and long-term task for now (Godfray et al. 2010). Furthermore, transferring all the N-fixing genes to a cereal would transform a cereal in a "legume" and, hence, spending a great deal of energy in the process with a likely loss in yield.

Therefore, symbiotic nitrogen fixation in agroecosystems should depend exclusively on legumes at least for some time. In order to increase knowledge of the symbiotic nitrogen fixation system, that is of particular importance in sustainable agriculture, a major emphasis should be laid on basic research. More work is needed on the genes responsible in rhizobia and legumes, the structural chemical bases of rhizobia-legume communication, and signal transduction pathways responsible for the finely orchestrated functioning of the symbiosis-specific genes involved in nodule development and nitrogen fixation. The mechanisms by which legumes choose their rhizobial partners operate independently from their ability to fix nitrogen. As result, symbiotic nitrogen fixation is often suboptimal. The initial recognition of the bacterial partner and the subsequent signal transduction in the host root utilizes components that are functionally conserved between legumes and probably actinorhiza host plants. In fact root nodule symbiosis uses genetic programs that are widely conserved in land plants. However, the later steps, which largely determine symbiotic performance, are subject to diversification of molecular mechanisms. Strategies towards increased nitrogen fixation of legumes in agriculture need to encompass these mechanisms, such as bacterial effector proteins, the occurrence of terminal bacteroid differentiation and the expression of bacterial hydrogenase (Den Herder and Parniske 2009; Markman and Parniske 2009).

In a world demanding more food from a limited amount of land, improved resistance to biotic stresses is a priority. Disease resistant cultivars can reduce both the risk of yield loss and the dependence on pesticides in high input systems, enabling a more stable production across diverse and adverse environments and soil conditions (Miklas et al. 2006). Abiotic stresses such as drought, heat and salinity add considerably to these losses, and are likely to increase with climate change.

Compared with legumes, cereals have received by far the strongest support from basic research on these areas leading to the availability of a large number of molecular markers, well saturated genetic and physical maps and sequencing projects of regions carrying specific genes. As a result, different genes of agronomic interest have been already identified and transferred to valuable genotypes (reviewed in Dwivedi 2007). In some cases, the knowledge of the structure and function of these genes have been studied thoroughly, and genetic engineering has produced transgenic cultivars resistant to some plagues and diseases (Narayanan et al. 2004; Vasil 2005). In legumes, although great efforts are underway (e.g., the recent multinational European Grain Legumes Integrated Project (GLIP) striving to develop new strategies to enhance the use of grain legumes crops in Europe and beyond), however in some cases the potential is still unexplored. This difference is largely proportional to the relative stage of development of genomics in these crops and to the number of trait mapping studies that has been completed (Dwivedi 2007).

In considering the legume-*Rhizobium* symbiosis, it is important to stress that legume yield is the result of this association, and hence why *Rhizobium* specialists blame legume breeders of considering only **a part** of the whole system, the legume host, in their breeding projects (see Chapter 11). In fact, the strong G x E interaction common in legumes could be partially explained by the fact that the rhizobium populations are very local in their characteristics so much so that a new improved variety developed in a certain area might not perform well in others because of the difference in the bacterium population in the soil.

Regardless of the nutritionally active phytochemicals mentioned above, legumes contain several non-nutritional factors (e.g., alkaloids, non-essential amino acids, phytic acid, tannins, etc.), formerly known as "antinutritional factors", which negatively affect legume digestibility and diminish the productivity of farm animals. Although, in modern animal husbandry the negative effects of these factors are reduced because the legume flour is mixed in low proportion with other ingredients, they are nevertheless a factor to be covered in plant breeding programs. Some of these compounds occasionally produce grave human diseases such as favism (see Chapter 2) or lathyrism (see Chapter 4).

In addition, legume allergy is a relatively common food allergy and legumes are causative agents of type I hypersensitivity in South Asia and Europe (Kumari et al. 2006; Ireneo et al. 2008). Seed storage proteins of the vicilin family have been characterized as major allergens in several seed legumes and tree nuts. Serological cross-reactivity among legumes is frequent and *in vitro* inhibition experiments demonstrated a high degree of cross-reactivity among lentils, chickpeas and peas. Fortunately, thermal treatments reduce immunoglobulin-E binding to legume proteins. However, several extremely resistant immunoreactive proteins still remained in legumes even after this treatment (Cuadrado et al. 2009). However, legume proteins are not often responsible for the allergic responses. For instance, there are patients who suffered from allergic symptoms (asthma and anaphylaxis) related to inhalation of lentil particles or ingestion of lentils in which sensitization to legume proteins was not clear. In these patients proteins from the lentil pest *Bruchus lentis* seem to be the cause of the allergic response since tests were positive to infested lentils and *B. lentis* in a group of these patients and negative to non-infested lentil extracts (Armentia et al. 2006).

Finally, the need of adequate mechanization in all phases of the culture is another reason explaining low yields in food legumes. Again there is a great difference between developed and developing countries, but even in the former, the required tools have been developed on the basis of the material already designed primarily for cereals. However, even in order to explain the differences in yield, this topic will not be discussed here.

The point remains that the key reason explaining all these facts is still insufficient research committed to legume crops. Apart from the lack of impact-oriented research and development approaches, until recently, legumes have also suffered from a dearth of genomic and molecular-genetic resources (Varshney et al. 2009). Significantly, many of the world's most important food legumes are grown in developing countries (arid and semi-arid regions of Africa and Asia), where crop productivity is hampered by biotic and abiotic stresses and other constraints, mentioned above. Except for two crops (soybean and common bean and to some extent pea), legume research is at the early stages and crops are "orphans" of the genome revolution. Only recently, the legume community has initiated a concerted effort to change this situation. Results have turned up in a series of international collaborations that benefit from recent advances in genome sequencing and genotyping technologies that can be used in high-throughput approaches to facilitate genomics-assisted breeding in these legumes (Varshney et al. 2009 and references herein).

The contents of this book summarize the recent progress achieved in five important crop legumes (pea, faba bean, chickpea, lentil and grass pea) in the development of genetic and genomics solutions to a range of agronomic constraints.

12.2 The Work Performed

Modern farming is not making the most of what legumes can offer. Being conscious of their many advantages, in the past two decades most research and development (R&D) programs have diverted significant portions of their investment towards legume activities. Apart from the funding devoted to the model species *Medicago truncatula* and *Lotus japonicus* and the soybean crop, most of the efforts have been focused on traditional cool season grain legumes with a significant role in temperate agriculture such as pea, faba bean, chickpea, lentils and grass pea (for details, see the corresponding chapters in the present book).

Germplasm and Genetics. A great number of conventional morphological markers have been described in, pea, chickpea, faba bean and lentil, at first from spontaneous and then from induced mutations. Besides induced mutagenesis, extensive variation has been reported to appear spontaneously in grass pea in the course of tissue culture studies (somaclonal variation). This type of genetic variability may also be successfully exploited to produce agronomically desirable types, providing a faster means of improvement than that allowed by conventional crossing and backcrossing methods.

Traditional breeding throughout the world has similar objectives with larger and more stable seed yield being one of the most important for all the considered species. Increasing yield often entails breeding for adaptation to environmental stress factors such as heat or cold, drought, and resistance to diseases caused by fungi, bacteria or viruses. Conventional breeding methods have been in general successful in improving the self-pollinating legume species (pea, chickpea and lentil) using genetic resources preserved in large international collections (ICARDA, ICRISAT, USDA, INIA, N I Vavilov Institute, IPK, etc.). Development of superior cultivars has been achieved through the introduction of novel traits from wild germplasm and landraces as well as pyramiding positive alleles in adapted genetic backgrounds. In case of the partially allogamous faba bean, whose wild ancestor has not been found, only a primary genepool well preserved in several collections, has been successfully used as a genetic resource for breeding programs. Notable achievements in both faba bean and lentil have been reached by ICARDA, which has released numerous improved cultivars (mainly in relation to resistance-tolerance to diseases and some abiotic factors) during the past three decades. Grass pea breeding effort throughout the world has been rather limited. The emphasis on grass pea breeding has been reduction in β-ODAP levels, a neurotoxin responsible for lathyrism. Nevertheless, some important programs exist to improve yield, resistance to stresses and to reduce the percentage of neurotoxin in the seed. Although, local landraces and cultivars are being lost as farmers switch to alternative crops, significant collections of cultivated and wild *Lathyrus* species have already been assembled and are maintained *ex situ* in a number of different institutes throughout the world.

Molecular breeding approaches. While traditional breeding methods have proved useful to breed traits of interest, conventional breeding could be greatly assisted by the availability of well developed genetic maps with robust and accurate markers for: 1) resistance to biotic stresses such as crenate broomrape and a variety of pathogenic fungi, 2) abiotic stresses such as frost and drought tolerance, and 3) traits related to the nutritional value of the seeds.

Detailed **genetic maps** have been generated in pea, chickpea, faba bean and lentil, placing many genes and QTLs of interest using a variety of molecular techniques.

From the preliminary linkage maps based on morphological traits, RFLPs (see Abbreviations for marker names), or anonymous markers (such as RAPDs or AFLPs), larger sets of transferable markers such as SSRs, or functional markers derived from ESTs have been incorporated, providing direct bridges among genetic maps. As mentioned above, there is limited literature on the generation of molecular linkage maps for grass pea and from the 1960s to now, only three maps have been developed, using RAPDs and SSRs.

Despite the available information, up to now, marker-assisted selection (MAS) is being more efficiently applied for **monogenic traits** tightly associated with DNA markers than for polygenic traits. Thus, in **pea** numerous major genes controlling pea disease resistance (*er1*, *er2*, *Er3*, *En*, BYMV, *Fwf*), seed composition (*Tr*), flowering time or plant architecture have been mapped and tagged (see Chapter 1 for details). Many results of gene tagging have been published in the *Pisum* gene list (http://data.jic.bbsrc.ac.uk/cgi-bin/pgene/) and are readily applicable in breeding programs. In case of **faba bean** (see Chapter 2), several markers (RAPDs, CAPs or SCARs) linked to a gene controlling the hypersensitive rust resistance or to select against traits affecting the nutritional value of seeds (e.g., low tannins, vicine and convicine content) have also been reported, which may facilitate a more efficient selection of new cultivars free of antinutritional compounds.

Although, the number of molecular markers available for genetic mapping and breeding purposes in *Lens* is increasing dramatically, the implementation of genetic markers for routine use in **lentil** breeding programs is still very limited and mainly focused on selecting resistance to a few major diseases (see Chapter 3). Molecular markers linked to *Fw* (conferring resistance to Fusarium wilt) and to a dominant locus for Ascochyta blight resistance (AbR_1) have been reported. Moreover, a SCAR markers linked to a single Ascochyta blight recessive gene (ral_2) is available to facilitate an efficient resistance selection. With regard to anthracnose, several RAPDs and AFLPs flanking a major resistance gene (*LCt-2*) have been documented as well. There is a limited number of references on the generation of molecular linkage maps for grass pea and the available maps are still insufficiently saturated with markers (see Chapter 4). However, the construction of a genetic map is not a prerequisite for identifying markers closely linked to a particular trait of interest. Thus, several markers have found to be linked to the gene responsible for the tendril trait of sweet pea (*Lathyrus odoratus*). In **chickpea,** the first traits mapped were related to the pigmentation of the flower (pink/white = *P/p* = *B/b*), brown/yellow seed testa (*T3/t3*) and purple/green epicotyl (*Gst/gst*), that appeared to be linked. Flower and seed color differentiate the two big cultivated chickpea types: kabuli and desi. Seed coat thickness (*Tt/tt*), considered as a qualitative trait is linked to flower color. Seed surface reticulation (*Rs/rs*) has been reported to be controlled by one or two genes and erect/prostrate plant growth habit (*Hg/hg*) is another trait with simple inheritance. Double pod is an interesting mutation controlled by a single recessive gene designated as "*s*" or "*sfl*" with a positive effect on yield stability.

Some resistances to fungal diseases in chickpea are controlled by oligogenic genes. Thus, genes controlling Fusarium wilt resistance to races 0, 1, 2, 3, 4 and 5 (*foc-0, foc-1, foc-2, foc-3, foc-4* and *foc-5*) have been mapped and arranged as a gene cluster in the same linkage group (LG). However, a new resistance gene for race 0 ($foc\text{-}0_1$) was found in a different LG confirming that the resistance is controlled by two independent genes ($foc\text{-}0_1$ and $foc\text{-}0_2$). Moreover, a gene controlling resistance to chickpea rusts (*Uca1/uca1*) has been located tightly flanked by two STMS markers. Microsatellite-based markers have shown to be the most useful in chickpea molecular breeding (see Chapter 5).

Knowledge of the genes underlying any trait is mandatory before transferring them into new lines. Gene discovery is now possible in pea, faba bean and lentil taking advantage of the wealth of mutations identified and enhanced by **candidate gene** and comparative mapping approaches with model species. Thus, in **pea,** *Le*, controlling internode length, was identified as an enzyme from the gibberellin biosynthetic pathway; the green cotyledon as an ortholog of a rice stay-green gene; the *r* mutation and four other independent mutations determining the wrinkled seed were identified in genes encoding enzymes involved in starch biosynthesis; major genes involved in the control of pea-*Rhizobium* symbiosis were identified by way of orthologous genes in *M. truncatula* (*sym9*) or *Lotus japonicus* (*sym29*). The *det* and *lf* genes were identified as homologs of the TFL1 family controlling flowering time in several plant species. *Stamina pistilloida* was identified as the ortholog of *Fim* and *UFO*. Comparative mapping of *L. japonicus* and pea *Cycloidea* genes helped to identify *K* and *LST1*, two genes controlling floral zygomorphy in pea. *Unifoliata*, a gene regulating leaf and flower morphogenesis and *Tendril-less* have been also already cloned. In **faba bean** only an ortholog of the *CEN/TFL1* responsible for the growth habit in the species has been identified so far using a candidate gene approach. Sequence alignment of *CEN/TFL1*-homolog or *CEN/TFL1*-like genes with the faba bean target sequence allowed the development of a diagnostic marker suitable for the selection of determinate growth habit plant in this crop. An attempt to saturate a QTL controlling resistance to Ascochyta blight has been performed in **chickpea** by developing markers from BAC clones. Microsynteny in this genomic area between chickpea and *M. truncatula*, allowed identifying candidate genes for this QTL. Up to now, no gene discovery has been reported in **lentil** or **grass pea**, nevertheless, several function-associated molecular maps (based on ESTs), are being developed in all the considered species that will facilitate direct selection of additional candidate genes in the near future.

The improvement of important agronomic characters such as **quantitative disease resistances and stability of seed yield and composition** are more difficult to breed. Breeding success will depend on availability of consistent resistance genes or QTLs within or outside the species, and adequate molecular markers flanking these genes or QTLs that can trace resistance in breeding populations. However, the long genetic distance existing in most cases between the identified genetic markers and the QTL, the common lack of codominant markers and the general lack of knowledge on biochemistry of the trait greatly limit the use of genetic markers in MAS approaches. Effectiveness of MAS might soon increase with the adoption of new improvement in marker technology together with the integration of comparative mapping and functional genomics.

QTLs have been mapped for a wide range of agronomic traits in **pea**. Concerning pathogen resistance, three QTLs controlling *Aphanomyces* partial resistance were consistently mapped in one pea population, in the field and the greenhouse after inoculation by two pathogen isolates. Ascochyta blight partial resistance and seed trait QTLs have been also mapped in six different pea populations, both in field and growth chamber experiments. The identification of genomic regions associated with resistance to *Orobanche crenata*, the major constraint for pea cultivation in the Mediterranean basin and the Middle East, has been also sought in this crop. An RIL population from a cross between a resistant accession of the wild pea *Pisum sativum* ssp. *syriacum* and a susceptible pea variety allowed identifying four QTLs associated with field resistance. Finally, the QTL approach has been used as well to identify the physiological determinants of seed yield and protein content in pea suggesting a predominant role of plant source capacity in determining seed protein content and yield (see Chapter 1 for details).

Faba bean research has mainly focused on developing molecular markers for selecting resistance against parasitic weeds (*Orobanche* ssp.) and other major fungal diseases (*Ascochyta* an rust). Three putative QTLs controlling resistance to *O. crenata* were first identified in an F_2 and later validated in the corresponding F_6 population. Two different F_2 populations have been used to detect QTLs responsible for Ascochyta blight resistance in faba bean. Polygenic control of the resistance was determined by at least two QTLs assigned to chromosomes 3 and 2, which explained 21.0 and 25.3% of the phenotypic variance, respectively. Validation studies are underway (see Chapter 2 for details).

Complex traits currently located on the **chickpea** genetic map are related to yield improvement, quality and resistance to Ascochyta blight. A QTL related with seed size has been found in LG4 by several authors. Other minor QTLs have been also reported, which should be validated in further studies. Early flowering time seems to have a positive effect on chickpea yield under the Mediterranean environment. QTLs controlling this trait have been found and validated in different environments and in both intra- and interspecific populations. Few QTLs related to quality components have been also published (two associated with β-carotene content and one associated with lutein concentration). Finally, several QTLs involved in blight resistance have been reported in chickpea genetic map. The most significant ones (located in LGs 2 and 4) have been corroborated by different authors. Microsynteny between chickpea and *M. truncatula*, has revealed several candidate genes for one of these QTLs (see Chapter 5 for details). In **lentil,** molecular markers have been used to locate QTLs related to winter hardiness and Ascochyta blight resistance. Moreover, several QTLs affecting earliness and plant height have also been reported (see Chapter 3 for details).

Molecular approaches to the improvement of **grass pea** identified one consistent putative QTL (QTL1) for Ascochyta blight resistance. A second potential QTL has also been reported. One of the markers mapped in the vicinity of QTL1 was generated from an STS primer pair designed from a *Lathyrus sativus* cDNA clone that showed sequence similarity to the *Cf-9* gene conferring resistance to *Cladosporium fulvum* in tomato (see Chapter 4 for details).

Breeding for resistance to abiotic stresses: drought stress and winter tolerance. Genetic maps have been also developed for other traits that are difficult to breed conventionally such as frost or drought tolerance. In all cool season legume crops, however, more common markers should be added to allow alignments among the maps so far reported, and to refine the position of the stable QTLs.

In **pea** QTLs for frost resistance have been identified, the most important of which co-segregate with the gene for delayed flowering, thus emphasizing the importance of appropriate flowering time for successful overwintering. More recently, QTLs of frost tolerance were associated with expressional QTLs of proteins potentially involved in frost response. No QTLs for drought stress have been reported so far in the species. The situation is similar in **faba bean** where only putative QTLs associated with frost tolerance and auxiliary traits have been reported so far. Five putative QTLs (three for unhardened and two for hardened frost tolerance) and three for fatty acid content, in unhardened leaves were detected. This fatty acid content was significantly correlated with unhardened frost tolerance.

Complex constraints such as drought and cold tolerance are essential to improve **chickpea and lentil** productivity in most areas where these crops are grown (particularly the Middle East and Mediterranean region). Development of early-maturing varieties and large root system is the most effective conventional approach to escape from terminal drought and ICARDA and ICRISAT have used promising accessions to developed cultivars tolerant to cold, adapted to winter sowing.

Compared with the previous legume species, grass pea is adapted to grow under drought stress conditions although grass pea accessions are generally very susceptible to cold. To date, no QTL analysis has been reported for any of the abiotic stresses in these three species.

Breeding for quality. Non-nutritional factors have also attracted attention and some progress has been made. In pea, both the natural and induced variation in seed quality traits is being used for improved end use, optimal feed use and general industrial application, using different genomic and transcriptomic approaches (e.g., TILLING to identify relevant genetic loci and candidate genes, transformation with RNAi, etc.). Markers to select for low tannin content and favism factors in faba bean are also known to be used in marker-assisted selection.

Since the discovery of ODAP 45 years ago, attempts to remove the **grass pea** neurotoxin producing lathyrism from the crop have been made worldwide, and there has been consistent lowering of ODAP levels in seed over time. ODAP has been shown to be independent of such characters as seed protein content and seed size and so it should be relatively easy to breed in conjunction with a number of other desirable plant characteristics. Transformation approaches to develop varieties with balanced composition of essential amino acids and low or zero-level ODAP are clearly underexploited. Genes controlling lathyrism have been investigated in grass pea. Quader (1985) postulated that ODAP content was controlled by a major gene. Tiwari and Campbell (1996), however, observed a normal distribution of the trait together with a continuous variation in F_2 progenies, indicating that ODAP content was quantitatively inherited. Grass pea mutation breeding can be a supplement to conventional plant breeding methods. Variations in the neurotoxic protein compound ODAP have been induced by different treatments and the results revealed the characteristic of a monogenic segregation. In addition to induced mutagenesis, extensive variation has appeared spontaneously in the course of tissue culture studies (somaclonal variation), that may also be successfully exploited for the improvement of the crop.

The work carried out in **rhizobia symbionts** has produced new bacterial genomes and the possibility of identifying important functions by using DNA arrays. Map-based gene cloning, transcriptomics, proteomics and metabolomics have helped in the identification of symbiotic genes thanks to the use of model legume species.

The number and quality of molecular markers is increasing and, as a consequence, **marker assisted selection** (MAS) is in progress. **Consensus gene maps** are now available for many food legume species including molecular markers as well as "true" genes. As stated above, several QTLs concerning important characters (resistance to stresses, seed and plant features) are being identified and mapped to study their stability and the possibility of their use in assisting selection. Although still in its infancy, studies on gene expression and chromosome "walking", are also in process.

There is also active work on **synteny** as gene maps of different food legumes (pea, faba bean, chickpea, lentil, etc.) as well as that of the model species *Medicago truncatula* (whose genome has already been sequenced) are also available. Genetic maps anchored with orthologous expressed markers (ESTs) from *M. truncatula* and other important legumes have been developed allowing, for the first time, to establish macrosyntenic relationship between *M. truncatula,* faba bean, pea and lentil. Synteny opens the door for devising new markers for a certain species based on the colinearity of related species maps as well as for the identification of candidate genes (see Chapter 7). *In vitro* **regeneration and transformation** is now possible in pea (Schröder et al. 1993, 1995) by *Agrobacterium*-mediated transformation of embryo axes and *in vitro* grafting to avoid the risk of non functional roots of the transformed plant. While pea transformation is based on a robust technology in chickpea and faba bean, transformation is still in its infancy. Nevertheless, recent publications on chickpea (Polowick et al. 2004; Tewari-Singh et al.

2004) and faba bean (Böttinger et al. 2001; Hanafy et al. 2005) also indicate that for this more recalcitrant legume species the avenue for routine genetic modifications is opening. In any case, the process is far from complete as the transformation efficiency is low and the whole process is time consuming. The efficiency has to be improved and many other important aspects (as for example the expression of transgenes) studied, but the door to that important process has been opened. All these achievements show that although the lag when compared with leading world crops is very large, it is no more at a "cosmic" distance (see Chapter 9 for additional information).

Some fields require more attention in the future. For example, functional genomics; most of the actual knowledge in food legumes derives from the model plants *Medicago truncatula* and *Lotus japonicus*. The transcriptomic and proteomic information for the cool season food legumes is still poor to say the least, without any comparison to the main crops such as soybean, rice and maize (see Chapters 6 and 8). The teams working on the former and the latter are not comparable either in number or in funding. Research in functional genomics on food legumes as a whole (some cases are real exceptions) require to develop expression markers, the study of the expression of QTLs (the case of resistance to *Orobanche* could be one of such expressions) and, perhaps not in the very near future, that of the epigenetic code.

12.3 The Prospects

There is still a huge unexplored genetic variation not used as yet . Leaves, seeds, pods, flowers, plant habits and cycles, etc., show a great variability in food legumes. Many new forms have been developed, some of them being of agronomic importance such as the determinate mutants in faba bean or the low antinutritional mutants in both faba bean and grass pea for example. High yielding cultivars have been obtained with little effort and many potentially valuable variants, including drought tolerance, have been identified in routine analysis of germplasm collections.

Mutation work, unfortunately almost abandoned in the current plant breeding practice (not for basic studies), produced many useful mutants as was the case in the systematic mutation work by J Sjödin in faba bean in the 1970s (Sjödin 1971); among the many variants produced were morphological characters as well as tetraploid and trisomic lines so successfully used in mapping. G Duc, in France, was also able to obtain by artificial mutation variants for rhizobium nodulation (Morandi et al. 2005). The potential still exists not only for theoretical but also for practical purposes.

Many studies have demonstrated that in food legumes it is possible to increase the seed protein content while maintaining a high yield. For instance, in faba bean it is possible to reach a protein content of up to 32%. However, to compensate the low level of lysine in cereals for mixtures in animal feeding, care must be taken to maintain the level of this amino acid in high-protein cultivars.

A recent avenue is the possibility of new uses for prebiotics and nutraceuticals. Soybean was the first legume to be used in this way, but there are many products derived from pea and in a few cases from faba bean. Not only proteins but also starch, fiber, shakes, powders, even gels and films to heal wounds have been produced and marketed in the USA and Canada (see also Section 3.1.2.1). The process allows a greater added value.

An additional advantage of food legumes is that its perception as healthy foods is increasing in developed countries pointing to a future increase in its consumption. Health benefits of legumes have been investigated in experimental, epidemiological and clinical studies and some conclusions are resulting from them. A low-fat, high-carbohydrate diet is often recommended as a part of a healthy life-style. Several research works have pointed out that inclusion of pulses in the daily diet has many beneficial effects in controlling and preventing various metabolic diseases, such as diabetes mellitus and coronary heart disease (Maki 2004).

Recent results reveal that chickpea, pea and lentil protein hydrolysates possess angiotensin converting enzyme (ACE) inhibitory properties, (Akillioglu and Karakaya 2009; Boye et al. 2010). Inhibition of ACE is important for the lowering of blood pressure. The total lentil protein isolate had the lowest half maximal inhibitory concentration (IC_{50}), followed by the legumin-rich, albumin-rich, and vicilin-rich fractions, respectively. Results further suggest that while each of the enriched protein fraction demonstrated ACE inhibitory activity, it is likely that lentil varieties having higher amounts of legumin and albumin proteins may have higher ACE inhibitory properties. ACE inhibitory activities are retained after heating treatments for 15 to 50 minutes as shown in *in vitro* experiments. It is also known that the legume proteins have a lowering effect on plasma cholesterol and triacylglycerols (TG) concentrations compared to animal proteins. Chickpea and lentil protein diets reduced liver TG and cholesterol by 31 and 45%, respectively, compared to a casein diet (Boulaga et al. 2009). Dry legume seeds are low glycemic index (GI) foods, which generate slow and moderate postprandial glucose and insulin response. This property of legumes has a beneficial effect in the management of diabetes and hyperlipidemia (Rizkalla et al. 2002).

The antioxidant effect of food components is highly valued nowadays. Research to evaluate the total antioxidant capacity in fiber-rich foods (i.e., cereals, legumes, and nuts) has shown that broad bean and lentil had the highest antioxidant capacity among pulses (Pellegrini et al. 2006). The journal Food Research International has published a review issue (Bassett et al. 2010) on current research being undertaken on pulses and pulse ingredients. This special issue contains several original manuscripts and six comprehensive reviews on the major and minor components of pulses including cool season grain legumes, techniques for their processing, their properties and potential health benefits, and current and novel applications. These results support the increasing role on legumes as healthy foods.

Finally, there are several reports on the anticarcinogenic effects of legume food components such as protease inhibitors and lectins (Guillamón et al. 2008; Caccialupi et al. 2010). For instance, Adebamowo et al. (2005) reported that there was a relationship between common bean and lentil consumption and a lower incidence of breast cancer. Bowman-Birk inhibitors (BBI) are effective in preventing or suppressing carcinogen-induced transformation *in vitro* and carcinogenesis in animal assays. The trypsin inhibitor of peas was significantly more effective in inhibiting cell proliferation (human colon adenocarcinoma cells), compared with BBI from soybean (Clemente et al. 2004). However, more research is needed in this field before definitive conclusions on legume food anticarcinogenic effects can be reached.

It is obvious that genomics and biotechnology tools could help in breeding new cool season grain legume healthier varieties with increased amounts of health positive components and reduced "antinutritional" factors. However this concept is becoming more imprecise as the example of trypsin inhibitors suggests. They are considered antinutritional factors but can help in preventing at least some cancer types.

Other open issues are the balance of the legume contribution to the global nitrogen cycle and the relationships with arbuscular mycorrhizal (AM) fungi. Nitrous oxide (N_2O) emissions and biological nitrogen (N_2) fixation by grain legumes are two major processes of nitrogen transformation in agroecosystems. However, the relationship between these two processes is not well understood. It has been suggested that rhizobia or rhizobia-legume symbiosis can produce nitrous oxide through denitrification but this has never been shown. It seems unlikely that a natural system would release nitrous oxide to the same degree that chemical fertilizers have been shown to do (Golding and Dong 2010). In an recent experiment with two cool season species, lentil and pea; Zong et al. (2009) conclude that N_2O emissions may not be directly related to biological nitrogen fixation by grain legumes under conditions comparable to the experimental conditions used. The rhizobial strains tested were neither capable of denitrification in pure culture, nor in symbiosis with lentil and pea in sterile Leonard jars, suggesting that introducing these *Rhizobium* strains into soils through rhizobial inoculation onto lentil and pea will not increase denitrification or N_2O emissions. Likewise, results indicted that that biological N_2 fixation by well nodulated lentil and pea was not a direct source of N_2O emissions. Although further work on N_2O emissions and nitrogen fixation by grain legumes are needed under field conditions to obtain more complete conclusions, these results support the positive role of cool season legumes in contributing to sustainable agriculture.

The roots of many plant species are naturally colonized by arbuscular mycorrhizal (AM) fungi, which are ubiquitous in soil. AM can directly take up inorganic nitrogen from the soil and transfer it to the host plant. Legumes are able to form symbiotic relationships with both rhizobial bacteria and AM fungi, responding to the non-rhizobial inoculants either through an effect on plant growth or, in addition, through an effect on the function of the legume–*Rhizobium* symbiosis. A review on this subject (Chalk et al. 2006) indicated that the beneficial effects that both indigenous and inoculated AM have on legume growth, nitrogen uptake and the proportional dependence of the legume on atmospheric N_2. The rhizobia provide nitrogen (N) and the AM provide phosphorous (P) in exchange for photosynthetically derived C from the host. Nitrogen and P are the two most limiting nutrients for plant growth, therefore the enhanced nutrition due to the respective symbionts can result in the improved growth and development of the legume, especially under nutrient limiting conditions. The C drain imposed by the respective symbionts can amount to a relatively large portion of the host fixed C. The AM fungus can receive between 10 and 23% of host photosynthate and the nodule between 6 and 30%. The positive effect of AM on the growth and performance of legumes and rhizobia has been established, especially under nutrient limiting conditions, in particular under low P availability (Mortimer et al. 2009). Species of the genus *Glomus* have been repeatedly tested as AM in cool season grain legumes (see examples with lentil in Section 3.2.3). Since the symbiotic relationships between the host plant and AM and rhizobia is controlled by genes, which are partially common to the two symbiotic pathways (Markmann and Parniske 2009) and probably to parasitic pathways (Fernández-Aparicio et al. 2010), a complete genomic understanding of these systems can help in improving the uptake of these limiting elements (N and P) and subsequently to a better crop yield. The use of symbiotic plant mutants in genome wide studies of fungal and plant gene responses will give more insight into the metabolic circuits controlled by *SYM* genes and driving symbiotic interactions. For instance, the pea *Sym* mutants not only control root colonization and morphological differentiation of an AM species (*Glomus intraradices*) but they also influence gene responses linked to symbiotic interactions between the partners (Kuznetsova et al. 2010).

12.4 Final Remarks

In spite of the convenient characteristics of food legumes in modern **sustainable agriculture**, the increase in surface on a world basis has been little or negative. Faba bean and chickpea, for example, have been grown in Canada and Australia, but the general trend has been towards a slight decrease of the cultivated area and no appreciable changes in yield and production (still on a global basis; there were significant increases at the regional level). Development of new varieties combining different resistances and agronomical and "marketing" characters continues to be a major task in food legumes as a group.

But if we have to deal with a substantial increase in food in the next 40 years, as stated at the beginning of this chapter, food legumes should occupy a significant area and a substantial part of plant breeding activities. Especially in developing countries, they are rightly considered the most promising and inexpensive sources of calories and proteins. Besides, it is still necessary to remember the important role of legumes in maintaining and improving the soil's physical and chemical properties. Legumes are largely the most important organisms fixing atmospheric nitrogen. The amount of fixed nitrogen are far from negligible; the data being very variable depending on sources, years and trials, the fixed atmospheric N in kg/ha/year can range from almost 100 to more than 200. At a time when reducing the use of nitrogen fertilizers is a must because of environmental concerns, the use of legumes in farming should be compulsory, and a lot of work is still left to improve the symbiosis.

Sustainable agriculture and sustainable food production will be impossible without legumes. In fact, the prospects for the future will not be reached without the wide use of legumes in farming. Food legumes can provide a natural N-fertilizer for the soil, increasing the organic matter content in the soil; they permit crop rotations and crop diversification; they produce food, dry and green feed and green manure; they produce products of high nutritional value and also new functional applications can be devised. To sum up it would provide agricultural and health benefits.

They only need more research.

References

Adebamowo CA, Cho EY, Sampson L, Katan MB, Spiegelman D, Willett WC, Holmes MD (2005) Dietary flavonols and flavonol-rich foods intake and the risk of breast cancer. Int J Cancer 114: 628–633.

Akillioglu HG, Karakaya S (2009) Effects of heat treatment and *in vitro* digestion on the Angiotensin converting enzyme inhibitory activity of some legume species. Eur Food Res Technol 229: 915–921.

Armentia A, Lombardero M, Blanco C, Fernandez S, Fernandez A, Sanchez-Monge R (2006) Allergic hypersensitivity to the lentil pest *Bruchus lentis*. Allergy 61: 1112–1116.

Bassett C, Boye J, Tyler R, Oomah BD (eds) (2010) Molecular, functional and processing characteristics of whole pulses and pulse fractions and their emerging food and nutraceutical applications. Food Res Int 43: 397–664.

Böttinger P, Steimetz A, Schieder O, Pickardt T (2001) *Agrobacterium*-mediated transformation of *Vicia faba*. Mol Breed 8: 243–254.
Boualga A, Prost J, Taleb-Senouci D, Krouf D, Kharoubi O, Lamri-Senhadji M, Belleville J, Bouchenak M (2009) Purified chickpea or lentil proteins impair VLDL metabolism and lipoprotein lipase activity in epididymal fat, but not in muscle, compared to casein, in growing rats. Eur J Nutr 48: 162–169.
Boye JI, Roufik S, Pesta N, Barbana C (2010) Angiotensin I-converting enzyme inhibitory properties and SDS-PAGE of red lentil protein hydrolysates. Food Sci Technol 43: 987–991.
Brujin FJ de, Jing Y, Dazzo FB (1995) Potential and pitfalls of trying to extend symbiotic interactions of nitrogen-fixing organisms to presently non-nodulated plants, such as rice. Plant Soil 174: 225–240.
Caccialupi P, Ceci LR, Siciliano RA, Pignone D, Clemente A, Sonnante G (2010) Bowman-Birk inhibitors in lentil: Heterologous expression, functional characterisation and anti-proliferative properties in human colon cancer cells. Food Chem 120: 1058–1066.
Chalk PM, Souza R, de F, Urquiaga S, Alves BJR, Boddey RM (2006) The role of arbuscular mycorrhiza in legume symbiotic performance. Soil Biol Biochem 38: 2944–2951.
Clemente A, MacKenzie DA, Jonson IT, Domoney C (2004) Investigation of legume seed protease inhibitors as potential anticarcinogenic proteins. In: M Muzquiz, GD Hill, C Cuadrado, MM Pedrosa, C Burbano (eds) Proc 4th Int Workshop on Antinutritional Factors in Legume Seeds and Oilseeds. EAA Publications, Wageningen, The Netherlands, pp 137–141.
Cuadrado C, Cabanillas B, Pedrosa MM, Varela A, Guillamón E, Muzquiz M, Crespo JF, Rodriguez J, Burbano C (2009) Influence of thermal processing on IgE reactivity to lentil and chickpea proteins. Mol Nutr Food Res 53: 1462–1468.
Davies HM (2010) Commercialization of whole-plant systems for biomanufacturing of protein products: evolution and prospects. Plant Biotechnol J 8: 845–861.
Den Herder G, Parniske M (2009) The unbearable naivety of legumes in symbiosis. Curr Opin Plant Biol 12: 491–499.
Den Herder G, Van Isterdael G, Beeckman T, De Smet I (2010) The roots of a new green revolution. Trends Plant Sci 15: 600–607.
Dixon R, Cheng Q, Shen G-F, Day A, Dawson-Day M (1997) *Nif* genes transfer and expression in chloroplast. Prospects and problems. Plant Soil 194: 193–203.
Dwivedi SL, Crouch JH, Mackill DJ, Xu Y, Blair MW, Ragot M, Upadhyaya HD, Ortiz R (2007) The molecularization of public sector crop breeding: progress, problems, and prospects. Adv Agron 95: 163–318.
Fernandez-Aparicio M, Rispail N, Prats E, Morandi D, García-Garrido JM, Dumas-Gaudot E, Duc G, Rubiales D (2010) Parasitic plant infection is partially controlled through symbiotic pathways. Weed Res 50: 76–82.
Godfray HCJ, Beddington JR, Crute IR, Haddad L, Lawrence D, Muir JF, Pretty J, Robinson S, Thomas SM, Toulmin C (2010) Food security: the challenge of feeding 9 billion people. Science 327: 812–818.
Golding A-L, Dong Z (2010) Hydrogen production by nitrogenase as a potential crop rotation benefit. Environ Chem Lett 8: 101–121.
Guillamón E, Pedrosa MM, Burbano C, Cuadrado C, Cortes Sánchez M, Muzquiz M (2008) The trypsin inhibitors present in seed of different grain legume species and cultivar. Food Chem 107: 68–74.
Hajjar R, Hodgkin T (2007) The use of wild relatives in crop improvement: a survey of developments over the last 20 years. Euphytica 156: 1–13.
Hanafy M, Pickardt T, Kiesecker H, Jacobsen HJ (2005) *Agrobacterium*-mediated transformation of faba bean (*Vicia faba* L.) using embryo axes. Euphytica 142: 227–236.
Heslop-Harrison JS (2002) Exploiting novel germplasm. Aust J Agri Res 53: 1–7.

Ireneo MMS, Ibanez MD, Fernandez-Caldas E, Carnes J (2008) *In vitro* and *in vivo* cross-reactivity studies of legume allergy in a Mediterranean population. Int Arch Allergy Immunol 147: 222–230.

Jellis GJ (2009) Crop plant resistance to biotic and abiotic factors: combating the pressures on production systems in a changing world. In: F Feldmann, DV Alford, C Furk (eds) Crop Plant Resistance to Biotic and Abiotic Factors. Deutsche Phytomedizinische Gesellschaft, Braunschweig, Germany, pp 15–20.

Kumari D, Kumar R, Sridhara S, Arora N, Gaur SN, Singh BP (2006) Sensitization to blackgram in patients with bronchial asthma and rhinitis: clinical evaluation and characterization of allergens. Allergy 61: 104–110.

Kuznetsova E, Seddas-Dozolme PMA, Arnould C, Tollot M, van Tuinen D, Borisov A, Gianinazzi S, Gianinazzi-Pearson V (2010) Symbiosis-related pea genes modulate fungal and plant gene expression during the arbuscule stage of mycorrhiza with *Glomus intraradices*. Mycorrhiza 20: 427–443.

Maki KC (2004) Dietary factors in the prevention of diabetes mellitus and coronary artery disease associated with the metabolic syndrome. Am J Cardiol 93: 12–17.

Markmann K, Parniske M (2009) Evolution of root endosymbiosis with bacteria: how novel are nodules? Trends Plant Sci 14: 77–86.

Messina MJ (1999) Legumes and soybeans: overview of their nutritional profiles and health effects. Am J Clin Nutr 70: 439S–450S.

Mehbood I, Naveed M, Zahir ZA (2009) Rhizobial association with non-legumes: mechanisms and applications. Crit Rev Plant Sci 28: 432–456.

Miklas PN, Kelly JD, Beebe SE, Blair MW (2006) Common bean breeding for resistance against biotic and abiotic stresses: from classical to MAS breeding. Euphytica 147: 105–131.

Montanez A, Abreu C, Gill PR, Hardarson G, Sicardi M (2009) Biological nitrogen fixation in maize (*Zea mays* L.) by N-15 isotope-dilution and identification of associated cultural diazotrophs. Biol Fert Soils 45: 253–263.

Morandi D, Prado E, Sagan M, Duc G (2005) Characterisation of new symbiotic *Medicago truncatula* (Gaertn.) mutants, and phenotypic or genotypic complementary information on previously described mutants. Mycorrhiza 15: 283–289.

Mortimer PE, Pérez-Fernández MA, Valentine AJ (2009) Arbuscular mycorrhizae affect the N and C economy of nodulated *Phaseolus vulgaris* (L.) during NH_4^+ nutrition. Soil Biol Biochem 41: 2115–2121.

Narayanan NN, Baisakh N, Oliva NP, VeraCruz CM, Gnanamanickam SS, Datta K, Datta SK (2004) Molecular breeding: marker-assisted selection combined with biolistic transformation for blast and bacterial blight resistance in indica rice (cv. CO39). Mol Breed 14: 61–71.

Pellegrini N, Serafini M, Salvatore S, Del Rio D, Bianchi M, Brighenti F (2006) Total antioxidant capacity of spices, dried fruits, nuts, pulses, cereals and sweets consumed in Italy assessed by three different *in vitro* assays. Mol Nutr Food Res 50: 1030–1038.

Polowick PL, Baliski DS, Mahon JD (2004) *Agrobacterium tumefaciens*-mediated transformation of chickpea (*Cicer arietinum* L.): gene integration, expression and inheritance. Plant Cell Rep 23: 485–491.

Quader MS (1985) Genetic analysis of neurotoxin content and some aspects of reproductive biology in *Lathyrus sativus* L. PhD Dissertation. Indian Agricultural Research Institute, New Delhi, India.

Rizkalla SW, Bellise F, Slama G (2002) Health benefits of low glycaemic index foods, such as pulses, in diabetic patients and healthy individuals. British J Nutr 88: S255–S262.

Rosenzweig C, Hillel D (1995) Potential impacts of climate change on agriculture and food supply. Consequences 1: 23–32.

Salunkhe DK, Jadhav SJ, Kadam SS, Chavan JK (1982) Chemical, biochemical, and biological significance of polyphenols in cereals and legumes. CRC Crit Rev Food Sci Nutr 17: 277–305.

Schroeder HE, Schotz AH, Wardley-Richardson T, Spencer D, Higgins T (1993) Transformation and regeneration of two cultivars of pea (*Pisum sativum* L.). Plant Physiol 101: 751–757.
Schroeder HE, Gollasch S, Moore A, Tabe LM, Craig S, Hardie DC, Chrispeels MJ, Spencer D, Higgins TJV (1995) Bean alpha-amylase inhibitor confers resistance to the pea weevil (*Bruchus pisorum*) in transgenic peas (*Pisum sativum* L.). Plant Physiol 107: 1233–1239.
Sjödin J (1971) Induced morphological variation in *Vicia faba* L. Hereditas 67: 155–180.
Tewari-Singh N, Sen J, Kiesecker H, Reddy VS, Jacobsen HJ, Guha-Mukherjee S, (2004) Use of a herbicide or lysine plus threonine for non-antibiotic selection of transgenic chickpea. Plant Cell Rep 22: 576–83.
Tiwari KR, Campbell CG (1996) Inheritance of neurotoxin (ODAP) content, flower and seed coat colour in grass pea (*Lathyrus sativus* L.) Euphytica 91: 195–203.
Varshney RK, Close TJ, Singh NK, Hoisington DA, Cook DR (2009) Orphan legume crops enter the genomics era! Curr Opin Plant Biol 12: 202–210.
Vasil IK (2005) The story of transgenic cereals: the challenge, the debate, and the solution—a historical perspective. *In Vitro* Cell Dev-Plant 41: 577–583.
Zhong Z, Lemke RL, Nelson LM (2009) Nitrous oxide emissions associated with nitrogen fixation by grain legumes. Soil Biol Biochem 41: 2283–2291.

Index

B

C

H

I

J

K

L

M

O

P

T